Optimization for Machine Learning

Neural Information Processing Series

Michael I. Jordan and Thomas Dietterich, editors

Optimization for Machine Learning

Edited by Suvrit Sra, Sebastian Nowozin, and Stephen J. Wright

The MIT Press
Cambridge, Massachusetts
London, England

This book was set in LaTeX by the authors and editors.

Library of Congress Cataloging-in-Publication Data

Optimization for machine learning / edited by Suvrit Sra, Sebastian Nowozin, and
Stephen J. Wright.
    p. cm. — (Neural information processing series)
Includes bibliographical references.
ISBN 978-0-262-53776-6 (paperback : alk. paper)
ISBN 978-0-262-01646-9 (hardcover : alk. paper) 1. Machine learning—
Mathematical models. 2. Mathematical optimization. I. Sra, Suvrit, 1976– II.
Nowozin, Sebastian, 1980– III. Wright, Stephen J., 1960–
Q325.5.O65 2012
006.3'1—c22
                                                            2011002059

Contents

Series Foreword

The yearly Neural Information Processing Systems (NIPS) workshops bring together scientists with broadly varying backgrounds in statistics, mathematics, computer science, physics, electrical engineering, neuroscience, and cognitive science, unified by a common desire to develop novel computational and statistical strategies for information processing and to understand the mechanisms for information processing in the brain. In contrast to conferences, these workshops maintain a flexible format that both allows and encourages the presentation and discussion of work in progress. They thus serve as an incubator for the development of important new ideas in this rapidly evolving field. The series editors, in consultation with workshop organizers and members of the NIPS Foundation Board, select specific workshop topics on the basis of scientific excellence, intellectual breadth, and technical impact. Collections of papers chosen and edited by the organizers of specific workshops are built around pedagogical introductory chapters, while research monographs provide comprehensive descriptions of workshop-related topics, to create a series of books that provides a timely, authoritative account of the latest developments in the exciting field of neural computation.

Michael I. Jordan and Thomas G. Dietterich

Preface

The intersection of interests between machine learning and optimization has engaged many leading researchers in both communities for some years now. Both are vital and growing fields, and the areas of shared interest are expanding too. This volume collects contributions from many researchers who have been a part of these efforts.

We are grateful first to the contributors to this volume. Their cooperation in providing high-quality material while meeting tight deadlines is highly appreciated. We further thank the many participants in the two workshops on Optimization and Machine Learning, held at the NIPS Workshops in 2008 and 2009. The interest generated by these events was a key motivator for this volume. Special thanks go to S. V. N. Vishawanathan (Vishy) for organizing these workshops with us, and to PASCAL2, MOSEK, and Microsoft Research for their generous financial support for the workshops.

S. S. thanks his father for his constant interest, encouragement, and advice regarding this book. S. N. thanks his wife and family. S. W. thanks all those colleagues who introduced him to machine learning, especially Partha Niyogi, to whose memory his efforts on this book are dedicated.

Suvrit Sra, Sebastian Nowozin, and Stephen J. Wright

1   Introduction: Optimization and Machine Learning

Suvrit Sra   suvrit.sra@tuebingen.mpg.de
Max Planck Institute for Intelligent Systems
Tübingen, Germany

Sebastian Nowozin   Sebastian.Nowozin@Microsoft.com
Microsoft Research
Cambridge, United Kingdom

Stephen J. Wright   swright@cs.wisc.edu
University of Wisconsin
Madison, Wisconsin, USA

Since its earliest days as a discipline, machine learning has made use of optimization formulations and algorithms. Likewise, machine learning has contributed to optimization, driving the development of new optimization approaches that address the significant challenges presented by machine learning applications. This cross-fertilization continues to deepen, producing a growing literature at the intersection of the two fields while attracting leading researchers to the effort.

Optimization approaches have enjoyed prominence in machine learning because of their wide applicability and attractive theoretical properties. While techniques proposed twenty years and more ago continue to be refined, the increased complexity, size, and variety of today's machine learning models demand a principled reassessment of existing assumptions and techniques. This book makes a start toward such a reassessment. Besides describing the resurgence in novel contexts of established frameworks such as first-order methods, stochastic approximations, convex relaxations, interior-point methods, and proximal methods, the book devotes significant attention to newer themes such as regularized optimization, robust optimization, a variety of gradient and subgradient methods, and the use of splitting techniques and second-order information. We aim to provide an up-to-date account of the optimization techniques useful to machine learning — those that are established and prevalent, as well as those that are rising in importance.

To illustrate our aim more concretely, we review in Section 1.1 and 1.2 two major paradigms that provide focus to research at the confluence of machine learning and optimization: support vector machines (SVMs) and regularized optimization. Our brief review charts the importance of these problems and discusses how both connect to the later chapters of this book. We then discuss other themes — applications, formulations, and algorithms — that recur throughout the book, outlining the contents of the various chapters and the relationship between them.

Audience.   This book is targeted to a broad audience of researchers and students in the machine learning and optimization communities; but the material covered is widely applicable and should be valuable to researchers in other related areas too. Some chapters have a didactic flavor, covering recent advances at a level accessible to anyone having a passing acquaintance with tools and techniques in linear algebra, real analysis, and probability. Other chapters are more specialized, containing cutting-edge material. We hope that from the wide range of work presented in the book, researchers will gain a broader perspective of the field, and that new connections will be made and new ideas sparked.

For background relevant to the many topics discussed in this book, we refer to the many good textbooks in optimization, machine learning, and related subjects. We mention in particular Bertsekas (1999), Nocedal and Wright (2006), and Nesterov (2004) for optimization over continuous variables, and Ben-Tal et al. (2009) for robust optimization. In machine learning, we refer for background to Vapnik (1999), Schölkopf and Smola (2002), Christianini and Shawe-Taylor (2000), and Hastie et al. (2009). Some fundamentals of graphical models and the use of optimization therein can be found in Wainwright and Jordan (2008) and Koller and Friedman (2009).

1.1   Support Vector Machines

The support vector machine (SVM) is the first contact that many optimization researchers had with machine learning, due to its classical formulation as a convex quadratic program — simple in form, though with a complicating constraint. It continues to be a fundamental paradigm today, with new algorithms being proposed for difficult variants, especially large-scale and nonlinear variants. Thus, SVMs offer excellent common ground on which to demonstrate the interplay of optimization and machine learning.

1.1.1 Background

The problem is one of learning a classification function from a set of labeled training examples. We denote these examples by $\{(\boldsymbol{x}_i, y_i),\ i = 1, \ldots, m\}$, where $\boldsymbol{x}_i \in \mathbb{R}^n$ are feature vectors and $y_i \in \{-1, +1\}$ are the labels. In the simplest case, the classification function is the signum of a linear function of the feature vector. That is, we seek a weight vector $\boldsymbol{w} \in \mathbb{R}^n$ and an intercept $b \in \mathbb{R}$ such that the predicted label of an example with feature vector $\boldsymbol{x}$ is $f(\boldsymbol{x}) = \operatorname{sgn}(\boldsymbol{w}^T \boldsymbol{x} + b)$. The pair $(\boldsymbol{w}, b)$ is chosen to minimize a weighted sum of: (a) a measure of the classification error on the training examples; and (b) $\|\boldsymbol{w}\|_2^2$, for reasons that will be explained in a moment. The formulation is thus

$$
\begin{aligned}
\underset{\boldsymbol{w}, b, \boldsymbol{\xi}}{\text{minimize}} \quad & \tfrac{1}{2}\boldsymbol{w}^T \boldsymbol{w} + C \sum\nolimits_{i=1}^{m} \xi_i \\
\text{subject to} \quad & y_i(\boldsymbol{w}^T \boldsymbol{x}_i + b) \geq 1 - \xi_i, \quad \xi_i \geq 0, \quad 1 \leq i \leq m.
\end{aligned}
\tag{1.1}
$$

Note that the summation term in the objective contains a penalty contribution from term i if $y_i = 1$ and $\boldsymbol{w}^T \boldsymbol{x}_i + b < 1$, or $y_i = -1$ and $\boldsymbol{w}^T \boldsymbol{x}_i + b > -1$. If the data are separable, it is possible to find a $(\boldsymbol{w}, b)$ pair for which this penalty is zero. Indeed, it is possible to construct two parallel hyperplanes in $\mathbb{R}^n$, both of them orthogonal to $\boldsymbol{w}$ but with different intercepts, that contain no training points between them. Among all such pairs of planes, the pair for which $\|\boldsymbol{w}\|_2$ is minimal is the one for which the separation is greatest. Hence, this $\boldsymbol{w}$ gives a robust separation between the two labeled sets, and is therefore, in some sense, most desirable. This observation accounts for the presence of the first term in the objective of (1.1).

Problem (1.1) is a convex quadratic program with a simple diagonal Hessian but general constraints. Some algorithms tackle it directly, but for many years it has been more common to work with its dual, which is

$$
\begin{aligned}
\underset{\boldsymbol{\alpha}}{\text{minimize}} \quad & \tfrac{1}{2}\boldsymbol{\alpha}^T \boldsymbol{Y} \boldsymbol{X}^T \boldsymbol{X} \boldsymbol{Y} \boldsymbol{\alpha} - \boldsymbol{\alpha}^T \mathbf{1} \\
\text{subject to} \quad & \sum\nolimits_i y_i \alpha_i = 0, \quad 0 \leq \alpha_i \leq C,
\end{aligned}
\tag{1.2}
$$

where $\boldsymbol{Y} = \operatorname{Diag}(y_1, \ldots, y_m)$ and $\boldsymbol{X} = [\boldsymbol{x}_1, \ldots, \boldsymbol{x}_m] \in \mathbb{R}^{n \times m}$. This dual is also a quadratic program. It has a positive semidefinite Hessian and simple bounds, plus a single linear constraint.

More powerful classifiers allow the inputs to come from an arbitrary set $\mathcal{X}$, by first mapping the inputs into a space $\mathcal{H}$ via a nonlinear (feature) mapping $\phi : \mathcal{X} \to \mathcal{H}$, and then solving the classification problem to find $(\boldsymbol{w}, b)$ with $\boldsymbol{w} \in \mathcal{H}$. The classifier is defined as $f(\boldsymbol{x}) := \operatorname{sgn}(\langle \boldsymbol{w}, \phi(\boldsymbol{x}) \rangle + b)$, and it can be found by modifying the Hessian from $\boldsymbol{Y} \boldsymbol{X}^T \boldsymbol{X} \boldsymbol{Y}$ to $\boldsymbol{Y} \boldsymbol{K} \boldsymbol{Y}$, where $\boldsymbol{K}_{ij} := \langle \phi(\boldsymbol{x}_i), \phi(\boldsymbol{x}_j) \rangle$ is the *kernel matrix*. The optimal weight vector can be recovered from the dual solution by setting $\boldsymbol{w} = \sum_{i=1}^{m} \alpha_i \phi(\boldsymbol{x}_i)$, so that the classifier is $f(\boldsymbol{x}) = \mathrm{sgn}\left[\sum_{i=1}^{m} \alpha_i \langle \phi(\boldsymbol{x}_i), \phi(\boldsymbol{x}) \rangle + b\right]$.

In fact, it is not even necessary to choose the mapping ϕ explicitly. We need only define a kernel mapping $k : \mathcal{X} \times \mathcal{X} \to \mathbb{R}$ and define the matrix $\boldsymbol{K}$ directly from this function by setting $\boldsymbol{K}_{ij} := k(\boldsymbol{x}_i, \boldsymbol{x}_j)$. The classifier can be written purely in terms of the kernel mapping k as follows: $f(\boldsymbol{x}) = \mathrm{sgn}\left[\sum_{i=1}^{m} \alpha_i k(\boldsymbol{x}_i, \boldsymbol{x}) + b\right]$.

1.1.2 Classical Approaches

There has been extensive research on algorithms for SVMs since at least the mid-1990s, and a wide variety of techniques have been proposed. Out-of-the-box techniques for convex quadratic programming have limited appeal because usually the problems have large size, and the Hessian in (1.2) can be dense and ill-conditioned. The proposed methods thus exploit the structure of the problem and the requirements on its (approximate) solution. We survey some of the main approaches here.

One theme that recurs across many algorithms is *decomposition* applied to the dual (1.2). Rather than computing a step in all components of $\boldsymbol{\alpha}$ at once, these methods focus on a relatively small subset and fix the other components. An early approach due to Osuna et al. (1997) works with a subset $B \subset \{1, 2, \ldots, s\}$, whose size is assumed to exceed the number of nonzero components of $\boldsymbol{\alpha}$ in the solution of (1.2); their approach replaces one element of B at each iteration and then re-solves the reduced problem (formally, a complete reoptimization is assumed, though heuristics are used in practice). The sequential minimal optimization (SMO) approach of Platt (1999) works with just two components of $\boldsymbol{\alpha}$ at each iteration, reducing each QP subproblem to triviality. A heuristic selects the pair of variables to relax at each iteration. LIBSVM[1] (see Fan et al., 2005) implements an SMO approach for (1.2) and a variety of other SVM formulations, with a particular heuristic based on second-order information for choosing the pair of variables to relax. This code also uses shrinking and caching techniques like those discussed below.

SVM[light][2] (Joachims, 1999) uses a linearization of the objective around the current point to choose the working set B to be the indices most likely to give descent, giving a fixed size limitation on B. Shrinking reduces the workload further by eliminating computation associated with components of $\boldsymbol{\alpha}$ that

---

1. http://www.csie.ntu.edu.tw/~cjlin/libsvm/
2. http://www.cs.cornell.edu/People/tj/svm_light/

seem to be at their lower or upper bounds. The method nominally requires computation of $|B|$ columns of the kernel $\boldsymbol{K}$ at each iteration, but columns can be saved and reused across iterations. Careful implementation of gradient evaluations leads to further computational savings. In early versions of SVM^{light}, the reduced QP subproblem was solved with an interior-point method (see below), but this was later changed to a coordinate relaxation procedure due to Hildreth (1957) and D'Esopo (1959). Zanni et al. (2006) use a similar method to select the working set, but solve the reduced problem using nonmontone gradient projection, with Barzilai-Borwein step lengths. One version of the gradient projection procedure is described by Dai and Fletcher (2006).

Interior-point methods have proved effective on convex quadratic programs in other domains, and have been applied to (1.2) (see Ferris and Munson, 2002; Gertz and Wright, 2003). However, the density, size, and ill-conditioning of the kernel matrix make achieving efficiency difficult. To ameliorate this difficulty, Fine and Scheinberg (2001) propose a method that replaces the Hessian with a low-rank approximation (of the form $\boldsymbol{V}\boldsymbol{V}^T$, where $\boldsymbol{V} \in \mathbb{R}^{m \times r}$ for $r \ll m$) and solves the resulting modified dual. This approach works well on problems of moderate scale, but may be too expensive for larger problems.

In recent years, the usefulness of the primal formulation (1.1) as the basis of algorithms has been revisited. We can rewrite this formulation as an unconstrained minimization involving the sum of a quadratic and a convex piecewise-linear function, as follows:

$$\operatorname*{minimize}_{\boldsymbol{w},b} \quad \tfrac{1}{2}\boldsymbol{w}^T\boldsymbol{w} + CR(\boldsymbol{w},b), \tag{1.3}$$

where the penalty term is defined by

$$R(\boldsymbol{w},b) := \sum_{i=1}^{m} \max(1 - y_i(\boldsymbol{w}^T\boldsymbol{x}_i + b), 0). \tag{1.4}$$

Joachims (2006) describes a cutting-plane approach that builds up a convex piecewise-linear lower bounding function for $R(\boldsymbol{w},b)$ based on subgradient information accumulated at each iterate. Efficient management of the inequalities defining the approximation ensures that subproblems can be solved efficiently, and convergence results are proved. Some enhancements are decribed in Franc and Sonnenburg (2008), and the approach is extended to nonlinear kernels by Joachims and Yu (2009). Implementations appear in the code SVM^{perf}.[3]

---

3. `http://www.cs.cornell.edu/People/tj/svm_light/svm_perf.html`

There has also been recent renewed interest in solving (1.3) by stochastic gradient methods. These appear to have been proposed originally by Bottou (see, for example, Bottou and LeCun, 2004) and are based on taking a step in the $(\boldsymbol{w}, b)$ coordinates, in a direction defined by the subgradient in a single term of the sum in (1.4). Specifically, at iteration k, we choose a steplength γ_k and an index $i_k \in \{1, 2, \ldots, m\}$, and update the estimate of $\boldsymbol{w}$ as follows:

$$\boldsymbol{w} \leftarrow \begin{cases} \boldsymbol{w} - \gamma_k(\boldsymbol{w} - mCy_{i_k}\boldsymbol{x}_{i_k}) & \text{if } 1 - y_{i_k}(\boldsymbol{w}^T\boldsymbol{x}_{i_k} + b) > 0, \\ \boldsymbol{w} - \gamma_k\boldsymbol{w} & \text{otherwise.} \end{cases}$$

Typically, one uses $\gamma_k \propto 1/k$. Each iteration is cheap, as it needs to observe just one training point. Thus, many iterations are needed for convergence; but in many large practical problems, approximate solutions that yield classifiers of sufficient accuracy can be found in much less time than is taken by algorithms that aim at an exact solution of (1.1) or (1.2). Implementations of this general approach include SGD[4] and Pegasos (see Shalev-Shwartz et al., 2007). These methods enjoy a close relationship with stochastic approximation methods for convex minimization; see Nemirovski et al. (2009) and the extensive literature referenced therein. Interestingly, the methods and their convergence theory were developed independently in the two communities, with little intersection until 2009.

1.1.3 Approaches Discussed in This Book

Several chapters of this book discuss the problem (1.1) or variants thereof. In Chapter 12, Gondzio gives some background on primal-dual interior-point methods for quadratic programming, and shows how structure can be exploited when the Hessian in (1.2) is replaced by an approximation of the form $\boldsymbol{Q}_0 + \boldsymbol{V}\boldsymbol{V}^T$, where $\boldsymbol{Q}_0$ is nonnegative diagonal and $\boldsymbol{V} \in \mathbb{R}^{m \times r}$ with $r \ll m$, as above. The key is careful design of the linear algebra operations that are used to form and solve the linear equations which arise at each iteration of the interior-point method. Andersen et al. in Chapter 3 also consider interior-point methods with low-rank Hessian approximations, but then go on to discuss robust and multiclass variants of (1.1). The robust variants, which replace each training vector $\boldsymbol{x}_i$ with an ellipsoid centered at $\boldsymbol{x}_i$, can be formulated as second-order cone programs and solved with an interior-point method.

A similar model for robust SVM is considered by Caramanis et al. in Chapter 14, along with other variants involving corrupted labels, missing

---

4. http://leon.bottou.org/projects/sgd.

data, nonellipsoidal uncertainty sets, and kernelization. This chapter also explores the connection between robust formulations and the regularization term $\|\boldsymbol{w}\|_2^2$ that appears in (1.1).

As Schmidt et al. note in Chapter 11, omission of the intercept term b from the formulation (1.1) (which can often be done without seriously affecting the quality of the classifier) leads to a dual (1.2) with no equality constraint — it becomes a bound-constrained convex quadratic program. As such, the problem is amenable to solution by gradient projection methods with second-order acceleration on the components of $\boldsymbol{\alpha}$ that satisfy the bounds.

Chapter 13, by Bottou and Bousquet, describes application of SGD to (1.1) and several other machine learning problems. It also places the problem in context by considering other types of errors that arise in its formulation, namely, the errors incurred by restricting the classifier to a finitely parametrized class of functions and by using an empirical, discretized approximation to the objective (obtained by sampling) in place of an assumed underlying continuous objective. The existence of these other errors obviates the need to find a highly accurate solution of (1.1).

1.2 Regularized Optimization

A second important theme of this book is finding *regularized* solutions of optimization problems originating from learning problems, instead of unregularized solutions. Though the contexts vary widely, even between different applications in the machine learning domain, the common thread is that such regularized solutions *generalize* better and provide a less complicated explanation of the phenomena under investigation. The principle of Occam's Razor applies: simple explanations of any given set of observations are generally preferable to more complicated explanations. Common forms of simplicity include sparsity of the variable vector $\boldsymbol{w}$ (that is, $\boldsymbol{w}$ has relatively few nonzeros) and low rank of a matrix variable $\boldsymbol{W}$.

One way to obtain simple approximate solutions is to modify the optimization problem by adding to the objective a *regularization function* (or *regularizer*), whose properties tend to favor the selection of unknown vectors with the desired structure. We thus obtain regularized optimization problems with the following *composite* form:

$$\underset{\boldsymbol{w}\in\mathbb{R}^n}{\text{minimize}}\quad \phi_\gamma(\boldsymbol{w}) := f(\boldsymbol{w}) + \gamma r(\boldsymbol{w}), \tag{1.5}$$

where f is the underlying objective, r is the regularizer, and γ is a nonnegative parameter that weights the relative importances of optimality and simplicity. (Larger values of γ promote simpler but less optimal solutions.) A desirable value of γ is often not known in advance, so it may be necessary to solve (1.5) for a range of values of γ.

The SVM problem (1.1) is a special case of (1.5) in which f represents the loss term (containing penalties for misclassified points) and r represents the regularizer $\boldsymbol{w}^T\boldsymbol{w}/2$, with weighting factor $\gamma = 1/C$. As noted above, when the training data are separable, a "simple" plane is the one that gives the largest separation between the two labeled sets. In the nonseparable case, it is not as intuitive to relate "simplicity" to the quantity $\boldsymbol{w}^T\boldsymbol{w}/2$, but we do see a trade-off between minimizing misclassification error (the f term) and reducing $\|\boldsymbol{w}\|_2$.

SVM actually stands in contrast to most regularized optimization problems in that the regularizer is smooth (though a nonsmooth regularization term $\|\boldsymbol{w}\|_1$ has also been considered, for example, by Bradley and Mangasarian, 2000). More frequently, r is a nonsmooth function with simple structure. We give several examples relevant to machine learning.

- In compressed sensing, for example, the regularizer $r(\boldsymbol{w}) = \|\boldsymbol{w}\|_1$ is common, as it tends to favor sparse vectors $\boldsymbol{w}$.

- In image denoising, r is often defined to be the total-variation (TV) norm, which has the effect of promoting images that have large areas of constant intensity (a cartoonlike appearance).

- In matrix completion, where $\boldsymbol{W}$ is a matrix variable, a popular regularizer is the *spectral norm*, which is the sum of singular values of $\boldsymbol{W}$. Analogously to the ℓ_1-norm for vectors, this regularizer favors matrices with low rank.

- Sparse inverse covariance selection, where we wish to find an approximation $\boldsymbol{W}$ to a given covariance matrix $\boldsymbol{\Sigma}$ such that $\boldsymbol{W}^{-1}$ is a sparse matrix. Here, f is a function that evaluates the fit between $\boldsymbol{W}$ and $\boldsymbol{\Sigma}$, and $r(\boldsymbol{W})$ is a sum of absolute values of components of $\boldsymbol{W}$.

- The well-known LASSO procedure for variable selection (Tibshirani, 1996) essentially uses an ℓ_1-norm regularizer along with a least-squares loss term.

- Regularized logistic regression instead uses logistic loss with an ℓ_1-regularizer; see, for example, Shi et al. (2008).

- Group regularization is useful when the components of $\boldsymbol{w}$ are naturally grouped, and where components in each group should be selected (or not selected) jointly rather than individually. Here, r may be defined as a sum of ℓ_2- or ℓ_∞-norms of subvectors of $\boldsymbol{w}$. In some cases, the groups are non-overlapping (see Turlach et al., 2005), while in others they are overlapping, for example, when there is a hierarchical relationship between components of $\boldsymbol{w}$ (see, for example, Zhao et al., 2009).

1.2.1 Algorithms

Problem (1.5) has been studied intensely in recent years largely in the context of the specific settings mentioned above; but some of the algorithms proposed can be extended to the general case. One elementary option is to apply gradient or subgradient methods directly to (1.5) without taking particular account of the structure. A method of this type would iterate $\boldsymbol{w}_{k+1} \leftarrow \boldsymbol{w}_k - \delta_k \boldsymbol{g}_k$, where $\boldsymbol{g}_k \in \partial \phi_\gamma(\boldsymbol{w}_k)$, and $\delta_k > 0$ is a steplength. When (1.5) can be formulated as a min-max problem; as is often the case with regularizers r of interest, the method of Nesterov (2005) can be used. This method ensures sublinear convergence, where the difference $\phi_\gamma(\boldsymbol{w}_k) - \phi_\gamma(\boldsymbol{w}^*) \leq O(1/k^2)$. Later work (Nesterov, 2009) expands on the min-max approach, and extends it to cases in which only noisy (but unbiased) estimates of the subgradient are available. For foundations of this line of work, see the monograph Nesterov (2004).

A fundamental approach that takes advantage of the structure of (1.5) solves the following subproblem (the *proximity problem*) at iteration k:

$$\boldsymbol{w}_{k+1} := \arg\min_{\boldsymbol{w}} \ (\boldsymbol{w} - \boldsymbol{w}_k)^T \nabla f(\boldsymbol{w}_k) + \gamma r(\boldsymbol{w}) + \frac{1}{2\mu} \|\boldsymbol{w} - \boldsymbol{w}_k\|_2^2, \quad (1.6)$$

for some $\mu > 0$. The function f (assumed to be smooth) is replaced by a linear approximation around the current iterate $\boldsymbol{w}_k$, while the regularizer is left intact and a quadratic damping term is added to prevent excessively long steps from being taken. The length of the step can be controlled by adjusting the parameter μ, for example to ensure a decrease in ϕ_γ at each iteration.

The solution to (1.6) is nothing but the *proximity operator* for $\gamma\mu r$, applied at the point $\boldsymbol{w}_k - \mu \nabla f(\boldsymbol{w}_k)$; (see Section 2.3 of Combettes and Wajs, 2005). Proximity operators are particularly attractive when the subproblem (1.6) is easy to solve, as happens when $r(\boldsymbol{w}) = \|\boldsymbol{w}\|_1$, for example. Approaches based on proximity operators have been proposed in numerous contexts under different guises and different names, such as "iterative shrinking and thresholding" and "forward-backward splitting." For early versions, see Figueiredo and Nowak (2003), Daubechies et al. (2004), and Combettes and Wajs (2005). A version for compressed sensing that adjusts μ to achieve global convergence is the SpaRSA algorithm of Wright et al. (2009). Nesterov (2007) describes enhancements of this approach that apply in the general setting, for f with Lipschitz continuous gradient. A simple scheme for adjusting μ (analogous to the classical Levenberg-Marquardt method for nonlinear least squares) leads to sublinear convergence of objective function values at rate $O(1/k)$ when ϕ_γ is convex, and at a linear rate when ϕ_γ is strongly convex. A more complex accelerated version improves the sublinear rate to $O(1/k^2)$.

The use of second-order information has also been explored in some settings. A method based on (1.6) for regularized logistic regression that uses second-order information on the reduced space of nonzero components of $\boldsymbol{w}$ is described in Shi et al. (2008), and inexact reduced Newton steps that use inexpensive Hessian approximations are described in Byrd et al. (2010).

A variant on subproblem (1.6) proposed by Xiao (2010) applies to problems of the form (1.5) in which $f(\boldsymbol{w}) = E_\xi F(\boldsymbol{w}; \xi)$. The gradient term in (1.6) is replaced by an average of unbiased subgradient estimates encountered at all iterates so far, while the final prox-term is replaced by one centered at a fixed point. Accelerated versions of this method are also described. Convergence analysis uses regret functions like those introduced by Zinkevich (2003).

Teo et al. (2010) describe the application of bundle methods to (1.5), with applications to SVM, ℓ_2-regularized logistic regression, and graph matching problems. Block coordinate relaxation has also been investigated; see, for example, Tseng and Yun (2009) and Wright (2010). Here, most of the components of $\boldsymbol{w}$ are fixed at each iteration, while a step is taken in the other components. This approach is most suitable when the function r is separable and when the set of components to be relaxed is chosen in accordance with the separability structure.

1.2.2   Approaches Discussed in This Book

Several chapters in this book discuss algorithms for solving (1.5) or its special variants. We outline these chapters below while relating them to the discussion of the algorithms above.

Bach et al. in Chapter 2 consider convex versions of (1.5) and describe the relevant duality theory. They discuss various algorithmic approaches, including proximal methods based on (1.6), active-set/pivoting approaches, block-coordinate schemes, and reweighted least-squares schemes. Sparsity-inducing norms are used as regularizers to induce different types of structure in the solutions. (Numerous instances of structure are discussed.) A computational study of the different methods is shown on the specific problem $\phi_\gamma(\boldsymbol{w}) = (1/2)\|\boldsymbol{A}\boldsymbol{w} - \boldsymbol{b}\|_2^2 + \gamma\|\boldsymbol{w}\|_1$, for various choices of the matrix $\boldsymbol{A}$ with different properties and for varying sparsity levels of the solution.

In Chapter 7, Franc et al. discuss cutting-plane methods for (1.5), in which a piecewise-linear lower bound is formed for f, and each iterate is obtained by minimizing the sum of this approximation with the *unaltered* regularizer $\gamma r(\boldsymbol{w})$. A line search enhancement is considered and application to multiple kernel learning is discussed.

Chapter 6, by Juditsky and Nemirovski, describes optimal first-order methods for the case in which (1.5) can be expressed in min-max form. The resulting saddle-point is solved for by a method that computes prox-steps similar to those from the scheme (1.6), but is adapted to the min-max form and uses generalized prox-terms. This "mirror-prox" algorithm is also distinguished by generating two sequences of primal-dual iterates and by its use of averaging. Accelerated forms of the method are also discussed.

In Chapter 18, Krishnamurthy et al. discuss an algorithm for sparse covariance selection, a particular case of (1.5). This method takes the dual and traces the path of solutions obtained by varying the regularization parameter γ, using a predictor-corrector approach. Scheinberg and Ma discuss the same problem in Chapter 17 but consider other methods, including a coordinate descent method and an alternating linearization method based on a reformulation of (1.5). This reformulation is then solved by a method based on augmented Lagrangians, with techniques customized to the application at hand. In Chapter 9, Tomioka et al. consider convex problems of the form (1.5) and highlight special cases. Methods based on variable splitting that use an augmented Lagrangian framework are described, and the relationship to proximal point methods is explored. An application to classification with multiple matrix-valued inputs is described.

Schmidt et al. in Chapter 11 consider special cases of (1.5) in which r is separable. They describe a minimum-norm subgradient method, enhanced with second-order information on the reduced subspace of nonzero components, as well as higher-order versions of methods based on (1.6).

1.3 Summary of the Chapters

The two motivating examples discussed above give an idea of the pervasiveness of optimization viewpoints and algorithms in machine learning. A confluence of interests is seen in many other areas, too, as can be gleaned from the summaries of individual chapters below. (We include additional comments on some of the chapters discussed above alongside a summary of those not yet discussed.)

Chapter 2 by Bach et al. has been discussed in Section 1.2.2.

We mentioned above that Chapter 3, by Andersen et al., describes solution of robust and multiclass variants of the SVM problem of Section 1.1, using interior-point methods. This chapter contains a wider discussion of conic programming over the three fundamental convex cones: the nonnegative orthant, the second-order cone, and the semidefinite cone. The linear algebra operations that dominate computation time are considered in detail, and the authors demonstrate how the Python software package CVXOPT[5] can be used to model and solve conic programs.

In Chapter 4, Bertsekas surveys incremental algorithms for convex optimization, especially gradient, subgradient, and proximal-point approaches. This survey offers an optimization perspective on techniques that have recently received significant attention in machine learning, such as stochastic gradients, online methods, and nondifferentiable optimization. Incremental methods encompass some online algorithms as special cases; the latter may be viewed as one "epoch" of an incremental method. The chapter connects many threads and offers a historical perspective along with sufficient technical details to allow ready implementation.

Chapters 5 and 6 by Juditsky and Nemirovski provide a broad and rigorous introduction to the subject of large-scale optimization for nonsmooth convex problems. Chapter 5 discusses state-of-the-art nonsmooth optimization methods, viewing them from a computation complexity framework that assumes only first-order oracle access to the nonsmooth convex objective of the problem. Particularly instructive is a discussion on the theoretical limits of performance of first-order methods; this discussion summarizes lower and upper bounds on the number of iterations needed to approximately minimize the given objective to within a desired accuracy. This chapter covers the basic theory for *mirror-descent* algorithms, and describes mirror descent in settings such as minimization over simple sets, minimization with nonlinear constraints, and saddle-point problems. Going beyond the "black-box" settings of Chapter 5, the focus of Chapter 6 is on settings where improved rates of convergence can be obtained by exploiting problem structure. A key property of the convergence rates is their near dimension independence. Potential speedups due to randomization (in the linear algebra operations, for instance) are also explored.

Chapter 7 and 8 both discuss inference problems involving discrete random variables that occur naturally in many structured models used in computer vision, natural language processing, and bioinformatics. The use of discrete variables allows the encoding of logical relations, constraints, and model assumptions, but poses significant challenges for inference and learning. In particular, solving for the exact maximum a posteriori probability state in these models is typically NP-hard. Moreover, the models can become very large, such as when each discrete variable represents an image pixel or Web user; problem sizes of a million discrete variables are not uncommon.

---

5. http://abel.ee.ucla.edu/cvxopt/.

As mentioned in Section 1.2, in Chapter 7 Franc et al., discuss cutting-plane methods for machine learning in a variety of contexts. Two continuous optimization problems are discussed — regularized risk minimization and multiple kernel learning — both of them solvable efficiently, using customized cutting-plane formulations. In the discrete case, the authors discuss the maximum a posteriori inference problem on Markov random fields, proposing a dual cutting-plane method.

Chapter 8, by Sontag et al., revisits the successful dual-decomposition method for linear programming relaxations of discrete inference problems that arise from Markov random fields and structured prediction problems. The method obtains its efficiency by exploiting exact inference over tractable substructures of the original problem, iteratively combining the partial inference results to reason over the full problem. As the name suggests, the method works in the Lagrangian dual of the original problem. Decoding a primal solution from the dual iterate is challenging. The authors carefully analyze this problem and provide a unified view on recent algorithms.

Chapter 9, by Tomioka et al., considers composite function minimization. This chapter also derives methods that depend on proximity operators, thus covering some standard choices such as ℓ_1-, ℓ_2-, and trace-norms. The key algorithmic approach shown in the chapter is a dual augmented Lagrangian method, which is shown under favorable circumstances to converge superlinearly. The chapter concludes with an application to brain-computer interface (BCI) data.

In Chapter 10, Hazan reviews online algorithms and regret analysis in the framework of convex optimization. He extracts the key tools essential to regret analysis, and casts the description using the regularized follow-the-leader framework. The chapter provides straightforward proofs for basic regret bounds, and proceeds to cover recent applications of convex optimization in regret minimization, for example, to bandit linear optimization and variational regret bounds.

Chapter 11, by Schmidt et al., considers Newton-type methods and their application to machine learning problems. For constrained optimization with a smooth objective (including bound-constrained optimization), two-metric projection and inexact Newton methods are described. For nonsmooth regularized minimization problems the form (1.5), the chapter sketches descent methods based on minimum-norm subgradients that use second-order information and variants of shrinking methods based on (1.6).

Chapter 12, by Gondzio, and Chapter 13, by Bottou and Bousquet have already been summarized in Section 1.1.3.

Chapter 14, by Caramanis et al., addresses an area of growing importance within machine learning: robust optimization. In such problems, solutions are identified that are robust to every possible instantiation of the uncertain data — even when the data take on their least favorable values. The chapter describes how to cope with adversarial or stochastic uncertainty arising in several machine-learning problems. SVM, for instance, allows for a number of uncertain variants, such as replacement of feature vectors with ellipsoidal regions of uncertainty. The authors establish connections between robustness and consistency of kernelized SVMs and LASSO, and conclude the chapter by showing how robustness can be used to control the generalization error of learning algorithms.

Chapter 15, by Le Roux et al., points out that optimization problems arising in machine learning are often proxies for the "real" problem of minimizing the generalization error. The authors use this fact to explicitly estimate the uncertain gradient of this true function of interest. Thus, a contrast between optimization and learning is provided by viewing the relationship between the Hessian of the objective function and the covariance matrix with respect to sample instances. The insight thus gained guides the authors' proposal for a more efficient learning method.

In Chapter 16, Audibert et al. describe algorithms for optimizing functions over finite sets where the function value is observed only stochastically. The aim is to identify the input that has the highest expected value by repeatedly evaluating the function for different inputs. This setting occurs naturally in many learning tasks. The authors discuss optimal strategies for optimization with a fixed budget of function evaluations, as well as strategies for minimizing the number of function evaluations while requiring a (ϵ, δ)-PAC optimality guarantee on the returned solution.

Chapter 17, by Scheinberg and Ma, focuses on sparse inverse covariance selection (SICS), an important problem that arises in learning with Gaussian Markov random fields. The chapter reviews several of the published approaches for solving SICS; it provides a detailed presentation of coordinate descent approaches to SICS and a technique called "alternating linearization" that is based on variable splitting (see also Chapter 9). Nesterov-style acceleration can be used to improve the theoretical rate of convergence. As is common for most methods dealing with SICS, the bottleneck lies in enforcing the positive definiteness constraint on the learned variable; some remarks on numerical performance are also provided.

Chapter 18, by Krishnamurthy et al., also studies SICS, but focuses on obtaining a full path of solutions as the regularization parameter varies over an interval. Despite a high theoretical complexity of $O(n^5)$, the methods are reported to perform well in practice, thanks to a combination of conjugate gradients, scaling, and warm restarting. The method could be a strong contender for small to medium-sized problems.

1.4 References

A. Ben-Tal, L. El Ghaoui, and A. Nemirovski. *Robust Optimization.* Princeton University Press, Princeton and Oxford, 2009.

D. P. Bertsekas. *Nonlinear Programming.* Athena Scientific, Belmont, Massachusetts, second edition, 1999.

L. Bottou and Y. LeCun. Large-scale online learning. In *Advances in Neural Information Processing Systems*, Cambridge, Massachusetts, 2004. MIT Press.

P. S. Bradley and O. L. Mangasarian. Massive data discrimination via linear support vector machines. *Optimization Methods and Software*, 13(1):1–10, 2000.

R. H. Byrd, G. M. Chin, W. Neveitt, and J. Nocedal. On the use of stochastic Hessian information in unconstrained optimization. Technical report, Optimization Technology Center, Northwestern University, June 2010.

N. Christianini and J. Shawe-Taylor. *An Introduction to Support Vector Machines and Other Kernel-Based Learning Methods.* Cambridge University Press, New York, NY, 2000.

P. L. Combettes and V. R. Wajs. Signal recovery by proximal forward-backward splitting. *Multiscale Modeling and Simulation*, 4(4):1168–1200, 2005.

Y. H. Dai and R. Fletcher. New algorithms for singly linearly constrained quadratic programs subject to lower and upper bounds. *Mathematical Programming, Series A*, 106:403–421, 2006.

I. Daubechies, M. Defriese, and C. De Mol. An iterative thresholding algorithm for linear inverse problems with a sparsity constraint. *Communications on Pure and Applied Mathematics*, 57(11):1413–1457, 2004.

D. A. D'Esopo. A convex programming procedure. *Naval Research Logistics Quarterly*, 6(1):33–42, 1959.

R. Fan, P. Chen, and C. Lin. Working set selection using second-order information for training SVM. *Journal of Machine Learning Research*, 6:1889–1918, 2005.

M. C. Ferris and T. S. Munson. Interior-point methods for massive support vector machines. *SIAM Journal on Optimization*, 13(3):783–804, 2002.

M. A. T. Figueiredo and R. D. Nowak. An EM algorithm for wavelet-based image restoration. *IEEE Transactions on Image Processing*, 12(8):906–916, 2003.

S. Fine and K. Scheinberg. Efficient SVM training using low-rank kernel representations. *Journal of Machine Learning Research*, 2:243–264, 2001.

V. Franc and S. Sonnenburg. Optimized cutting plane algorithm for support vector machines. In *Proceedings of the 25th International Conference on Machine Learning*, pages 320–327, New York, NY, 2008. ACM.

E. M. Gertz and S. J. Wright. Object-oriented software for quadratic programming. *ACM Transactions on Mathematical Software*, 29(1):58–81, 2003.

T. Hastie, R. Tibshirani, and J. Friedman. *The Elements of Statistical Learning Theory.* Series in Statistics. Springer, second edition, 2009.

C. Hildreth. A quadratic programming procedure. *Naval Research Logistics Quarterly*, 4(1):79–85, 1957.

T. Joachims. Making large-scale support vector machine learning practical. In B. Schölkopf, C. J. C. Burges, and A. J. Smola, editors, *Advances in Kernel Methods: Support Vector Learning*, chapter 11, pages 169–184. MIT Press, Cambridge, Massachusetts, 1999.

T. Joachims. Training linear SVMs in linear time. In *Proceedings of the 12th ACM SIGKDD International Conference on Knowledge Discovery and Data Mining*, pages 217–226, New York, NY, 2006. ACM Press.

T. Joachims and C.-N. J. Yu. Sparse kernel SVMs via cutting-plane training. *Machine Learning Journal*, 76(2–3):179–193, 2009. Special Issue for the European Conference on Machine Learning.

D. Koller and N. Friedman. *Probabilistic Graphical Models: Principles and Techniques*. MIT Press, Cambridge, Massachusetts, 2009.

A. Nemirovski, A. Juditsky, G. Lan, and A. Shapiro. Robust stochastic approximation approach to stochastic programming. *SIAM Journal on Optimization*, 19 (4):1574–1609, 2009.

Y. Nesterov. *Introductory Lectures on Convex Optimization: A Basic Course*. Kluwer Academic Publishers, 2004.

Y. Nesterov. Smooth minimization of nonsmooth functions. *Mathematical Programming, Series A*, 103:127–152, 2005.

Y. Nesterov. Gradient methods for minimizing composite objective function. CORE Discussion Paper 2007/76, CORE, Catholic University of Louvain, September 2007. Revised May 2010.

Y. Nesterov. Primal-dual subgradient methods for convex programs. *Mathematical Programming, Series B*, 120(1):221–259, 2009.

J. Nocedal and S. J. Wright. *Numerical Optimization*. Springer, New York, second edition, 2006.

E. E. Osuna, R. Freund, and F. Girosi. Support vector machines: Training and applications. A. I. Memo 1602, Artificial Intelligence Laboratory, Massachusetts Institute of Technology, March 1997.

J. C. Platt. Fast training of support vector machines using sequential minimal optimization. In B. Schölkopf, C. J. C. Burges, and A. J. Smola, editors, *Advances in Kernel Methods: Support Vector Learning*, pages 185–208, Cambridge, Massachusetts, 1999. MIT Press.

B. Schölkopf and A. J. Smola. *Learning with Kernels*. MIT Press, Cambridge, Massachusetts, 2002.

S. Shalev-Shwartz, Y. Singer, and N. Srebro. Pegasos: Primal Estimated sub-GrAdient SOlver for SVM. In *Proceedings of the 24th International Conference on Machine Learning*, pages 807–814, 2007.

W. Shi, G. Wahba, S. J. Wright, K. Lee, R. Klein, and B. Klein. LASSO-Patternsearch algorithm with application to ophthalmology data. *Statistics and its Interface*, 1:137–153, January 2008.

C. H. Teo, S. V. N. Vishwanathan, A. J. Smola, and Q. V. Le. Bundle methods for regularized risk minimization. *Journal of Machine Learning Research*, 11: 311–365, 2010.

R. Tibshirani. Regression shrinkage and selection via the LASSO. *Journal of the Royal Statistical Society, Series B*, 58(1):267–288, 1996.

P. Tseng and S. Yun. A coordinate gradient descent method for nonsmooth separable minimization. *Mathematical Programming, Series B*, 117:387–423, June 2009.

B. Turlach, W. N. Venables, and S. J. Wright. Simultaneous variable selection.

Technometrics, 47(3):349–363, 2005.

V. N. Vapnik. *The Nature of Statistical Learning Theory*. Statistics for Engineering and Information Science. Springer, second edition, 1999.

M. J. Wainwright and M. I. Jordan. *Graphical Models, Exponential Families, and Variational Inference*. Now Publishers, 2008.

S. J. Wright. Accelerated block-coordinate relaxation for regularized optimization. Technical report, Computer Sciences Department, University of Wisconsin-Madison, August 2010.

S. J. Wright, R. D. Nowak, and M. A. T. Figueiredo. Sparse reconstruction by separable approximation. *IEEE Transactions on Signal Processing*, 57:2479–2493, August 2009.

L. Xiao. Dual averaging methods for regularized stochastic learning and online optimization. *Journal of Machine Learning Research*, 11:2543–2596, 2010.

L. Zanni, T. Serafini, and G. Zanghirati. Parallel software for training large scale support vector machines on multiprocessor systems. *Journal of Machine Learning Research*, 7:1467–1492, 2006.

P. Zhao, G. Rocha, and B. Yu. The composite absolute penalties family for grouped and hierarchical model selection. *Annals of Statistics*, 37(6A):3468–3497, 2009.

M. Zinkevich. Online convex programming and generalized infinitesimal gradient ascent. In *Proceedings of the 20th International Conference on Machine Learning*, pages 928–936, 2003.

2   Convex Optimization with Sparsity-Inducing Norms

Francis Bach
INRIA - Willow Project-Team
23, avenue d'Italie, 75013 PARIS

francis.bach@inria.fr

Rodolphe Jenatton
INRIA - Willow Project-Team
23, avenue d'Italie, 75013 PARIS

rodolphe.jenatton@inria.fr

Julien Mairal
INRIA - Willow Project-Team
23, avenue d'Italie, 75013 PARIS

julien.mairal@inria.fr

Guillaume Obozinski
INRIA - Willow Project-Team
23, avenue d'Italie, 75013 PARIS

guillaume.obozinski@inria.fr

2.1   Introduction

The principle of parsimony is central to many areas of science: the simplest explanation of a given phenomenon should be preferred over more complicated ones. In the context of machine learning, it takes the form of variable or feature selection, and it is commonly used in two situations. First, to make the model or the prediction more interpretable or computationally cheaper to use, that is, even if the underlying problem is not sparse, one looks for the best sparse approximation. Second, sparsity can also be used given prior knowledge that the model should be sparse.

For variable selection in linear models, parsimony may be achieved directly by penalization of the empirical risk or the log-likelihood by the cardinality of the support of the weight vector. However, this leads to hard combinatorial problems (see, e.g., Tropp, 2004). A traditional convex approximation of the problem is to replace the cardinality of the support with the ℓ_1-norm. Estimators may then be obtained as solutions of convex programs.

Casting sparse estimation as convex optimization problems has two main benefits. First, it leads to efficient estimation algorithms—and this chapter focuses primarily on these. Second, it allows a fruitful theoretical analysis answering fundamental questions related to estimation consistency, prediction efficiency (Bickel et al., 2009; Negahban et al., 2009), or model consistency (Zhao and Yu, 2006; Wainwright, 2009). In particular, when the sparse model is assumed to be well specified, regularization by the ℓ_1-norm is adapted to high-dimensional problems, where the number of variables to learn from may be exponential in the number of observations.

Reducing parsimony to finding the model of lowest cardinality turns out to be limiting, and *structured parsimony* has emerged as a natural extension, with applications to computer vision (Jenatton et al., 2010b), text processing (Jenatton et al., 2010a) and bioinformatics (Kim and Xing, 2010; Jacob et al., 2009). Structured sparsity may be achieved through regularizing by norms other than the ℓ_1-norm. In this chapter, we focus primarily on norms which can be written as linear combinations of norms on subsets of variables (section 2.1.1). One main objective of this chapter is to present methods which are adapted to most sparsity-inducing norms with loss functions potentially beyond least squares.

Finally, similar tools are used in other communities such as signal processing. While the objectives and the problem setup are different, the resulting convex optimization problems are often very similar, and most of the techniques reviewed in this chapter also apply to sparse estimation problems in signal processing.

This chapter is organized as follows. In section 2.1.1, we present the optimization problems related to sparse methods, and in section 2.1.2, we review various optimization tools that will be needed throughout the chapter. We then quickly present in section 2.2 generic techniques that are not best suited to sparse methods. In subsequent sections, we present methods which are well adapted to regularized problems: proximal methods in section 2.3, block coordinate descent in section 2.4, reweighted ℓ_2-methods in section 2.5, and working set methods in section 2.6. We provide quantitative evaluations of all of these methods in section 2.7.

2.1.1 Loss Functions and Sparsity-Inducing Norms

We consider in this chapter convex optimization problems of the form

$$\min_{\boldsymbol{w}\in\mathbb{R}^p} f(\boldsymbol{w}) + \lambda\Omega(\boldsymbol{w}), \qquad (2.1)$$

where $f : \mathbb{R}^p \to \mathbb{R}$ is a convex differentiable function and $\Omega : \mathbb{R}^p \to \mathbb{R}$ is a sparsity-inducing—typically nonsmooth and non-Euclidean—norm.

In supervised learning, we predict outputs y in $\mathcal{Y}$ from observations $\boldsymbol{x}$ in $\mathcal{X}$; these observations are usually represented by p-dimensional vectors, so that $\mathcal{X} = \mathbb{R}^p$. In this supervised setting, f generally corresponds to the empirical risk of a loss function $\ell : \mathcal{Y} \times \mathbb{R} \to \mathbb{R}_+$. More precisely, given n pairs of data points $\{(\boldsymbol{x}^{(i)}, y^{(i)}) \in \mathbb{R}^p \times \mathcal{Y}; \; i = 1, \ldots, n\}$, we have for linear models $f(\boldsymbol{w}) := \frac{1}{n}\sum_{i=1}^n \ell(y^{(i)}, \boldsymbol{w}^T\boldsymbol{x}^{(i)})$. Typical examples of loss functions are the square loss for least squares regression, that is, $\ell(y, \hat{y}) = \frac{1}{2}(y - \hat{y})^2$ with y in $\mathbb{R}$, and the logistic loss $\ell(y, \hat{y}) = \log(1 + e^{-y\hat{y}})$ for logistic regression, with y in $\{-1, 1\}$. We refer the reader to Shawe-Taylor and Cristianini (2004) for a more complete description of loss functions.

When one knows a priori that the solutions $\boldsymbol{w}^\star$ of problem (2.1) have only a few non-zero coefficients, Ω is often chosen to be the ℓ_1-norm, that is, $\Omega(\boldsymbol{w}) = \sum_{j=1}^p |\boldsymbol{w}_j|$. This leads, for instance, to the Lasso (Tibshirani, 1996) with the square loss and to the ℓ_1-regularized logistic regression (see, for instance, Shevade and Keerthi, 2003; Koh et al., 2007) with the logistic loss. Regularizing by the ℓ_1-norm is known to induce sparsity in the sense that a number of coefficients of $\boldsymbol{w}^\star$, depending on the strength of the regularization, will be *exactly* equal to zero.

In some situations, for example, when encoding categorical variables by binary dummy variables, the coefficients of $\boldsymbol{w}^\star$ are naturally partitioned in subsets, or *groups*, of variables. It is then natural to *simultaneously* select or remove all the variables forming a group. A regularization norm explicitly exploiting this group structure can be shown to improve the prediction performance and/or interpretability of the learned models (Yuan and Lin, 2006; Roth and Fischer, 2008; Huang and Zhang, 2010; Obozinski et al., 2010; Lounici et al., 2009). Such a norm might, for instance, take the form

$$\Omega(\boldsymbol{w}) := \sum_{g\in\mathcal{G}} d_g\|\boldsymbol{w}_g\|_2, \qquad (2.2)$$

where $\mathcal{G}$ is a partition of $\{1, \ldots, p\}$, $(d_g)_{g\in\mathcal{G}}$ are positive weights, and $\boldsymbol{w}_g$ denotes the vector in $\mathbb{R}^{|g|}$ recording the coefficients of $\boldsymbol{w}$ indexed by g in $\mathcal{G}$. Without loss of generality we may assume all weights $(d_g)_{g\in\mathcal{G}}$ to be equal to one. As defined in Eq. (2.2), Ω is known as a mixed ℓ_1/ℓ_2-norm. It behaves like an ℓ_1-norm on the vector $(\|\boldsymbol{w}_g\|_2)_{g \in \mathcal{G}}$ in $\mathbb{R}^{|\mathcal{G}|}$, and therefore Ω induces group sparsity. In other words, each $\|\boldsymbol{w}_g\|_2$, and equivalently each $\boldsymbol{w}_g$, is encouraged to be set to zero. On the other hand, within the groups g in $\mathcal{G}$, the ℓ_2-norm does not promote sparsity. Combined with the square loss, it leads to the group Lasso formulation (Yuan and Lin, 2006). Note that when $\mathcal{G}$ is the set of singletons, we retrieve the ℓ_1-norm. More general mixed ℓ_1/ℓ_q-norms for $q > 1$ are also used in the literature (Zhao et al., 2009):

$$\Omega(\boldsymbol{w}) = \sum_{g \in \mathcal{G}} \|\boldsymbol{w}_g\|_q := \sum_{g \in \mathcal{G}} \left\{ \sum_{j \in g} |\boldsymbol{w}_j|^q \right\}^{1/q}.$$

In practice, though, the ℓ_1/ℓ_2- and ℓ_1/ℓ_∞-settings remain the most popular ones.

In an attempt to better encode structural links between variables at play (e.g., spatial or hierarchical links related to the physics of the problem at hand), recent research has explored the setting where $\mathcal{G}$ can contain groups of variables that *overlap* (Zhao et al., 2009; Bach, 2008a; Jenatton et al., 2009; Jacob et al., 2009; Kim and Xing, 2010; Schmidt and Murphy, 2010). In this case, Ω is still a norm, and it yields sparsity in the form of specific patterns of variables. More precisely, the solutions $\boldsymbol{w}^\star$ of problem (2.1) can be shown to have a set of zero coefficients, or simply *zero pattern*, that corresponds to a union of some groups g in $\mathcal{G}$ (Jenatton et al., 2009). This property makes it possible to control the sparsity patterns of $\boldsymbol{w}^\star$ by appropriately defining the groups in $\mathcal{G}$. This form of *structured sparsity* has proved to be useful notably in the context of hierarchical variable selection (Zhao et al., 2009; Bach, 2008a; Schmidt and Murphy, 2010), multitask regression of gene expressions (Kim and Xing, 2010), and the design of localized features in face recognition (Jenatton et al., 2010b).

2.1.2   Optimization Tools

The tools used in this chapter are relatively basic and should be accessible to a broad audience. Most of them can be found in classic books on convex optimization (Boyd and Vandenberghe, 2004; Bertsekas, 1999; Borwein and Lewis, 2006; Nocedal and Wright, 2006), but for self-containedness, we present here a few of them related to nonsmooth unconstrained optimization.

2.1.2.1 Subgradients

Given a convex function $g : \mathbb{R}^p \rightarrow \mathbb{R}$ and a vector $\boldsymbol{w}$ in $\mathbb{R}^p$, let us define the *subdifferential* of g at $\boldsymbol{w}$ as

$$\partial g(\boldsymbol{w}) := \{\boldsymbol{z} \in \mathbb{R}^p \mid g(\boldsymbol{w}) + \boldsymbol{z}^T(\boldsymbol{w}' - \boldsymbol{w}) \leq g(\boldsymbol{w}') \text{ for all vectors } \boldsymbol{w}' \in \mathbb{R}^p\}.$$

The elements of $\partial g(\boldsymbol{w})$ are called the *subgradients* of g at $\boldsymbol{w}$. This definition admits a clear geometric interpretation: any subgradient $\boldsymbol{z}$ in $\partial g(\boldsymbol{w})$ defines an affine function $\boldsymbol{w}' \mapsto g(\boldsymbol{w}) + \boldsymbol{z}^T(\boldsymbol{w}' - \boldsymbol{w})$ which is tangent to the graph of the function g. Moreover, there is a bijection (one-to-one correspondence) between such tangent affine functions and the subgradients. Let us now illustrate how subdifferentials can be useful for studying nonsmooth optimization problems with the following proposition:

Proposition 2.1 (subgradients at optimality).
For any convex function $g : \mathbb{R}^p \rightarrow \mathbb{R}$, a point $\boldsymbol{w}$ in $\mathbb{R}^p$ is a global minimum of g if and only if the condition $0 \in \partial g(\boldsymbol{w})$ holds.

Note that the concept of a subdifferential is useful mainly for nonsmooth functions. If g is differentiable at $\boldsymbol{w}$, the set $\partial g(\boldsymbol{w})$ is indeed the singleton $\{\nabla g(\boldsymbol{w})\}$, and the condition $0 \in \partial g(\boldsymbol{w})$ reduces to the classical first-order optimality condition $\nabla g(\boldsymbol{w}) = 0$. As a simple example, let us consider the following optimization problem:

$$\min_{w \in \mathbb{R}} \frac{1}{2}(x - w)^2 + \lambda |w|.$$

Applying proposition 2.1 and noting that the subdifferential $\partial | \cdot |$ is $\{+1\}$ for $w > 0$, $\{-1\}$ for $w < 0$, and $[-1, 1]$ for $w = 0$, it is easy to show that the unique solution admits a closed form called the *soft-thresholding* operator, following a terminology introduced by Donoho and Johnstone (1995); it can be written

$$w^\star = \begin{cases} 0 & \text{if } |x| \leq \lambda \\ (1 - \frac{\lambda}{|x|})x & \text{otherwise.} \end{cases} \tag{2.3}$$

This operator is a core component of many optimization techniques for sparse methods, as we shall see later.

2.1.2.2 Dual Norm and Optimality Conditions

The next concept we introduce is the dual norm, which is important to the study of sparsity-inducing regularizations (Jenatton et al., 2009; Bach, 2008a; Negahban et al., 2009). It arises notably in the analysis of estimation bounds (Negahban et al., 2009) and in the design of working-set strategies, as will be shown in section 2.6. The dual norm Ω^* of the norm Ω is defined for any vector $\boldsymbol{z}$ in $\mathbb{R}^p$ by

$$\Omega^*(\boldsymbol{z}) := \max_{\boldsymbol{w} \in \mathbb{R}^p} \boldsymbol{z}^T \boldsymbol{w} \text{ such that } \Omega(\boldsymbol{w}) \leq 1.$$

Moreover, the dual norm of Ω^* is Ω itself, and as a consequence, the formula above also holds if the roles of Ω and Ω^* are exchanged. It is easy to show that in the case of an ℓ_q-norm, $q \in [1; +\infty]$, the dual norm is the $\ell_{q'}$-norm, with q' in $[1; +\infty]$ such that $\frac{1}{q} + \frac{1}{q'} = 1$. In particular, the ℓ_1- and ℓ_∞-norms are dual to each other, and the ℓ_2-norm is self-dual (dual to itself).

The dual norm plays a direct role in computing optimality conditions of sparse regularized problems. By applying proposition 2.1 to equation (2.1), a little calculation shows that a vector $\boldsymbol{w}$ in $\mathbb{R}^p$ is optimal for equation (2.1) if and only if $-\frac{1}{\lambda} \nabla f(\boldsymbol{w}) \in \partial \Omega(\boldsymbol{w})$ with

$$\partial \Omega(\boldsymbol{w}) = \begin{cases} \{\boldsymbol{z} \in \mathbb{R}^p; \; \Omega^*(\boldsymbol{z}) \leq 1\} \text{ if } \boldsymbol{w} = 0, \\ \{\boldsymbol{z} \in \mathbb{R}^p; \; \Omega^*(\boldsymbol{z}) \leq 1 \text{ and } \boldsymbol{z}^T \boldsymbol{w} = \Omega(\boldsymbol{w})\} \text{ otherwise.} \end{cases} \tag{2.4}$$

As a consequence, the vector 0 is a solution if and only if $\Omega^*\big(\nabla f(0)\big) \leq \lambda$.

These general optimality conditions can be specified to the Lasso problem (Tibshirani, 1996), also known as basis pursuit (Chen et al., 1999):

$$\min_{\boldsymbol{w} \in \mathbb{R}^p} \frac{1}{2} \|\boldsymbol{y} - \boldsymbol{X}\boldsymbol{w}\|_2^2 + \lambda \|\boldsymbol{w}\|_1, \tag{2.5}$$

where $\boldsymbol{y}$ is in $\mathbb{R}^n$, and $\boldsymbol{X}$ is a design matrix in $\mathbb{R}^{n \times p}$. From equation (2.4) and since the ℓ_∞-norm is the dual of the ℓ_1-norm, we obtain that necessary and sufficient optimality conditions are

$$\forall j = 1, \ldots, p, \quad \begin{cases} |\boldsymbol{X}_j^T(\boldsymbol{y} - \boldsymbol{X}\boldsymbol{w})| \; \leq \; \lambda & \text{if } \boldsymbol{w}_j = 0, \\ \boldsymbol{X}_j^T(\boldsymbol{y} - \boldsymbol{X}\boldsymbol{w}) \; = \; \lambda \operatorname{sgn}(\boldsymbol{w}_j) & \text{if } \boldsymbol{w}_j \neq 0, \end{cases} \tag{2.6}$$

where $\boldsymbol{X}_j$ denotes the jth column of $\boldsymbol{X}$, and $\boldsymbol{w}_j$ the jth entry of $\boldsymbol{w}$. As we will see in section 2.6.1, it is possible to derive interesting properties of the Lasso from these conditions, as well as efficient algorithms for solving it. We have presented a useful duality tool for norms. More generally, there exists a related concept for convex functions, which we now introduce.

2.1.2.3 *Fenchel Conjugate and Duality Gaps*

Let us denote by f^* the Fenchel conjugate of f (Rockafellar, 1997), defined by

$$f^*(\boldsymbol{z}) := \sup_{\boldsymbol{w} \in \mathbb{R}^p} [\boldsymbol{z}^T \boldsymbol{w} - f(\boldsymbol{w})].$$

The Fenchel conjugate is related to the dual norm. Let us define the indicator function ι_Ω such that $\iota_\Omega(\boldsymbol{w})$ is equal to 0 if $\Omega(\boldsymbol{w}) \le 1$ and $+\infty$ otherwise. Then ι_Ω is a convex function and its conjugate is exactly the dual norm Ω^*.

For many objective functions, the Fenchel conjugate admits closed forms, and therefore can be computed efficiently (Borwein and Lewis, 2006). Then it is possible to derive a duality gap for problem (2.1) from standard Fenchel duality arguments (see Borwein and Lewis, 2006), as shown below.

Proposition 2.2 (duality for problem (2.1)).
If f^* and Ω^* are respectively the Fenchel conjugate of a convex and differentiable function f, and the dual norm of Ω, then we have

$$\max_{\boldsymbol{z}\in\mathbb{R}^p:\,\Omega^*(\boldsymbol{z})\le\lambda} -f^*(\boldsymbol{z}) \;\le\; \min_{\boldsymbol{w}\in\mathbb{R}^p} f(\boldsymbol{w}) + \lambda\Omega(\boldsymbol{w}). \tag{2.7}$$

Moreover, equality holds as soon as the domain of f has a non-empty interior.

Proof. This result is a specific instance of theorem 3.3.5 in Borwein and Lewis (2006). In particular, we use the facts that (a) the conjugate of a norm Ω is the indicator function ι_{Ω^*} of the unit ball of the dual norm Ω^*, and that (b) the subdifferential of a differentiable function (here, f) reduces to its gradient.

If $\boldsymbol{w}^\star$ is a solution of equation (2.1), and $\boldsymbol{w}, \boldsymbol{z}$ in $\mathbb{R}^p$ are such that $\Omega^*(\boldsymbol{z}) \le \lambda$, this proposition implies that we have

$$f(\boldsymbol{w}) + \lambda\Omega(\boldsymbol{w}) \ge f(\boldsymbol{w}^\star) + \lambda\Omega(\boldsymbol{w}^\star) \ge -f^*(\boldsymbol{z}). \tag{2.8}$$

The difference between the left and right terms of equation (2.8) is called a duality gap. It represents the difference between the value of the primal objective function $f(\boldsymbol{w}) + \lambda\Omega(\boldsymbol{w})$ and a dual objective function $-f^*(\boldsymbol{z})$, where $\boldsymbol{z}$ is a dual variable. The proposition says that the duality gap for a pair of optima $\boldsymbol{w}^\star$ and $\boldsymbol{z}^\star$ of the primal and dual problem is equal to zero. When the optimal duality gap is zero, we say that *strong duality* holds.

Duality gaps are important in convex optimization because they provide an upper bound on the difference between the current value of an objective function and the optimal value which allows setting proper stopping criteria for iterative optimization algorithms. Given a current iterate $\boldsymbol{w}$, computing a duality gap requires choosing a "good" value for $\boldsymbol{z}$ (and in particular a feasible one). Given that at optimality, $\boldsymbol{z}(\boldsymbol{w}^\star) = \nabla f(\boldsymbol{w}^\star)$ is the unique solution to the dual problem, a natural choice of dual variable is $\boldsymbol{z} = \min\left(1, \frac{\lambda}{\Omega^*(\nabla f(\boldsymbol{w}))}\right)\nabla f(\boldsymbol{w})$, which reduces to $\boldsymbol{z}(\boldsymbol{w}^\star)$ at the optimum and therefore yields a zero duality gap at optimality.

Note that in most formulations we will consider,* the function f is of the form $f(\boldsymbol{w}) = \psi(\boldsymbol{X}\boldsymbol{w})$ with $\psi : \mathbb{R}^n \to \mathbb{R}$, and $\boldsymbol{X}$ is a design matrix; typically, the Fenchel conjugate of ψ is easy to compute, whereas the design matrix $\boldsymbol{X}$ makes it hard[1] to compute f^*. In that case, (2.1) can be rewritten as

$$\min_{\boldsymbol{w}\in\mathbb{R}^p, \boldsymbol{u}\in\mathbb{R}^n} \quad \psi(\boldsymbol{u}) + \lambda\,\Omega(\boldsymbol{w}) \qquad \text{s.t.} \ \ \boldsymbol{u} = \boldsymbol{X}\boldsymbol{w}, \tag{2.9}$$

and equivalently as the optimization of the Lagrangian

$$\min_{\boldsymbol{w}\in\mathbb{R}^p, \boldsymbol{u}\in\mathbb{R}^n} \quad \max_{\boldsymbol{\alpha}\in\mathbb{R}^n} \quad \left(\psi(\boldsymbol{u}) - \lambda\boldsymbol{\alpha}^T\boldsymbol{u}\right) + \lambda\left(\Omega(\boldsymbol{w}) + \boldsymbol{\alpha}^T\boldsymbol{X}\boldsymbol{w}\right), \tag{2.10}$$

which is obtained by introducing the Lagrange multiplier $\boldsymbol{\alpha}$. The corresponding Fenchel dual[2] is then

$$\max_{\boldsymbol{\alpha}\in\mathbb{R}^n} \quad -\psi^*(\lambda\boldsymbol{\alpha}) \quad \text{such that} \quad \Omega^*(\boldsymbol{X}^T\boldsymbol{\alpha}) \leq \lambda, \tag{2.11}$$

which does not require any inversion of $\boldsymbol{X}$.

2.2   Generic Methods

The problem defined in equation (2.1) is convex as soon as both the loss f and the regularizer Ω are convex functions. In this section, we consider optimization strategies which are essentially blind to problem structure, namely, subgradient descent (e.g., see Bertsekas, 1999), which is applicable under weak assumptions, and interior-point methods solving reformulations such as linear programs (LP), quadratic programs (QP) or, more generally, second-order cone programming (SOCP) or semidefinite programming (SDP) problems (e.g., see Boyd and Vandenberghe, 2004). The latter strategy is usually possible only with the square loss and makes use of general-purpose optimization toolboxes.

2.2.1   Subgradient descent

For all convex unconstrained problems, subgradient descent can be used as soon as one subgradient can be computed efficiently. In our setting, this is possible when a subgradient of the loss f, and a subgradient of the regularizer Ω can be computed. This is true for all classical settings, and leads to the

---

1. It would require computing the pseudo-inverse of $\boldsymbol{X}$.
2. Fenchel conjugacy naturally extends to this case (for more details see Borwein and Lewis, 2006, theorem 3.3.5).

iterative algorithm

$$\boldsymbol{w}_{t+1} = \boldsymbol{w}_t - \frac{\alpha}{t}(\boldsymbol{s} + \lambda \boldsymbol{s}'), \text{ where } \boldsymbol{s} \in \partial f(\boldsymbol{w}_t), \ \boldsymbol{s}' \in \partial \Omega(\boldsymbol{w}_t)$$

with α a positive parameter. These updates are globally convergent. More precisely, we have, from Nesterov (2004), $F(\boldsymbol{w}_t) - \min_{\boldsymbol{w} \in \mathbb{R}^p} F(\boldsymbol{w}) = O(\frac{1}{\sqrt{t}})$. However, the convergence is in practice slow (i.e., many iterations are needed), and the solutions obtained are usually not sparse. This is to be contrasted with the proximal methods presented in the next section, which are less generic but more adapted to sparse problems.

2.2.2   Reformulation as LP, QP, SOCP, or SDP

For all the sparsity-inducing norms we consider in this chapter, the corresponding regularized least-squares problem can be represented by standard mathematical programming problems, all of them being SDPs, and often simpler (e.g., QP). For example, for the ℓ_1-norm regularized least-squares regression, we can reformulate $\min_{w \in \mathbb{R}^p} \frac{1}{2n}\|\boldsymbol{y} - \boldsymbol{X}\boldsymbol{w}\|_2^2 + \lambda \Omega(\boldsymbol{w})$ as

$$\min_{\boldsymbol{w}_+, \boldsymbol{w}_- \in \mathbb{R}_+^p} \frac{1}{2n}\|\boldsymbol{y} - \boldsymbol{X}\boldsymbol{w}_+ + \boldsymbol{X}\boldsymbol{w}_-\|_2^2 + \lambda(1^\top \boldsymbol{w}_+ + 1^\top \boldsymbol{w}_-),$$

which is a quadratic program. Other problems can be cast similarly (for the trace-norm, see Fazel et al., 2001; Bach, 2008b).

General-purpose toolboxes can then be used to get solutions with high precision (low duality gap). However, in the context of machine learning, this is inefficient for two reasons: (1) these toolboxes are generic and blind to problem structure and tend to be too slow, or cannot even run because of memory problems; (2) as outlined by Bottou and Bousquet (2007), high precision is not necessary for machine learning problems, and a duality gap of the order of machine precision (which would be a typical result from toolboxes) is not necessary.

2.3   Proximal Methods

2.3.1   Principle of Proximal Methods

Proximal methods are specifically tailored to optimize an objective of the form (2.1), that is, an objective which can be written as the sum of a generic differentiable function f with Lipschitz gradient, and a non-differentiable function $\lambda \Omega$. They have drawn increasing attention in the machine learning community, especially because of their convergence rates (optimal for the class of first-order techniques) and their ability to deal with large nonsmooth convex problems (e.g., Nesterov 2007; Beck and Teboulle 2009; Wright et al. 2009; Combettes and Pesquet 2010).

Proximal methods can be described as follows. At each iteration the function f is linearized around the current point and a problem of the form

$$\min_{\boldsymbol{w}\in\mathbb{R}^p} \ f(\boldsymbol{w}^t)+\nabla f(\boldsymbol{w}^t)^T(\boldsymbol{w}-\boldsymbol{w}^t) + \lambda\Omega(\boldsymbol{w}) + \frac{L}{2}\|\boldsymbol{w}-\boldsymbol{w}^t\|_2^2 \qquad (2.12)$$

is solved. The quadratic term, called the proximal term, keeps the update in a neighborhood of the current iterate $\boldsymbol{w}^t$ where f is close to its linear approximation; $L > 0$ is a parameter which should essentially be an upper bound on the Lipschitz constant of ∇f and is typically set with a line search. This problem can be rewritten as

$$\min_{\boldsymbol{w}\in\mathbb{R}^p} \ \frac{1}{2}\|\boldsymbol{w} - \left(\boldsymbol{w}^t - \tfrac{1}{L}\nabla f(\boldsymbol{w}^t) \right)\|_2^2 + \tfrac{\lambda}{L}\Omega(\boldsymbol{w}). \qquad (2.13)$$

It should be noted that when the nonsmooth term Ω is not present, the solution of (2.13) just yields the standard gradient update rule $\boldsymbol{w}^{t+1} \leftarrow \boldsymbol{w}^t - \frac{1}{L}\nabla f(\boldsymbol{w}^t)$. Furthermore, if Ω is the indicator function of a set ι_C, that is, defined by $\iota_C(x) = 0$ for $x \in C$ and $\iota_C(x) = +\infty$ otherwise, then solving (2.13) yields the projected gradient update with projection on the set C. This suggests that the solution of the proximal problem provides an interesting generalization of gradient updates, and motivates the introduction of the notion of a *proximal operator* associated with the regularization term $\lambda\Omega$.

The proximal operator, which we will denote as $\mathrm{Prox}_{\mu\Omega}$, was defined by Moreau (1962) as the function that maps a vector $\boldsymbol{u} \in \mathbb{R}^p$ to the unique[3] solution of

$$\min_{\boldsymbol{w}\in\mathbb{R}^p} \ \frac{1}{2}\|\boldsymbol{u} - \boldsymbol{w}\|^2 + \mu\,\Omega(\boldsymbol{w}). \qquad (2.14)$$

This operator is clearly central to proximal methods since their main step consists in computing $\mathrm{Prox}_{\frac{\lambda}{L}\Omega}\big(\boldsymbol{w}^t - \frac{1}{L}\nabla f(\boldsymbol{w}^t)\big)$.

In section 2.3.3, we present analytical forms of proximal operators associated with simple norms and algorithms to compute them in some more elaborate cases.

2.3.2 Algorithms

The basic proximal algorithm uses the solution of problem (2.13) as the next update $\boldsymbol{w}^{t+1}$; however, fast variants such as the accelerated algorithm

---

3. Because the objective is strongly convex.

presented in Nesterov (2007) or FISTA (Beck and Teboulle, 2009) maintain two variables and use them to combine the solution of (2.13) with information about previous steps. Often, an upper bound on the Lipschitz constant of ∇f is not known, and even if it is, it is often better to obtain a local estimate. A suitable value for L can be obtained by iteratively increasing L by a constant factor until the condition

$$f(\boldsymbol{w}_L^\star) \leq M_f^L(\boldsymbol{w}^t, \boldsymbol{w}_L^\star) := f(\boldsymbol{w}^t) + \nabla f(\boldsymbol{w}^t)^T(\boldsymbol{w}_L^\star - \boldsymbol{w}^t) + \tfrac{L}{2}\|\boldsymbol{w}_L^\star - \boldsymbol{w}^t\|_2^2 \quad (2.15)$$

is met, where $\boldsymbol{w}_L^\star$ denotes the solution of (2.13).

For functions f whose gradients are Lipschitz, the basic proximal algorithm has a global convergence rate in $O(\frac{1}{t})$ where t is the number of iterations of the algorithm. Accelerated algorithms like FISTA can be shown to have global convergence rate in $O(\frac{1}{t^2})$. Perhaps more important, both basic (ISTA) and accelerated (Nesterov, 2007) proximal methods are adaptive in the sense that if f is strongly convex—and the problem is therefore better conditioned—the convergence is actually linear (i.e., with rates in $O(C^t)$ for some constant $C < 1$; see Nesterov 2007). Finally, it should be noted that accelerated schemes are not necessarily descent algorithms, in the sense that the objective does not necessarily decrease at each iteration in spite of the global convergence properties.

2.3.3   Computing the Proximal Operator

Computing the *proximal operator efficiently* and *exactly* is crucial to enjoying the fast convergence rates of proximal methods. We therefore focus here on properties of this operator and on its computation for several sparsity-inducing norms.

■ *Dual proximal operator.* In the case where Ω is a norm, by Fenchel duality the following problem is dual (see proposition 2.2) to problem (2.13):

$$\max_{\boldsymbol{v}\in\mathbb{R}^p} -\frac{1}{2}\left[\|\boldsymbol{v}-\boldsymbol{u}\|_2^2 - \|\boldsymbol{u}\|^2\right] \quad \text{such that} \quad \Omega^*(\boldsymbol{v}) \leq \mu. \quad (2.16)$$

Lemma 2.3 (Relation to dual proximal operator). *Let $Prox_{\mu\Omega}$ be the proximal operator associated with the regularization $\mu\Omega$, where Ω is a norm, and let $Proj_{\{\Omega^*(\cdot)\leq\mu\}}$ be the projector on the ball of radius μ of the dual norm Ω^*. Then $Proj_{\{\Omega^*(\cdot)\leq\mu\}}$ is the proximal operator for the dual problem (2.16) and, denoting the identity I_d, these two operators satisfy the relation*

$$Prox_{\mu\Omega} = I_d - Proj_{\{\Omega^*(\cdot)\leq\mu\}}. \quad (2.17)$$

Proof. By proposition 2.2, if $\boldsymbol{w}^\star$ is optimal for (2.14) and $\boldsymbol{v}^\star$ is optimal for

(2.16), we have[4] $-\boldsymbol{v}^\star = \nabla f(\boldsymbol{w}^\star) = \boldsymbol{w}^\star - \boldsymbol{u}$. Since $\boldsymbol{v}^\star$ is the projection of $\boldsymbol{u}$ on the ball of radius μ of the norm Ω^*, the result follows.

This lemma shows that the proximal operator can always be computed as the residual of a projection on a convex set.

- ℓ_1-*norm regularization.* Using optimality conditions for (2.16) and then (2.17) or subgradient condition (2.4) applied to (2.14), it is easy to check that $\mathrm{Proj}_{\{\|\cdot\|_\infty \leq \mu\}}$ and $\mathrm{Prox}_{\mu\|\cdot\|_1}$ respectively satisfy

$$\left[\mathrm{Proj}_{\{\|\cdot\|_\infty \leq \mu\}}(\boldsymbol{u})\right]_j = \min\left(1, \tfrac{\mu}{|u_j|}\right)u_j \quad \text{and} \quad \left[\mathrm{Prox}_{\mu\|\cdot\|_1}(\boldsymbol{u})\right]_j = \left(1 - \tfrac{\mu}{|u_j|}\right)_+ u_j,$$

for $j \in \{1,\ldots,p\}$, with $(x)_+ := \max(x,0)$. Note that $\mathrm{Prox}_{\mu\|\cdot\|_1}$ is componentwise the *soft-thresholding operator* of Donoho and Johnstone (1995) presented in section 2.1.2.

- ℓ_1-*norm constraint.* Sometimes, the ℓ_1-norm is used as a hard constraint and, in that case, the optimization problem is

$$\min_{\boldsymbol{w}} f(\boldsymbol{w}) \quad \text{such that} \quad \|\boldsymbol{w}\|_1 \leq C.$$

This problem can still be viewed as an instance of (2.1), with Ω defined by $\Omega(\boldsymbol{u}) = 0$ if $\|\boldsymbol{u}\|_1 \leq C$ and $\Omega(\boldsymbol{u}) = +\infty$ otherwise. Proximal methods thus apply, and the corresponding proximal operator is the projection on the ℓ_1-ball, for which efficient pivot algorithms with linear complexity have been proposed (Brucker, 1984; Maculan and Galdino de Paula Jr, 1989).

- ℓ_1/ℓ_q-*norm ("group Lasso").* If $\mathcal{G}$ is a partition of $\{1,\ldots,p\}$, the dual norm of the ℓ_1/ℓ_q-norm is the $\ell_\infty/\ell_{q'}$-norm, with $\frac{1}{q} + \frac{1}{q'} = 1$. It is easy to show that the orthogonal projection on a unit $\ell_\infty/\ell_{q'}$ ball is obtained by projecting each subvector $\boldsymbol{u}_g$ separately on a unit $\ell_{q'}$-ball in $\mathbb{R}^{|g|}$. For the ℓ_1/ℓ_2-norm $\Omega: \boldsymbol{w} \mapsto \sum_{g\in\mathcal{G}} \|\boldsymbol{w}_g\|_2$ we have

$$[\mathrm{Prox}_{\mu\Omega}(\boldsymbol{u})]_g = \left(1 - \frac{\lambda}{\|\boldsymbol{u}_g\|_2}\right)_+ \boldsymbol{u}_g, \quad g \in \mathcal{G}.$$

This is shown easily by considering that the subgradient of the ℓ_2-norm is $\partial\|\boldsymbol{w}\|_2 = \{\frac{\boldsymbol{w}}{\|\boldsymbol{w}\|_2}\}$ if $\boldsymbol{w} \neq \boldsymbol{0}$ or $\partial\|\boldsymbol{w}\|_2 = \{\boldsymbol{z} \mid \|\boldsymbol{z}\|_2 \leq 1\}$ if $\boldsymbol{w} = 0$ and by applying the result of (2.4).

For the ℓ_1/ℓ_∞-norm, whose dual norm is the ℓ_∞/ℓ_1-norm, an efficient algorithm to compute the proximal operator is based on (2.17). Indeed,

---

this equation indicates that the proximal operator can be computed on each group g as the residual of a projection on an ℓ_1-norm ball in $\mathbb{R}^{|g|}$; the latter is done efficiently with the previously mentioned linear-time algorithms.

In general, the case where groups overlap is more complicated because the regularization is no longer separable. Nonetheless, in some cases it is still possible to compute the proximal operator efficiently.

■ *Hierarchical ℓ_1/ℓ_q-norms.* Hierarchical norms were proposed by Zhao et al. (2009). Following Jenatton et al. (2010a), we focus on the case of a norm $\Omega : \boldsymbol{w} \mapsto \sum_{g \in \mathcal{G}} \|\boldsymbol{w}_g\|_q$, with $q \in \{2, \infty\}$, where the set of groups $\mathcal{G}$ is *tree-structured*, meaning either that two groups are disjoint or that one is included in the other. Let $\preceq$ be a total order such that $g_1 \preceq g_2$ if and only if either $g_1 \subset g_2$ or $g_1 \cap g_2 = \emptyset$.[5] Then, if $g_1 \preceq \ldots \preceq g_m$ with $m = |\mathcal{G}|$, and if we define Π_g as (a) the proximal operator $\boldsymbol{w}_g \mapsto \mathrm{Prox}_{\mu\|\cdot\|_q}(\boldsymbol{w}_g)$ on the subspace corresponding to group g, and (b) as the identity on the orthogonal, it can be shown (Jenatton et al., 2010a) that

$$\mathrm{Prox}_{\mu\Omega} = \Pi_{g_m} \circ \ldots \circ \Pi_{g_1}. \tag{2.18}$$

In other words, the proximal operator associated with the norm can be obtained as the composition of the proximal operators associated to individual groups, provided that the ordering of the groups is well chosen. Note that this result does not hold for $q \notin \{1, 2, \infty\}$.

■ *Combined $\ell_1 + \ell_1/\ell_q$-norm (sparse group Lasso).* The possibility of combining an ℓ_1/ℓ_q-norm that takes advantage of sparsity at the group level with an ℓ_1-norm that induces sparsity within the groups is quite natural (Friedman et al., 2010; Sprechmann et al., 2010). Such regularizations are in fact a special case of the hierarchical ℓ_1/ℓ_q-norms presented above, and the corresponding proximal operator is therefore readily computed by applying soft-thresholding and then group soft-thresholding.

■ *Overlapping ℓ_1/ℓ_∞-norms.* When the groups overlap but do not have a tree structure, computing the proximal operator has proved to be more difficult, but it can still be done efficiently when $q = \infty$. Indeed, as shown by Mairal et al. (2010), there exists a dual relation between such an operator and a quadratic min-cost flow problem on a particular graph, which can be tackled using network flow optimization techniques.

---

5. For a tree-structured $\mathcal{G}$ such an order exists.

2.4   (Block) Coordinate Descent Algorithms

Coordinate descent algorithms solving ℓ_1-regularized learning problems go back to Fu (1998). They optimize (exactly or approximately) the objective with respect to one variable at a time while all others are kept fixed.

2.4.1   Coordinate Descent for ℓ_1-Regularization

We first consider the following special case of an ℓ_1-regularized problem:

$$\min_{w \in \mathbb{R}} \frac{1}{2}(w - w_0)^2 + \lambda|w|. \tag{2.19}$$

As shown in (2.3), $w^\star$ can be obtained by *soft-thresholding*:

$$w^\star = \text{Prox}_{\lambda|\cdot|}(w_0) := \left(1 - \frac{\lambda}{|w_0|}\right)_+ w_0 \tag{2.20}$$

2.4.1.1   Lasso Case

In the case of the least-square loss, the minimization with respect to a single coordinate can be written as

$$\min_{\boldsymbol{w}_j \in \mathbb{R}} \nabla_j f(\boldsymbol{w}^t)\,(\boldsymbol{w}_j - \boldsymbol{w}_j^t) + \frac{1}{2}\nabla_{jj}^2 f(\boldsymbol{w}^t)(\boldsymbol{w}_j - \boldsymbol{w}_j^t)^2 + \lambda|\boldsymbol{w}_j|,$$

with $\nabla_j f(\boldsymbol{w}) = \boldsymbol{X}_j^T(\boldsymbol{X}\boldsymbol{w} - \boldsymbol{y})$ and $\nabla_{jj}^2 f(\boldsymbol{w}) = \boldsymbol{X}_j^T \boldsymbol{X}_j$ independent of $\boldsymbol{w}$. Since the above equation is of the form (2.19), it is solved in closed form:

$$\boldsymbol{w}_j^\star = \text{Prox}_{\lambda|\cdot|}\left(\boldsymbol{w}_j^t - \nabla_j f(\boldsymbol{w}_j^t)/\nabla_{jj}^2 f\right). \tag{2.21}$$

In words, $\boldsymbol{w}_j^\star$ is obtained by solving the unregularized problem with respect to coordinate j and soft-thresholding the solution.

This is the update proposed in the shooting algorithm of Fu (1998), which cycles through all variables in a fixed order.[6]

An efficient implementation is obtained if the quantity $\boldsymbol{X}\boldsymbol{w} - \boldsymbol{y}$ or even better $\nabla f(\boldsymbol{w}^t) = \boldsymbol{X}^T\boldsymbol{X}\boldsymbol{w} - \boldsymbol{X}^T\boldsymbol{y}$ is kept updated.[7]

---

6. Coordinate descent with a cyclic order is sometimes called the Gauss-Seidel procedure.
7. In the former case, at each iteration, $\boldsymbol{X}\boldsymbol{w} - \boldsymbol{y}$ can be updated in $\Theta(n)$ operations if w_j changes and $\nabla_{j^{t+1}} f(\boldsymbol{w})$ can always be updated in $\Theta(n)$ operations. The complexity of one cycle is therefore $O(pn)$. However, a better complexity is obtained in the latter case, provided the matrix $\boldsymbol{X}^T\boldsymbol{X}$ is precomputed (with complexity $O(p^2n)$). Indeed, $\nabla f(\boldsymbol{w}^t)$ is updated in $\Theta(p)$ iterations only if w_j does not stay at 0. Otherwise, if w_j stays at 0, the step costs $O(1)$; the complexity of one cycle is therefore $\Theta(ps)$ where s is the number of non-zero variables at the end of the cycle.

2.4.1.2   Smooth loss

For more general smooth losses, such as the logistic loss, the optimization with respect to a single variable cannot be solved in closed form. It is possible to solve it numerically, using a sequence of modified Newton steps as proposed by Shevade and Keerthi (2003). We present here a fast algorithm of Tseng and Yun (2009) based on solving just a quadratic approximation of f with an inexact line search at each iteration.

Given $d = w_j^\star - w_j^t$ where $w_j^\star$ is the solution of (2.21), a line search is performed to choose the largest step of the form $\alpha^k d$ with $\alpha \in (0,1)$, $k \in \mathbb{N}$, such that the following modified Armijo condition is satisfied:

$$F(w^t + \alpha d e_j) - F(w^t) \le \sigma\alpha\big(\nabla_j f(w)d + |w_j^t + d| - |w_j^t|\big)$$

where $F(w) := f(w) + \lambda\,\Omega(w)$ and $\sigma < 1$. Tseng and Yun (2009) show that if f is continuously differentiable and if H^t has a uniformly upper and lower bounded spectrum, the sequence generated by the algorithm is decreasing and its cluster points are stationary points of F. It should be noted that the algorithm generalizes to separable regularizations other than the ℓ_1-norm.

Variants of coordinate descent algorithms have also been considered by Genkin et al. (2007), by Krishnapuram et al. (2005), and by Wu and Lange (2008). Generalizations based on the Gauss-Southwell rule have been considered by Tseng and Yun (2009).

2.4.2   Block Coordinate Descent for ℓ_1/ℓ_2-Regularization

When $\Omega(w)$ is the ℓ_1/ℓ_2-norm with groups $g \in \mathcal{G}$ forming a partition of $\{1,\dots,p\}$, the previous methods are generalized by block coordinate descent (BCD) algorithms, and in particular the algorithm of Tseng and Yun (2009) generalizes easily to that case.

Specifically, at each iteration the BCD generalization solves a problem of the form

$$\min_{w_g \in \mathbb{R}^{|g|}} \nabla_g f(w^t)^T(w_g - w_g^t) + \frac{1}{2}(w_g - w_g^t)^T H_{gg}(w_g - w_g^t) + \lambda\|w_g\|_2, \quad (2.22)$$

where H_{gg} equals or approximates[8] $\nabla_{gg}^2 f(w^t)$. The above problem is solved in closed form if $H_{gg} = h_{gg}I_{|g|}$, in which case the solution $w_g^\star$ is obtained by

---

8. It is, however, not required to have good approximation properties of H_{gg} to obtain convergence guarantees for the algorithm.

group soft-thresholding of the Newton step:

$$\boldsymbol{w}_g^\star = \mathrm{Prox}_{\lambda\|\cdot\|_2}\big(\boldsymbol{w}_g^t - h_{gg}^{-1}\nabla_g f(\boldsymbol{w}_g^t)\big) \quad \text{with} \quad \mathrm{Prox}_{\lambda\|\cdot\|_2}(\boldsymbol{w}) = \left(1 - \frac{\lambda}{\|\boldsymbol{w}\|_2}\right)_+ \boldsymbol{w}.$$

In univariate learning problems regularized by the ℓ_1/ℓ_2-norm, and for the square loss, it is common to orthonormalize the set of variables belonging to a given group (Yuan and Lin, 2006; Wu and Lange, 2008), in which case it is natural to choose $H_{gg} = \nabla_{gg}^2 f(\boldsymbol{w}^t) = I_{|g|}$. If H_{gg} is not a multiple of the identity, the solution of (2.22) can be found by replacing $\lambda\|\boldsymbol{w}_g\|_2$ with $\lambda'\|\boldsymbol{w}_g\|_2^2$ in (2.22), which yields an analytic solution; it is then a standard result in optimization that there exists a value of λ'—which can be found by binary search—such that the obtained solution also solves (2.22). More simply, it is sufficient to choose $H_{gg} = h_{gg}I_{|g|}$ with h_{gg} an approximation of the largest eigenvalue of $\nabla_{gg}^2 f(\boldsymbol{w}^t)$.[9]

In the case of general smooth losses, the descent direction is given by $\boldsymbol{d} = \boldsymbol{w}_g^\star - \boldsymbol{w}_g^t$ with $\boldsymbol{w}_g^\star$ as above and with a stepsize chosen to satisfy the modified Armijo rule

$$F(\boldsymbol{w}^t + \alpha\boldsymbol{d}) - F(\boldsymbol{w}^t) \le \sigma\alpha\big(\nabla_g f(\boldsymbol{w})^T\boldsymbol{d} + \|\boldsymbol{w}_g^t + \boldsymbol{d}\|_2 - \|\boldsymbol{w}_g^t\|\big).$$

2.5 Reweighted-ℓ_2 Algorithms

Approximating a nonsmooth or constrained optimization problem by a series of smooth unconstrained problems is common in optimization (see, e.g., Nesterov, 2005; Boyd and Vandenberghe, 2004; Nocedal and Wright, 2006). In the context of objective functions regularized by sparsity-inducing norms, it is natural to consider variational formulations of these norms in terms of squared ℓ_2-norms, since many efficient methods are available to solve ℓ_2-regularized problems (e.g., linear system solvers for least-squares regression).

In this section, we show on our motivating example of sums of ℓ_2-norms of subsets how such formulations arise (see, e.g., Argyriou et al., 2007; Rakotomamonjy et al., 2008; Jenatton et al., 2010b; Daubechies et al., 2010).

---

9. This can be done easily for joint feature selection in multitask learning, since in that case the Hessian $\nabla_{gg}^2 f(\boldsymbol{w}^t)$ is diagonal (Obozinski et al., 2010).

2.5.1 Variational Formulation for Sums of ℓ_2-Norms

A simple application of the Cauchy-Schwarz inequality and the inequality $\sqrt{ab} \leq \frac{1}{2}(a+b)$ leads to

$$
\Omega(\boldsymbol{w}) \;=\; \sum_{g \in \mathcal{G}} \|\boldsymbol{w}_g\|_2 = \frac{1}{2} \min_{\forall g \in \mathcal{G},\, \boldsymbol{\eta}_g \geqslant 0} \sum_{g \in \mathcal{G}} \left\{ \frac{\|\boldsymbol{w}_g\|_2^2}{\eta_g} + \eta_g \right\}
$$

$$
= \frac{1}{2} \min_{\forall g \in \mathcal{G},\, \boldsymbol{\eta}_g \geqslant 0} \left\{ \sum_{j=1}^{p} \Big(\sum_{g \in \mathcal{G},\, j \in g} \eta_g^{-1} \Big) \boldsymbol{w}_j^2 + \sum_{g \in \mathcal{G}} \eta_g \right\},
$$

with equality if and only if $\forall g \in \mathcal{G}$, $\boldsymbol{\eta}_g = \|\boldsymbol{w}_g\|_2$ (Argyriou et al., 2007; Rakotomamonjy et al., 2008; Jenatton et al., 2010b). In the case of the ℓ_1-norm, it simplifies to $\sum_{j=1}^{p} |\boldsymbol{w}_j| = \frac{1}{2} \min_{\eta \geqslant 0} \sum_{j=1}^{p} \left\{ \frac{w_j^2}{\eta_j} + \eta_j \right\}$.

The variational formulation we presented in the previous proposition allows us to consider the following function $H(\boldsymbol{w}, \boldsymbol{\eta})$ defined as

$$
H(\boldsymbol{w}, \boldsymbol{\eta}) = f(\boldsymbol{w}) + \frac{\lambda}{2} \sum_{j=1}^{p} \left\{ \sum_{g \in \mathcal{G},\, j \in g} \eta_g^{-1} \right\} \boldsymbol{w}_j^2 + \frac{\lambda}{2} \sum_{g \in \mathcal{G}} \eta_g.
$$

It is jointly convex in $(\boldsymbol{w}, \boldsymbol{\eta})$; the minimization with respect to $\boldsymbol{\eta}$ can be done in closed form, and the optimum is equal to $F(\boldsymbol{w}) = f(\boldsymbol{w}) + \lambda\Omega(\boldsymbol{w})$; as for the minimization with respect to $\boldsymbol{w}$, it is an ℓ_2-regularized problem.

Unfortunately, the alternating minimization algorithm that is immediately suggested is not convergent in general, because the function H is not continuous (in particular around $\boldsymbol{\eta}$, which has zero coordinates). In order to make the algorithm convergent, two strategies are usually used:

■ *Smoothing*: we can add a term of the form $\frac{\varepsilon}{2} \sum_{g \in \mathcal{G}} \eta_g^{-1}$, which yields a joint cost function with compact level sets on the set of positive numbers. Alternating minimization algorithms are then convergent (as a consequence of general results on block coordinate descent), and have two different iterations: (1) minimization with respect to $\boldsymbol{\eta}$ in closed form, through $\eta_g = (\|\boldsymbol{w}_g\|_2 + \varepsilon)$, and (2) minimization with respect to $\boldsymbol{w}$, which is an ℓ_2-regularized problem which can be, for example, solved in closed form for the square loss. Note, however, that the second problem does not need to be optimized exactly at all iterations.

■ *First-order method in $\boldsymbol{\eta}$*: While the joint cost function $H(\boldsymbol{\eta}, \boldsymbol{w})$ is not continuous, the function $I(\boldsymbol{\eta}) = \min_{\boldsymbol{w} \in \mathbb{R}^p} H(\boldsymbol{w}, \boldsymbol{\eta})$ is continuous and, under general assumptions, is continuously differentiable; it is thus amenable to first-order methods (e.g., proximal methods, gradient descent). When the groups in $\mathcal{G}$ do not overlap, one sufficient condition is that the function $f(\boldsymbol{w})$ is of the form $f(\boldsymbol{w}) = \psi(\boldsymbol{X}\boldsymbol{w})$, where $\boldsymbol{X} \in \mathbb{R}^{n \times p}$ is any matrix (typically the design matrix), and ψ is a strongly convex function on $\mathbb{R}^n$. This strategy is particularly interesting when evaluating $I(\boldsymbol{\eta})$ is computationally cheap.

2.6 Working-Set Methods

Working-set algorithms address optimization problems by solving an increasing sequence of small subproblems of (2.1). The working set, which we will denote as J, refers to the subset of variables involved in the optimization of these subproblems.

Working-set algorithms proceed as follows: after computing a solution to the problem restricted to the variables in J, global optimality is checked to determine whether the algorithm has to continue. If it does, new variables enter the working set J according to a strategy that has to be defined. Note that we consider only *forward* algorithms, that is, those where the working set grows monotonically. In other words, there are no *backward* steps where variables would be allowed to leave the set J. Provided this assumption is met, it is easy to see that these procedures stop in a finite number of iterations.

This class of algorithms takes advantage of sparsity from a computational point of view (Lee et al., 2007; Szafranski et al., 2007; Bach, 2008a; Roth and Fischer, 2008; Obozinski et al., 2010; Jenatton et al., 2009; Schmidt and Murphy, 2010), since the subproblems that need to be solved are typically much smaller than the original one.

Working-set algorithms require three ingredients:

- *Inner-loop solver*: At each iteration of the working-set algorithm, problem (2.1) has to be solved on J, that is, subject to the additional equality constraint that $\boldsymbol{w}_j = 0$ for all j in J^c:

$$\min_{\boldsymbol{w} \in \mathbb{R}^p} f(\boldsymbol{w}) + \lambda \Omega(\boldsymbol{w}), \text{ such that } \boldsymbol{w}_{J^c} = 0. \qquad (2.23)$$

The computation can be performed by any of the methods presented in this chapter. Working-set algorithms should therefore be viewed as "meta-algorithms". Since solutions for successive working sets are typically close to each other, the approach is efficient if the method chosen can use *warm-restarts*.

- *Computing the optimality conditions*: Given a solution $\boldsymbol{w}^\star$ of problem (2.23), it is then necessary to check whether $\boldsymbol{w}^\star$ is also a solution for the original problem (2.1). This test relies on the duality gaps of problems (2.23) and (2.1). In particular, if $\boldsymbol{w}^\star$ is a solution of problem (2.23), it follows from proposition 2.2 in section 2.1.2 that

$$f(\boldsymbol{w}^\star) + \lambda\Omega(\boldsymbol{w}^\star) + f^*(\nabla f(\boldsymbol{w}^\star)) = 0.$$

In fact, the Lagrangian parameter associated with the equality constraint ensures the feasibility of the dual variable formed from the gradient of f at $\boldsymbol{w}^\star$. In turn, this guarantees that the duality gap of problem (2.23) vanishes. The candidate $\boldsymbol{w}^\star$ is now a solution of the full problem (2.1), that is, without the equality constraint, if and only if

$$\Omega^*(\nabla f(\boldsymbol{w}^\star)) \leq \lambda. \tag{2.24}$$

Condition (2.24) points out that the dual norm Ω^* is a key quantity to monitor the progress of the working-set algorithm (Jenatton et al., 2009). In simple settings, for instance, when Ω is the ℓ_1-norm, checking condition (2.24) can be easily computed since Ω^* is just the ℓ_∞-norm. In this case, condition (2.24) becomes

$$|[\nabla f(\boldsymbol{w}^\star)]_j| \leq \lambda, \text{ for all } j \text{ in } \{1,\dots,p\}.$$

Note that by using the optimality of problem (2.23), the components of the gradient of f indexed by J are already guaranteed to be no greater than λ. For more general sparsity-inducing norms with overlapping groups of variables (see section 2.1.1), the dual norm Ω^* can no longer be computed easily, prompting the need for approximations and upper bounds of Ω^* (Bach, 2008a; Jenatton et al., 2009; Schmidt and Murphy, 2010).

- *Strategy for the growth of the working set*: If condition (2.24) is not satisfied for the current working set J, some inactive variables in J^c have to become active. This point raises the questions of *how many* variables and *how* these variables should be chosen.
First, depending on the structure of Ω, *one* or a *group* of inactive variables have to be considered to enter the working set. Furthermore, one natural way to proceed is to look at the variables that violate condition (2.24) most. In the example of ℓ_1-regularized least-squares regression with normalized predictors, this strategy amounts to selecting the inactive variable that has the highest correlation with the current residual.

The working-set algorithms we have described so far aim at solving problem (2.1) for a fixed value of the regularization parameter λ. However, for specific types of loss and regularization functions, the set of solutions of problem (2.1) can be obtained efficiently for all possible values of λ, which is the topic of the next section.

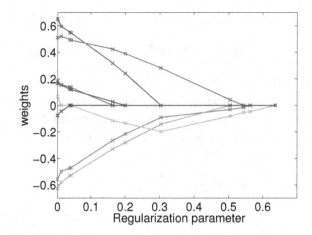

Figure 2.1: The weights $w^\star(\lambda)$ are represented as functions of the regularization parameter λ. When λ increases, more and more coefficients are set to zero. These functions are all piecewise linear.

2.6.1   LARS - Homotopy

We present in this section an active-set method for solving the Lasso problem (Tibshirani, 1996) of equation (2.5). Active-set and working-set methods are very similar; they differ in that active-set methods allow variables returning to zero to exit the set. The problem of the Lasso is, again,

$$\min_{w \in \mathbb{R}^p} \frac{1}{2}\|y - Xw\|_2^2 + \lambda\|w\|_1, \tag{2.25}$$

where y is in $\mathbb{R}^n$, and X is a design matrix in $\mathbb{R}^{n \times p}$. Even though generic working-set methods introduced above could be used to solve this formulation, a specific property of the ℓ_1-norm associated with a quadratic loss makes it possible to address it more efficiently.

Under mild assumptions (which we will detail later), the solution of equation (2.25) is unique, and we denote it by $w^\star(\lambda)$. We apply the term *regularization path* to the function $\lambda \mapsto w^\star(\lambda)$ that associates to a regularization parameter λ the corresponding solution. We will show that this function is piecewise linear, a behavior illustrated in figure 2.1, where the entries of $w^\star(\lambda)$ for a particular instance of the Lasso are represented as functions of λ.

An efficient algorithm can thus be constructed by choosing a particular value of λ for which finding this solution is trivial, and by following the piecewise linear path, computing the directions of the current linear parts and the points where the direction changes (also known as kinks). This piecewise linearity was first discovered and exploited by Markowitz (1952) in the context of portfolio selection; revisited by Osborne et al. (2000), who described a *homotopy* algorithm; and popularized by Efron et al. (2004) with the LARS algorithm.

Let us show how to construct the path. From the optimality conditions presented in equation (2.6), denoting the set of active variables by $J := \{j; |\boldsymbol{X}_j^T(\boldsymbol{y} - \boldsymbol{X}\boldsymbol{w}^\star)| = \lambda\}$, and defining the vector $\boldsymbol{\epsilon}$ in $\{-1; 0; 1\}^p$ as $\boldsymbol{\epsilon} := \mathrm{sgn}\left(\boldsymbol{X}^T(\boldsymbol{y} - \boldsymbol{X}\boldsymbol{w}^\star)\right)$, we have the closed form

$$\begin{cases} \boldsymbol{w}_J^\star(\lambda) & = (\boldsymbol{X}_J^T\boldsymbol{X}_J)^{-1}(\boldsymbol{X}_J^T\boldsymbol{y} - \lambda\boldsymbol{\epsilon}_J) \\ \boldsymbol{w}_{J^c}^\star(\lambda) & = 0, \end{cases}$$

where we have assumed the matrix $\boldsymbol{X}_J^T\boldsymbol{X}_J$ to be invertible (which is a sufficient condition to guarantee the uniqueness of $\boldsymbol{w}^\star$). This is an important point: if one knows the set J and the signs $\boldsymbol{\epsilon}_J$ in advance, then $\boldsymbol{w}^\star(\lambda)$ admits a simple closed form. Moreover, when J and $\boldsymbol{\epsilon}_J$ are fixed, the function $\lambda \mapsto (\boldsymbol{X}_J^T\boldsymbol{X}_J)^{-1}(\boldsymbol{X}_J^T\boldsymbol{y} - \lambda\boldsymbol{\epsilon}_J)$ is affine in λ. With this observation in hand, we can now present the main steps of the path-following algorithm. It basically starts from a trivial solution of the regularization path, then follows the path by exploiting this formula, updating J and $\boldsymbol{\epsilon}_J$ whenever needed so that optimality conditions (2.6) remain satisfied. This procedure requires some assumptions—namely, that (a) the matrix $\boldsymbol{X}_J^T\boldsymbol{X}_J$ is always invertible, and (b) that updating J along the path consists of adding or removing from this set a single variable at the same time. Concretely, we proceed as follows:

1. Set λ to $\|\boldsymbol{X}^T\boldsymbol{y}\|_\infty$ for which it is easy to show from equation (2.6) that $\boldsymbol{w}^\star(\lambda) = 0$ (trivial solution on the regularization path).

2. Set $J := \{j; |\boldsymbol{X}_j^T\boldsymbol{y}| = \lambda\}$.

3. Follow the regularization path by decreasing the value of λ, with the formula $\boldsymbol{w}_J^\star(\lambda) = (\boldsymbol{X}_J^T\boldsymbol{X}_J)^{-1}(\boldsymbol{X}_J^T\boldsymbol{y} - \lambda\boldsymbol{\epsilon}_J)$ keeping $\boldsymbol{w}_{J^c}^\star = 0$, until one of the following events occurs:

   - There exists j in J^c such that $|\boldsymbol{X}_j^T(\boldsymbol{y} - \boldsymbol{X}\boldsymbol{w}^\star)| = \lambda$. Then, add j to the set J.

   - There exists j in J such that a non-zero coefficient $\boldsymbol{w}_j^\star$ hits zero. Then, remove j from J.

We suppose that only one such event can occur at the same time. It is also easy to show that the value of λ corresponding to the next event can be obtained in closed form.

4. Go back to 3.

Let us now briefly discuss assumptions (a) and (b). When the matrix $\boldsymbol{X}_J^T\boldsymbol{X}_J$

is not invertible, the regularization path is non-unique, and the algorithm fails. This can easily be fixed by addressing a slightly modified formulation. It is possible to consider instead the elastic-net formulation of Zou and Hastie (2005) that uses $\Omega(\boldsymbol{w}) = \lambda \|\boldsymbol{w}\|_1 + \frac{\gamma}{2} \|\boldsymbol{w}\|_2^2$. Indeed, it amounts to replacing the matrix $\boldsymbol{X}_J^T \boldsymbol{X}_J$ by $\boldsymbol{X}_J^T \boldsymbol{X}_J + \gamma \boldsymbol{I}$, which is positive definite and therefore always invertible, even with a small value for γ, and applying the same algorithm in practice. The second assumption (b) can be unsatisfied in practice because of machine precision. To the best of our knowledge, the algorithm will fail in such cases, but we consider this scenario unlikely with real data.

The complexity of the above procedure depends on the number of kinks of the regularization path (which is also the number of iterations of the algorithm). Even though it is possible to build examples where this number is large, we often observe in practice that the event where one variable leaves the active set is rare. The complexity also depends on the implementation. By maintaining the computations of $\boldsymbol{X}_j^T(\boldsymbol{y} - \boldsymbol{X}\boldsymbol{w}^\star)$ and a Cholesky decomposition of $(\boldsymbol{X}_J^T \boldsymbol{X}_J)^{-1}$, it is possible to obtain an implementation in $O(psn + ps^2 + s^3)$ operations, where s is the sparsity of the solution when the algorithm is stopped (which we consider approximately equal to the number of iterations). The product psn corresponds to the computation of the matrices $\boldsymbol{X}_J^T \boldsymbol{X}_J$; ps^2, to the updates of the correlations $\boldsymbol{X}_j^T(\boldsymbol{y} - \boldsymbol{X}\boldsymbol{w}^\star)$ along the path; and s^3, to the Cholesky decomposition.

2.7   Quantitative Evaluation

To illustrate and compare the methods presented in this chapter, we consider in this section three benchmarks. These benchmarks are chosen to be representative of problems regularized with sparsity-inducing norms, involving different norms and different loss functions. To make comparisons that are as fair as possible, each algorithm is implemented in C/C++, using efficient BLAS and LAPACK libraries for basic linear algebra operations. All subsequent simulations are run on a single core of a 3.07Ghz CPU, with 8GB of memory. In addition, we take into account several criteria which strongly influence the convergence speed of the algorithms. In particular, we consider (a) different problem scales, (b) different levels of correlations, and (c) different strengths of regularization. We also show the influence of the required precision by monitoring the time of computation as a function of the objective function.

For the convenience of the reader, we list here the algorithms compared and the acronyms we use to refer to them throughout this section: the LARS algorithm (LARS), coordinate descent (CD), reweighted-ℓ_2 schemes (Re-ℓ_2), the simple proximal method (ISTA), and its accelerated version (FISTA); we will also include in the comparisons generic algorithms such as a subgradient descent algorithm (SG), and a commercial software (Mosek) for cone (CP), quadratic (QP), and second-order cone programming (SOCP) problems.

2.7.1 Speed Benchmarks

We first present a large benchmark evaluating the performance of various optimization methods for solving the Lasso.

We perform small-scale ($n = 200, p = 200$) and medium-scale ($n = 2000, p = 10,000$) experiments. We generate design matrices as follows. For the scenario with low correlation, all entries of $\boldsymbol{X}$ are independently drawn from a Gaussian distribution $\mathcal{N}(0, 1/n)$, which is a setting often used to evaluate optimization algorithms in the literature. For the scenario with large correlation, we draw the rows of the matrix $\boldsymbol{X}$ from a multivariate Gaussian distribution for which the *average absolute value* of the correlation between two different columns is eight times the one of the scenario with low correlation. Test data vectors $\boldsymbol{y} = \boldsymbol{X}\boldsymbol{w} + \boldsymbol{n}$ where $\boldsymbol{w}$ are randomly generated, with two levels of sparsity to be used with the two different levels of regularization. $\boldsymbol{n}$ is a noise vector whose entries are i.i.d. samples from a Gaussian distribution $\mathcal{N}(0, 0.01\|\boldsymbol{X}\boldsymbol{w}\|_2^2/n)$. In the low regularization setting, the sparsity of the vectors $\boldsymbol{w}$ is $s = 0.5\min(n,p)$, and in the high regularization one, $s = 0.01\min(n,p)$, corresponding to fairly sparse vectors. For SG, we take the step size to be equal to $a/(k+b)$, where k is the iteration number and (a,b) are the best[10] parameters selected on a logarithmic grid $(a,b) \in \{10^3, \dots, 10\} \times \{10^2, 10^3, 10^4\}$; we proceeded this way so as not to disadvantage SG by an arbitrary choice of stepsize.

To sum up, we make a comparison for 8 different conditions (2 scales $\times$ 2 levels of correlation $\times$ 2 levels of regularization). All results are reported in figures 2.2 and 2.3, by averaging 5 runs for each experiment. Interestingly, we observe that the relative performance of the different methods change significantly with the scenario.

Our conclusions for the different methods are as follows.

- *LARS:* For the small-scale problem, LARS outperforms all other methods for almost every scenario and precision regime. It is therefore *definitely the right choice for the small-scale setting.*

---

10. "The best step size" is understood here as being the step size leading to the smallest objective function after 500 iterations.

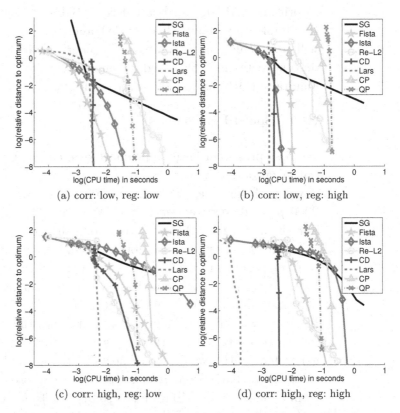

Figure 2.2: Benchmarks for solving the Lasso for the small-scale experiment ($n = 200$, $p = 200$), for the two levels of correlation and two levels of regularization, and 8 optimization methods (see main text for details). The curves represent the relative value of the objective function as a function of the computational time in seconds on a $\log_{10} / \log_{10}$ scale.

Unlike first-order methods, its performance does not depend on the correlation of the design matrix $\boldsymbol{X}$, but on the sparsity s of the solution. In our larger-scale setting, it has been competitive either when the solution is *very sparse* (high regularization) or when there is *high correlation* in $\boldsymbol{X}$ (in that case, other methods do not perform as well). More important, LARS gives an exact solution and computes the regularization path.

• *Proximal methods (ISTA, FISTA):* FISTA outperforms ISTA in all scenarios but one. The methods are close for high regularization or low correlation, but FISTA is significantly better for high correlation and/or low regularization. These methods are almost always outperformed by LARS in the small-scale setting, except for *low precision and low correlation*.
Both methods *suffer from correlated features*, which is consistent with the fact that their convergence rate is proportional to the Lipschitz constant of

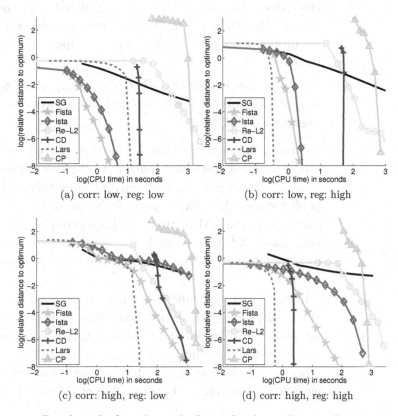

Figure 2.3: Benchmarks for solving the Lasso for the medium-scale experiment $n = 2000$, $p = 10,000$, for the two levels of correlation and two levels of regularization, and 8 optimization methods (see main text for details). The curves represent the relative value of the objective function as a function of the computational time in seconds on a $\log_{10} / \log_{10}$ scale.

the gradient of f, which grows with the amount of correlation. They are *well adapted to large-scale settings with low or medium correlation.*

- *Coordinate descent (CD):* To the best of our knowledge, no theoretical convergence rate is available for this method. Empirically, we have observed that the behavior of CD often translates into a "warm-up" stage followed by a fast convergence phase.
Its performance in the *small-scale setting is competitive* (though always behind LARS), but *less efficient in the large-scale one*. For a reason we cannot explain, *it suffers less than proximal methods do from correlated features.*

- *Reweighted-ℓ_2:* This method was outperformed in all our experiments by other dedicated methods.[11] We considered only the smoothed alternating scheme of section 2.5 and not first-order methods in η such as that of Rakotomamonjy et al. (2008). A more exhaustive comparison should include these as well.

■ *Generic methods (SG, QP, CP):* As expected, generic methods are not adapted for solving the Lasso and are always outperformed by dedicated ones such as LARS.

Among the methods that we have presented, some require an overhead computation of the Gram matrix $X^T X$: this is the case for coordinate descent and reweighted-ℓ_2 methods. We took this overhead time into account in all figures, which explains the behavior of the corresponding convergence curves. Like the LARS, these methods could benefit from an offline precomputation of $X^T X$, and would therefore be more competitive if the solutions corresponding to several values of the regularization parameter have to be computed.

In the above experiments we have considered the case of the square loss. Obviously, some of the conclusions drawn above would not be valid for other smooth losses. On the one hand, the LARS no longer applies; on the other hand, proximal methods are clearly still available, and coordinate descent schemes, which were dominated by the LARS in our experiments, would most likely turn out to be very good contenders in that setting.

2.7.2   Structured Sparsity

In this second series of experiments, the optimization techniques of the previous sections are further evaluated when applied to other types of loss and sparsity-inducing functions. Instead of the ℓ_1-norm previously studied, we focus on the particular *hierarchical* ℓ_1/ℓ_2-norm Ω introduced in section 2.3. From an optimization standpoint, although Ω shares some similarities with the ℓ_1-norm (e.g., the convexity and the non-smoothness), it differs in that it cannot be decomposed into independent parts (because of the overlapping structure of $\mathcal{G}$). CD schemes hinge on this property, and as a result, they cannot be straightforwardly applied in this case.

---

11. Note that the reweighted-ℓ_2 scheme requires iteratively solving large-scale linear systems that are badly conditioned. Our implementation uses LAPACK Cholesky decompositions, but a better performance might be obtained using a preconditioned conjugate gradient, especially in the very large-scale setting.

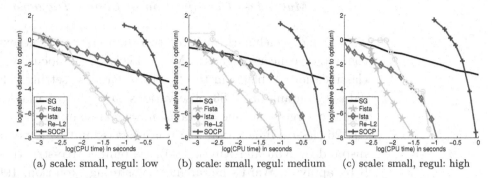

(a) scale: small, regul: low   (b) scale: small, regul: medium   (c) scale: small, regul: high

Figure 2.4: Benchmarks for solving a least-squares regression problem regularized by the hierarchical norm Ω. The experiment is small-scale ($n = 256, p = 151$) and shows the performances of five optimization methods (see main text for details) for three levels of regularization. The curves represent the relative value of the objective function as a function of the computational time in seconds on a $\log_{10} / \log_{10}$ scale.

2.7.2.1 *Denoising of Natural Image Patches*

In this first benchmark, we consider a least-squares regression problem regularized by Ω that arises in the context of the denoising of natural image patches (Jenatton et al., 2010a). In particular, based on a hierarchical set of features that accounts for different types of edge orientations and frequencies in natural images, we seek to reconstruct noisy 16×16 patches. Although the problem involves a small number of variables ($p = 151$), it has to be solved repeatedly for thousands of patches, at moderate precision. It is therefore crucial to be able to solve this problem efficiently.

The algorithms involved in the comparisons are ISTA, FISTA, Re-ℓ_2, SG, and SOCP. All results are reported in figure 2.4, by averaging five runs.

We can draw several conclusions from the simulations. First, we observe that across all levels of sparsity, the accelerated proximal scheme performs better than, or similarly to the other approaches. In addition, as opposed to FISTA, ISTA seems to suffer in non-sparse scenarios. In the least sparse setting, the reweighted-ℓ_2 scheme matches the performance of FISTA. However, this scheme does not yield truly sparse solutions, and would therefore require a subsequent thresholding operation, which can be difficult to motivate in a principled way. As expected, the generic techniques such as SG and SOCP do not compete with the dedicated algorithms.

2.7.2.2   Multi-class Classification of Cancer Diagnosis

The second benchmark involves two datasets[12] of gene expressions in the context of cancer diagnosis. More precisely, we focus on two multi-class classification problems in the "small n, large p" setting. The medium-scale dataset contains $n = 83$ observations, $p = 4615$ variables and 4 classes, and the large-scale one contains $n = 308$ samples, $p = 30017$ variables and 26 classes. In addition, both datasets exhibit highly correlated features. Inspired by Kim and Xing (2010), we built a tree-structured set of groups $\mathcal{G}$ by applying Ward's hierarchical clustering (Johnson, 1967) on the gene expressions. The norm Ω built that way aims at capturing the hierarchical structure of gene expression networks (Kim and Xing, 2010).

Instead of the square loss function, we consider the multinomial logistic loss function, which is better suited for multi-class classification problems. As a direct consequence, the algorithms whose applicability crucially depends on the choice of the loss function are removed from the benchmark. This is, for instance, the case for reweighted-ℓ_2 schemes that have closed-form updates available only with the square loss (see section 2.5). Importantly, the choice of the multinomial logistic loss function requires optimizing over a matrix with dimensions p times the number of classes (i.e., a total of $4615 \times 4 \approx 18,000$ and $30,017 \times 26 \approx 780,000$ variables). Also, for lack of scalability, generic interior-point solvers could not be considered here. To summarize, the following comparisons involve ISTA, FISTA, and SG.

All the results are reported in figure 2.5. The benchmark especially points out that overall the accelerated proximal scheme performs better than the two other methods. Again, it is important to note that both proximal algorithms yield sparse solutions, which is not the case for SG. More generally, this experiment illustrates the flexibility of proximal algorithms with respect to the choice of the loss function.

We conclude this section with general remarks on the experiments that we presented. First, the use of proximal methods is often advocated because of their optimal worst-case complexities in $O(\frac{1}{k^2})$. In practice, in our experiments these and several other methods empirically exhibit convergence rates that are at least linear, if not better, which suggests that the adaptivity of the method (e.g., its ability to take advantage of local curvature) might be more crucial to its practical success. Second, our experiments concentrated on regimes that are of interest for sparse methods in machine learning, where typically p is larger than n and where it is possible to find

---

12. The two datasets we used are *SRBCT* and *14_Tumors*, which are freely available at http://www.gems-system.org/.

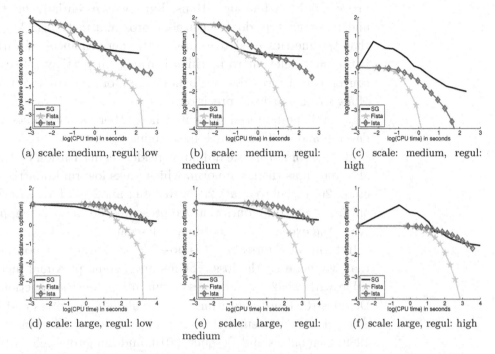

Figure 2.5: Medium- and large-scale multi-class classification problems for three optimization methods (see details about the datasets and the methods in the main text). Three levels of regularization are considered. The curves represent the relative value of the objective function as a function of the computation time in second on a $\log_{10} / \log_{10}$ scale. In the highly regularized setting, the tuning of the stepsize for the subgradient turned out to be difficult, which explains the behavior of SG in the first iterations.

good sparse solutions. The setting where n is much larger than p was out of scope here, but would be worth a separate study, and should involve methods from stochastic optimization. Also, even though it might make sense from an optimization viewpoint, we did not consider problems with low levels of sparsity, that is, with more dense solution vectors, since it would be a more difficult regime for many of the algorithms that we presented (namely, LARS, CD, or proximal methods).

2.8   Extensions

We obviously could not exhaustively cover the literature on algorithms for sparse methods in this chapter.

Surveys and comparisons of algorithms for sparse methods have been proposed by Schmidt et al. (2007) and Yuan et al. (2010). These papers present quite a few algorithms, but focus essentially on ℓ_1-regularization and unfortunately do not consider proximal methods. Also, it is not clear that the metrics used to compare the performances of various algorithms is the most relevant to machine learning; in particular, we present the full convergence curves that we believe are more informative than the ordering of algorithms at fixed precision.

Beyond the material presented here, there a few topics that we did not develop and that are worth mentioning.

In terms of norms, we did not consider regularization by the nuclear norm, also known as the trace-norm, which seeks low-rank matrix solutions (Fazel et al., 2001; Srebro et al., 2005; Recht et al., 2007; Bach, 2008b). Most of the optimization techniques that we presented do, however, apply to this norm (with the exception of coordinate descent).

In terms of algorithms, it is possible to relax the smoothness assumptions that we made on the loss. For instance, some proximal methods are applicable with weaker smoothness assumptions on the function f, such as the Douglas-Rachford algorithm (see details in Combettes and Pesquet, 2010). The related augmented Lagrangian techniques (Glowinski and Le Tallec, 1989; Combettes and Pesquet, 2010, and numerous references therein), also known as alternating-direction methods of multipliers, are also relevant in that setting. These methods are applicable in particular to cases where several regularizations are mixed.

In the context of proximal methods, the metric used to define the proximal operator can be (1) modified by judicious rescaling operations, in order to better fit the geometry of the data (Duchi et al., 2010), or even (2) replaced with norms associated with functional spaces, in the context of kernel methods (Rosasco et al., 2009).

Finally, from a broader outlook, our—a priori deterministic—optimization problem (2.1) may also be tackled with stochastic optimization approaches, which has been the focus of much research (Bottou, 1998; Bottou and LeCun, 2003; Shapiro et al., 2009).

2.9   Conclusion

We presented and compared four families of algorithms for sparse methods: proximal methods, block coordinate descent algorithms, reweighted-ℓ_2 algorithms, and the LARS that are representative of the state of the art. We did not aim at being exhaustive. The properties of these methods can be summarized as follows:

- Proximal methods provide efficient and scalable algorithms that are applicable to a wide family of loss functions, that are simple to implement, that are compatible with many sparsity-inducing norms, and, that are often competitive with the other methods considered.

- For the square loss, the LARS remains the fastest algorithm for (a) small- and medium-scale problems, since its complexity depends essentially on the size of the active sets, and (b) cases with very correlated designs. It computes the whole path up to a certain sparsity level.

- For smooth losses, block coordinate descent provides one of the fastest algorithms, but it is limited to separable regularizers.

- For the square-loss and possibly sophisticated sparsity-inducing regularizers, ℓ_2-reweighted algorithms provide generic algorithms that are still pretty competitive compared with subgradient and interior-point methods. For general losses, these methods currently require solving ℓ_2-regularized problems iteratively, and it would be desirable to relax this constraint.

2.10   References

A. Argyriou, T. Evgeniou, and M. Pontil.   Multi-task feature learning.   In B. Schölkopf, J. Platt, and T. Hoffman, editors, *Advances in Neural Information Processing Systems 19*, pages 41–48. MIT Press, 2007.

F. Bach. Exploring large feature spaces with hierarchical multiple kernel learning. In D. Koller, D. Schuurmans, Y. Bengio, and L. Bottou, editors, *Advances in Neural Information Processing Systems 21*, pages 105–112. MIT Press, 2008a.

F. Bach. Consistency of trace norm minimization. *Journal of Machine Learning Research*, 9:1019–1048, 2008b.

A. Beck and M. Teboulle. A fast iterative shrinkage-thresholding algorithm for linear inverse problems. *SIAM Journal on Imaging Sciences*, 2(1):183–202, 2009.

D. P. Bertsekas. *Nonlinear programming*. Athena Scientific, Belmont, MA, second edition, 1999.

P. Bickel, Y. Ritov, and A. Tsybakov. Simultaneous analysis of Lasso and Dantzig selector. *Annals of Statistics*, 37(4):1705–1732, 2009.

J. M. Borwein and A. S. Lewis. *Convex Analysis and Nonlinear Optimization: Theory and Examples*. Springer-Verlag, second edition, 2006.

L. Bottou. Online algorithms and stochastic approximations. In D. Saad, editor, *Online Learning and Neural Networks*. Cambridge University Press, Cambridge, UK, 1998.

L. Bottou and O. Bousquet. The tradeoffs of large scale learning. In J. C. Platt, D. Koller, Y. Singer, and S. Roweis, editors, *Advances in Neural Information Processing Systems 20*, pages 161–168. MIT Press, 2007.

L. Bottou and Y. LeCun. Large scale online learning. In S. Thrun, L. Saul, and B. Schölkopf, editors, *Advances in Neural Information Processing Systems 16*, pages 217–224. MIT Press, 2003.

S. Boyd and L. Vandenberghe. *Convex Optimization*. Cambridge University Press, 2004.

P. Brucker. An $O(n)$ algorithm for quadratic knapsack problems. *Operations Research Letters*, 3(3):163–166, 1984.

S. S. Chen, D. L. Donoho, and M. A. Saunders. Atomic decomposition by basis pursuit. *SIAM Journal on Scientific Computing*, 20(1):33–61, 1999.

P. Combettes and J. Pesquet. *Fixed-Point Algorithms for Inverse Problems in Science and Engineering*, chapter Proximal Splitting Methods in Signal Processing. Springer-Verlag, New York, 2010.

I. Daubechies, R. DeVore, M. Fornasier, and C. S. Güntürk. Iteratively reweighted least squares minimization for sparse recovery. *Communications on Pure and Applied Mathematics*, 63(1):1–38, 2010.

D. L. Donoho and I. M. Johnstone. Adapting to unknown smoothness via wavelet shrinkage. *Journal of the American Statistical Association*, 90(432):1200–1224, 1995.

J. Duchi, E. Hazan, and Y. Singer. Adaptive subgradient methods for online learning and stochastic optimization. In A. T. Kalai and M. Mohri, editors, *Proceedings of the 23rd Conference on Learning Theory*, pages 257–269. Omnipress, 2010.

B. Efron, T. Hastie, I. Johnstone, and R. Tibshirani. Least angle regression. *Annals of Statistics*, 32(2):407–499, 2004.

M. Fazel, H. Hindi, and S. P. Boyd. A rank minimization heuristic with application to minimum order system approximation. In *Proceedings of the American Control Conference*, volume 6, pages 4734–4739, 2001.

J. Friedman, T. Hastie, and R. Tibshirani. A note on the group lasso and a sparse group lasso. *preprint*, 2010. arXiv:1001.0736.

W. J. Fu. Penalized regressions: The bridge versus the lasso. *Journal of Computational and Graphical Statistics*, 7(3):397–416, 1998.

A. Genkin, D. D. Lewis, and D. Madigan. Large-scale bayesian logistic regression for text categorization. *Technometrics*, 49(3):291–304, 2007.

R. Glowinski and P. Le Tallec. *Augmented Lagrangian and Operator-Splitting Methods in Nonlinear Mechanics*. Studies in Applied Mathematics. SIAM, 1989.

J. Huang and T. Zhang. The benefit of group sparsity. *Annals of Statistics*, 38(4):1978–2004, 2010.

L. Jacob, G. Obozinski, and J.-P. Vert. Group Lasso with overlaps and graph Lasso. In *Proceedings of the 26th International Conference on Machine Learning*, pages 433–440. ACM Press, 2009.

R. Jenatton, J.-Y. Audibert, and F. Bach. Structured variable selection with sparsity-inducing norms. Technical report, 2009. Preprint arXiv:0904.3523v1.

R. Jenatton, J. Mairal, G. Obozinski, and F. Bach. Proximal methods for sparse hierarchical dictionary learning. In *Proceedings of the 27th International Conference on Machine Learning*, 2010a.

R. Jenatton, G. Obozinski, and F. Bach. Structured sparse principal component analysis. In *Proceedings of International Conference on Artificial Intelligence and Statistics*, pages 366–373, 2010b.

S. C. Johnson. Hierarchical clustering schemes. *Psychometrika*, 32(3):241–254, 1967.

S. Kim and E. P. Xing. Tree-guided group lasso for multi-task regression with structured sparsity. In *Proceedings of the 27th International Conference on Machine Learning*, pages 543–550, 2010.

K. Koh, S. J. Kim, and S. Boyd. An Interior-Point Method for Large-Scale l 1-Regularized Logistic Regression. *Journal of Machine Learning Research*, 8:1555, 2007.

B. Krishnapuram, L. Carin, M. A. T. Figueiredo, and A. J. Hartemink. Sparse multinomial logistic regression: Fast algorithms and generalization bounds. *IEEE Transactions Pattern Analysis and Machine Intelligence*, 27(6):957–968, 2005.

H. Lee, A. Battle, R. Raina, and A. Y. Ng. Efficient sparse coding algorithms. In J. C. Platt, D. Koller, Y. Singer, and S. Roweis, editors, *Advances in Neural Information Processing Systems 20*, pages 801–808. MIT Press, 2007.

K. Lounici, M. Pontil, A. B. Tsybakov, and S. van de Geer. Taking advantage of sparsity in multi-task learning. Technical report, Preprint arXiv:0903.1468, 2009.

N. Maculan and G. Galdino de Paula Jr. A linear-time median-finding algorithm for projecting a vector on the simplex of $\mathbb{R}^n$. *Operations Research Letters*, 8(4): 219–222, 1989.

J. Mairal, R. Jenatton, G. Obozinski, and F. Bach. Network flow algorithms for structured sparsity. In J. Lafferty, C. K. I. Williams, J. Shawe-Taylor, R. S. Zemel, and A. Culotta, editors, *Advances in Neural Information Processing Systems 23*. MIT Press, 2010.

H. Markowitz. Portfolio selection. *Journal of Finance*, 7(1):77–91, 1952.

J. Moreau. Fonctions convexes duales et points proximaux dans un espace hilbertien. *Comptes Rendus de l'Académie des Sciences, Paris, Série A, Mathématique*, 255:2897–2899, 1962.

S. Negahban, P. Ravikumar, M. J. Wainwright, and B. Yu. A unified framework for high-dimensional analysis of M-estimators with decomposable regularizers. In Y. Bengio, D. Schuurmans, J. Lafferty, C. K. I. Williams, and A. Culotta, editors, *Advances in Neural Information Processing Systems 22*. MIT Press, 2009.

Y. Nesterov. *Introductory Lectures on Convex Optimization: A Basic Course*. Kluwer Academic Publishers, 2004.

Y. Nesterov. Smooth minimization of non-smooth functions. *Mathematical Programming*, 103(1):127–152, 2005.

Y. Nesterov. Gradient methods for minimizing composite objective function. Technical report, Center for Operations Research and Econometrics, Catholic University of Louvain, 2007. revised 2010.

J. Nocedal and S. J. Wright. *Numerical Optimization*. Springer-Verlag, second edition, 2006.

G. Obozinski, B. Taskar, and M. I. Jordan. Joint covariate selection and joint subspace selection for multiple classification problems. *Statistics and Computing*, 20(2):231–252, 2010.

M. R. Osborne, B. Presnell, and B. A. Turlach. On the Lasso and its dual. *Journal of Computational and Graphical Statistics*, 9(2):319–337, 2000.

A. Rakotomamonjy, F. Bach, S. Canu, and Y. Grandvalet. SimpleMKL. *Journal of Machine Learning Research*, 9:2491–2521, 2008.

B. Recht, M. Fazel, and P. A. Parrilo. Guaranteed minimum-rank solutions of linear matrix equations via nuclear norm minimization. Technical report, 2007. Preprint arXiv:0706.4138.

R. T. Rockafellar. *Convex analysis.* Princeton University Press, 1997.

L. Rosasco, S. Mosci, M. Santoro, A. Verri, and S. Villa. Iterative Projection Methods for Structured Sparsity Regularization. Technical report, Computer Science and Artificial Intelligence Laboratory, MIT, 2009. CBCL-282.

V. Roth and B. Fischer. The Group-Lasso for generalized linear models: uniqueness of solutions and efficient algorithms. In *Proceedings of the 25th International Conference on Machine Learning*, pages 848–855, 2008.

M. Schmidt and K. Murphy. Convex structure learning in log-linear models: Beyond pairwise potentials. In *Proceedings of the 13th International Conference on Artificial Intelligence and Statistics*, 2010.

M. Schmidt, G. Fung, and R. Rosales. Fast optimization methods for L1 regularization: A comparative study and two new approaches. *Machine Learning: ECML 2007*, pages 286–297, 2007.

A. Shapiro, D. Dentcheva, A. Ruszczyński, and A. P. Ruszczyński. *Lectures on Stochastic Programming: Modeling and Theory.* SIAM, 2009.

J. Shawe-Taylor and N. Cristianini. *Kernel Methods for Pattern Analysis.* Cambridge University Press, 2004.

S. K. Shevade and S. S. Keerthi. A simple and efficient algorithm for gene selection using sparse logistic regression. *Bioinformatics*, 19(17):2246–2253, 2003.

P. Sprechmann, I. Ramirez, G. Sapiro, and Y. Eldar. Collaborative hierarchical sparse modeling. In *Proceedings of the 44th Annual Conference on Information Sciences and Systems*, 2010.

N. Srebro, J. D. M. Rennie, and T. S. Jaakkola. Maximum-margin matrix factorization. In L. K. Saul, Y. Weiss, and L. Bottou, editors, *Advances in Neural Information Processing Systems 17*, pages 1329–1336. MIT Press, 2005.

M. Szafranski, Y. Grandvalet, and P. Morizet-Mahoudeaux. Hierarchical penalization. In J. C. Platt, D. Koller, Y. Singer, and S. Roweis, editors, *Advances in Neural Information Processing Systems 20*. MIT Press, 2007.

R. Tibshirani. Regression shrinkage and selection via the Lasso. *Journal of the Royal Statistical Society, series B*, 58(1):267–288, 1996.

J. A. Tropp. Greed is good: Algorithmic results for sparse approximation. *IEEE Transactions on Information Theory*, 50(10):2231–2242, 2004.

P. Tseng and S. Yun. A coordinate gradient descent method for nonsmooth separable minimization. *Mathematical Programming, series B*, 117(1):387–423, 2009.

M. J. Wainwright. Sharp thresholds for noisy and high-dimensional recovery of sparsity using ℓ_1-constrained quadratic programming. *IEEE Transactions on Information Theory*, 55(5):2183–2202, 2009.

S. Wright, R. Nowak, and M. Figueiredo. Sparse reconstruction by separable approximation. *IEEE Transactions on Signal Processing*, 57(7):2479–2493, 2009.

T. Wu and K. Lange. Coordinate descent algorithms for lasso penalized regression. *Annals of Statistics*, 2(1):224–244, 2008.

G. Yuan, K. Chang, C. Hsieh, and C. Lin. A comparison of optimization methods for large-scale l1-regularized linear classification. Technical report, Department of Computer Science, National University of Taiwan, 2010.

M. Yuan and Y. Lin. Model selection and estimation in regression with grouped variables. *Journal of the Royal Statistical Society, series B*, 68:49–67, 2006.

P. Zhao and B. Yu. On model selection consistency of Lasso. *Journal of Machine Learning Research*, 7:2541–2563, 2006.

P. Zhao, G. Rocha, and B. Yu. The composite absolute penalties family for grouped and hierarchical variable selection. *Annals of Statistics*, 37(6A):3468–3497, 2009.

H. Zou and T. Hastie. Regularization and variable selection via the elastic net. *Journal of the Royal Statistical Society Series B*, 67(2):301–320, 2005.

3   # Interior-Point Methods for Large-Scale Cone Programming

Martin Andersen   msa@ee.ucla.edu
University of California, Los Angeles
Los Angeles, CA 90095-1594, USA

Joachim Dahl   dahl.joachim@gmail.com
MOSEK ApS
Fruebjergvej 3, 2100 København Ø, Denmark

Zhang Liu   zhang.liu@gmail.com
Northrop Grumman Corporation
San Diego, CA 92127-2412, USA

Lieven Vandenberghe   vandenbe@ee.ucla.edu
University of California, Los Angeles
Los Angeles, CA 90095-1594, USA

In the conic formulation of a convex optimization problem the constraints are expressed as linear inequalities with respect to a possibly non-polyhedral convex cone. This makes it possible to formulate elegant extensions of interior-point methods for linear programming to general nonlinear convex optimization. Recent research on cone programming algorithms has focused particularly on three convex cones for which symmetric primal-dual methods have been developed: the nonnegative orthant, the second-order cone, and the positive semidefinite matrix cone. Although not all convex constraints can be expressed in terms of the three standard cones, cone programs associated with these cones are sufficiently general to serve as the basis of convex modeling packages. They are also widely used in machine learning.

The main difficulty in the implementation of interior-point methods for cone programming is the complexity of the linear equations that need to be solved at each iteration. These equations are usually dense, unlike the equations that arise in linear programming, and it is therefore difficult to develop general-purpose strategies for exploiting problem structure based solely on sparse matrix methods. In this chapter we give an overview of ad hoc techniques that can be used to exploit nonsparse structure in specific classes of applications. We illustrate the methods with examples from machine learning and present numerical results with CVXOPT, a software package that supports the rapid development of customized interior-point methods.

3.1 Introduction

3.1.1 Cone Programming

The cone programming formulation has been popular in the recent literature on convex optimization. In this chapter we define a *cone linear program* (cone LP or conic LP) as an optimization problem of the form

$$\begin{aligned}
&\text{minimize} && c^T x \\
&\text{subject to} && Gx \preceq_C h \\
&&& Ax = b
\end{aligned} \tag{3.1}$$

with optimization variable x. The inequality $Gx \preceq_C h$ is a *generalized inequality*, which means that $h - Gx \in C$, where C is a closed, pointed, convex cone with nonempty interior. We will also encounter *cone quadratic programs* (cone QPs),

$$\begin{aligned}
&\text{minimize} && (1/2)x^T P x + c^T x \\
&\text{subject to} && Gx \preceq_C h \\
&&& Ax = b,
\end{aligned} \tag{3.2}$$

with P positive semidefinite.

If $C = \mathbb{R}^p_+$ (the nonnegative orthant in $\mathbb{R}^p$), the generalized inequality is a componentwise vector inequality, equivalent to p scalar linear inequalities, and problem (3.1) reduces to a linear program (LP). If C is a nonpolyhedral cone, the problem is substantially more general than an LP, in spite of the similarity in notation. In fact, as Nesterov and Nemirovskii (1994) point out, any convex optimization problem can be reformulated as a cone LP by a simple trick: a general constraint $x \in Q$, where Q is a closed convex set with nonempty interior, can be reformulated in a trivial way as $(x, t) \in C, t = 1$, if we define C as the *conic hull* of Q, that is, $C = \mathbf{cl}\{(x, t) \mid t > 0, (1/t)x \in Q\}$. More important in practice, it turns out that a surprisingly small number of cones is sufficient to express the convex constraints that are most commonly encountered in applications. In addition to the nonnegative orthant, the most common cones are the *second-order cone*,

$$\mathcal{Q}_p = \{(y_0, y_1) \in \mathbb{R} \times \mathbb{R}^{p-1} \mid \|y_1\|_2 \le y_0\},$$

and the *positive semidefinite cone*,

$$\mathcal{S}_p = \big\{ \mathbf{vec}(U) \mid U \in S_+^p \big\}.$$

Here S_+^p denotes the positive semidefinite matrices of order p and $\mathbf{vec}(U)$ is the symmetric matrix U stored as a vector:

$$\mathbf{vec}(U) =$$
$$\sqrt{2}\left(\frac{U_{11}}{\sqrt{2}}, U_{21}, \ldots, U_{p1}, \frac{U_{22}}{\sqrt{2}}, U_{32}, \ldots, U_{p2}, \ldots, \frac{U_{p-1,p-1}}{\sqrt{2}}, U_{p,p-1}, \frac{U_{pp}}{\sqrt{2}}\right).$$

(The scaling of the off-diagonal entries ensures that the standard trace inner product of symmetric matrices is preserved, that is, $\text{Tr}(UV) = \mathbf{vec}(U)^T \mathbf{vec}(V)$ for all U, V.) Since the early 1990s a great deal of research has been directed at developing a comprehensive theory and software for modeling optimization problems as cone programs involving the three "canonical" cones (Nesterov and Nemirovskii, 1994; Boyd et al., 1994; Ben-Tal and Nemirovski, 2001; Alizadeh and Goldfarb, 2003; Boyd and Vandenberghe, 2004). YALMIP and CVX, two modeling packages for general convex optimization, use cone LPs with the three canonical cones as their standard format (Löfberg, 2004; Grant and Boyd, 2007, 2008).

In this chapter we assume that the cone C in (3.1) is a direct product

$$C = C_1 \times C_2 \times \cdots \times C_K, \tag{3.3}$$

where each cone C_i is of one of the three canonical types (nonnegative orthant, second-order cone, or positive semidefinite cone). These cones are self-dual, and the dual of the cone LP therefore involves an inequality with respect to the same cone:

$$\begin{aligned}
\text{maximize} \quad & -h^T z - b^T y \\
\text{subject to} \quad & G^T z + A^T y + c = 0 \\
& z \succeq_C 0.
\end{aligned} \tag{3.4}$$

The cone LP (3.1) is called a *second-order cone program* (SOCP) if C is a direct product of one or more second-order cones. (The nonnegative orthant can be written as a product of second-order cones $\mathcal{Q}_1$ of order 1.) A common and more explicit standard form of an SOCP is

$$
\begin{array}{ll}
\text{minimize} & c^T x \\
\text{subject to} & \|F_i x + g_i\|_2 \le d_i^T x + f_i, \quad i = 1, \dots, K \\
& Ax = b.
\end{array}
\tag{3.5}
$$

This corresponds to choosing

$$
G = \begin{bmatrix} -d_1^T \\ -F_1 \\ \vdots \\ -d_K^T \\ -F_K \end{bmatrix}, \qquad
h = \begin{bmatrix} f_1 \\ g_1 \\ \vdots \\ f_K \\ g_K \end{bmatrix}, \qquad
C = \mathcal{Q}_{p_1} \times \cdots \times \mathcal{Q}_{p_K}
$$

in (3.1), if the row dimensions of the matrices F_k are equal to $p_k - 1$.

The cone LP (3.1) is called a *semidefinite program* (SDP) if C is a direct product of positive semidefinite matrix cones. For purposes of exposition, a simple standard form with one matrix inequality is sufficient:

$$
\begin{array}{ll}
\text{minimize} & c^T x \\
\text{subject to} & \displaystyle\sum_{i=1}^{n} x_i F_i \preceq F_0 \\
& Ax = b,
\end{array}
\tag{3.6}
$$

where the coefficients F_i are symmetric matrices of order p and the inequality denotes matrix inequality. This can be seen as the special case of (3.1) obtained by choosing

$$
G = \begin{bmatrix} \mathbf{vec}(F_1) & \cdots & \mathbf{vec}(F_n) \end{bmatrix}, \qquad h = \mathbf{vec}(F_0), \qquad C = \mathcal{S}_p. \tag{3.7}
$$

The SDP (3.6) is in fact as general as the cone LP (3.1) with an arbitrary combination of the three cone types. A componentwise vector inequality $Gx \preceq h$ can be represented as a diagonal matrix inequality $\mathrm{Diag}(Gx) \preceq \mathrm{Diag}(h)$. A second-order cone constraint $\|Fx + g\|_2 \le d^T x + f$ is equivalent to the linear matrix inequality

$$
\begin{bmatrix} d^T x + f & (Fx + g)^T \\ Fx + g & (d^T x + f)I \end{bmatrix} \succeq 0.
$$

Multiple matrix inequalities can be represented by choosing block-diagonal matrices F_i. For algorithmic purposes, however, it is better to handle the three types of cones separately.

3.1.2 Interior-Point Methods

Interior-point algorithms dominated the research on convex optimization methods from the early 1990s until recently. They are popular because they reach a high accuracy in a small number (10–50) of iterations, almost independent of problem size, type, and data. Each iteration requires the solution of a set of linear equations with fixed dimensions and known structure. As a result, the time needed to solve different instances of a given problem family can be estimated quite accurately. Interior-point methods can be extended to handle infeasibility gracefully (Nesterov et al., 1999; Andersen, 2000), by returning a certificate of infeasibility if a problem is primal or dual infeasible. Finally, interior-point methods depend on only a small number of algorithm parameters, which can be set to values that work well for a wide range of data, and do not need to be tuned for a specific problem.

The key to efficiency of an interior-point solver is the set of linear equations solved in each iteration. These equations are sometimes called Newton equations, because they can be interpreted as a linearization of the nonlinear equations that characterize the central path, or Karush-Kuhn-Tucker (KKT) equations, because they can be interpreted as optimality (or KKT) conditions of an equality-constrained quadratic optimization problem. The cost of solving the Newton equations determines the size of the problems that can be solved by an interior-point method. General-purpose convex optimization packages rely on sparse matrix factorizations to solve the Newton equations efficiently. This approach is very successful in linear programming, where problems with several hundred thousand variables and constraints are solved routinely. The success of general-purpose sparse linear programming solvers can be attributed to two facts. First, the Newton equations of a sparse LP can usually be reduced to sparse positive definite sets of equations, which can be solved very effectively by sparse Cholesky factorization methods. Second, dense linear programs, which of course are not uncommon in practice, can often be converted into sparse problems by introducing auxiliary variables and constraints. This increases the problem dimensions, but if the resulting problem is sufficiently sparse, the net gain in efficiency is often significant.

For other classes of cone optimization problems (for example, semidefinite programming), the sparse linear programming approach to exploiting problem structure is less effective, either because the Newton equations are not sufficiently sparse or because the translation of problem structure into sparsity requires an excessive number of auxiliary variables. For these problem classes, it is difficult to develop *general-purpose* techniques that are as efficient and scalable as linear programming solvers. Nevertheless, the recent literature contains many examples of large-scale convex optimization problems that were solved successfully by scalable *customized* implementations of interior-point algorithms (Benson et al., 1999; Roh and Vandenberghe, 2006; Gillberg and Hansson, 2003; Koh et al., 2007; Kim et al., 2007; Joshi and Boyd, 2008; Liu and Vandenberghe, 2009; Wallin et al., 2009). These results were obtained by a variety of direct and iterative linear algebra techniques that take advantage of non-sparse problem structure. The purpose of this chapter is to survey some of these techniques and illustrate them with applications from machine learning. There is of course a trade-off in how much effort one is prepared to make to optimize performance of an interior-point method for a specific application. We will present results for a software package, CVXOPT (Dahl and Vandenberghe, 2009), that was developed to assist in the development of custom interior-point solvers for specific problem families. It allows the user to specify an optimization problem via an operator description, that is, by providing functions for evaluating the linear mappings in the constraints and for supplying a custom method for solving the Newton equations. This makes it possible to develop efficient solvers that exploit various types of problem structure in a fraction of the time needed to write a custom interior-point solver from scratch. Other interior-point software packages that allow customization include the QP solver OOQP (Gertz and Wright, 2003) and the Matlab-based conic solver SDPT3 (Tütüncü et al., 2003).

3.2   Primal-Dual Interior-Point Methods

We first describe some implementation details for primal-dual interior-point methods based on the *Nesterov-Todd scaling* (Nesterov and Todd, 1997, 1998). However, much of the following discussion also applies to other types of primal-dual interior-point methods for second-order cone and semidefinite programming (Helmberg et al., 1996; Kojima et al., 1997; Monteiro and Zhang, 1998).

3.2.1   Newton Equations

Consider the cone LP (3.1) and cone QP (3.2). The Newton equations for a primal-dual interior-point method based on the Nesterov-Todd scaling have the form

$$\begin{bmatrix} P & A^T & G^T \\ A & 0 & 0 \\ G & 0 & -W^TW \end{bmatrix} \begin{bmatrix} \Delta x \\ \Delta y \\ \Delta z \end{bmatrix} = \begin{bmatrix} r_x \\ r_y \\ r_z \end{bmatrix} \tag{3.8}$$

(with $P = 0$ for the cone LP). The right-hand sides r_x, r_y, r_z change at each iteration and are defined differently in different algorithms. The matrix W is a *scaling matrix* that depends on the current primal and dual iterates. If the inequalities in (3.1) and (3.4) are generalized inequalities with respect to a cone of the form (3.3), then the scaling matrix W is block-diagonal with K diagonal blocks W_k, defined as follows:

- If C_k is a nonnegative orthant of dimension p ($C_k = \mathbb{R}^p_+$), then W_k is a positive diagonal matrix,

$$W_k = \mathrm{Diag}(d),$$

for some $d \in \mathbb{R}^p_{++}$.

- If C_k is a second-order cone of dimension p ($C_k = \mathcal{Q}_p$), then W_k is a positive multiple of a hyperbolic Householder matrix

$$W_k = \beta(2vv^T - J), \qquad J = \begin{bmatrix} 1 & 0 \\ 0 & -I \end{bmatrix}, \tag{3.9}$$

where $\beta > 0$, $v \in \mathbb{R}^p$ satisfies $v^T Jv = 1$, and I is the identity matrix of order $p - 1$. The inverse of W_k is given by

$$W_k^{-1} = \frac{1}{\beta}(2Jvv^T J - J).$$

- If C_k is a positive semidefinite cone of order p ($C_k = \mathcal{S}_p$), then W_k is the matrix representation of a congruence operation: W_k and its transpose are defined by the identities

$$W_k\,\mathbf{vec}(U) = \mathbf{vec}(R^T U R), \qquad W_k^T\,\mathbf{vec}(U) = \mathbf{vec}(R U R^T), \tag{3.10}$$

for all U, where $R \in \mathbb{R}^{p \times p}$ is a nonsingular matrix. The inverses of W_k and W_k^T are defined by

$$W_k^{-1}\,\mathbf{vec}(U) = \mathbf{vec}(R^{-T} U R^{-1}), \qquad W_k^{-T}\,\mathbf{vec}(U) = \mathbf{vec}(R^{-1} U R^{-T}).$$

The values of the parameters d, β, v, R (or R^{-1}) in these definitions depend on the current primal and dual iterates, and are updated after each iteration of the interior-point algorithm.

The number of Newton equations solved per iteration varies with the type of algorithm. It is equal to two in a predictor-corrector method, three in a predictor-corrector method that uses a self-dual embedding, and it can be higher than three if iterative refinement is used. However, since the scaling W is identical for all the Newton equations solved in a single iteration, only one factorization step is required per iteration, and the cost per iteration is roughly equal to the cost of solving one Newton equation.

By eliminating Δz, the Newton equation can be reduced to a smaller system:

$$\begin{bmatrix} P + G^T W^{-1} W^{-T} G & A^T \\ A & 0 \end{bmatrix} \begin{bmatrix} \Delta x \\ \Delta y \end{bmatrix} = \begin{bmatrix} r_x + G^T W^{-1} W^{-T} r_z \\ r_y \end{bmatrix}. \quad (3.11)$$

The main challenge in an efficient implementation is to exploit structure in the matrices P, G, A when assembling the matrix

$$P + G^T W^{-1} W^{-T} G = P + \sum_{k=1}^{K} G_k^T W_k^{-1} W_k^{-T} G_k, \quad (3.12)$$

(where G_k is the block row of G corresponding to the kth inequality) and when solving equation (3.11).

General-purpose solvers for cone programming rely on sparsity in P, G, and A to solve large-scale problems. For example, if the problem does not include equality constraints, one can solve (3.11) by a Cholesky factorization of the matrix (3.12). For pure LPs or QPs (W diagonal) this matrix is typically sparse if P and G are sparse, and a sparse Cholesky factorization can be used. In problems that involve all three types of cones it is more difficult to exploit sparsity. Even when P and G are sparse, the matrix (3.12) is often dense. In addition, forming the matrix can be expensive.

3.2.2   Customized Implementations

In the following sections we will give examples of techniques for exploiting certain types of non-sparse problem structure in the Newton equations (3.8). The numerical results are obtained using the Python software package CVXOPT, which provides two mechanisms for customizing the interior-point solvers.

- Users can specify the matrices G, A, P in (3.1) and (3.2) as operators by providing Python functions that evaluate the matrix-vector products and their adjoints.

- Users can provide a Python function for solving the Newton equation (3.8).

This is made straightforward by certain elements of the Python syntax, as the following example illustrates. Suppose we are interested in solving several equations of the form

$$
\begin{bmatrix} -I & A^T \\ A & 0 \end{bmatrix} \begin{bmatrix} x_1 \\ x_2 \end{bmatrix} = \begin{bmatrix} b_1 \\ b_2 \end{bmatrix},
\tag{3.13}
$$

with the same matrix $A \in \mathbb{R}^{m \times n}$ and different right-hand sides b_1, b_2. (We assume $m \leq n$ and $\mathrm{rank}(A) = m$.) The equations can be solved by first solving

$$
AA^T x_2 = b_2 + Ab_1,
$$

using a Cholesky factorization of AA^T and then computing x_1 from $x_1 = A^T x_2 - b_1$. The following code defines a Python function `factor()` that computes the Cholesky factorization of $C = AA^T$, and returns a function `solve()` that calculates x_1 and x_2 for a given right-hand side b. A function call `f = factor(A)` therefore returns a function `f` that can be used to compute the solution for a particular right-hand side b as `x1, x2 = f(b)`.

```
from cvxopt import blas, lapack, matrix

def factor(A):
    m, n = A.size
    C = matrix(0.0, (m, m))
    blas.syrk(A, C)                 # C := A * A^T.
    lapack.potrf(C)                 # Factor C = L * L^T and set C := L.
    def solve(b):
        x2 = b[-m:] + A * b[:n]
        lapack.potrs(C, x2)         # x2 := L^-T * L^-1 * x2.
        x1 = A.T * x2 - b[:n]
        return x1, x2
    return solve
```

Note that the Python syntax proves very useful in this type of application. For example, Python treats functions as other objects, so the factor function can simply return a solve function. Note also that the symbols `A` and `C` are used in the body of the function `solve()` but are not defined there. To resolve these names, Python therefore looks at the enclosing scope (the function block with the definition of `factor()`). These scope rules make it possible to pass problem-dependent parameters to functions without using global variables.

3.3   Linear and Quadratic Programming

In the case of a (non-conic) LP or QP the scaling matrix W in the Newton equation (3.8) and (3.11) is a positive diagonal matrix. As already mentioned, general-purpose interior-point codes for linear and quadratic programming are very effective at exploiting sparsity in the data matrices P, G, A. Moreover, many types of non-sparse problem structures can be translated into sparsity by adding auxiliary variables and constraints. Nevertheless, even in the case of LPs or QPs, it is sometimes advantageous to exploit problem structure directly by customizing the Newton equation solver. In this section we discuss a few examples.

3.3.1 ℓ_1-Norm Approximation

The basic idea is illustrated by the ℓ_1-norm approximation problem

$$\text{minimize}\quad \|Xu - d\|_1, \tag{3.14}$$

with $X \in \mathbb{R}^{m \times n}$, $d \in \mathbb{R}^m$, and variable $u \in \mathbb{R}^n$. This is equivalent to an LP with $m + n$ variables and $2m$ constraints:

$$
\begin{aligned}
\text{minimize}\quad & \mathbf{1}^T v \\
\text{subject to}\quad & \begin{bmatrix} X & -I \\ -X & -I \end{bmatrix} \begin{bmatrix} u \\ v \end{bmatrix} \preceq \begin{bmatrix} d \\ -d \end{bmatrix},
\end{aligned}
\tag{3.15}
$$

with $\mathbf{1}$ the m-vector with entries equal to one. The reduced Newton equation (3.11) for this LP is

$$
\begin{bmatrix} X^T(W_1^{-2} + W_2^{-2})X & X^T(W_2^{-2} - W_1^{-2}) \\ (W_2^{-2} - W_1^{-2})X & W_1^{-2} + W_2^{-2} \end{bmatrix} \begin{bmatrix} \Delta u \\ \Delta v \end{bmatrix} = \begin{bmatrix} r_u \\ r_v \end{bmatrix} \tag{3.16}
$$

where W_1 and W_2 are positive diagonal matrices. (To simplify the notation, we do not propagate the expressions for the right-hand sides when applying block elimination.) By eliminating the variable Δv the Newton equation can be further reduced to the equation

$$X^T D X \Delta u = r,$$

where D is the positive diagonal matrix

$$D = 4W_1^{-2}W_2^{-2}(W_1^{-2} + W_2^{-2})^{-1} = 4(W_1^2 + W_2^2)^{-1}.$$

The cost of solving the ℓ_1-norm approximation problem is therefore equal to a small multiple (10–50) of the cost of solving the same problem in the ℓ_2-norm, that is, solving the normal equations $X^T X u = X^T d$ of the corresponding least-squares problem (Boyd and Vandenberghe, 2004, page 617).

The Python code shown below exploits this fact. The matrix

$$G = \begin{bmatrix} X & -I \\ -X & -I \end{bmatrix}$$

is specified via a Python function G that evaluates the matrix-vector products with G and G^T. The function F factors the matrix $X^T D X$ and returns a solve routine f that takes the right-hand side of (3.8) as its input argument and replaces it with the solution. The input argument of F is the scaling matrix W stored as a Python dictionary W containing the various parameters of W. The last line calls the CVXOPT cone LP solver. The code can be further optimized by a more extensive use of the BLAS.

Table 3.1 shows the result of an experiment with six randomly generated dense matrices X. We compare the speed of the customized CVXOPT solver shown above, the same solver with further BLAS optimizations, and the general-purpose LP solver in MOSEK (MOSEK ApS, 2010), applied to the LP (3.15). The last column shows the results for MOSEK applied to the equivalent formulation

$$
\begin{aligned}
\text{minimize} \quad & \mathbf{1}^T v + \mathbf{1}^T w \\
\text{subject to} \quad & Xu - d = v - w \\
& v \succeq 0, \quad w \succeq 0.
\end{aligned}
\tag{3.17}
$$

The times are in seconds on an Intel Core 2 Quad Q9550 (2.83 GHz) with 4GB of memory.

The table shows that a customized solver, implemented in Python with a modest programming effort, can be competitive with one of the best general-purpose sparse linear programming codes. In this example, the customized solver takes advantage of the fact that the dense matrix X appears in two positions of the matrix G. This property is not exploited by a general-purpose sparse solver.

3.3.2 Least-Squares with ℓ_1-Norm Regularization

As a second example, we consider a least-squares problem with ℓ_1-norm regularization,

$$\text{minimize} \quad \|Xu - d\|_2^2 + \|u\|_1,$$

```
from cvxopt import lapack, solvers, matrix, mul, div

m, n = X.size

def G(x, y, alpha = 1.0, beta = 0.0, trans = 'N'):
    if trans == 'N':  # y := alpha * G * x + beta * y
        u = X * x[:n]
        y[:m] = alpha * ( u - x[n:] ) + beta * y[:m]
        y[m:] = alpha * (-u - x[n:] ) + beta * y[m:]
    else:             # y := alpha * G' * x + beta * y
        y[:n] =  alpha * X.T * (x[:m] - x[m:]) + beta * y[:n]
        y[n:] = -alpha * (x[:m] + x[m:]) + beta * y[n:]

def F(W):
    d1, d2 = W['d'][:m]**2, W['d'][m:]**2
    D = 4*(d1 + d2)**-1
    A = X.T * spdiag(D) * X
    lapack.potrf(A)
    def f(x, y, z):
        x[:n] += X.T * ( mul( div(d2 - d1, d1 + d2), x[n:] ) +
            mul( .5*D, z[:m] - z[m:] ) )
        lapack.potrs(A, x)
        u = X * x[:n]
        x[n:] = div( x[n:] - div(z[:m], d1) - div(z[m:], d2) +
            mul(d1**-1 - d2**-1, u), d1**-1 + d2**-1 )
        z[:m] = div(u - x[n:] - z[:m], W['d'][:m])
        z[m:] = div(-u - x[n:] - z[m:], W['d'][m:])
    return f

c = matrix(n*[0.0] + m*[1.0])
h = matrix([d, -d])
sol = solvers.conelp(c, G, h, kktsolver = F)
```

m	n	CVXOPT	CVXOPT/BLAS	MOSEK (3.15)	MOSEK (3.17)
500	100	0.12	0.06	0.75	0.40
1000	100	0.22	0.11	1.53	0.81
1000	200	0.52	0.29	1.95	1.06
2000	200	1.23	0.60	3.87	2.19
1000	500	2.44	1.32	3.63	2.38
2000	500	5.00	2.68	7.44	5.11
2000	1000	17.1	9.52	32.4	12.8

Table 3.1: Solution times (seconds) for six randomly generated dense ℓ_1-norm approximation problems of dimension $m \times n$. Column 3 gives the CPU times for the customized CVXOPT code. Column 4 gives the CPU times for a customized CVXOPT code with more extensive use of the BLAS for matrix-vector and matrix-matrix multiplications. Columns 5 and 6 show the times for the interior-point solver in MOSEK v6 (with basis identification turned off) applied to the LPs (3.15) and (3.17), respectively.

with $X \in \mathbb{R}^{m \times n}$. The problem is equivalent to a QP

$$
\begin{aligned}
\text{minimize} \quad & (1/2)\|Xu - d\|_2^2 + \mathbf{1}^T v \\
\text{subject to} \quad & -v \preceq u \preceq v,
\end{aligned}
\tag{3.18}
$$

with $2n$ variables and $2n$ constraints. The reduced Newton equation (3.11) for this QP is

$$
\begin{bmatrix} X^T X + W_1^{-2} + W_2^{-2} & W_2^{-2} - W_1^{-2} \\ W_2^{-2} - W_1^{-2} & W_1^{-2} + W_2^{-2} \end{bmatrix} \begin{bmatrix} \Delta u \\ \Delta v \end{bmatrix} = \begin{bmatrix} r_u \\ r_v \end{bmatrix}
$$

where W_1 and W_2 are diagonal. Eliminating Δv, as in the example of section 3.3.1, results in a positive definite equation of order n:

$$
(X^T X + D)\Delta u = r,
$$

where $D = 4(W_1^2 + W_2^2)^{-1}$. Alternatively, we can apply the matrix inversion lemma and convert this to an equation of order m:

$$
(X D^{-1} X^T + I)\Delta \tilde{u} = \tilde{r}.
\tag{3.19}
$$

The second option is attractive when $n \gg m$, but requires a customized interior-point solver, since the matrix D depends on the current iterates. A general-purpose QP solver applied to (3.18), on the other hand, is expensive if $n \gg m$, since it does not recognize the low-rank structure of the matrix $X^T X$ in the objective.

Table 3.2 shows the result of an experiment with randomly generated

m	n	CVXOPT	MOSEK (3.18)	MOSEK (3.20)
50	200	0.02	0.35	0.32
50	400	0.03	1.06	0.59
100	1000	0.12	9.57	1.69
100	2000	0.24	66.5	3.43
500	1000	1.19	10.1	7.54
500	2000	2.38	68.6	17.6

Table 3.2: Solution times (seconds) for six randomly generated dense least-squares problems with ℓ_1-norm regularization. The matrix X has dimension $m \times n$. Column 3 gives the CPU times for the customized CVXOPT code. Column 4 shows the times for MOSEK applied to (3.18). Column 5 shows the times for MOSEK applied to (3.20).

dense matrices X. We compare the speed of a customized QP solver with the general-purpose QP solver in MOSEK applied to the QP (3.18) and the equivalent QP

$$
\begin{aligned}
\text{minimize} \quad & (1/2)w^T w + \mathbf{1}^T v \\
\text{subject to} \quad & -v \preceq x \preceq v \\
& Xu - w = d
\end{aligned}
\tag{3.20}
$$

with variables u, v, w. Although this last formulation has more variables and constraints than (3.18), MOSEK solves it more efficiently because it is sparser. For the custom solver the choice between (3.18) and (3.20) is irrelevant because the Newton equations for both QPs reduce to an equation of the form (3.19).

3.3.3   Support Vector Machine Training

A well-known example of the technique in the previous section arises in the training of support vector machine classifiers via the QP:

$$
\begin{aligned}
\text{minimize} \quad & (1/2)u^T Q u - d^T u \\
\text{subject to} \quad & 0 \preceq \mathrm{Diag}(d)u \preceq \gamma \mathbf{1} \\
& \mathbf{1}^T u = 0.
\end{aligned}
\tag{3.21}
$$

In this problem Q is the kernel matrix and has entries $Q_{ij} = k(x_i, x_j)$, $i, j = 1, \ldots, N$, where $x_1, \ldots, x_N \in \mathbb{R}^n$ are the training examples and $k : \mathbb{R}^n \times \mathbb{R}^n \to \mathbb{R}$ is a positive definite kernel function. The vector $d \in \{-1, +1\}^N$ contains the labels of the training vectors. The parameter γ is given. The reduced Newton equation for (3.21) is

$$\begin{bmatrix} Q + W_1^{-2} + W_2^{-2} & \mathbf{1} \\ \mathbf{1}^T & 0 \end{bmatrix} \begin{bmatrix} \Delta u \\ \Delta y \end{bmatrix} = \begin{bmatrix} r_u \\ r_y \end{bmatrix}. \qquad (3.22)$$

This equation is expensive to solve when N is large because the kernel matrix Q is generally dense. If the linear kernel $k(v, \tilde{v}) = v^T \tilde{v}$ is used, the kernel matrix can be written as $Q = XX^T$ where $X \in \mathbb{R}^{N \times n}$ is the matrix with rows x_i^T. If $N \gg n$, we can apply the matrix inversion lemma as in the previous example, and reduce the Newton equation to an equation

$$\left(I + X^T (W_1^{-2} + W_2^{-2})^{-1} X \right) \Delta w = r$$

of order n. This method for exploiting low-rank structure or diagonal-plus-low-rank structure in the kernel matrix Q is well known in machine learning (Ferris and Munson, 2002; Fine and Scheinberg, 2002).

Crammer and Singer (2001) extended the binary SVM classifier to classification problems with more than two classes. The training problem of the Crammer-Singer multiclass SVM can be expressed as a QP

$$\begin{array}{ll} \text{minimize} & (1/2)\operatorname{Tr}(U^T Q U) - \operatorname{Tr}(E^T U) \\ \text{subject to} & U \preceq \gamma E \\ & U \mathbf{1}_m = 0 \end{array} \qquad (3.23)$$

with a variable $U \in \mathbb{R}^{N \times m}$, where N is the number of training examples and m is the number of classes. As in the previous section, Q is a kernel matrix with entries $Q_{ij} = k(x_i, x_j)$, $i, j = 1, \ldots, N$. The matrix $E \in \mathbb{R}^{N \times m}$ is defined as

$$E_{ij} = \begin{cases} 1 & \text{training example } i \text{ belongs to class } j \\ 0 & \text{otherwise.} \end{cases}$$

The inequality $U \preceq \gamma E$ denotes componentwise inequality between matrices. From the optimal solution U one obtains the multiclass classifier, which maps a test point x to the class number

$$\operatorname*{argmax}_{j=1,\ldots,m} \sum_{i=1}^{N} U_{ij} k(x_i, x).$$

An important drawback of this formulation, compared with multiclass classifiers based on a combination of binary classifiers, is the high cost of solving the QP (3.23), which has Nm variables, Nm inequality constraints, and N equality constraints. Let us therefore examine the reduced Newton equations

$$
\begin{bmatrix}
Q + W_1^{-2} & 0 & \cdots & 0 & I \\
0 & Q + W_2^{-2} & \cdots & 0 & I \\
\vdots & \vdots & \ddots & \vdots & \vdots \\
0 & 0 & \cdots & Q + W_m^{-2} & I \\
I & I & \cdots & I & 0
\end{bmatrix}
\begin{bmatrix}
\Delta u_1 \\
\Delta u_2 \\
\vdots \\
\Delta u_m \\
\Delta y
\end{bmatrix}
=
\begin{bmatrix}
r_{u_1} \\
r_{u_2} \\
\vdots \\
r_{u_m} \\
r_y
\end{bmatrix}
$$

with variables $\Delta u_k, \Delta y \in \mathbb{R}^N$. The variables Δu_k are the columns of the search direction ΔU corresponding to the variable U in (3.23). Eliminating the variables Δu_k gives the equation $H \Delta y = r$ with

$$
H = \sum_{k=1}^m (Q + W_k^{-2})^{-1}.
$$

Now suppose the linear kernel is used, and $Q = XX^T$ with $X \in \mathbb{R}^{N \times n}$ and N large (compared to mn). Then we can exploit the low rank structure in Q and write H as

$$
\begin{aligned}
H &= \sum_{k=1}^m \left(W_k^2 - W_k^2 X (I + X^T W_k^2 X)^{-1} X^T W_k^2 \right) \\
&= D - YY^T
\end{aligned}
$$

where $D = \sum_k W_k^2$ is diagonal and Y is an $N \times mn$ matrix, and

$$
Y = \begin{bmatrix} W_1^2 X L_1^{-1} & W_2^2 X L_2^{-1} & \cdots & W_m^2 X L_m^{-1} \end{bmatrix}
$$

where L_k is a Cholesky factor of $I + X^T W_k^2 X = L_k L_k^T$. A second application of the matrix inversion lemma gives

$$
\begin{aligned}
\Delta y &= (D - YY^T)^{-1} r \\
&= \left(D^{-1} + D^{-1} Y (I + Y^T D^{-1} Y)^{-1} Y^T D^{-1} \right) r.
\end{aligned}
$$

The largest dense matrix that needs to be factored in this method is the $mn \times mn$ matrix $I + Y^T D^{-1} Y$. For large N the cost is dominated by the matrix products $X^T W_i^2 D^{-1} W_j^2 X$, $i, j = 1, \ldots, m$, needed to compute $Y^T D^{-1} Y$. This takes $O(m^2 n^2 N)$ operations.

In table 3.3 we show computational results for the multiclass classifier applied to the MNIST handwritten digit data set (LeCun and Cortes, 1998). The images are 28×28. We add a constant feature to each training example, so the dimension of the feature space is $n = 1 + 28^2 = 785$. We use $\gamma = 10^5/N$. For the largest N, the QP (3.23) has $600,000$ variables and inequality constraints, and $60,000$ equality constraints.

N	time	iterations	test error
10000	5699	27	8.6%
20000	12213	33	4.0%
30000	35738	38	2.7%
40000	47950	39	2.0%
50000	63592	42	1.6%
60000	82810	46	1.3%

Table 3.3: Solution times (seconds) and numbers of iterations for the multiclass SVM training problem applied to the MNIST set of handwritten digits ($m = 10$ classes, $n = 785$ features)

3.4   Second-Order Cone Programming

Several authors have provided detailed studies of techniques for exploiting sparsity in SOCPs (Andersen et al., 2003; Goldfarb and Scheinberg, 2005). The coefficient matrix (3.12) of the reduced Newton equation of a linear and quadratic cone program with K second-order cone constraints of dimension $p_1, \ldots, p_K$ is

$$P + \sum_{k=1}^{K} G_k^T W_k^{-2} G_k, \qquad W_k^{-1} = \frac{1}{\beta_k}(2 J v_k v_k^T J - J). \qquad (3.24)$$

The scaling matrices are parameterized by parameters $\beta_k > 0$ and $v_k \in \mathbb{R}^{p_k}$ with $v_k^T J v_k = 1$ and J the sign matrix defined in (3.9). Note that

$$W_k^{-2} = \frac{1}{\beta^2}(2 w_k w_k^T - J) = \frac{1}{\beta^2}(I + 2 w_k w_k^T - 2 e_0 e_0^T), \qquad w_k = \begin{bmatrix} v_k^T v_k \\ -2 v_{k0} v_{k1} \end{bmatrix}$$

where e_0 is the first unit vector in $\mathbb{R}^p$, v_{k0} is the first entry of v_k, and v_{k1} is the $(p-1)$-vector of the other entries. Therefore

$$G_k^T W_k^{-2} G_k = \frac{1}{\beta^2} \left(G_k^T G_k + 2(G_k^T w_k)(G_k^T w_k)^T - 2(G_k^T e_0)(G_k^T e_0)^T \right),$$

that is, a multiple of $G_k^T G_k$ plus a rank-two term.

We can distinguish two cases when examining the sparsity of the sum (3.24). If the dimensions p_k of the second-order cones are small, then the matrices G_k are likely to have many zero columns and the vectors $G_k^T w_k$ will be sparse (for generic dense w_k). Therefore the products $G_k^T W_k^{-2} G_k$ and the entire matrix (3.24) are likely to be sparse. At the extreme end ($p_k = 1$) this reduces to the situation in linear programming where the matrix (3.12) has the sparsity of $P + G^T G$.

The second case arises when the dimensions p_k are large. Then $G_k^T w_k$ is likely to be dense, which results into a dense matrix (3.24). If $K \ll n$, we can still separate the sum (3.24) in a sparse part and a few dense rank-one terms, and apply techniques for handling dense rows in sparse linear programs (Andersen et al., 2003; Goldfarb and Scheinberg, 2005).

3.4.1   Robust Support Vector Machine Training

Second-order cone programming has found wide application in *robust optimization*. As an example, we discuss the robust SVM formulation of Shivaswamy et al. (2006). This problem can be expressed as a cone QP with second-order cone constraints:

$$
\begin{aligned}
& \text{minimize} && (1/2)w^T w + \gamma \mathbf{1}^T v \\
& \text{subject to} && \mathrm{Diag}(d)(Xw + b\mathbf{1}) \succeq \mathbf{1} - v + Eu \\
& && v \succeq 0 \\
& && \|S_j w\|_2 \leq u_j, \quad j = 1, \ldots, t.
\end{aligned}
\tag{3.25}
$$

The variables are $w \in \mathbb{R}^n$, $b \in \mathbb{R}$, $v \in \mathbb{R}^N$, and $u \in \mathbb{R}^t$. The matrix $X \in \mathbb{R}^{N \times n}$ has as its rows the training examples x_i^T, and the vector $d \in \{-1, 1\}^N$ contains the training labels. For $t = 0$, the term Eu and the norm constraints are absent, and the problem reduces to the standard linear SVM

$$
\begin{aligned}
& \text{minimize} && (1/2)w^T w + \gamma \mathbf{1}^T v \\
& \text{subject to} && d_i(x_i^T w + b) \geq 1 - v_i, \quad i = 1, \ldots, N \\
& && v \succeq 0.
\end{aligned}
\tag{3.26}
$$

In problem (3.25) the inequality constraints in (3.26) are replaced by a robust version that incorporates a model of the uncertainty in the training examples. The uncertainty is described by t matrices S_j, with t ranging from 1 to N, and an $N \times n$-matrix E with 0-1 entries and exactly one entry equal to one in each row. The matrices S_j can be assumed to be symmetric positive semidefinite. To interpret the constraints, suppose $E_{ij} = 1$. Then the constraint in (3.25) that involves training example x_i can be written as a second-order cone constraint:

$$
d_i(x_i^T w + b) - \|S_j w\|_2 \geq 1 - v_i.
$$

This is equivalent to

$$
\inf_{\|\eta\|_2 \leq 1} \left(d_i(x_i + S_j \eta)^T w + b \right) \geq 1 - v_i.
$$

In other words, we replace the training example x_i with an ellipsoid $\{x_i + S_j\eta \mid \|\eta\|_2 \leq 1\}$ and require that $d_i(x^T w + b) \geq 1 - v_i$ holds for all x in the ellipsoid. The matrix S_j defines the shape and magnitude of the uncertainty about training example i. If we take $t = 1$, we assume that all training examples are subject to the same type of uncertainty. Values of t larger than one allow us to use different uncertainty models for different subsets of the training examples.

To evaluate the merits of the robust formulation, it is useful to compare the costs of solving the robust and non-robust problems. Recall that the cost per iteration of an interior-point method applied to the QP (3.26) is of order Nn^2 if $N \geq n$, and is dominated by an equation of the form $(I + X^T D X)\Delta w = r$ with D positive diagonal. To determine the cost of solving the robust problem, we write it in the standard cone QP form (3.2) by choosing $x = (w, b, v, u) \in \mathbb{R}^n \times \mathbb{R} \times \mathbb{R}^N \times \mathbb{R}^t$, $K = 1 + t$, $C = \mathbb{R}_+^{2N} \times \mathcal{Q}_{n+1} \times \cdots \mathcal{Q}_{n+1}$. We have

$$
P = \begin{bmatrix} I & 0 & 0 & 0 \\ 0 & 0 & 0 & 0 \\ 0 & 0 & 0 & 0 \\ 0 & 0 & 0 & 0 \end{bmatrix}, \qquad
G = \begin{bmatrix} -\operatorname{Diag}(d)X & -d & -I & E \\ 0 & 0 & -I & 0 \\ 0 & 0 & 0 & -e_1^T \\ -S_1 & 0 & 0 & 0 \\ \vdots & \vdots & \vdots & \vdots \\ 0 & 0 & 0 & -e_t^T \\ -S_t & 0 & 0 & 0 \end{bmatrix}
$$

where e_k is the kth unit vector in $\mathbb{R}^t$. Note that $E^T D E$ is diagonal for any diagonal matrix D, and this property makes it inexpensive to eliminate the extra variable Δu from the Newton equations. As in the nonrobust case, the Newton equations can then be further reduced to an equation in n variables Δw. The cost of forming the reduced coefficient matrix is of order $Nn^2 + tn^3$. When $n \leq N$ and for modest values of t, the cost of solving the robust counterpart of the linear SVM training problem is therefore comparable to the standard non-robust linear SVM.

Table 3.4 shows the solution times for a customized CVXOPT interior-point method applied to randomly generated test problems with $n = 200$ features. Each training vector is assigned to one of t uncertainty models. For comparison, the general-purpose solver SDPT3 v.4 called from CVX takes about 130 seconds for $t = 50$ and $N = 4000$ training vectors.

N	$t = 2$	$t = 10$	$t = 50$	$t = 100$
4000	2.5	2.8	4.1	5.0
8000	5.4	5.3	6.0	6.9
16000	12.5	12.5	12.7	13.7

Table 3.4: Solution times (seconds) for customized interior-point method for robust SVM training ($n = 200$ features and t different uncertainty models)

3.5   Semidefinite Programming

We now turn to the question of exploiting problem structure in cone programs that include linear matrix inequalities. To simplify the notation, we explain the ideas for the inequality form SDP (3.6).

Consider the coefficient matrix $H = G^T W^{-1} W^{-T} G$ of the reduced Newton equations, with G defined in (3.7) and the scaling matrix W defined in (3.10). The entries of H are

$$H_{ij} = \text{Tr}\left(R^{-1} F_i R^{-T} R^{-1} F_j R^{-T}\right), \quad i, j = 1, \ldots, n. \qquad (3.27)$$

The matrix R is generally dense, and therefore the matrix H is usually dense, so the equation $H \Delta x = r$ must be solved by a dense Cholesky factorization. The cost of evaluating the expressions (3.27) is also significant, and often exceeds the cost of solving the system. For example, if $p = O(n)$ and the matrices F_i are dense, then it takes $O(n^4)$ operations to compute the entire matrix H and $O(n^3)$ operations to solve the system.

Efforts to exploit problem structure in SDPs have focused on using sparsity and low-rank structure in the coefficient matrices F_i to reduce the cost of assembling H. Sparsity is exploited, in varying degrees, by all general-purpose SDP solvers (Sturm, 1999, 2002; Tütüncü et al., 2003; Yamashita et al., 2003; Benson and Ye, 2005; Borchers, 1999). Several of these techniques use ideas from the theory of chordal sparse matrices and positive definite matrix completion theory to reduce the problem size or speed up critical calculations (Fukuda et al., 2000; Nakata et al., 2003; Burer, 2003; Andersen et al., 2010). It was also recognized early on that low-rank structure in the coefficients F_i can be very useful to reduce the complexity of interior-point methods (Gahinet and Nemirovski, 1997; Benson et al., 1999). For example, if $F_i = a_i a_i^T$, then it can be verified that

$$H = (A^T R^{-T} R^{-1} A) \circ (A^T R^{-T} R^{-1} A)$$

where A is the matrix with columns a_i and $\circ$ is componentwise matrix multiplication. This expression for H takes only $O(n^3)$ operations to evaluate if $p = O(n)$. Low-rank structure is exploited in the LMI Control Toolbox (Gahinet et al., 1995), DSDP (Benson and Ye, 2005), and SDPT3 (Tütüncü et al., 2003). Recent applications of dense, low-rank structure include SDPs derived from sum-of-squares formulations of nonnegative polynomials (Löfberg and Parrilo, 2004; Roh and Vandenberghe, 2006; Roh et al., 2007; Liu and Vandenberghe, 2007). Kandola et al. (2003) describe an application in machine learning.

Sparsity and low-rank structure do not exhaust the useful types of problem structure that can be exploited in SDP interior-point methods, as demonstrated by the following two examples.

3.5.1 SDPs with Upper Bounds

A simple example from Toh et al. (2007) and Nouralishahi et al. (2008) will illustrate the limitations of techniques based on sparsity. Consider a standard form SDP with an added upper bound:

$$
\begin{aligned}
&\text{minimize} && \operatorname{Tr}(CX) \\
&\text{subject to} && \operatorname{Tr}(A_i X) = b_i, \quad i = 1, \dots, m \\
& && 0 \preceq X \preceq I.
\end{aligned}
\tag{3.28}
$$

The variable X is a symmetric matrix of order p. Since general-purpose SDP solvers do not accept this format directly, the problem needs to be reformulated as one without upper bounds. An obvious reformulation is to introduce a slack variable S and solve the standard form SDP

$$
\begin{aligned}
&\text{minimize} && \operatorname{Tr}(CX) \\
&\text{subject to} && \operatorname{Tr}(A_i X) = b_i, \quad i = 1, \dots, m \\
& && X + S = I \\
& && X \succeq 0, \quad S \succeq 0.
\end{aligned}
\tag{3.29}
$$

This is the semidefinite programming analog of converting an LP with variable bounds,

$$
\begin{aligned}
&\text{minimize} && c^T x \\
&\text{subject to} && Ax = b \\
& && 0 \preceq x \preceq 1,
\end{aligned}
$$

into a standard form LP,

$$\begin{aligned}
\text{minimize} \quad & c^T x \\
\text{subject to} \quad & Ax = b, \quad x + s = \mathbf{1} \\
& x \succeq 0, \quad s \succeq 0.
\end{aligned} \tag{3.30}$$

Even though this is unnecessary in practice (LP solvers usually handle variable upper bounds directly), the transformation to (3.30) would have only a minor effect on the complexity. In (3.30) we add n extra variables (assuming the dimension of x is n) and n extremely sparse equality constraints. A good LP solver that exploits the sparsity will solve the LP at roughly the same cost as the corresponding problem without upper bounds. The situation is very different for SDPs. In (3.29) we increase the number of equality constraints from m to $m + p(p+1)/2$. SDP solvers are not as good at exploiting sparsity as LP solvers, so (3.29) is much harder to solve using general-purpose solvers than the corresponding problem without upper bound.

Nevertheless, the SDP with upper bounds can be solved at a cost comparable to the standard form problem, via a technique proposed by Toh et al. (2007) and Nouralishahi et al. (2008). The reduced Newton equations (3.11) for the SDP with upper bounds (3.29) are

$$T_1 \Delta X T_1 + T_2 \Delta X T_2 + \sum_{i=1}^{m} \Delta y_i A_i = r_X \tag{3.31a}$$

$$\text{Tr}(A_i \Delta X) = r_{y_i}, \quad i = 1, \dots, m \tag{3.31b}$$

where $T_1 = R_1^{-T} R_1^{-1}$ and $T_2 = R_2^{-T} R_2^{-1}$ are positive definite matrices. (The Newton equations for the standard form problem (3.28) are similar, but have only one term $T \Delta X T$ in the first equation, making it easy to eliminate ΔX.)

To solve (3.31) we first determine a congruence transformation that simultaneously diagonalizes T_1 and T_2,

$$V^T T_1 V = I, \qquad V^T T_2 V = \text{Diag}(\gamma),$$

where γ is a positive vector (see (Golub and Van Loan, 1996, section 8.7.2)). If we define $\Delta \tilde{X} = V^{-1} \Delta X V^{-T}$, $\tilde{A}_i = V^T A_i V$, the equations reduce to

$$\begin{aligned}
\Delta \tilde{X} + \text{Diag}(\gamma) \Delta \tilde{X} \, \text{Diag}(\gamma) + \sum_{i=1}^{m} \Delta y_i \tilde{A}_i &= V^T r_X V \\
\text{Tr}(\tilde{A}_i \Delta \tilde{X}) &= r_{y_i}, \quad i = 1, \dots, m.
\end{aligned}$$

From the first equation, we can express $\Delta \tilde{X}$ in terms of Δy:

$$\Delta \tilde{X} = (V^T r_X V) \circ \Gamma - \sum_{i=1}^{m} \Delta y_i (\tilde{A}_i \circ \Gamma) \tag{3.32}$$

$m = p$	time per iteration
50	0.05
100	0.33
200	2.62
300	10.5
400	30.4
500	70.8

Table 3.5: Time (seconds) per iteration of a customized interior-point method for SDPs with upper bounds

where Γ is the symmetric matrix with entries $\Gamma_{ij} = 1/(1+\gamma_i\gamma_j)$. Substituting this in the second equation gives a set of equations $H\Delta y = r$ where

$$H_{ij} = \text{Tr}(\tilde{A}_i(\tilde{A}_j \circ \Gamma)) = \text{Tr}((\tilde{A}_i \circ \tilde{A}_j)\Gamma)), \quad i,j = 1,\ldots,m.$$

After solving for Δy, one easily obtains ΔX from (3.32). The cost of this method is dominated by the cost of computing the matrices $\tilde{A}_i$ ($O(p^4)$ flops if $m = O(p)$), the cost of assembling H ($O(p^4)$ flops), and the cost of solving for Δy ($O(p^3)$ flops). For dense coefficient matrices A_i, the overall cost is comparable to the cost of solving the Newton equations for the standard form SDP (3.28) without upper bound.

Table 3.5 shows the time per iteration of a CVXOPT implementation of the method described above. The test problems are randomly generated, with $m = p$ and dense coefficient matrices A_i. The general-purpose SDP solver SDPT3 v.4, called from CVX, and applied to problem (3.29) with $m = p = 100$, takes about 23 seconds per iteration.

3.5.2 Nuclear Norm Approximation

In section 3.3.1 we discussed the ℓ_1-norm approximation problem (3.14) and showed that the cost per iteration of an interior-point method is comparable to the cost of solving the corresponding least-squares problem (that is, $O(mn^2)$ operations). We can ask the same question about the matrix counterpart of ℓ_1-norm approximation, the *nuclear norm* approximation problem:

$$\text{minimize} \quad \|X(u) - D\|_*. \tag{3.33}$$

Here $\|\cdot\|_*$ denotes the nuclear matrix norm (sum of singular values) and $X(u) = \sum_{i=1}^{n} u_i X_i$ is a linear mapping from $\mathbb{R}^n$ to $\mathbb{R}^{p \times q}$. The nuclear norm is popular in convex heuristics for rank minimization problems in system theory and machine learning (Fazel et al., 2001; Fazel, 2002; Fazel et al., 2004; Recht et al., 2010; Candès and Plan, 2010). These heuristics extend ℓ_1-norm heuristics for sparse optimization.

Problem (3.33) is equivalent to an SDP

$$
\begin{aligned}
\text{minimize} \quad & (\operatorname{Tr} V_1 + \operatorname{Tr} V_2)/2 \\
\text{subject to} \quad & \begin{bmatrix} V_1 & X(u) - D \\ (X(u) - D)^T & V_2 \end{bmatrix} \succeq 0,
\end{aligned}
\tag{3.34}
$$

with auxiliary symmetric matrix variables V_1, V_2. The presence of the extra variables V_1 and V_2 clearly makes solving (3.34) using a general-purpose SDP solver very expensive unless p and q are small, and much more expensive than solving the corresponding least-squares approximation problem (that is, problem (3.33) with the Frobenius norm replacing the nuclear norm).

A specialized interior-point method is described in Liu and Vandenberghe (2009). The basic idea can be summarized as follows. The Newton equations for (3.34) are

$$
\Delta Z_{11} = r_{V_1}, \qquad \Delta Z_{22} = r_{V_2}, \qquad \operatorname{Tr}(X_i^T \Delta Z_{12}) = r_{u_i}, \quad i = 1, \ldots, n
$$

and

$$
\begin{bmatrix} \Delta V_1 & X(\Delta u) \\ X(\Delta u)^T & \Delta V_2 \end{bmatrix} + T \begin{bmatrix} \Delta Z_{11} & \Delta Z_{12} \\ \Delta Z_{12}^T & \Delta Z_{22} \end{bmatrix} T = r_Z,
$$

where $T = RR^T$. The variables ΔZ_{11}, ΔZ_{22}, ΔV_1, ΔV_2 are easily eliminated, and the equations reduce to

$$
\begin{aligned}
X(\Delta u) + T_{11} \Delta Z_{12} T_{22} + T_{12} \Delta Z_{12}^T T_{12} &= r_{Z_{12}} \\
\operatorname{Tr}(X_i^T \Delta Z_{12}) &= r_{u_i}, \quad i = 1, \ldots, n,
\end{aligned}
$$

where T_{ij} are subblocks of T partitioned as the matrix in the constraint (3.34). The method of Liu and Vandenberghe (2009) is based on applying a transformation that reduces T_{11} and T_{22} to identity matrices and T_{12} to a (rectangular) diagonal matrix, and then eliminating ΔZ_{12} from the first equation, to obtain a dense linear system in Δu. The cost of solving the Newton equations is $O(n^2 pq)$ operations if $n \geq \max\{p, q\}$. For dense X_i this is comparable to the cost of solving the approximation problem in the least-squares (Frobenius norm) sense.

Table 3.6 shows the time per iteration of a CVXOPT code for (3.34). The problems are randomly generated with $n = p = 2q$. Note that the SDP (3.34) has $n + p(p+1)/2 + q(q+1)/2$ variables and is very expensive to solve by general-purpose interior-point codes. CVX/SDPT3 applied to (3.33) takes

$n = p = 2q$	time per iteration
100	0.30
200	2.33
300	8.93
400	23.9
500	52.4

Table 3.6: Time (seconds) per iteration of a customized interior-point method for the nuclear norm approximation problem

22 seconds per iteration for the first problem ($n = p = 100$, $q = 50$).

3.6   Conclusion

Interior-point algorithms for conic optimization are attractive in machine learning and other applications because they converge to a high accuracy in a small number of iterations and are quite robust with respect to data scaling. The main disadvantages are the high memory requirements and the linear algebra complexity associated with the linear equations that are solved at each iteration. It is therefore critical to exploit problem structure when solving large problems. For linear and quadratic programming, sparse matrix techniques provide a general and effective approach to handling problem structure. For nonpolyhedral cone programs, and semidefinite programs in particular, the sparse approach is less effective for two reasons. First, translating non-sparse problem structure into a sparse model may require introducing a very large number of auxiliary variables and constraints. Second, techniques for exploiting sparsity in SDPs are less well developed than for LPs. It is therefore difficult to develop *general-purpose* techniques for exploiting problem structure in cone programs that are as scalable as sparse linear programming solvers. However, it is sometimes quite straightforward to find special-purpose techniques that exploit various types of problem structure. When this is the case, customized implementations can be developed that are orders of magnitude more efficient than general-purpose interior-point implementations.

3.7   References

F. Alizadeh and D. Goldfarb.   Second-order cone programming.   *Mathematical Programming, series B*, 95:3–51, 2003.

E. D. Andersen. On primal and dual infeasibility certificates in a homogeneous model for convex optimization. *SIAM Journal on Optimization*, 11(2):380–388, 2000.

E. D. Andersen, C. Roos, and T. Terlaky. On implementing a primal-dual interior-point method for conic quadratic optimization. *Mathematical Programming, series B*, 95(2):249–277, 2003.

M. S. Andersen, J. Dahl, and L. Vandenberghe. Implementation of nonsymmetric interior-point methods for linear optimization over sparse matrix cones. *Mathematical Programming Computation*, 2(3–4):167–201, 2010.

A. Ben-Tal and A. Nemirovski. *Lectures on Modern Convex Optimization: Analysis, Algorithms, and Engineering Applications*. SIAM, Philadelphia, 2001.

S. J. Benson and Y. Ye. DSDP5: Software for semidefinite programming. Technical Report ANL/MCS-P1289-0905, Mathematics and Computer Science Division, Argonne National Laboratory, Argonne, IL, 2005.

S. J. Benson, Y. Ye, and X. Zhang. Solving large-scale sparse semidefinite programs for combinatorial optimization. *SIAM Journal on Optimization*, 10:443–461, 1999.

B. Borchers. CSDP, a C library for semidefinite programming. *Optimization Methods and Software*, 11(1):613–623, 1999.

S. Boyd and L. Vandenberghe. *Convex Optimization*. Cambridge University Press, 2004.

S. Boyd, L. El Ghaoui, E. Feron, and V. Balakrishnan. *Linear Matrix Inequalities in System and Control Theory*, volume 15 of *SIAM Studies in Applied Mathematics*. Philadelphia, 1994.

S. Burer. Semidefinite programming in the space of partial positive semidefinite matrices. *SIAM Journal on Optimization*, 14(1):139–172, 2003.

E. J. Candès and Y. Plan. Matrix completion with noise. *Proceedings of the IEEE*, 98(6):925–936, 2010.

K. Crammer and Y. Singer. On the algorithmic implementation of the multiclass kernel-based vector machines. *Journal of Machine Learning Research*, 2:265–292, 2001.

J. Dahl and L. Vandenberghe. *CVXOPT: A Python Package for Convex Optimization*. http://abel.ee.ucla.edu/cvxopt, 2009.

M. Fazel. *Matrix Rank Minimization with Applications*. PhD thesis, Stanford University, 2002.

M. Fazel, H. Hindi, and S. Boyd. A rank minimization heuristic with application to minimum order system approximation. In *Proceedings of the American Control Conference*, volume 6, pages 4734–4739, 2001.

M. Fazel, H. Hindi, and S. Boyd. Rank minimization and applications in system theory. In *Proceedings of the American Control Conference*, pages 3273–3278, 2004.

M. C. Ferris and T. S. Munson. Interior-point methods for massive support vector machines. *SIAM Journal on Optimization*, 13(3):783–804, 2002.

S. Fine and K. Scheinberg. Efficient SVM training using low-rank kernel representations. *Journal of Machine Learning Research*, 2:243–264, 2002.

M. Fukuda, M. Kojima, K. Murota, and K. Nakata. Exploiting sparsity in semidefinite programming via matrix completion I: general framework. *SIAM Journal* on *Optimization*, 11(3):647–674, 2000.

P. Gahinet and A. Nemirovski. The projective method for solving linear matrix inequalities. *Mathematical Programming*, 77(2):163–190, May 1997.

P. Gahinet, A. Nemirovski, A. J. Laub, and M. Chilali. *LMI Control Toolbox*. The MathWorks, 1995.

E. M. Gertz and S. J. Wright. Object-oriented software for quadratic programming. *ACM Transactions on Mathematical Software*, 29(1):58–81, 2003.

J. Gillberg and A. Hansson. Polynomial complexity for a Nesterov-Todd potential-reduction method with inexact search directions. In *Proceedings of the 42nd IEEE Conference on Decision and Control*, volume 3, pages 3824–3829, 2003.

D. Goldfarb and K. Scheinberg. Product-form Cholesky factorization in interior point methods for second-order cone programming. *Mathematical Programming Series A*, 103(1):153–179, 2005.

G. H. Golub and C. F. Van Loan. *Matrix Computations*. John Hopkins University Press, third edition, 1996.

M. Grant and S. Boyd. *CVX: Matlab Software for Disciplined Convex Programming (Web Page and Software)*. http://stanford.edu/~boyd/cvx, 2007.

M. Grant and S. Boyd. Graph implementations for nonsmooth convex programs. In V. Blondel, S. Boyd, and H. Kimura, editors, *Recent Advances in Learning and Control (a Tribute to M. Vidyasagar)*, pages 95–110. Springer, 2008.

C. Helmberg, F. Rendl, R. J. Vanderbei, and H. Wolkowicz. An interior-point method for semidefinite programming. *SIAM Journal on Optimization*, 6(2): 342–361, 1996.

S. Joshi and S. Boyd. An efficient method for large-scale gate sizing. *IEEE Transactions on Circuits and Systems I*, 55(9):2760–2773, 2008.

J. Kandola, T. Graepel, and J. Shawe-Taylor. Reducing kernel matrix diagonal dominance using semi-definite programming. In B. Schölkopf and M. Warmuth, editors, *Learning Theory and Kernel Machines, Proceedings of the 16th Annual Conference on Learning Theory and 7th Kernel Workshop*, pages 288–302. Springer-Verlag, 2003.

S.-J. Kim, K. Koh, M. Lustig, S. Boyd, and D. Gorinevsky. An interior-point method for large-scale ℓ_1-regularized least squares. *IEEE Journal on Selected Topics in Signal Processing*, 1(4):606–617, 2007.

K. Koh, S.-J. Kim, and S. Boyd. An interior-point method for large-scale ℓ_1-regularized logistic regression. *Journal of Machine Learning Research*, 8:1519–1555, 2007.

M. Kojima, S. Shindoh, and S. Hara. Interior-point methods for the monotone semidefinite linear complementarity problem in symmetric matrices. *SIAM Journal on Optimization*, 7:86–125, 1997.

Y. LeCun and C. Cortes. The MNIST Database of Handwritten Digits. Available at http://yann.lecun.com/exdb/mnist/, 1998.

Z. Liu and L. Vandenberghe. Low-rank structure in semidefinite programs derived from the KYP lemma. In *Proceedings of the 46th IEEE Conference on Decision and Control*, pages 5652–5659, 2007.

Z. Liu and L. Vandenberghe. Interior-point method for nuclear norm approximation with application to system identification. *SIAM Journal on Matrix Analysis and Applications*, 31(3):1235–1256, 2009.

J. Löfberg. YALMIP: A Toolbox for Modeling and Optimization in MATLAB. In *Proceedings of the International Symposium on Computer Aided Control Systems Design*, pages 284–289, 2004.

J. Löfberg and P. A. Parrilo. From coefficients to samples: A new approach to SOS optimization. In *Proceedings of the 43rd IEEE Conference on Decision and Control*, volume 3, pages 3154–3159, 2004.

R. D. C. Monteiro and Y. Zhang. A unified analysis for a class of long-step primal-dual path-following interior-point algorithms for semidefinite programming. *Mathematical Programming*, 81:281–299, 1998.

MOSEK ApS. *The MOSEK Optimization Tools Manual. Version 6.0.*, 2010. Available from www.mosek.com.

K. Nakata, K. Fujisawa, M. Fukuda, M. Kojima, and K. Murota. Exploiting sparsity in semidefinite programming via matrix completion II: Implementation and numerical details. *Mathematical Programming, series B*, 95(2):303–327, 2003.

Y. Nesterov and A. Nemirovskii. *Interior-Point Polynomial Methods in Convex Programming*, volume 13 of *Studies in Applied Mathematics*. SIAM, Philadelphia, 1994.

Y. Nesterov, M. J. Todd, and Y. Ye. Infeasible-start primal-dual methods and infeasibility detectors for nonlinear programming problems. *Mathematical Programming*, 84(2):227–267, 1999.

Y. E. Nesterov and M. J. Todd. Self-scaled barriers and interior-point methods for convex programming. *Mathematics of Operations Research*, 22(1):1–42, 1997.

Y. E. Nesterov and M. J. Todd. Primal-dual interior-point methods for self-scaled cones. *SIAM Journal on Optimization*, 8(2):324–364, May 1998.

M. Nouralishahi, C. Wu, and L. Vandenberghe. Model calibration for optical lithography via semidefinite programming. *Optimization and Engineering*, 9: 19–35, 2008.

B. Recht, M. Fazel, and P. A. Parrilo. Guaranteed minimum-rank solutions of linear matrix equations via nuclear norm minimization. *SIAM Review*, 52(3):471–501, 2010.

T. Roh and L. Vandenberghe. Discrete transforms, semidefinite programming, and sum-of-squares representations of nonnegative polynomials. *SIAM Journal on Optimization*, 16(4):939–964, 2006.

T. Roh, B. Dumitrescu, and L. Vandenberghe. Multidimensional FIR filter design via trigonometric sum-of-squares optimization. *IEEE Journal of Selected Topics in Signal Processing*, 1(4):641–650, 2007.

P. K. Shivaswamy, C. Bhattacharyya, and A. J. Smola. Second order cone programming approaches for handling missing and uncertain data. *Journal of Machine Learning Research*, 7:1283–1314, 2006.

J. F. Sturm. Using SEDUMI 1.02, a Matlab toolbox for optimization over symmetric cones. *Optimization Methods and Software*, 11-12:625–653, 1999.

J. F. Sturm. Implementation of interior point methods for mixed semidefinite and second order cone optimization problems. *Optimization Methods and Software*, 17(6):1105–1154, 2002.

K. C. Toh, R. H. Tütüncü, and M. J. Todd. Inexact primal-dual path-following algorithms for a special class of convex quadratic SDP and related problems. *Pacific Journal of Optimization*, 3, 2007.

R. H. Tütüncü, K. C. Toh, and M. J. Todd. Solving semidefinite-quadratic-linear programs using SDPT3. *Mathematical Programming, series B*, 95:189–217, 2003.

R. Wallin, A. Hansson, and J. H. Johansson. A structure exploiting preprocessor for semidefinite programs derived from the Kalman-Yakubovich-Popov lemma. *IEEE Transactions on Automatic Control*, 54(4):697–704, 2009.

M. Yamashita, K. Fujisawa, and M. Kojima. Implementation and evaluation of SDPA 6.0 (Semidefinite Programming Algorithm 6.0). *Optimization Methods and Software*, 18(4):491–505, 2003.

4   Incremental Gradient, Subgradient, and Proximal Methods for Convex Optimization: A Survey

Dimitri P. Bertsekas   dimitri@mit.edu
Dept. of Electr. Engineering and Comp. Science, M.I.T.
Cambridge, MA, 02139

We survey incremental methods for minimizing a sum $\sum_{i=1}^{m} f_i(x)$ consisting of a large number of convex component functions f_i. Our methods consist of iterations applied to single components, and have proved very effective in practice. We introduce a unified algorithmic framework for a variety of such methods, some involving gradient and subgradient iterations, which are known, and some involving combinations of subgradient and proximal methods, which are new and offer greater flexibility in exploiting the special structure of f_i. We provide an analysis of the convergence and rate of convergence properties of these methods, including the advantages offered by randomization in the selection of components. We also survey applications in inference/machine learning, signal processing, and large-scale and distributed optimization.

4.1   Introduction

We consider optimization problems with a cost function consisting of a large number of component functions, such as

$$
\text{minimize} \quad \sum_{i=1}^{m} f_i(x) \\
\text{subject to} \;\; x \in X,
\tag{4.1}
$$

where $f_i : \Re^n \mapsto \Re$, $i = 1, \ldots, m$ are real-valued functions, and X is a closed convex set.[1] We focus on the case where the number of components m is very large, and there is an incentive to use incremental methods that operate on a single component f_i at each iteration, rather than on the entire cost function. If each incremental iteration tends to make reasonable progress in some "average" sense, then, depending on the value of m, an incremental method may significantly outperform (by orders of magnitude) its nonincremental counterpart, as extensive experience has shown.

In this chapter, we survey the algorithmic properties of incremental methods in a unified framework, based on the author's recent work on incremental proximal methods (Bertsekas, 2010). In this section, we first provide an overview of representative applications, and then we discuss three types of incremental methods: gradient, subgradient, and proximal. We unify these methods into a combined method, which we use as a vehicle for analysis in Sections 4.2, 4.3, and 4.4. Finally, we discuss in greater detail some illustrative applications in Section 4.5. Some of the proofs of propositions have been omitted and can be found in the report (Bertsekas, 2010).

4.1.1   Some Examples of Additive Cost Problems

Additive cost problems of the form (4.1) arise in a variety of contexts. Let us provide a few examples where the incremental approach may have an advantage over alternatives.

Example 4.1 (Least Squares and Inference). An important context where cost functions of the form $\sum_{i=1}^{m} f_i(x)$ arise is inference/machine learning, where each term $f_i(x)$ corresponds to error between some data and

---

1. Throughout the chapter, we will operate within the n-dimensional space $\Re^n$ with the standard Euclidean norm, denoted $\| \cdot \|$. All vectors are considered column vectors and a prime denotes transposition, so $x'x = \|x\|^2$. We will be using standard terminology of convex optimization throughout, as given, for example, in textbooks such as Rockafellar (1970), or the author's recent book (Bertsekas, 2009).

the output of a parametric model, with x being the vector of parameters. An example is *linear least-squares* problems, where f_i has quadratic structure, except for a regularization function, which may be differentiable/quadratic, as in the classical regression problem

$$\sum_{i=1}^{m}(a_i'x - b_i)^2 + \gamma\|x - \overline{x}\|^2, \quad \text{s.t.} \quad x \in \Re^n,$$

where $\overline{x}$ is given, or nondifferentiable, as in the ℓ_1-regularization problem

$$\sum_{i=1}^{m}(a_i'x - b_i)^2 + \gamma\sum_{j=1}^{n}|x_j|, \quad \text{s.t.} \quad (x_1,\dots,x_n) \in \Re^n,$$

which will be discussed further in Section 4.5.

A more general class of additive cost problems is *nonlinear least squares*. Here

$$f_i(x) = \big(h_i(x)\big)^2,$$

where $h_i(x)$ represents the difference between the ith measurement (out of m) from a physical system and the output of a parametric model whose parameter vector is x. Problems of nonlinear curve fitting and regression, as well as problems of training neural networks, fall in this category, and they are typically nonconvex.

Another possibility is to use a nonquadratic function to penalize the error between some data and the output of the parametric model. For example, in place of the squared error $(a_i'x - b_i)^2$, we may use

$$f_i(x) = \ell(a_i'x - b_i),$$

where ℓ is a convex function. This is a common approach in robust estimation and some support vector machine formulations.

Still another example is *maximum likelihood estimation*, where f_i is a log-likelihood function of the form

$$f_i(x) = -\log P_Y(y_i; x),$$

where $y_1,\dots,y_m$ represents values of independent samples of a random vector whose distribution $P_Y(\cdot; x)$ depends on an unknown parameter vector $x \in \Re^n$ that one wishes to estimate. Related contexts include "incomplete" data cases, where the expectation-maximization (EM) approach is used.

Example 4.2 (Dual Optimization in Separable Problems). Consider the problem

$$\text{maximize} \quad \sum_{i=1}^{m} c_i(y_i)$$

$$\text{subject to} \quad \sum_{i=1}^{m} g_i(y_i) \geq 0, \quad y_i \in Y_i, \quad i = 1, \ldots, m,$$

where $c_i : \Re \mapsto \Re$ and $g_i : \Re \mapsto \Re^n$ are functions of the single scalar coordinate y_i, and Y_i are given sets of scalars. Then, by assigning a dual vector/multiplier $x \in \Re^n$ to the n-dimensional constraint function, we obtain the dual problem

$$\text{minimize} \quad \sum_{i=1}^{n} f_i(x), \quad \text{subject to } x \geq 0,$$

where

$$f_i(x) = \sup_{y_i \in Y_i} \left\{ c_i(y_i) + x' g_i(y_i) \right\},$$

which has the additive form (4.1). Note that Y_i is not assumed to be convex, so integer programming and other discrete optimization problems are included. However, the dual cost function components f_i are always convex, and their values and subgradients can often be computed either analytically or with a one-dimensional maximization.

Example 4.3 (Minimization of an Expected Value: Stochastic Programming). Consider the minimization of an expected value

$$\begin{aligned} \text{minimize} \quad & E\{F(x, w)\} \\ \text{subject to} \quad & x \in X, \end{aligned} \tag{4.2}$$

where w is a random variable taking a finite but very large number of values w_i, $i = 1, \ldots, m$, with corresponding probabilities π_i. Then the cost function consists of the sum of the m functions $\pi_i F(x, w_i)$.

An example is *stochastic programming*, a classical model of two-stage optimization under uncertainty. A vector $x \in X$ is selected, a random event occurs that has m possible outcomes $w_1, \ldots, w_m$, and then another vector y is selected from some set Y with knowledge of the outcome that occurred. Then, for optimization purposes, we need to specify a different vector $y_i \in Y$ for each outcome w_i. The problem is to minimize the expected cost

$$F(x) + \sum_{i=1}^{m} \pi_i G_i(y_i),$$

where $G_i(y_i)$ is the cost associated with the occurrence of w_i, and π_i is the corresponding probability. This is a problem with an additive cost function. Furthermore, if there are separable (e.g., linear) constraints coupling the vectors x and y_i, the problem has a separable form.

Additive cost function problems also arise from problem (4.2) in a different way: when the expected value $E\{F(x, w)\}$ is approximated by an m-sample average

$$f(x) = \frac{1}{m} \sum_{i=1}^{m} F(x, w_i),$$

where w_i are independent samples of the random variable w. The minimum of the sample average $f(x)$ is then taken as an approximation of the minimum of $E\{F(x, w)\}$.

Example 4.4 (Problems with Many Constraints). Problems of the form

$$
\begin{aligned}
&\text{minimize} \quad f(x) \\
&\text{subject to} \quad g_j(x) \le 0, \quad j = 1, \dots, r, \ x \in X,
\end{aligned}
\tag{4.3}
$$

where the number r of constraints is very large, often arise in practice, either directly or via reformulation from other problems. They can be handled in a variety of ways. One possibility is to adopt a penalty function approach, and replace problem (4.3) with

$$
\begin{aligned}
&\text{minimize} \quad f(x) + c \sum_{j=1}^{r} P\big(g_j(x)\big) \\
&\text{subject to} \quad x \in X,
\end{aligned}
\tag{4.4}
$$

where $P(\cdot)$ is a scalar penalty function satisfying $P(t) = 0$ if $t \le 0$, and $P(t) > 0$ if $t > 0$, and c is a positive penalty parameter. For example, one may use the quadratic penalty $P(t) = \big(\max\{0, t\}\big)^2$. An interesting alternative is to use $P(t) = \max\{0, t\}$, in which case it can be shown that the optimal solutions of problems (4.3) and (4.4) coincide when c is sufficiently large (see, for example, Bertsekas et al. (2003, Section 7.3) for the case in which f is convex). The cost function of the penalized problem (4.4) is of the additive form (4.1).

The idea of replacing constraints with penalties can be extended to the case where the constraint $x \in X$ in problem (4.3) has the form $x \in \cap_{j=1}^{m} X_j$. Then, under relatively mild conditions, problem (4.3) is equivalent to the unconstrained minimization of

$$f(x) + c \sum_{j=1}^{r} P\big(g_j(x)\big) + \gamma \sum_{j=1}^{m} \text{dist}(x; X_j),$$

where $\text{dist}(x; X_j) = \inf_{y \in X_j} \|y - x\|$ and γ is a sufficiently large penalty parameter. We discuss this possibility in Section 4.5.

Example 4.5 (Distributed Incremental Optimization in Sensor Networks). Consider a network of m sensors where data are collected and used to solve some inference problem involving a parameter vector x. If $f_i(x)$ represents an error penalty for the data collected by the ith sensor, the inference problem is of the form (4.1). While it is possible to collect all the data at a fusion center where the problem will be solved in centralized manner, it may be preferable to adopt a distributed approach in order to save data communication overhead and/or take advantage of parallelism in computation. In such an approach the current iterate x_k is passed from one sensor to another, with each sensor i performing an incremental iteration involving just its local component function f_i, and the entire cost function need not be known at any one location. We refer to Blatt et al. (2007), and Rabbat and Nowak (2004, 2005) for further discussion.

Example 4.6 (Weber Problem in Location Theory). We want to find a point x in the plane whose sum of weighted distances from a given set of points $y_1, \ldots, y_m$ is minimized. Mathematically, the problem is

$$\sum_{i=1}^{m} w_i \|x - y_i\|, \quad \text{s.t.} \quad x \in \Re^n,$$

where $w_1, \ldots, w_m$ are given positive scalars. This problem descends from the famous Fermat-Torricelli-Viviani problem (see (Boltyanski et al., 1999) for an account of the history; Fermat's formulation was for the case of a triangle, where $m = 3$). It is a basic problem in location theory, and has received a lot of attention. The algorithmic approaches of this chapter would be of potential interest when the number of points m is large. We refer to Beck and Teboulle (2010) for a discussion that is relevant to our context.

4.1.2 Incremental Gradient Methods: Differentiable Problems

In the case where the components f_i are differentiable (not necessarily convex), we may use incremental gradient methods, which have the form

$$x_{k+1} = P_X\big(x_k - \alpha_k \nabla f_{i_k}(x_k)\big), \tag{4.5}$$

where α_k is a positive stepsize, $P_X(\cdot)$ denotes projection on X, and i_k is the index of the cost component that is iterated on. Such methods have a long history, particularly for the unconstrained case ($X = \Re^n$), starting with the Widrow-Hoff least-mean-squares (LMS) method (Widrow and Hoff, 1960) for positive semidefinite quadratic component functions (see e.g., (Luo, 1993), (Bertsekas and Tsitsiklis, 1996, Section 3.2.5), (Bertsekas, 1999, Section 1.5.2)). They have also been used extensively for the training of neural networks, a case of nonquadratic/nonconvex cost components, under the generic name "backpropagation methods." There are several variants of these methods, which differ in the stepsize selection scheme, and iin the order in which components are taken up for iteration (it could be deterministic or randomized). They are supported by convergence analyses under various conditions; see Luo (1993), Grippo (1994), Grippo (2000), Luo and Tseng (1994), Mangasarian and Solodov (1994), Bertsekas (1997), Solodov (1998), and Tseng (1998).

When comparing the incremental gradient method with its classical non-incremental gradient counterpart (where $m = 1$ and all components are lumped into a single function $f(x) = \sum_{i=1}^m f_i(x)$), it is important to realize that there are two complementary performance issues to consider.

1. *Progress when far from convergence.* Here the incremental method can be much faster. For an extreme case let $X = \Re^n$ (no constraints), and take m very large and all components f_i identical to each other. Then an incremental iteration requires m times less computation than a classical gradient iteration, but gives exactly the same result. While this is an extreme example, it reflects the essential mechanism by which incremental methods can be far superior: when the components f_i are not too dissimilar, far from the minimum a single component gradient will point to, "more or less," the right direction (see also the discussion of Bertsekas (1997) and Bertsekas (1999, Example 1.5.5 and Exercise 1.5.5).)

2. *Progress when close to convergence.* Here the incremental method is generally inferior. As we will discuss shortly, it converges at a sublinear rate because it requires a diminishing stepsize α_k, compared with the typically linear rate achieved with the classical gradient method when a small, constant stepsize is used ($\alpha_k \equiv \alpha$). One may use a constant stepsize with the incremental method - and indeed this may be the preferred mode of implementation - but then the method typically oscillates in the neighborhood of a solution, with the size of the oscillation roughly proportional to α, as examples and theoretical analysis show.

To understand the convergence mechanism of incremental gradient methods, let us consider the case $X = \Re^n$, and assume that the component functions f_i are selected for iteration according to a cyclic order (i.e., for every ℓ, $i_{\ell m} = 1, i_{\ell m+1} = 2, \ldots, i_{\ell m+m-1} = m$), and let us assume that α_k is constant within a cycle (i.e., $\alpha_{\ell m} = \alpha_{\ell m+1} = \cdots = \alpha_{\ell m+m-1}$). Then, viewing the iteration (4.5) in terms of cycles, we have, for every k that marks the beginning of a cycle ($i_k = 1$),

$$x_{k+m} = x_k - \alpha_k \sum_{i=1}^{m} \nabla f_i(x_{k+i-1}) = x_k - \alpha_k \big(\nabla f(x_k) + e_k \big),$$

where f is the cost function/sum of components, $f(x) = \sum_{i=1}^{m} f_i(x)$, and e_k is given by

$$e_k = \sum_{i=1}^{m} \big(\nabla f_i(x_k) - \nabla f_i(x_{k+i-1}) \big),$$

and may be viewed as an error in the calculation of the gradient $\nabla f(x_k)$. For Lipschitz continuous gradient functions ∇f_i, the error e_k is proportional to α_k, and this shows two fundamental properties of incremental gradient methods, which hold generally for the other incremental methods of this chapter as well.

1. A constant stepsize ($\alpha_k \equiv \alpha$) typically cannot guarantee convergence, since then the size of the gradient error $\|e_k\|$ is typically bounded away from 0. Instead, a peculiar form of convergence takes place for constant but sufficiently small α, whereby the iterates within cycles converge to corresponding points of a limit cycle. This is true even in the most favorable case of a linear least squares problem (see Luo (1993), or the textbook analysis of Bertsekas (1999, Section 1.5.1)).

2. A diminishing stepsize (such as $\alpha_k = O(1/k)$) leads to a diminishing error e_k, so (under the appropriate Lipschitz condition) it can result in convergence to a stationary point of f.

A corollary of these properties is that the price for achieving convergence is the slow (sublinear) asymptotic rate of convergence associated with a diminishing stepsize, which compares unfavorably with the often linear rate of convergence associated with a constant stepsize and the nonincremental gradient method. However, in practical terms this argument does not tell the entire story, since in the early iterations, the incremental gradient method often achieves a much faster convergence rate than its nonincremental counterpart. In practice, the incremental method is usually operated with a stepsize that either is constant or is gradually reduced up to a positive value small enough that the resulting asymptotic oscillation is of no essential concern. An alternative is to use a constant stepsize throughout, but to reduce over time the degree of incrementalism, so that ultimately the method becomes nonincremental and achieves a linear convergence rate (see Bertsekas (1997) and Solodov (1998)).

Aside from extensions to nondifferentiable cost problems, for $X = \Re^n$ there is an important variant of the incremental gradient method that involves extrapolation along the direction of the difference of the preceding two iterates:

$$x_{k+1} = x_k - \alpha_k \nabla f_{i_k}(x_k) + \beta(x_k - x_{k-1}), \tag{4.6}$$

where β is a scalar in $[0, 1)$ and $x_{-1} = x_0$ (see e.g., Mangasarian and Solodov (1994), Tseng (1998), Bertsekas (1996, Section 3.2)). This is sometimes called the *incremental gradient method with momentum*. The nonincremental version of this method is the *heavy ball* method of Poljak (1964), which can be shown to have a faster convergence rate than the corresponding gradient method (see Polyak (1987, Section 3.2.1)). A nonincremental method of this type, but with variable and suitably chosen value of β, has been proposed by Nesterov (1983), and has received a lot of attention recently because it has optimal iteration complexity properties under certain conditions (see Nesterov (2004, 2005), Lu et al. (2008), Tseng (2008), and Beck and Teboulle (2009, 2010)). However, no incremental analogs of this method with favorable complexity properties are currently known.

Another variant of the incremental gradient method for the case $X = \Re^n$ has been proposed by Blatt et al. (2007), which (after the first m iterates are computed) has the form

$$x_{k+1} = x_k - \alpha \sum_{\ell=0}^{m-1} \nabla f_{i_{k-\ell}}(x_{k-\ell}). \tag{4.7}$$

(For $k < m$, the summation should go up to $\ell = \min\{k, m - 1\}$, and α should be replaced by a corresponding larger value, such as $\alpha_k = m\alpha/(k + 1)$.) This method also computes the gradient incrementally, one component per iteration, but in place of the single component gradient $\nabla f_{i_k}(x_k)$ in (4.5), it uses an approximation to the total cost gradient $\nabla f(x_k)$, which is an aggregate of the component gradients computed in the past m iterations. A cyclic order of component function selection ($i_k = k$ modulo m plus 1) is assumed in (Blatt et al., 2007), and a convergence analysis is given, including a linear convergence rate result for a sufficiently small constant stepsize α and quadratic component functions f_i. It is not clear how iterations (4.5) and (4.7) compare in terms of rate of convergence, although the latter seems likely to make faster progress when close to convergence. Note that iteration (4.7) bears similarity to the incremental gradient iteration with momentum

(4.6) where $\beta \approx 1$. In particular, when $\alpha_k \equiv \alpha$, the sequence generated by (4.6) satisfies

$$x_{k+1} = x_k - \alpha \sum_{\ell=0}^{k} \beta^\ell \, \nabla f_{i_{k-\ell}}(x_{k-\ell}),$$

which resembles (4.7). There are no known analogs of iterations (4.6) and (4.7) for nondifferentiable cost problems.

Among alternative incremental methods for differentiable cost problems, we also mention versions of the Gauss-Newton method for nonlinear least-squares problems, based on the extended Kalman filter ((Davidon, 1976), (Bertsekas, 1996), and (Moriyama et al., 2003)). They are mathematically equivalent to the ordinary Gauss-Newton method for linear least squares, which they solve exactly after a single pass through the component functions f_i, but they often perform much faster in the nonlinear case, particularly when m is large.

Let us finally note that incremental gradient methods are related to stochastic gradient methods, which aim to minimize an expected value $E\{F(x,w)\}$ (cf. Example 1.3) by using the iteration

$$x_{k+1} = x_k - \alpha_k \nabla F(x_k, w_k),$$

where w_k is a sample of the random variable w. These methods also have a long history (see Polyak and Tsypkin (1973), Ljung (1977), Kushner and Clark (1978), Tsitsiklis et al. (1986), Polyak (1987), Bertsekas and Tsit-siklis (1989, 1996, 2000), Gaivoronski (1994), Pflug (1996), Kushner and Yin (1997), Bottou (2005), Meyn (2007), Borkar (2008), Nemirovski et al. (2009), Lee and Wright (2010)), and are strongly connected with stochastic approximation algorithms. The main difference between stochastic and de-terministic formulations is that the former involve sequentially sampling cost components from an infinite population under some statistical assumptions, while in the latter the set of cost components is predetermined and finite. However, it is possible to view the incremental gradient method (4.5), with a randomized selection of the component function f_i (i.e., with i_k chosen to be any one of the indexes $1, \ldots, m$, with equal probability $1/m$), as a stochas-tic gradient method (see Bertsekas and Tsitsiklis (1996, Example 4.4) and (Bertsekas and Tsitsiklis, 2000, Section 5)).

The stochastic formulation of incremental methods just discussed high-lights an important application context where the component functions f_i are not given a priori, but become known sequentially through some obser-vation process. Then it often makes sense to use an incremental method to process the component functions as they become available, and to obtain approximate solutions as early as possible. In fact, this may be essential in time-sensitive and possibly time-varying environments, where solutions are needed "online." In such cases, one may hope that an adequate estimate of the optimal solution will be obtained before all the functions f_i are processed for the first time.

4.1.3   Incremental Subgradient Methods - Nondifferentiable Problems

Incremental subgradient methods apply to the case where the component functions f_i are convex and nondifferentiable at some points. They are similar to their gradient counterparts (4.5) except that an arbitrary subgradient $\tilde{\nabla} f_{i_k}(x_k)$ of the cost component f_{i_k} is used in place of the gradient:[2]

$$x_{k+1} = P_X\big(x_k - \alpha_k \tilde{\nabla} f_{i_k}(x_k)\big). \tag{4.8}$$

Such methods were first proposed in the general form (4.8) in the Soviet Union by Kibardin (1980), following the earlier paper by Litvakov (1966) (which considered convex/nondifferentiable extensions of linear least-squares problems) and related subsequent proposals.[3] These works remained unnoticed until about 2005 in the Western literature, where incremental methods were often reinvented in different contexts and with different lines of analysis. See Ben-Tal et al. (2001), Nedić and Bertsekas (2000, 2001, 2010), Nedić et al. (2001), Kiwiel (2004), Rabbat and Nowak (2004, 2005), Gaudioso et al. (2006), Shalev-Shwartz et al. (2007), Neto and De Pierro (2009), Johansson et al. (2009), Predd et al. (2009), Ram et al. (2009a,b), and Duchi et al. (2010).

Incremental subgradient methods have convergence characteristics that are similar in many ways to their gradient counterparts, the most important similarity being the necessity for a diminishing stepsize α_k for convergence. The lines of analysis, however, tend to be different, since incremental gradient methods rely for convergence on arguments based on decrease of the cost function value, while incremental subgradient methods rely on argu-

---

2. In this chapter, we use $\tilde{\nabla} f(x)$ to denote a subgradient of a convex real-valued function f at a vector x. The choice of $\tilde{\nabla} f(x)$ from within the subdifferential $\partial f(x)$ at x will be clear from the context.

3. Generally, in those times, algorithmic ideas relating to simple gradient methods with and without deterministic and stochastic errors were popular in the Soviet scientific community, partly due to an emphasis on stochastic iterative algorithms, such as pseudogradient and stochastic approximation; the works of Ermoliev, Polyak, and Tsypkin, to name a few of the principal contributors, are representative (Ermoliev, 1969; Polyak and Tsypkin, 1973; Ermoliev, 1976; Polyak, 1978, 1987). By contrast, the emphasis in the Western literature at the time was on more complex Newton-like and conjugate direction methods.

ments based on decrease of the iterates' distance from the optimal solution set. The line of analysis of this chapter is of the latter type, and is similar to earlier works of the author and his collaborators (see Nedić and Bertsekas (2000), Nedić and Bertsekas (2001), Nedić et al. (2001), and the textbook presentations in Bertsekas (1999) and Bertsekas et al. (2003)).

Note two important ramifications of the lack of differentiability of the component functions f_i:

1. Convexity of f_i becomes essential, since the notion of subgradient is connected with convexity (subgradient-like algorithms for nondifferentiable / nonconvex problems have been suggested in the literature, but tend to be complicated and have not found much application thus far).

2. There is more reason to favor the incremental over the nonincremental methods, since (contrary to the differentiable case) nonincremental subgradient methods also require a diminishing stepsize for convergence, and typically achieve a sublinear rate of convergence. Thus the one theoretical advantage of the nonincremental gradient method discussed earlier is not shared by its subgradient counterpart.

Finally, just as in the differentiable case, there is a substantial literature for stochastic versions of subgradient methods. In fact, as we will discuss in this chapter, there is a potentially significant advantage in turning the method into a stochastic one by randomizing the order of selection of the components f_i for iteration.

4.1.4   Incremental Proximal Methods

We now consider an extension of the incremental approach to proximal algorithms. The simplest one for problem (4.1) is of the form

$$x_{k+1} = \arg\min_{x \in X} \left\{ f_{i_k}(x) + \frac{1}{2\alpha_k} \|x - x_k\|^2 \right\}, \tag{4.9}$$

which relates to the proximal minimization algorithm ((Martinet, 1970), (Rockafellar, 1976)) in the same way that the incremental subgradient method (4.8) relates to the classical nonincremental subgradient method.[4] Here $\{\alpha_k\}$ is a positive scalar sequence, and we will assume that each $f_i : \Re^n \mapsto \Re$ is a convex function and X is a nonempty closed convex set.

---

4. In this chapter, we restrict our attention to proximal methods with the quadratic regularization term $\|x - x_k\|^2$. Our approach is applicable in principle when a nonquadratic term is used instead, in order to match the structure of the given problem. The discussion of such alternative algorithms is beyond our scope.

The motivation for this type of method, which was considered only recently in Bertsekas (2010), is that with a favorable structure of the components, the proximal iteration (4.8) may be obtained in closed form or be relatively simple, in which case it may be preferable to a gradient or subgradient iteration. In this connection, we note that, generally, proximal iterations are considered more stable than gradient iterations; for example, in the nonincremental case, they converge essentially for any choice of α_k, while this is not so for gradient methods.

While some cost function components may be well suited for a proximal iteration, others may not be because the minimization (4.9) is inconvenient, so it makes sense to consider combinations of gradient/subgradient and prox-imal iterations. In fact, in the past this has motivated nonincremental com-binations of gradient and proximal methods for minimizing the sum of two functions (or more generally, finding a zero of the sum of two nonlinear oper-ators). These methods have a long history, dating to the splitting algorithms of Lions and Mercier (1979) and Passty (1979), and have become popular more recently (see Beck and Teboulle (2009, 2010), and the references they cite for specialized algorithms, such as shrinkage/thresholding, cf. Section 5.1).

With similar motivation in mind, we adopt in this paper a unified algorith-mic framework that includes incremental gradient, subgradient, and proxi-mal methods and their combinations, and highlights their common structure and behavior. We focus on problems of the form

$$\text{minimize} \quad F(x) \overset{\text{def}}{=} \sum_{i=1}^{m} F_i(x)$$

$$\text{subject to} \ \ x \in X,$$

(4.10)

where for all i,

$$F_i(x) = f_i(x) + h_i(x),$$

(4.11)

$f_i : \Re^n \mapsto \Re$ and $h_i : \Re^n \mapsto \Re$ are real-valued convex functions, and X is a nonempty closed convex set.

In Section 4.2, we consider several incremental algorithms that iterate on the components f_i with a proximal iteration, and on the components h_i with a subgradient iteration. By choosing all the f_i or all the h_i to be identically zero, we obtain the subgradient and proximal iterations (4.8) and (4.9), respectively, as special cases. However, our methods offer greater flexibility, and may exploit the special structure of problems where the functions f_i are suitable for a proximal iteration, while the components h_i are not suitable, and thus may be preferably treated with a subgradient iteration.

In Section 4.3, we discuss the convergence and rate of convergence properties of methods that use a cyclic rule for component selection, and in Section 4.4, we discuss a randomized component selection rule. In summary, the convergence behavior of our incremental methods is similar to the one outlined earlier for the incremental subgradient method (4.8). This includes convergence within a certain error bound for a constant stepsize, exact convergence to an optimal solution for an appropriately diminishing stepsize, and improved convergence rate/iteration complexity when randomization is used to select the cost component for iteration. In Section 4.5, we illustrate our methods for some example applications.

4.2   Incremental Subgradient-Proximal Methods

In this section, we consider problems (4.10) and (4.11), and introduce several incremental algorithms that involve a combination of a proximal and a subgradient iteration. One of our algorithms has the form

$$z_k = \arg\min_{x \in X} \left\{ f_{i_k}(x) + \frac{1}{2\alpha_k} \|x - x_k\|^2 \right\}, \tag{4.12}$$

$$x_{k+1} = P_X\big(z_k - \alpha_k \tilde{\nabla} h_{i_k}(z_k)\big), \tag{4.13}$$

where $\tilde{\nabla} h_{i_k}(z_k)$ is an arbitrary subgradient of h_{i_k} at z_k. The iteration is well defined because the minimum in (4.12) is uniquely attained since f_i is continuous and $\|x - x_k\|^2$ is real-valued, strictly convex, and coercive, while the subdifferential $\partial h_i(z_k)$ is nonempty since h_i is real-valued. Also, by choosing all the f_i or all the h_i to be identically zero, we obtain the subgradient and proximal iterations (4.8) and (4.9), respectively, as special cases.

The iterations (4.12) and (4.13) maintain both sequences $\{z_k\}$ and $\{x_k\}$ within the constraint set X, but it may be convenient to relax this constraint for either the proximal or the subgradient iteration, thereby requiring a potentially simpler computation. This leads to the algorithm

$$z_k = \arg\min_{x \in \Re^n} \left\{ f_{i_k}(x) + \frac{1}{2\alpha_k} \|x - x_k\|^2 \right\}, \tag{4.14}$$

$$x_{k+1} = P_X\big(z_k - \alpha_k \tilde{\nabla} h_{i_k}(z_k)\big), \tag{4.15}$$

where the restriction $x \in X$ has been omitted from the proximal iteration, and to the algorithm

$$z_k = x_k - \alpha_k \tilde{\nabla} h_{i_k}(x_k), \tag{4.16}$$

$$x_{k+1} = \arg\min_{x \in X} \left\{ f_{i_k}(x) + \frac{1}{2\alpha_k}\|x - z_k\|^2 \right\}, \tag{4.17}$$

where the projection onto X has been omitted from the subgradient iteration. It is also possible to use different stepsize sequences in the proximal and subgradient iterations, but for notational simplicity we will not discuss this type of algorithm.

All of the incremental proximal algorithms given above are new to our knowledge, having first been proposed by Bertsekas (2010). The closest connection to the existing proximal methods is the "proximal gradient" method, which has been analyzed and discussed recently in the context of several machine-learning applications by Beck and Teboulle (2009, 2010). (It can also be interpreted in terms of splitting algorithms (Lions and Mercier, 1979), (Passty, 1979).) This method is nonincremental, applies to differentiable h_i and, contrary to subgradient and incremental methods, it does not require a diminishing stepsize for convergence to the optimum. In fact, the line of convergence analysis of Beck and Teboulle (2009, 2010) relies on the differentiability of h_i and the nonincremental character of the proximal gradient method, and thus is different from ours.

Part (a) of the following proposition is a key fact about incremental proximal iterations. It shows that they are closely related to incremental subgradient iterations, the only difference being that the subgradient is evaluated at the end point of the iteration rather than at the starting point. Part (b) of the proposition provides an inequality that is well known in the theory of proximal methods, and will be useful for our convergence analysis. In the following method, we denote by ri(S) the relative interior of a convex set S, and by dom(f) the effective domain $\{x \mid f(x) < \infty\}$ of a function $f : \Re^n \mapsto (-\infty, \infty]$.

Proposition 4.1. *Let X be a nonempty closed convex set, and let $f : \Re^n \mapsto (-\infty, \infty]$ be a closed proper convex function such that* ri$(X) \cap$ri$(\mathrm{dom}(f)) \neq \emptyset$. *For any $x_k \in \Re^n$ and $\alpha_k > 0$, consider the proximal iteration*

$$x_{k+1} = \arg\min_{x \in X} \left\{ f(x) + \frac{1}{2\alpha_k}\|x - x_k\|^2 \right\}. \tag{4.18}$$

(a) The iteration can be written as

$$x_{k+1} = P_X\big(x_k - \alpha_k \tilde{\nabla} f(x_{k+1})\big), \qquad i = 1, \ldots, m, \tag{4.19}$$

where $\tilde{\nabla} f(x_{k+1})$ is some subgradient of f at x_{k+1}.

(b) For all $y \in X$, we have

$$\|x_{k+1} - y\|^2 \leq \|x_k - y\|^2 - 2\alpha_k(f(x_{k+1}) - f(y)) - \|x_k - x_{k+1}\|^2$$
$$\leq \|x_k - y\|^2 - 2\alpha_k(f(x_{k+1}) - f(y)). \tag{4.20}$$

Proof. (a) We use the formula for the subdifferential of the sum of the three functions f, $(1/2\alpha_k)\|x - x_k\|^2$, and the indicator function of X (cf. (Bertsekas, 2009, Proposition 5.4.6)), together with the condition that 0 should belong to this subdifferential at the optimum x_{k+1}. We obtain that (4.18) holds if and only if

$$\frac{1}{\alpha_k}(x_k - x_{k+1}) \in \partial f(x_{k+1}) + N_X(x_{k+1}), \tag{4.21}$$

where $N_X(x_{k+1})$ is the normal cone of X at x_{k+1} (which is the set of vectors y such that $y'(x - x_{k+1}) \leq 0$ for all $x \in X$, and also the subdifferential of the indicator function of X at x_{k+1}; see (Bertsekas, 2009, p. 185)). This is true if and only if

$$x_k - x_{k+1} - \alpha_k \tilde{\nabla} f(x_{k+1}) \in N_X(x_{k+1})$$

for some $\tilde{\nabla} f(x_{k+1}) \in \partial f(x_{k+1})$, which in turn is true if and only if (4.19) holds (cf. Bertsekas (2009, Proposition 5.4.6)).

(b) We have

$$\|x_k - y\|^2 = \|x_k - x_{k+1} + x_{k+1} - y\|^2$$
$$= \|x_k - x_{k+1}\|^2 - 2(x_k - x_{k+1})'(y - x_{k+1}) + \|x_{k+1} - y\|^2. \tag{4.22}$$

Also since from (4.21), $\frac{1}{\alpha_k}(x_k - x_{k+1})$ is a subgradient at x_{k+1} of the sum of f and the indicator function of X, we have (also using the assumption $y \in X$) that

$$f(x_{k+1}) + \frac{1}{\alpha_k}(x_k - x_{k+1})'(y - x_{k+1}) \leq f(y).$$

Combining this relation with (4.22), the result follows. $\square$

Based on the preceding proposition, we see that all the preceding iterations can be written in an incremental subgradient format:

(a) Iteration (4.12)-(4.13) can be written as

$$z_k = P_X\big(x_k - \alpha_k \tilde{\nabla} f_{i_k}(z_k)\big), \qquad x_{k+1} = P_X\big(z_k - \alpha_k \tilde{\nabla} h_{i_k}(z_k)\big). \tag{4.23}$$

(b) Iteration (4.14)-(4.15) can be written as

$$z_k = x_k - \alpha_k \tilde{\nabla} f_{i_k}(z_k), \qquad x_{k+1} = P_X\big(z_k - \alpha_k \tilde{\nabla} h_{i_k}(z_k)\big). \tag{4.24}$$

(c) Iteration (4.16)-(4.17) can be written as

$$z_k = x_k - \alpha_k \tilde{\nabla} h_{i_k}(x_k), \qquad x_{k+1} = P_X\big(z_k - \alpha_k \tilde{\nabla} f_{i_k}(x_{k+1})\big). \qquad (4.25)$$

In all the preceding updates, the subgradient $\tilde{\nabla} h_{i_k}$ can be *any* vector in the subdifferential of h_{i_k}, while the subgradient $\tilde{\nabla} f_{i_k}$ must be a *specific* vector in the subdifferential of f_{i_k}, specified according to Proposition 4.1(a). Also, iteration (4.24) can be written as

$$x_{k+1} = P_X\big(x_k - \alpha_k \tilde{\nabla} F_{i_k}(z_k)\big),$$

and resembles the incremental subgradient method for minimizing over X the cost $F(x) = \sum_{i=1}^{m} F_i(x)$ (cf. (4.10)), the only difference being that the subgradient of F_{i_k} is taken at z_k rather than x_k.

An important issue which affects the methods' effectiveness is the order in which the components $\{f_i, h_i\}$ are chosen for iteration. We consider two possibilities:

1. A *cyclic order*, whereby $\{f_i, h_i\}$ are taken up in the fixed deterministic order $1, \ldots, m$, so that i_k is equal to (k modulo m) plus 1. A contiguous block of iterations involving $\{f_1, h_1\}, \ldots, \{f_m, h_m\}$ in this order and exactly once is called a *cycle*. We assume that the stepsize α_k is constant within a cycle (for all k with $i_k = 1$ we have $\alpha_k = \alpha_{k+1} \ldots = \alpha_{k+m-1}$).

2. A *randomized order*, whereby at each iteration a component pair $\{f_i, h_i\}$ is chosen randomly by sampling over all component pairs with a uniform distribution, independently of the past history of the algorithm.

It is essential to include all components in a cycle in the cyclic case, and to sample according to the uniform distribution in the randomized case, for otherwise some components will be sampled more often than others, leading to a bias in the convergence process.

For the remainder of the chapter, we denote the optimal value of problem (4.10) by F^* :

$$F^* = \inf_{x \in X} F(x),$$

and the set of optimal solutions (which could be empty) by X^*:

$$X^* = \big\{ x^* \mid x^* \in X,\, F(x^*) = F^* \big\}.$$

Also, for a nonempty closed set X, we denote by $\mathrm{dist}(\cdot; X)$ the distance function, defined as follows:

$$\mathrm{dist}(x; X) = \min_{z \in X} \|x - z\|, \qquad x \in \Re^n.$$

4.3   Convergence for Methods with Cyclic Order

In this section, we discuss convergence under the cyclic order. We consider a randomized order in the next section. We focus on the sequence $\{x_k\}$ rather than $\{z_k\}$, which need not lie within X in the case of iterations (4.24) and (4.25) when $X \neq \Re^n$. In summary, the idea is to show that the effect of taking subgradients of f_i or h_i at points near x_k (e.g., at z_k rather than at x_k) is inconsequential, and diminishes as the stepsize α_k becomes smaller, as long as some subgradients relevant to the algorithms are uniformly bounded in norm by some constant. This is similar to the convergence mechanism of incremental gradient methods described in Section 4.2. We use the following assumptions throughout the present section.

Assumption 4.1 (For iterations (4.23) and (4.24)). *There is a constant $c \in \Re$ such that for all k*

$$\max \left\{ \|\tilde{\nabla} f_{i_k}(z_k)\|, \|\tilde{\nabla} h_{i_k}(z_k)\| \right\} \leq c. \tag{4.26}$$

Furthermore, for all k that mark the beginning of a cycle (i.e., all $k > 0$ with $i_k = 1$), we have for all $j = 1, \dots, m$:

$$\max \left\{ f_j(x_k) - f_j(z_{k+j-1}), \, h_j(x_k) - h_j(z_{k+j-1}) \right\} \leq c \, \|x_k - z_{k+j-1}\|. \tag{4.27}$$

Assumption 4.2 (For iteration (4.25)). *There is a constant $c \in \Re$ such that for all k*

$$\max \left\{ \|\tilde{\nabla} f_{i_k}(x_{k+1})\|, \|\tilde{\nabla} h_{i_k}(x_k)\| \right\} \leq c. \tag{4.28}$$

Furthermore, for all k that mark the beginning of a cycle (i.e., all $k > 0$ with $i_k = 1$), we have for all $j = 1, \dots, m$:

$$\max \left\{ f_j(x_k) - f_j(x_{k+j-1}), \, h_j(x_k) - h_j(x_{k+j-1}) \right\} \leq c \, \|x_k - x_{k+j-1}\|, \tag{4.29}$$

$$f_j(x_{k+j-1}) - f_j(x_{k+j}) \leq c \, \|x_{k+j-1} - x_{k+j}\|. \tag{4.30}$$

The condition (4.27) is satisfied if for each i and k, there is a subgradient of f_i at x_k and a subgradient of h_i at x_k, whose norms are bounded by c. Conditions that imply the preceding assumptions are:

(a) For algorithm (4.23): f_i and h_i are Lipschitz continuous over the set X.

(b) For algorithms (4.24) and (4.25): f_i and h_i are Lipschitz continuous over the entire space $\Re^n$.

(c) For algorithms (4.23), (4.24), and (4.25): f_i and h_i are polyhedra (this is a special case of (a) and (b)).

(d) The sequences $\{x_k\}$ and $\{z_k\}$ are bounded, since then, f_i and h_i, being real-valued and convex, are Lipschitz continuous over any bounded set that contains $\{x_k\}$ and $\{z_k\}$ (see, e.g., Bertsekas (2009, Proposition 5.4.2))].

The following proposition provides a key estimate that reveals the convergence mechanism of our methods.

Proposition 4.2. *Let $\{x_k\}$ be the sequence generated by any one of the algorithms (4.23)-(4.25), with a cyclic order of component selection. Then for all $y \in X$ and all k that mark the beginning of a cycle (i.e., all k with $i_k = 1$), we have*

$$\|x_{k+m} - y\|^2 \le \|x_k - y\|^2 - 2\alpha_k\big(F(x_k) - F(y)\big) + \alpha_k^2 \beta m^2 c^2, \qquad (4.31)$$

where $\beta = \frac{1}{m} + 4$ in the case of (4.23) and (4.24), and $\beta = \frac{5}{m} + 4$ in the case of (4.25).

Proof. We first prove the result for algorithms (4.23) and (4.24), and then indicate the modifications necessary for algorithm (4.25). Using Proposition 4.1(b), we have for all $y \in X$ and k,

$$\|z_k - y\|^2 \le \|x_k - y\|^2 - 2\alpha_k\big(f_{i_k}(z_k) - f_{i_k}(y)\big). \qquad (4.32)$$

Also, using the nonexpansion property of the projection (i.e., $\big\|P_X(u) - P_X(v)\big\| \le \|u - v\|$ for all $u, v \in \Re^n$), the definition of subgradient, and (4.26), we obtain for all $y \in X$ and k:

$$
\begin{aligned}
\|x_{k+1} - y\|^2 &= \big\|P_X\big(z_k - \alpha_k \tilde{\nabla} h_{i_k}(z_k)\big) - y\big\|^2 \\
&\le \big\|z_k - \alpha_k \tilde{\nabla} h_{i_k}(z_k) - y\big\|^2 \\
&\le \|z_k - y\|^2 - 2\alpha_k \tilde{\nabla} h_{i_k}(z_k)'(z_k - y) + \alpha_k^2 \big\|\tilde{\nabla} h_{i_k}(z_k)\big\|^2 \\
&\le \|z_k - y\|^2 - 2\alpha_k\big(h_{i_k}(z_k) - h_{i_k}(y)\big) + \alpha_k^2 c^2.
\end{aligned}
\qquad (4.33)
$$

Combining (4.32) and (4.33), and using the definition $F_j = f_j + h_j$, we have

$$
\begin{aligned}
\|x_{k+1} - y\|^2 &\le \|x_k - y\|^2 - 2\alpha_k\big(f_{i_k}(z_k) + h_{i_k}(z_k) - f_{i_k}(y) - h_{i_k}(y)\big) + \alpha_k^2 c^2 \\
&= \|x_k - y\|^2 - 2\alpha_k\big(F_{i_k}(z_k) - F_{i_k}(y)\big) + \alpha_k^2 c^2.
\end{aligned}
$$
$$(4.34)$$

Now let k mark the beginning of a cycle (i.e., $i_k = 1$). Then, at iteration $k + j - 1$, $j = 1, \dots, m$, the selected components are $\{f_j, h_j\}$, in view of the assumed cyclic order. We may thus replicate the preceding inequality with

k replaced by $k+1, \dots, k+m-1$, and add to obtain

$$\|x_{k+m} - y\|^2 \le \|x_k - y\|^2 - 2\alpha_k \sum_{j=1}^{m} \big(F_j(z_{k+j-1}) - F_j(y)\big) + m\alpha_k^2 c^2$$

or, equivalently,

$$\|x_{k+m} - y\|^2 \le \|x_k - y\|^2 - 2\alpha_k \big(F(x_k) - F(y)\big) + m\alpha_k^2 c^2$$
$$+ 2\alpha_k \sum_{j=1}^{m} \big(F_j(x_k) - F_j(z_{k+j-1})\big). \quad (4.35)$$

The remainder of the proof deals with appropriately bounding the last term above.

From (4.27), we have for $j = 1, \dots, m$ that

$$F_j(x_k) - F_j(z_{k+j-1}) \le 2c \, \|x_k - z_{k+j-1}\|. \qquad (4.36)$$

We also have

$$\|x_k - z_{k+j-1}\| \le \|x_k - x_{k+1}\| + \dots + \|x_{k+j-2} - x_{k+j-1}\| + \|x_{k+j-1} - z_{k+j-1}\|, \qquad (4.37)$$

and by the definition of algorithms (4.23) and (4.24), the nonexpansion property of the projection, and (4.26), each of the terms in the right-hand side above is bounded by $2\alpha_k c$, except for the last, which is bounded by $\alpha_k c$. Thus (4.37) yields $\|x_k - z_{k+j-1}\| \le \alpha_k(2j-1)c$ which, together with (4.36), shows that

$$F_j(x_k) - F_j(z_{k+j-1}) \le 2\alpha_k c^2 (2j-1). \qquad (4.38)$$

Combining (4.35) and (4.38), we have

$$\|x_{k+m} - y\|^2 \le \|x_k - y\|^2 - 2\alpha_k \big(F(x_k) - F(y)\big) + m\alpha_k^2 c^2 + 4\alpha_k^2 c^2 \sum_{j=1}^{m} (2j-1),$$

and finally

$$\|x_{k+m} - y\|^2 \le \|x_k - y\|^2 - 2\alpha_k \big(F(x_k) - F(y)\big) + m\alpha_k^2 c^2 + 4\alpha_k^2 c^2 m^2,$$

which is of the form (4.31) with $\beta = \frac{1}{m} + 4$.

For algorithm (4.25), a similar argument goes through using Assumption 4.2. In place of (4.32), using the nonexpansion property of the projection, the definition of subgradient, and (4.28), we obtain, for all $y \in X$ and $k \ge 0$,

$$\|z_k - y\|^2 \le \|x_k - y\|^2 - 2\alpha_k \big(h_{i_k}(x_k) - h_{i_k}(y)\big) + \alpha_k^2 c^2. \qquad (4.39)$$

In place of (4.33), using Proposition 4.1(b), we have

$$\|x_{k+1} - y\|^2 \leq \|z_k - y\|^2 - 2\alpha_k\big(f_{i_k}(x_{k+1}) - f_{i_k}(y)\big). \tag{4.40}$$

Combining these equations, in analogy to (4.34), we obtain

$$
\begin{aligned}
\|x_{k+1} - y\|^2 &\leq \|x_k - y\|^2 - 2\alpha_k\big(f_{i_k}(x_{k+1}) + h_{i_k}(x_k) - f_{i_k}(y) - h_{i_k}(y)\big) + \alpha_k^2 c^2 \\
&= \|x_k - y\|^2 - 2\alpha_k\big(F_{i_k}(x_k) - F_{i_k}(y)\big) \\
&\quad + \alpha_k^2 c^2 + 2\alpha_k\big(f_{i_k}(x_k) - f_{i_k}(x_{k+1})\big).
\end{aligned}
\tag{4.41}
$$

and, similar to (4.35),

$$
\begin{aligned}
\|x_{k+m} - y\|^2 &\leq \|x_k - y\|^2 - 2\alpha_k\big(F(x_k) - F(y)\big) + m\alpha_k^2 c^2 \\
&\quad + 2\alpha_k \sum_{j=1}^{m} \big(F_j(x_k) - F_j(x_{k+j-1})\big) \\
&\quad + 2\alpha_k \sum_{j=1}^{m} \big(f_j(x_{k+j-1}) - f_j(x_{k+j})\big).
\end{aligned}
\tag{4.42}
$$

We now bound the last two terms in the preceding relation, using Assumption 4.2. From (4.29), we have

$$
\begin{aligned}
F_j(x_k) - F_j(x_{k+j-1}) &\leq 2c\|x_k - x_{k+j-1}\| \\
&\leq 2c\big(\|x_k - x_{k+1}\| + \cdots + \|x_{k+j-2} - x_{k+j-1}\|\big),
\end{aligned}
$$

and since by (4.28) and the definition of the algorithm, each norm term in the right-hand side above is bounded by $2\alpha_k c$:

$$F_j(x_k) - F_j(x_{k+j-1}) \leq 4\alpha_k c^2(j-1).$$

Also, from (4.28) and (4.30) and the nonexpansion property of the projection, we have

$$f_j(x_{k+j-1}) - f_j(x_{k+j}) \leq c\|x_{k+j-1} - x_{k+j}\| \leq 2\alpha_k c^2.$$

Combining the preceding relations and adding, we obtain

$$
\begin{aligned}
2\alpha_k \sum_{j=1}^{m} \big(F_j(x_k) - F_j(x_{k+j-1})\big) &+ 2\alpha_k \sum_{j=1}^{m} \big(f_j(x_{k+j-1}) - f_j(x_{k+j})\big) \\
&\leq 8\alpha_k^2 c^2 \sum_{j=1}^{m}(j-1) + 4\alpha_k^2 c^2 m \\
&= 4\alpha_k^2 c^2 m^2 + 4\alpha_k^2 c^2 m = \left(4 + \frac{4}{m}\right)\alpha_k^2 c^2 m^2,
\end{aligned}
$$

which, together with (4.42), yields (4.31) with $\beta = 4 + \frac{5}{m}$. $\square$

Among other things, Proposition 4.2 guarantees that with a cyclic order, given the iterate x_k at the start of a cycle and any point $y \in X$ having lower cost than x_k (for example an optimal point), the algorithm yields a point x_{k+m} at the end of the cycle that will be closer to y than x_k, provided the stepsize α_k is less than $\frac{2\left(F(x_k)-F(y)\right)}{\beta m^2 c^2}$. In particular, for any $\epsilon > 0$ and assuming that there exists an optimal solution x^*, either we are within $\frac{\alpha_k \beta m^2 c^2}{2} + \epsilon$ of the optimum,

$$F(x_k) \leq F(x^*) + \frac{\alpha_k \beta m^2 c^2}{2} + \epsilon,$$

or the squared distance to the optimum will be strictly decreased by at least $2\alpha_k \epsilon$:

$$\|x_{k+m} - x^*\|^2 < \|x_k - x^*\|^2 - 2\alpha_k \epsilon.$$

Thus, using Proposition 4.2, we can provide various types of convergence results. As an example, for a constant stepsize ($\alpha_k \equiv \alpha$), convergence can be established to a neighborhood of the that which shrinks to 0 as $\alpha \to 0$, as stated in the following proposition. Its proof and all the proofs of propositions that follow are given in (Bertsekas, 2010).

Proposition 4.3. *Let $\{x_k\}$ be the sequence generated by any one of the algorithms (4.23)-(4.25), with a cyclic order of component selection, and let the stepsize α_k be fixed at some positive constant α.*

(a) If $F^* = -\infty$, then

$$\liminf_{k \to \infty} F(x_k) = F^*.$$

(b) If $F^* > -\infty$, then

$$\liminf_{k \to \infty} F(x_k) \leq F^* + \frac{\alpha \beta m^2 c^2}{2},$$

where c and β are the constants of Proposition 4.2.

The next proposition gives an estimate of the number of iterations needed to guarantee a given level of optimality up to the threshold tolerance $\alpha \beta m^2 c^2 / 2$ of the preceding proposition.

Proposition 4.4. *Assume that X^* is nonempty. Let $\{x_k\}$ be a sequence generated as in Proposition 4.3. Then, for $\epsilon > 0$ we have*

$$\min_{0 \leq k \leq N} F(x_k) \leq F^* + \frac{\alpha \beta m^2 c^2 + \epsilon}{2}, \tag{4.43}$$

where N is given by

$$N = m \left\lfloor \frac{dist(x_0; X^*)^2}{\alpha \epsilon} \right\rfloor. \tag{4.44}$$

According to Proposition 4.4, to achieve a cost function value within $O(\epsilon)$ of the optimal, the term $\alpha \beta m^2 c^2$ must also be of order $O(\epsilon)$, so α must be of order $O(\epsilon/m^2 c^2)$, and from (4.44), the number of necessary iterations N is $O(m^3 c^2/\epsilon^2)$ and the number of necessary cycles is $O\big((mc)^2/\epsilon^2\big)$. This is the same type of estimate as for the nonincremental subgradient method (i.e., $O(1/\epsilon^2)$, counting a cycle as one iteration of the nonincremental method, and viewing mc as a Lipschitz constant for the entire cost function F), and does not reveal any advantage for the incremental methods given here. However, in the next section, we demonstrate a much more favorable iteration complexity estimate for the incremental methods that use a randomized order of component selection.

Exact Convergence for a Diminishing Stepsize

We can also obtain an exact convergence result for the case where the stepsize α_k diminishes to zero. The idea is that with a constant stepsize α we can get to within an $O(\alpha)$-neighborhood of the optimum, as shown above, so with a diminishing stepsize α_k, we should be able to reach an arbitrarily small neighborhood of the optimum. However, for this to happen, α_k should not be reduced too fast, and should satisfy $\sum_{k=0}^{\infty} \alpha_k = \infty$ (so that the method can "travel" infinitely far if necessary).

Proposition 4.5. *Let $\{x_k\}$ be the sequence generated by any one of the algorithms (4.23)-(4.25), with a cyclic order of component selection, and let the stepsize α_k satisfy*

$$\lim_{k \to \infty} \alpha_k = 0, \qquad \sum_{k=0}^{\infty} \alpha_k = \infty.$$

Then,

$$\liminf_{k \to \infty} F(x_k) = F^*.$$

Furthermore, if X^* is nonempty and

$$\sum_{k=0}^{\infty} \alpha_k^2 < \infty,$$

then $\{x_k\}$ converges to some $x^* \in X^*$.

4.4   Convergence for Methods with Randomized Order

In this section, we discuss convergence for the randomized component selection order and a constant stepsize α. The randomized versions of iterations (4.23), (4.24), and (4.25), are

$$z_k = P_X\big(x_k - \alpha\tilde{\nabla} f_{\omega_k}(z_k)\big), \qquad x_{k+1} = P_X\big(z_k - \alpha\tilde{\nabla} h_{\omega_k}(z_k)\big), \qquad (4.45)$$

$$z_k = x_k - \alpha\tilde{\nabla} f_{\omega_k}(z_k), \qquad x_{k+1} = P_X\big(z_k - \alpha\tilde{\nabla} h_{\omega_k}(z_k)\big), \qquad (4.46)$$

$$z_k = P_X\big(x_k - \alpha\tilde{\nabla} f_{\omega_k}(z_k)\big), \qquad x_{k+1} = z_k - \alpha\tilde{\nabla} h_{\omega_k}(z_k), \qquad (4.47)$$

respectively, where $\{\omega_k\}$ is a sequence of random variables taking values from the index set $\{1,\ldots,m\}$.

We assume the following throughout the present section.

Assumption 4.3 (For iterations (4.45) and (4.46)).

(a) $\{\omega_k\}$ is a sequence of random variables, each uniformly distributed over $\{1,\ldots,m\}$, and such that for each k, ω_k is independent of the history $\{x_k, z_{k-1}, \ldots, z_0, x_0\}$.

(b) There is a constant $c \in \Re$ such that for all k, we have with probability 1

$$\max\{\|\tilde{\nabla} f_i(z_k^i)\|, \|\tilde{\nabla} h_i(z_k^i)\|\} \le c, \qquad \forall\, i = 1,\ldots,m, \qquad (4.48)$$

$$\max\{f_i(x_k) - f_i(z_k^i),\, h_i(x_k) - h_i(z_k^i)\} \le c\|x_k - z_k^i\|, \qquad \forall\, i = 1,\ldots,m, \qquad (4.49)$$

where z_k^i is the result of the proximal iteration starting at x_k, if ω_k were i, that is,

$$z_k^i = \arg\min_{x\in X}\left\{f_i(x) + \frac{1}{2\alpha_k}\|x - x_k\|^2\right\} \qquad (4.50)$$

in the case of iteration (4.45), and

$$z_k^i = \arg\min_{x\in\Re^n}\left\{f_i(x) + \frac{1}{2\alpha_k}\|x - x_k\|^2\right\} \qquad (4.51)$$

in the case of iteration (4.46).

Assumption 4.4 (For iteration (4.47)).

(a) $\{\omega_k\}$ is a sequence of random variables, each uniformly distributed over $\{1,\ldots,m\}$, and such that for each k, ω_k is independent of the history $\{x_k, z_{k-1}, \ldots, z_0, x_0\}$.

(b) *There is a constant* $c \in \Re$ *such that for all* k, *we have with probability 1*

$$\max \left\{ \|\tilde{\nabla} f_i(x_{k+1}^i)\|, \|\tilde{\nabla} h_i(x_k)\| \right\} \leq c, \qquad \forall \, i = 1, \ldots, m, \qquad (4.52)$$

$$f_i(x_k) - f_i(x_{k+1}^i) \leq c \|x_k - x_{k+1}^i\|, \qquad \forall \, i = 1, \ldots, m, \qquad (4.53)$$

where x_{k+1}^i *is the result of the iteration, starting at* x_k *if* ω_k *would be* i, *that is,*

$$x_{k+1}^i = P_X \left(z_k^i - \alpha_k \tilde{\nabla} f_i(x_{k+1}^i) \right), \qquad (4.54)$$

with

$$z_k^i = x_k - \alpha_k \tilde{\nabla} h_i(x_k). \qquad (4.55)$$

Note that condition (4.49) is satisfied if there exist subgradients of f_i and h_i at x_k with norms less than or equal to c. Thus the conditions (4.48) and (4.49) are similar, the main difference being that the first applies to slopes of f_i and h_i at z_k^i while the second applies to the slopes of f_i and h_i at x_k. As in the case of Assumption 4.1, these conditions are guaranteed by Lipschitz continuity assumptions on f_i and h_i. The convergence analysis of the randomized algorithms of this section is somewhat more complicated than the one of the cyclic order counterparts, and relies on the Supermartingale convergence theorem (see Bertsekas (2010)). The following proposition deals with the case of a constant stepsize, and parallels Proposition 4.3 for the cyclic-order case.

Proposition 4.6. *Let* $\{x_k\}$ *be the sequence generated by one of the randomized incremental methods (4.45)-(4.47), and let the stepsize* α_k *be fixed at some positive constant* α.

(a) *If* $F^* = -\infty$, *then with probability 1*

$$\inf_{k \geq 0} F(x_k) = F^*.$$

(b) *If* $F^* > -\infty$, *then with probability 1*

$$\inf_{k \geq 0} F(x_k) \leq F^* + \frac{\alpha \beta m c^2}{2},$$

where $\beta = 5$.

By comparing Proposition 4.6(b) with Proposition 4.3(b), we see that when $F^* > -\infty$ and the stepsize α is constant, the randomized methods (4.45), (4.46), and (4.47), have a better error bound (by a factor m) than their nonrandomized counterparts. In fact, an example given in (Bertsekas et al., 2003, p. 514) for the incremental subgradient method can be adapted to show that the bound of Proposition 4.3(b) is tight in the sense that for a bad problem/cyclic order we have $\liminf_{k\to\infty} F(x_k) - F^* = O(\alpha m^2 c^2)$. By contrast, the randomized method will get to within $O(\alpha mc^2)$ with probability 1 for any problem, according to Proposition 4.6(b). Thus, with the randomized algorithm we do not run the risk of accidentally choosing a bad cyclic order. A related result is provided by the following proposition, which should be compared with Proposition 4.4 for the nonrandomized methods.

Proposition 4.7. *Assume that X^* is nonempty. Let $\{x_k\}$ be a sequence generated as in Proposition 4.6. Then, for any positive scalar ϵ, we have with probability 1*

$$\min_{0\le k\le N} F(x_k) \le F^* + \frac{\alpha\beta mc^2 + \epsilon}{2}, \tag{4.56}$$

where N is a random variable with

$$E\{N\} \le m\,\frac{dist(x_0; X^*)^2}{\alpha\epsilon}. \tag{4.57}$$

Like Proposition 4.6, a comparison of Propositions 4.4 and 4.7 again suggests an advantage for the randomized methods: compared to their deterministic counterparts, they achieve a much smaller error tolerance (a factor of m) in the same *expected* number of iterations. Note, however, that the preceding assessment is based on upper bound estimates, which may not be sharp on a given problem (although the bound of Proposition 4.3(b) is tight with a worst-case problem selection as mentioned earlier; see Bertsekas et al. (2003, p. 514)). Moreover, the comparison based on worst-case values versus expected values may not be strictly valid. In particular, while Proposition 4.4 provides an upper bound estimate on N, Proposition 4.7 provides an upper bound estimate on $E\{N\}$, which is not quite the same.

Finally, for the case of a diminishing stepsize, the following proposition parallels Proposition 4.5, for the cyclic order.

Proposition 4.8. *Let $\{x_k\}$ be the sequence generated by one of the randomized incremental methods (4.45)-(4.47), and let the stepsize α_k satisfy*

$$\lim_{k\to\infty} \alpha_k = 0, \qquad \sum_{k=0}^{\infty} \alpha_k = \infty.$$

Then, with probability 1,

$$\liminf_{k\to\infty} F(x_k) = F^*.$$

Furthermore, if X^* is nonempty and

$$\sum_{k=0}^{\infty} \alpha_k^2 < \infty,$$

then $\{x_k\}$ converges to some $x^* \in X^*$ with probability 1.

4.5   Some Applications

In this section we illustrate our methods in the context of two types of practical applications, and discuss relations with known algorithms.

4.5.1   Regularized Least Squares

Let us consider least-squares problems involving minimization of a sum of quadratic component functions $f_i(x)$ that correspond to errors between data and the output of a model that is parameterized by a vector x. Often a convex regularization function $R(x)$ is added to the least-squares objective, to induce desirable properties of the solution. This gives rise to problems of the form

$$\text{minimize} \quad \gamma R(x) + \frac{1}{2} \sum_{i=1}^{m} (c_i' x - d_i)^2 \tag{4.58}$$

$$\text{subject to } x \in \Re^n,$$

where c_i and d_i are given vectors and scalars, respectively, and γ is a positive scalar. When R is differentiable (e.g., quadratic), and either m is very large or the data (c_i, d_i) become available sequentially over time, it makes sense to consider incremental gradient methods, which have a history of applications since the 1960s, starting with the Widrow-Hoff least mean-squares (LMS) method (Widrow and Hoff, 1960).

The classical type of regularization involves a quadratic function R (as in classical regression and the LMS method), but recently nondifferentiable regularization functions have become increasingly important. On the other hand, to apply our incremental methods, a quadratic R is not essential. What is important is that R has a simple form that facilitates the use of proximal algorithms, such as a separable form, so that the proximal iteration on R is simplified through decomposition. As an example, consider the ℓ_1-

regularization problem, where

$$R(x) = \|x\|_1 = \sum_{j=1}^{n} |x^j| \qquad (4.59)$$

and x^j is the jth coordinate of x. Then the proximal iteration

$$z_k = \arg\min_{x \in \Re^n} \left\{ \gamma \|x\|_1 + \frac{1}{2\alpha_k} \|x - x_k\|^2 \right\}$$

decomposes into the n one-dimensional minimizations

$$z_k^j = \arg\min_{x^j \in \Re} \left\{ \gamma |x^j| + \frac{1}{2\alpha_k} |x^j - x_k^j|^2 \right\}, \qquad j = 1, \ldots, n,$$

and can be done in closed form: For each component $j = 1, 2, \ldots, n$, we have

$$z_k^j = \begin{cases} x_k^j - \gamma\alpha_k & \text{if } \gamma\alpha_k \le x_k^j, \\ x_k^j & \text{if } -\gamma\alpha_k < x_k^j < \gamma\alpha_k, \\ x_k^j + \gamma\alpha_k & \text{if } x_k^j \le -\gamma\alpha_k. \end{cases} \qquad (4.60)$$

We refer to Figueiredo et al. (2007), Wright et al. (2009), Beck and Teboulle (2010), and the references given there for a discussion of a broad variety of applications in estimation and signal-processing problems, where nondifferentiable regularization functions play an important role.

The incremental algorithms of this chapter are well suited for solution of ℓ_1-regularization problems of the form (4.58)-(4.59). For example, the kth incremental iteration may consist of selecting a data pair (c_{i_k}, d_{i_k}) and performing a proximal iteration of the form (4.60) to obtain z_k, followed by a gradient iteration on the component $\frac{1}{2}(c_{i_k}' x - d_{i_k})^2$, starting at z_k:

$$x_{k+1} = z_k - \alpha_k c_{i_k}(c_{i_k}' z_k - d_{i_k}).$$

This algorithm is the special case of algorithms (4.23)-(4.25) (here $X = \Re^n$, and all three algorithms coincide), with $f_i(x)$ being $\gamma\|x\|_1$ (we use m copies of this function) and $h_i(x) = \frac{1}{2}(c_i' x - d_i)^2$. It can be viewed as an incremental version of a popular class of algorithms in signal processing, known as iterative shrinkage/thresholding (see Chambolle et al. (1998); Figueiredo and Nowak (2003); Daubechies et al. (2004); Combettes and Wajs (2005); Elad et al. (2006); Bioucas-Dias and Figueiredo (2007); and Beck and Teboulle (2009, 2010)). Our methods bear the same relation to this class of algorithms as the LMS method bears to gradient algorithms for the classical least-squares problem with quadratic regularization function.

Finally, as an alternative, the proximal iteration (4.60) could be replaced by a proximal iteration on $\gamma |x^j|$ for some selected index j, with all indexes selected cyclically in incremental iterations. Randomized selection of the data pair (c_{i_k}, d_{i_k}) would also be interesting, particularly in contexts where the data have a natural stochastic interpretation.

4.5.2   Iterated Projection Algorithms

A feasibility problem that arises in many contexts involves finding a point with certain properties within a set intersection $\cap_{i=1}^m X_i$, where each X_i is a closed convex set. For the case where m is large and each of the sets X_i has a simple form, incremental methods that make successive projections on the component sets X_i have a long history (see, e.g., Gubin et al. (1967), and recent works such as (Bauschke, 2001), (Bauschke et al., 2006), and (Cegielski and Suchocka, 2008), and their bibliographies). We may consider the following generalized version of the classical feasibility problem,

$$\begin{aligned} \text{minimize} \quad & f(x) \\ \text{subject to} \quad & x \in \cap_{i=1}^m X_i, \end{aligned} \tag{4.61}$$

where $f : \Re^n \mapsto \Re$ is a convex cost function, and the method is

$$x_{k+1} = P_{X_{i_k}}\big(x_k - \alpha_k \tilde{\nabla} f(x_k)\big), \tag{4.62}$$

where the index i_k is chosen from $\{1, \ldots, m\}$ according to a randomized rule. The incremental approach is particularly well suited for problems of the form (4.61), where the sets X_i are not known in advance, but are revealed as the algorithm progresses.

While (4.61) does not involve a sum of component functions, it may be converted into one that does by using an exact penalty function. In particular, consider the problem

$$\begin{aligned} \text{minimize} \quad & f(x) + \gamma \sum_{i=1}^m \text{dist}(x; X_i) \\ \text{subject to} \quad & x \in \Re^n, \end{aligned} \tag{4.63}$$

where γ is a positive penalty parameter. Then for f Lipschitz continuous and γ sufficiently large, problems (4.61) and (4.63) are equivalent, as shown in the following proposition, the proof of which may be found in (Bertsekas, 2010).

Proposition 4.9. *Let $f : Y \mapsto \Re$ be a function defined on a subset Y of $\Re^n$, and let X_i, $i = 1, \ldots, m$ be closed subsets of Y with nonempty intersection. Assume that f is Lipschitz continuous over Y. Then there is a scalar $\overline{\gamma} > 0$ such that for all $\gamma \geq \overline{\gamma}$, the set of minima of f over $\cap_{i=1}^m X_i$ coincides with* the set of minima of

$$f(x) + \gamma \sum_{i=1}^{m} dist(x; X_i)$$

over Y.

From Proposition 4.9, it follows that we may consider in place of the original problem (4.61) the additive cost problem (4.63) to which our algorithms apply. In particular, let us consider algorithms (4.23)-(4.25), with $X = \Re^n$, which involve a proximal iteration on one of the functions $c\,dist(x; X_i)$ followed by a subgradient iteration on f. A key fact here is that the proximal iteration

$$z_k = \arg \min_{x \in \Re^n} \left\{ \gamma\,dist(x; X_{i_k}) + \frac{1}{2\alpha_k} \|x - x_k\|^2 \right\} \tag{4.64}$$

involves a projection on of x_k onto X_{i_k}, followed by an interpolation. In particular, it can be shown (see, e.g., Bertsekas (2010)) that the vector z_k produced by the proximal iteration (4.64) is $z_k = x_k$ if $x_k \in X_{i_k}$. It is otherwise given by

$$z_k = \begin{cases} (1 - \beta_k)x_k + \beta_k P_{X_{i_k}}(x_k) & \text{if } \beta_k < 1, \\ P_{X_{i_k}}(x_k) & \text{if } \beta_k \geq 1, \end{cases} \tag{4.65}$$

where

$$\beta_k = \frac{\alpha_k \gamma}{dist(x_k; X_{i_k})}.$$

Finally, our incremental methods also apply to the case where f has an additive form:

$$\text{minimize} \quad \sum_{i=1}^{m} f_i(x)$$

$$\text{subject to} \quad x \in \cap_{i=1}^{m} X_i.$$

In this case the interpolated projection iterations (4.65) on the sets X_i are followed by subgradient or proximal iterations on the components f_i.

4.6   Conclusions

We have surveyed incremental algorithms, which can deal with many of the challenges posed by large data sets in machine learning applications. We have used a unified analytical framework that includes incremental proximal algorithms and their combinations with the more established incremental gradient and subgradient methods. This allows the flexibility to separate the cost function into the parts that are conveniently handled by proximal iterations (e.g., in essentially closed form) and the remaining parts to be handled by subgradient iterations. We have outlined the convergence properties of these methods, and we have shown that our algorithms apply to some important problems that have been the focus of recent research.

4.7 References

H. H. Bauschke. Projection algorithms: Results and open problems. In D. Butnariu, Y. Censor, and S. Reich, editors, *Inherently Parallel Algorithms in Feasibility and Optimization and their Applications*. Elsevier, Amsterdam, 2001.

H. H. Bauschke, P. L. Combettes, and S. G. Kruk. Extrapolation algorithm for affine-convex feasibility problems. *Numerical Algorithms*, 41:239–274, 2006.

A. Beck and M. Teboulle. A fast iterative shrinkage-threshold algorithm for linear inverse problems. *SIAM Journal on Imaging Sciences*, 2:183–202, 2009.

A. Beck and M. Teboulle. Gradient-based algorithms with applications to signal-recovery problems. In Y. Eldar and D. P. Palomar, editors, *Convex Optimization in Signal Processing and Communications*, pages 42–88. Cambridge University Press, 2010.

A. Ben-Tal, T. Margalit, and A. Nemirovski. The ordered subsets mirror descent optimization method and its use for positron emission tomography reconstruction. In D. Butnariu, Y. Censor, and S. Reich, editors, *Inherently Parallel Algorithms in Feasibility and Optimization and their Applications*. Elsevier, Amsterdam, 2001.

D. P. Bertsekas. Incremental least squares methods and the extended Kalman filter. *SIAM Journal on Optimization*, 6:807–822, 1996.

D. P. Bertsekas. A hybrid incremental gradient method for least squares. *SIAM Journal on Optimization*, 7:913–926, 1997.

D. P. Bertsekas. *Nonlinear Programming*. Athena Scientific, Belmont, MA, second edition, 1999.

D. P. Bertsekas. *Convex Optimization Theory*. Athena Scientific, Belmont, MA, 2009.

D. P. Bertsekas. Incremental proximal methods for large scale convex optimization. Report LIDS-P-2847, Laboratory for Information and Decision Sciences, MIT, Cambridge, MA, 2010.

D. P. Bertsekas and J. N. Tsitsiklis. *Parallel and Distributed Computation: Numerical Methods*. Prentice-Hall, Englewood Cliffs, NJ, 1989.

D. P. Bertsekas and J. N. Tsitsiklis. *Neuro-Dynamic Programming*. Athena Scientific, Belmont, MA, 1996.

D. P. Bertsekas and J. N. Tsitsiklis. Gradient convergence in gradient methods. *SIAM Journal on Optimization*, 10:627–642, 2000.

D. P. Bertsekas, A. Nedić, and A. E. Ozdaglar. *Convex Analysis*. Athena Scientific, Belmont, MA, 2003.

J. M. Bioucas-Dias and M. A. T. Figueiredo. A new TwIST: Two-step iterative shrinking/thresholding algorithms for image restoration. *IEEE Transactions on*

Image Processing, 16(12):2992–3004, 2007.

D. Blatt, A. O. Hero, and H. Gauchman. A convergent incremental gradient method with a constant step size. *SIAM Journal on Optimization*, 18(1):29–51, 2007.

V. Boltyanski, H. Martini, and V. Soltan. *Geometric Methods and Optimization Problems*. Kluwer Academic, Boston, 1999.

V. S. Borkar. *Stochastic Approximation: A Dynamical Systems Viewpoint*. Cambridge University Press, 2008.

L. Bottou. SGD: stochastic gradient descent, 2005. URL http://leon.bottou.org/projects/sgd.

A. Cegielski and A. Suchocka. Relaxed alternating projection methods. *SIAM Journal on Optimization*, 19(3):1093–1106, 2008.

A. Chambolle, R. DeVore, N. Y. Lee, and B. J. Lucier. Nonlinear wavelet image processing: Variational problems, compression, and noise removal through wavelet shrinkage. *IEEE Transactions on Image Processing*, 7(3):319–335, 1998.

P. L. Combettes and V. R. Wajs. Signal recovery by proximal forward-backward splitting. *Multiscale Modeling and Simulation*, 4(4):1168–1200, 2005.

I. Daubechies, M. Defriese, and C. De Mol. An iterative thresholding algorithm for linear inverse problems with a sparsity constraint. *Communications on Pure and Applied Mathematics*, 57:1413–1457, 2004.

W. C. Davidon. New least squares algorithms. *Journal of Optimization Theory and Applications*, 18:187–197, 1976.

J. Duchi, E. Hazan, and Y. Singer. Adaptive subgradient methods for online learning and stochastic optimization. EECS Technical Report 2010-24, UC Berkeley, 2010. To appear in Journal of Machine Learning Research.

M. Elad, B. Matalon, and M. Zibulevsky. Coordinate and subspace optimization methods for linear least squares with non-quadratic regularization. *Journal on Applied and Computational Harmonic Analysis*, 23:346–367, 2006.

Y. Ermoliev. On the stochastic quasi-gradient method and stochastic quasi-Feyer sequences. *Kibernetika*, 2:73–83, 1969.

Y. Ermoliev. *Stochastic Programming Methods*. Nauka, Moscow, 1976.

M. A. T. Figueiredo and R. D. Nowak. An EM algorithm for wavelet-based image restoration. *IEEE Transactions on Image Processing*, 12(8):906–916, 2003.

M. A. T. Figueiredo, R. D. Nowak, and S. J. Wright. Gradient projection for sparse reconstruction: Application to compressed sensing and other inverse problems. *IEEE Journal on Selected Topics in Signal Processing*, 1(4):586–597, 2007.

A. A. Gaivoronski. Convergence of parallel backpropagation algorithm for neural networks. *Optimization Methods and Software*, 4:117–134, 1994.

M. Gaudioso, G. Giallombardo, and G. Miglionico. An Incremental method for solving convex finite min-max problems. *Mathematics of Operations Research*, 31:173–187, 2006.

L. Grippo. A Class of unconstrained minimization methods for neural network training. *Optimization Methods and Software*, 4:135–150, 1994.

L. Grippo. Convergent on-line algorithms for supervised learning in neural networks. *IEEE Transactions on Neural Networks*, 11:1284–1299, 2000.

L. G. Gubin, B. T. Polyak, and E. V. Raik. The method of projection for finding the common point in convex sets. *U.S.S.R. Computational Mathematics and*

Mathematical Physics (English Translation), 7:1–24, 1967.

B. Johansson, M. Rabi, and M. Johansson. A Randomized incremental subgradient method for distributed optimization in networked systems. *SIAM Journal on Optimization*, 20:1157–1170, 2009.

V. M. Kibardin. Decomposition into functions in the minimization problem. *Automation and Remote Control*, 40:1311–1323, 1980.

K. C. Kiwiel. Convergence of approximate and incremental subgradient methods for Convex Optimization. *SIAM Journal on Optimization*, 14(3):807–840, 2004.

H. J. Kushner and D. S. Clark. *Stochastic Approximation Methods for Constrained and Unconstrained Systems*. Springer-Verlag, New York, NY, 1978.

H. J. Kushner and G. Yin. *Stochastic Approximation Methods and Applications*. Springer-Verlag, New York, NY, 1997.

S. Lee and S. J. Wright. Sparse nonlinear support vector machines via stochastic approximation. Technical report, Computer Sciences Department, University of Wisconsin, 2010. submitted.

P. L. Lions and B. Mercier. Splitting algorithms for the sum of two nonlinear operators. *SIAM Journal on Numerical Analysis*, 16:964–979, 1979.

B. M. Litvakov. On an iteration method in the problem of approximating a function from a finite number of observations. *Avtom. Telemech.*, 4:104–113, 1966.

L. Ljung. Analysis of recursive stochastic algorithms. *IEEE Transactions on Automatic Control*, 22:551–575, 1977.

Z. Lu, R. D. C. Monteiro, and M. Yuan. Convex optimization methods for dimension reduction and coefficient estimation in multivariate linear regression. Technical report, School of Industrial and Systems Engineering, Georgia Institute of Technology, Atlanta, 2008. To appear in Mathematical Programming.

Z. Q. Luo. On the convergence of the LMS algorithm with adaptive learning rate for linear feedforward networks. *Neural Computation*, 3(2):226–245, 1993.

Z. Q. Luo and P. Tseng. Analysis of an approximate gradient projection method with applications to the backpropagation algorithm. *Optimization Methods and Software*, 4:85–101, 1994.

O. L. Mangasarian and M. V. Solodov. Serial and parallel backpropagation convergence via nonmonotone perturbed minimization. *Optimization Methods and Software*, 4:103–116, 1994.

B. Martinet. Régularisation d'inéquations variationelles par approximations successives. *Rev. Française Information Recherche Opérationelle*, 4:154–159, 1970.

S. Meyn. *Control Techniques for Complex Networks*. Cambridge University Press, New York, NY, 2007.

H. Moriyama, Y. N., and M. Fukushima. The incremental Gauss-Newton algorithm with adaptive stepsize rule. *Computational Optimization and Applications*, 26(2): 107–141, 2003.

A. Nedić and D. P. Bertsekas. Convergence rate of the incremental subgradient algorithm. In *Stochastic Optimization: Algorithms and Applications*, pages 263–304. Kluwer Academic, 2000.

A. Nedić and D. P. Bertsekas. Incremental subgradient methods for nondifferentiable optimization. *SIAM Journal on Optimization*, 12:109–138, 2001.

A. Nedić and D. P. Bertsekas. The Effect of deterministic noise in subgradient methods. *Mathematical Programming*, 125(1):75–99, 2010.

A. Nedić, D. P. Bertsekas, and V. Borkar. Distributed asynchronous incremental subgradient methods. In *Inherently Parallel Algorithms in Feasibility and Optimization and Their Applications*. Elsevier, Amsterdam, 2001.

A. Nemirovski, A. Juditsky, G. Lan, and A. Shapiro. Robust stochastic approximation approach to stochastic programming. *SIAM Journal on Optimization*, 19: 1574–1609, 2009.

Y. Nesterov. A Method for unconstrained convex minimization problem with the rate of convergence $O(1/k^2)$. *Doklady AN SSSR*, 269:543–547, 1983. translated as Soviet Math. Dokl.

Y. Nesterov. *Introductory Lectures on Convex Optimization*. Kluwer Academic, Dordrecht, The Netherlands, 2004.

Y. Nesterov. Smooth minimization of nonsmooth functions. *Mathematical Programming, Series A*, 103:127–152, 2005.

E. S. Neto and A. R. De Pierro. Incremental subgradients for constrained convex optimization: A Unified framework and new methods. *SIAM Journal on Optimization*, 20:1547–1572, 2009.

G. B. Passty. Ergodic convergence to a zero of the sum of monotone operators in Hilbert Space. *Journal of Mathematical Analysis and Applications*, 72:383–390, 1979.

G. Pflug. *Optimization of Stochastic Models. The Interface Between Simulation and Optimization*. Kluwer Academic, Boston, 1996.

B. T. Poljak. Some methods of speeding up the convergence of iteration methods. *Z. VyČisl. Mat. i Mat. Fiz.*, 4:1–17, 1964.

B. T. Polyak. Nonlinear programming methods in the presence of noise. *Mathematical Programming*, 14:87–97, 1978.

B. T. Polyak. *Introduction to Optimization*. Optimization Software Inc., NY, 1987.

B. T. Polyak and Y. Z. Tsypkin. Pseudogradient adaptation and training algorithms. *Automation and Remote Control*, 12:83–94, 1973.

J. B. Predd, S. R. Kulkarni, and H. V. Poor. A Collaborative training algorithm for distributed learning. *IEEE Transactions on Information Theory*, 55:1856–1871, 2009.

M. G. Rabbat and R. D. Nowak. Distributed optimization in sensor networks. In *Proceedings of the International Conference on Information Processing in Sensor Networks*, pages 20–27, 2004.

M. G. Rabbat and R. D. Nowak. Quantized incremental algorithms for distributed optimization. *IEEE Journal on Selected Areas in Communications*, 23:798–808, 2005.

S. S. Ram, A. Nedić, and V. V. Veeravalli. Incremental stochastic subgradient algorithms for convex optimization. *SIAM Journal on Optimization*, 20:691–717, 2009a.

S. S. Ram, A. Nedić, and V. V. Veeravalli. Distributed stochastic subgradient projection algorithms for convex optimization. *Submitted*, 2009b.

R. T. Rockafellar. *Convex Analysis*. Princeton Univ. Press, 1970.

R. T. Rockafellar. Monotone operators and the proximal point algorithm. *SIAM Journal on Control and Optimization*, 14:877–898, 1976.

S. Shalev-Shwartz, Y. Singer, and N. Srebro. Pegasos: Primal Estimated Subgradient Solver for SVM. In *Proceedings of the 24th International Conference on*

Machine Learning, pages 807–814, New York, NY, 2007.

M. V. Solodov. Incremental gradient algorithms with stepsizes bounded away from zero. *Computational Optimization and Applications*, 11:28–35, 1998.

P. Tseng. An Incremental gradient(-projection) method with momentum term and adaptive stepsize rule. *SIAM Journal on Optimization*, 8:506–531, 1998.

P. Tseng. On Accelerated proximal gradient methods for convex-concave optimization. Technical report, Mathematics Department, University of Washington, 2008.

J. N. Tsitsiklis, D. P. Bertsekas, and M. Athans. Distributed asynchronous deterministic and stochastic gradient optimization algorithms. *IEEE Transactions on Automatic Control*, AC-31:803–812, 1986.

B. Widrow and M. E. Hoff. Adaptive switching circuits. In *Institute of Radio Engineers, Western Electronic Show and Convention, Convention Record*, volume 4, pages 96–104, 1960.

S. J. Wright, R. D. Nowak, and M. A. T. Figueiredo. Sparse reconstruction by separable approximation. *IEEE Transations on Signal Processing*, 57:2479–2493, 2009.

5 First-Order Methods for Nonsmooth Convex Large-Scale Optimization, I: General Purpose Methods

Anatoli Juditsky   Anatoli.Juditsky@imag.fr
Laboratoire Jean Kuntzmann , Université J. Fourier
B. P. 53 38041 Grenoble Cedex, France

Arkadi Nemirovski   nemirovs@isye.gatech.edu
School of Industrial and Systems Engineering, Georgia Institute of Technology
765 Ferst Drive NW, Atlanta Georgia 30332, USA

We discuss several state-of-the-art computationally cheap, as opposed to the polynomial time interior-point algorithms, first-order methods for minimizing convex objectives over simple large-scale feasible sets. Our emphasis is on the general situation of a nonsmooth convex objective represented by deterministic/stochastic first-order oracle and on the methods which, under favorable circumstances, exhibit a (nearly) dimension-independent convergence rate.

5.1 Introduction

At present, almost all of convex programming is within the grasp of polynomial time interior-point methods (IPMs) capable of solving convex programs to high accuracy at a low iteration count. However, the iteration cost of all known polynomial methods grows nonlinearly with a problem's design dimension n (number of decision variables), something like n^3. As a result, as the design dimension grows, polynomial time methods eventually become impractical—roughly speaking, a single iteration lasts forever. What "eventually" means in fact depends on a problem's structure. For instance, typical linear programming programs of decision-making origin have extremely sparse constraint matrices, and IPMs are able to solve programs of this type with tens and hundreds of thousands variables and constraints in reasonable time. In contrast to this, linear programming programs arising in machine learning and signal processing often have dense constraint matrices. Such programs with "just" few thousand variables and constraints can become very difficult for an IPM. At the present level of our knowledge, the methods of choice when solving convex programs which, because of their size, are beyond the practical grasp of IPMs, are the *first-order methods* (FOMs) with computationally cheap iterations. In this chapter, we present several state-of-the-art FOMs for large-scale convex optimization, focusing on the most general *nonsmooth unstructured* case, where the convex objective f to be minimized can be nonsmooth and is represented by a black box, a routine able to compute the values and subgradients of f.

5.1.1   First-Order Methods: Limits of Performance

We start by explaining what can and can*not* be expected from FOMs, restricting ourselves for the time being to convex programs of the form

$$\text{Opt}(f) = \min_{x \in \mathcal{X}} f(x), \tag{5.1}$$

where $\mathcal{X}$ is a compact convex subset of $\mathbb{R}^n$, and f is known to belong to a given family $\mathcal{F}$ of convex and (at least) Lipschitz continuous functions on $\mathcal{X}$. Formally, an FOM is an algorithm $\mathcal{B}$ which knows in advance what $\mathcal{X}$ and $\mathcal{F}$ are, but does not know exactly what $f \in \mathcal{F}$ is. It is restricted to learning f via subsequent calls to a *first-order oracle*—a routine which, given a point $x \in \mathcal{X}$ on input, returns on output a value $f(x)$ and a (sub)gradient $f'(x)$ of f at x (informally speaking, this setting implicitly assumes that $\mathcal{X}$ is simple (like box, or ball, or standard simplex), while f can be complicated). Specifically, as applied to a particular objective $f \in \mathcal{F}$ and given on input a required accuracy $\epsilon > 0$, the method $\mathcal{B}$, after generating a finite sequence of *search points* $x_t \in \mathcal{X}$, $t = 1, 2, ...$, where the first-order oracle is called, terminates and outputs an approximate solution $\widehat{x} \in \mathcal{X}$ which should be ϵ-optimal: $f(\widehat{x}) - \text{Opt}(f) \leq \epsilon$. In other words, the method itself is a collection of rules for generating subsequent search points, identifying the terminal step, and building the approximate solution.

These rules, in principle, can be arbitrary, with the only limitation of being *nonanticipating*, meaning that the output of a rule is uniquely defined by $\mathcal{X}$ and the first-order information on f accumulated before the rule is applied. As a result, for a given $\mathcal{B}$ and $\mathcal{X}$, x_1 is independent of f, x_2 depends solely on $f(x_1), f'(x_1)$, and so on. Similarly, the decision to terminate after a particular number t of steps, as well as the resulting approximate solution $\hat{x}$, are uniquely defined by the first-order information $f(x_1), f'(x_1), ..., f(x_t), f'(x_t)$ accumulated in the course of these t steps. Performance limits of FOMs are given by *information-based complexity theory*, which says what, for given $\mathcal{X}, \mathcal{F}, \epsilon$, may be the minimal number of steps of an FOM solving all problems (5.1) with $f \in \mathcal{F}$ within accuracy ϵ. Here are several instructive examples (see Nemirovsky and Yudin, 1983).

(a) Let $\mathcal{X} \subset \{x \in \mathbb{R}^n : \|x\|_p \leq R\}$, where $p \in \{1, 2\}$, and let $\mathcal{F} = \mathcal{F}_p$ comprise all convex functions f which are Lipschitz continuous, with a given constant L, w.r.t. $\|\cdot\|_p$. When $\mathcal{X} = \{x \in \mathbb{R}^n : \|x\|_p \leq R\}$, the number N of steps of *any* FOM able to solve every problem from the outlined family within accuracy ϵ is *at least* $O(1) \min[n, L^2 R^2/\epsilon^2]$. [1] When $p = 2$, this lower complexity bound remains true when $\mathcal{F}$ is restricted to being the family of all functions of the type $f(x) = \max_{1 \leq i \leq n} [\epsilon_i L x_i + a_i]$ with $\epsilon_i = \pm 1$. Moreover, the bound is nearly achievable: whenever $\mathcal{X} \subset \{x \in \mathbb{R}^n : \|x\|_p \leq R\}$, there exist quite transparent (and simple to implement when $\mathcal{X}$ is simple) FOMs able to solve all problems (5.1) with $f \in \mathcal{F}_p$ within accuracy ϵ in $O(1)(\ln(n))^{2/p-1} L^2 R^2/\epsilon^2$ steps.

It should be stressed that the outlined nearly dimension-independent performance of FOMs depends heavily on the assumption $p \in \{1, 2\}$. [2] With p set to $+\infty$ (i.e., when minimizing convex functions that are Lipschitz continuous with constant L w.r.t. $\|\cdot\|_\infty$ over the box $\mathcal{X} = \{x \in \mathbf{R}^n : \|x\|_\infty \leq R\}$), the lower and upper complexity bounds are $O(1)n \ln(LR/\epsilon)$, provided that $LR/\epsilon \geq 2$; these bounds depend heavily on the problem's dimension.

(b) Let $\mathcal{X} = \{x \in \mathbb{R}^n : \|x\|_2 \leq R\}$, and let $\mathcal{F}$ comprise all differentiable convex functions, Lipschitz continuous with constant L w.r.t. $\|\cdot\|_2$, gradient. Then the number N of steps of any FOM able to solve every problem from the outlined family within accuracy ϵ is *at least* $O(1) \min[n, \sqrt{LR^2/\epsilon}]$. This lower complexity bound remains true when $\mathcal{F}$ is restricted to be the family of convex quadratic forms $\frac{1}{2}x^T A x + b^T x$ with positive semidefinite symmetric matrices A of spectral norm (maximal singular value) not exceeding L. Here again the lower complexity bound is nearly achievable. Whenever $\mathcal{X} \subset \{x \in \mathbb{R}^n : \|x\|_2 \leq R\}$, there exists a simple implementation when $\mathcal{X}$ is simple (although by far not transparent) FOM: *Nesterov's optimal algorithm for smooth convex minimization* (Nesterov, 1983, 2005), which allows one to

---

1. From now on, all $O(1)$'s are appropriate positive *absolute* constants.
2. In fact, it can be relaxed to $1 \leq p \leq 2$.

solve within accuracy ϵ all problems (5.1) with $f \in \mathcal{F}$ in $O(1)\sqrt{LR^2/\epsilon}$ steps.
(c) Let $\mathcal{X}$ be as in (b), and let $\mathcal{F}$ comprise all functions of the form
$f(x) = \|Ax - b\|_2$, where the spectral norm of A (which is no longer
positive semidefinite) does not exceed a given L. Let us slightly extend
the power of the first-order oracle and assume that at a step of an FOM
we observe b (but not A) and are allowed to carry out $O(1)$ matrix-vector
multiplications involving A and A^T. In this case, the number of steps of any
method capable to solve all problems in question within accuracy ϵ is at
least $O(1)\min[n, LR/\epsilon]$, and there exists a method (specifically, Nesterov's
optimal algorithm as applied to the quadratic form $\|Ax - b\|_2^2$), which
achieves the desired accuracy in $O(1)LR/\epsilon$ steps.

The outlined results bring us both bad and good news on FOMs as applied
to large-scale convex programs. The bad news is that *unless the number of
steps of the method exceeds the problem's design dimension n* (which is of
no interest when n is really large), *and without imposing severe additional
restrictions on the objectives to be minimized, an FOM can exhibit only a
sublinear rate of convergence*: specifically denoting by t the number of steps,
the rate $O(1)(\ln(n))^{1/p-1/2}LR/t^{1/2}$ in the case of (a) (better than nothing,
but really slow), $O(1)LR^2/t^2$ in the case of (b) (much better, but simple
$\mathcal{X}$ along with smooth f is a rare commodity), and $O(1)LR/t$ in the case of
(c) (in-between (a) and (b)). As a consequence, *FOMs are poorly suited for
building high-accuracy solutions to large-scale convex problems.*

The good news is that *for problems with favorable geometry* (e.g., those in
(a)-(c)), *good FOMs exhibit a dimension-independent, or nearly so, rate of
convergence*, which is of paramount importance in large-scale applications.
Another bit of good news (not declared explicitly in the above examples)
is that *when $\mathcal{X}$ is simple, typical FOMs have cheap iterations—modulo
computations hidden in the oracle, an iteration costs just $O(\dim \mathcal{X})$ a.o.*
The bottom line is that *FOMs are well suited for finding medium-accuracy
solutions to large-scale convex problems*, at least when the latter possess
favorable geometry.

Another conclusion of the presented results is that the performance limits
of FOMs depend heavily on the size R of the feasible domain and on the
Lipschitz constant L (of f in the case of (a), and of f' in the case of (b)).
This is in a sharp contrast to IPMs, where the complexity bounds depend
logarithmically on the magnitudes of an optimal solution and of the data
(the analogies of R and L, respectively), which, practically speaking, allows
one to handle problems with unbounded domains (one may impose an upper
bound of 10^6 or 10^{100} on the variables) and not to bother much about how

the data are scaled.[3] Strong dependence of the complexity of FOMs on L and R implies a number of important consequences. In particular:

• Boundedness of $\mathcal{X}$ is of paramount importance, at least theoretically. In this respect, unconstrained settings, as in Lasso: $\min_{x}\{\lambda\|x\|_1 + \|Ax - b\|_2^2\}$ are less preferable than their bounded domain counterparts, as in $\min\{\|Ax - b\|_2 : \|x\|_1 \leq R\}$[4] in full accordance with common sense—however difficult it is to find a needle in a haystack, a small haystack in this respect is better than a large one!

• For a given problem (5.1), the size R of the feasible domain and the Lipschitz constant L of the objective depend on the norm $\|\cdot\|$ used to quantify these quantities: $R = R_{\|\cdot\|}$, $L = L_{\|\cdot\|}$. When $\|\cdot\|$ varies, the product $L_{\|\cdot\|}R_{\|\cdot\|}$ (this product is all that matters) changes,[5] and this phenomenon should be taken into account when choosing an FOM for a particular problem.

5.1.2 What Is Ahead

Literature on FOMs, which has always been huge, is now growing explosively—partly due to rapidly increasing demand for large-scale optimization, and partly due to endogenous reasons stemming primarily from discovering ways (Nesterov, 2005) to accelerate FOMs by exploiting problems' structure (for more details on the latter subject, see Chapter 6). Even a brief overview of this literature in a single chapter would be completely unrealistic. Our primary selection criteria were (a) to focus on techniques for large-scale *nonsmooth* convex programs (these are the problems arising in most applications known to us), (b) to restrict ourselves to FOMs possessing state-of-the-art (in some cases—even provably optimal) nonasymptotic efficiency estimates, and (c) the possibility for self-contained presentation of the methods, given space limitations. Last, but not least, we preferred to focus on the situations of which we have first-hand (or nearly so) knowledge. As a result, our presentation of FOMs is definitely incomplete. As for citation policy, we restrict ourselves to referring to works directly related to what we are pre-

---

3. In IPMs, scaling of the data affects stability of the methods w.r.t. rounding errors, but this is another story.

4. We believe that the desire to end up with unconstrained problems stems from the common belief that the unconstrained convex minimization is simpler than the constrained one. To the best of our understanding, this belief is misleading, and the actual distinction is between optimization over simple and over sophisticated domains; what is simple depends on the method in question.

5. For example, the ratio $[L_{\|\cdot\|_2}R_{\|\cdot\|_2}]/L_{\|\cdot\|_1}R_{\|\cdot\|_1}$ can be as small as $1/\sqrt{n}$ and as large as $\sqrt{n}$

senting, with no attempt to give even a nearly exhaustive list of references to FOM literature. We apologize in advance for potential omissions even on this reduced list.

In this chapter, we focus on the simplest general-purpose FOMs, *mirror descent* (MD) *methods* aimed at solving nonsmooth convex minimization problems, specifically, general-type problems (5.1) (Section 5.2), problems (5.1) with strongly convex objectives (Section 5.4), convex problems with functional constraints $\min_{x \in \mathcal{X}} \{f_0(x) : f_i(x) \leq 0, 1 \leq i \leq m\}$ (Section 5.3), and stochastic versions of problems (5.1), where the first-order oracle is replaced with its stochastic counterpart, thus providing unbiased random estimates of the subgradients of the objective rather than the subgradients themselves (Section 5.5). Finally, Section 5.6 presents extensions of the mirror descent scheme from problems of convex minimization to the convex-concave saddle-point problems.

As we have already said, this chapter is devoted to general-purpose FOMs, meaning that the methods in question are fully black-box-oriented—they do not assume any a priori knowledge of the structure of the objective (and the functional constraints, if any) aside from convexity and Lipschitz continuity. By itself, this generality is redundant: convex programs arising in applications always possess a lot of known in advance structure, and utilizing a priori knowledge of this structure can accelerate the solution process dramatically. Acceleration of FOMs by utilizing a problems' structure is the subject of Chapter 6.

5.2 Mirror Descent Algorithm: Minimizing over a Simple Set

5.2.1 Problem of Interest

We focus primarily on solving an optimization problem of the form

$$\text{Opt} = \min_{x \in \mathcal{X}} f(x), \tag{5.2}$$

where $\mathcal{X} \subset E$ is a closed convex set in a finite-dimensional Euclidean space E, and $f : \mathcal{X} \to \mathbb{R}$ is a Lipschitz continuous convex function represented by a *first-order oracle*. This oracle is a routine which, given a point $x \in \mathcal{X}$ on input, returns the value $f(x)$ and a subgradient $f'(x)$ of f at x. We always assume that $f'(x)$ is bounded on $\mathcal{X}$. We also assume that (5.2) is solvable.

5.2.2 Mirror Descent setup

We set up the MD method with two entities:

- a norm $\|\cdot\|$ on the space E embedding $\mathcal{X}$, and the conjugate norm $\|\cdot\|_*$ on E^*: $\|\xi\|_* = \max_x \{\langle \xi, x \rangle : \|x\| \leq 1\}$;
- a *distance-generating function* (d.-g.f. for short) *for* $\mathcal{X}$ *compatible with the norm* $\|\cdot\|$, that is, a continuous convex function $\omega(x) : \mathcal{X} \to \mathbb{R}$ such that
  —$\omega(x)$ admits a selection $\omega'(x)$ of a subgradient which is continuous on the set $\mathcal{X}^o = \{x \in \mathcal{X} : \partial\omega(x) \neq \emptyset\}$;
  —$\omega(\cdot)$ is strongly convex, with modulus 1, w.r.t. $\|\cdot\|$:

$$\forall(x, x' \in \mathcal{X}^o) : \langle \omega'(x) - \omega'(x'), x - x' \rangle \geq \|x - y\|^2. \tag{5.3}$$

For $x \in \mathcal{X}^o$, $u \in \mathcal{X}$, let

$$V_x(u) = \omega(u) - \omega(x) - \langle \omega'(x), u - x \rangle. \tag{5.4}$$

Denote $x_c = \operatorname{argmin}_{u \in \mathcal{X}} \omega(u)$ (the existence of a minimizer is given by continuity and strong convexity of ω on $\mathcal{X}$ and by closedness of $\mathcal{X}$, and its uniqueness by strong convexity of ω). When $\mathcal{X}$ is bounded, we define $\omega(\cdot)$-diameter $\Omega = \max_{u \in \mathcal{X}} V_{x_c}(u) \leq \max_{\mathcal{X}} \omega(u) - \min_{\mathcal{X}} \omega(u)$ of $\mathcal{X}$. Given $x \in \mathcal{X}^o$, we define the *prox-mapping* $\operatorname{Prox}_x(\xi) : E \to \mathcal{X}^o$ as

$$\operatorname{Prox}_x(\xi) = \operatorname{argmin}_{u \in \mathcal{X}} \{\langle \xi, u \rangle + V_x(u)\}. \tag{5.5}$$

From now on we make the

Simplicity Assumption. $\mathcal{X}$ and ω are simple and fit each other. Specifically, given $x \in \mathcal{X}^o$ and $\xi \in E$, it is easy to compute $\operatorname{Prox}_x(\xi)$.

5.2.3 Basic Mirror Descent algorithm

The MD algorithm associated with the outlined setup, as applied to problem (5.2), is the recurrence

$$
\begin{aligned}
&(a) \quad x_1 = \operatorname{argmin}_{x \in \mathcal{X}} \omega(x) \\
&(b) \quad x_{t+1} = \operatorname{Prox}_{x_t}(\gamma_t f'(x_t)), \ t = 1, 2, \ldots \\
&(c) \quad x^t = \left[\sum_{\tau=1}^t \gamma_\tau\right]^{-1} \sum_{\tau=1}^t \gamma_\tau x_\tau \\
&(d) \quad \widehat{x}^t = \operatorname{argmin}_{x \in \{x_1, \ldots, x_t\}} f(x)
\end{aligned}
\tag{5.6}
$$

Here, x_t are subsequent *search points*, and x^t (or $\widehat{x}^t$—the error bounds that follow work for both these choices) are subsequent *approximate solutions* generated by the algorithm. Note that $x_t \in \mathcal{X}^o$ and $x^t, \widehat{x}^t \in \mathcal{X}$ for all t.

The convergence properties of MD stem from the following simple observation:

Proposition 5.1. *Suppose that f is Lipschitz continuous on $\mathcal{X}$ with $L :=$*

$\sup_{x \in \mathcal{X}} \|f'(x)\|_* < \infty$. *Let* $\overline{f}_t = \max[f(x^t), f(\widehat{x}^t)]$. *Then*

(i) *for all* $u \in \mathcal{X}$, $t \geq 1$ *one has*

$$
\begin{aligned}
\sum_{\tau=1}^{t} \gamma_\tau \langle f'(x_\tau), x_\tau - u \rangle &\leq V_{x_1}(u) + \tfrac{1}{2} \sum_{\tau=1}^{t} \gamma_\tau^2 \|f'(x_\tau)\|_*^2 \\
&\leq V_{x_1}(u) + \tfrac{L^2}{2} \sum_{\tau=1}^{t} \gamma_\tau^2.
\end{aligned}
\tag{5.7}
$$

As a result, for all $t \geq 1$,

$$
\overline{f}_t - \mathrm{Opt} \leq \epsilon_t := \frac{V_{x_1}(x_*) + \frac{L^2}{2} \sum_{\tau=1}^{t} \gamma_\tau^2}{\sum_{\tau=1}^{t} \gamma_\tau},
\tag{5.8}
$$

where x_* *is an optimal solution to* (5.2). *In particular, in the divergent series case* $\gamma_t \to 0$, $\sum_{\tau=1}^{t} \gamma_\tau \to +\infty$ *as* $t \to \infty$, *the algorithm converges:* $\overline{f}_t - \mathrm{Opt} \to 0$ *as* $t \to \infty$. *Moreover, with the stepsizes*

$$
\gamma_t = \gamma / [\|f'(x_t)\|_* \sqrt{t}]
$$

for all t, *one has*

$$
\overline{f}_t - \mathrm{Opt} \leq O(1) \left[\frac{V_{x_1}(x_*)}{\gamma} + \frac{\ln(t+1)\gamma}{2} \right] L t^{-1/2}.
\tag{5.9}
$$

(ii) *Let* $\mathcal{X}$ *be bounded so that the* $\omega(\cdot)$-*diameter* Ω *of* $\mathcal{X}$ *is finite. Then, for every number* N *of steps, the* N-*step MD algorithm with constant stepsizes,*

$$
\gamma_t = \frac{\sqrt{2\Omega}}{L\sqrt{N}}, \quad 1 \leq t \leq N,
\tag{5.10}
$$

ensures that

$$
\begin{aligned}
\underline{f}_N &= \min_{u \in \mathcal{X}} \tfrac{1}{N} \sum_{\tau=1}^{N} [f(x_\tau) + \langle f'(x_\tau), u - x_\tau \rangle] \leq \mathrm{Opt}, \\
\overline{f}_N &- \mathrm{Opt} \leq \overline{f}_N - \underline{f}_N \leq \frac{\sqrt{2\Omega}L}{\sqrt{N}}.
\end{aligned}
\tag{5.11}
$$

In other words, the quality of approximate solutions $(x^N$ *or* $\widehat{x}^N)$ *can be certified by the easy-to-compute online lower bound* $\underline{f}_N$ *on* Opt, *and the certified level of nonoptimality of the solutions can only be better than the one given by the worst-case upper bound in the right-hand side of* (5.11).

Proof. From the definition of the prox-mapping,

$$
x_{\tau+1} = \operatorname*{argmin}_{z \in \mathcal{X}} \left\{ \langle \gamma_\tau f'(x_\tau) - \omega'(x_\tau), z \rangle + \omega(z) \right\},
$$

whence, by optimality conditions,

$$
\langle \gamma_\tau f'(x_\tau) - \omega'(x_\tau) + \omega'(x_{\tau+1}), u - x_{\tau+1} \rangle \geq 0 \; \forall u \in \mathcal{X}.
$$

When rearranging terms, this inequality can be rewritten as

$$
\begin{aligned}
\gamma_\tau \langle f'(x_\tau), x_\tau - u \rangle \;\leq\; & [\omega(u) - \omega(x_\tau) - \langle \omega'(x_\tau), u - x_\tau \rangle] \\
& -[\omega(u) - \omega(x_{\tau+1}) - \langle \omega'(x_{\tau+1}), u - x_{\tau+1} \rangle] \\
& +\gamma_\tau \langle f'(x_\tau), x_\tau - x_{\tau+1} \rangle \\
& -[\omega(x_{\tau+1}) - \omega(x_\tau) - \langle \omega'(x_\tau), x_{\tau+1} - x_\tau \rangle] \\
=\; & V_{x_\tau}(u) - V_{x_{\tau+1}}(u) + \underbrace{[\gamma_\tau \langle f'(x_\tau), x_\tau - x_{\tau+1} \rangle - V_{x_\tau}(x_{\tau+1})]}_{\delta_\tau}. \quad (5.12)
\end{aligned}
$$

From the strong convexity of V_{x_τ} it follows that

$$
\begin{aligned}
\delta_\tau \;\leq\; & \gamma_\tau \langle f'(x_\tau), x_\tau - x_{\tau+1} \rangle - \tfrac{1}{2}\|x_\tau - x_{\tau+1}\|^2 \\
\leq\; & \gamma_\tau \|f'(x_\tau)\|_* \|x_\tau - x_{\tau+1}\| - \tfrac{1}{2}\|x_\tau - x_{\tau+1}\|^2 \\
\leq\; & \max_s [\gamma_\tau \|f'(x_\tau)\|_* s - \tfrac{1}{2}s^2] = \tfrac{\gamma_\tau^2}{2}\|f'(x_\tau)\|_*^2,
\end{aligned}
$$

and we get

$$
\gamma_\tau \langle f'(x_\tau), x_\tau - u \rangle \leq V_{x_\tau}(u) - V_{x_{\tau+1}}(u) + \gamma_\tau^2 \|f'(x_\tau)\|_*^2 / 2. \quad (5.13)
$$

Summing these inequalities over $\tau = 1, ..., t$ and taking into account that $V_x(u) \geq 0$, we arrive at (5.7). With $u = x_*$, (5.7), when taking into account that $\langle f'(x_\tau), x_\tau - x_* \rangle \geq f(x_\tau) - \mathrm{Opt}$ and setting $f^t = [\sum_{\tau=1}^t \gamma_\tau]^{-1} \sum_{\tau=1}^t \gamma_\tau f(x_\tau)$ results in

$$
f^t - \mathrm{Opt} \leq \frac{V_{x_1}(x_*) + L^2 \left[\sum_{\tau=1}^t \gamma_\tau^2\right]/2}{\sum_{\tau=1}^t \gamma_\tau}.
$$

Since, clearly, $\overline{f}_t = \max[f(x^t), f(\widehat{x}^t)] \leq f^t$, we have arrived at (5.8). This inequality straightforwardly implies the remaining results of (i).

To prove (ii), note that by the definition of Ω and due to $x_1 = \operatorname{argmin}_X \omega$, (5.7) combines with (5.10) to imply that

$$
f^N - \underline{f}_N = \max_{u \in X} \left[f^N - \frac{1}{N} \sum_{\tau=1}^N [f(x_\tau) + \langle f'(x_\tau), u - x_\tau \rangle] \right] \leq \frac{\sqrt{2\Omega}L}{\sqrt{N}}. \quad (5.14)
$$

Since f is convex, the function $\frac{1}{N}\sum_{\tau=1}^N [f(x_\tau) + \langle f'(x_\tau), u - x_\tau \rangle]$ underestimates $f(u)$ everywhere on X, that is, $\underline{f}_N \leq \mathrm{Opt}$. And, as we have seen, $f^N \geq \overline{f}_N$, therefore (ii) follows from (5.14). $\qquad\square$

5.3   Problems with Functional Constraints

The MD algorithm can be extended easily from the case of problem (5.2) to the case of problem

$$\text{Opt} = \min_{x \in \mathcal{X}} \left\{ f_0(x) : f_i(x) \leq 0, \ 1 \leq i \leq m \right\}, \tag{5.15}$$

where f_i, $0 \leq f_i \leq m$, are Lipschitz continuous convex functions on $\mathcal{X}$ given by the first-order oracle which, given $x \in \mathcal{X}$ on input, returns the values $f_i(x)$ and subgradients $f_i'(x)$ of f_i at x, with selections of the subgradients $f_i'(\cdot)$ bounded on $\mathcal{X}$. Consider the N-step algorithm:

1. *Initialization:* Set $x_1 = \text{argmin}_{\mathcal{X}}\, \omega$.

2. *Step t, $1 \leq t \leq N$:* Given $x_t \in \mathcal{X}$, call the first-order oracle (x_t being the input) and check whether

$$f_i(x_t) \leq \gamma \|f_i'(x_t)\|_*, \ i = 1, ..., m. \tag{5.16}$$

If it is the case (productive step), set $i(t) = 0$; otherwise (nonproductive step) choose $i(t) \in \{1, ..., m\}$ such that $f_{i(t)}(x) > \gamma \|f_{i(t)}'(x_t)\|_*$. Set

$$\gamma_t = \gamma/\|f_{i(t)}'(x_t)\|_*, \ x_{t+1} = \text{Prox}_{x_t}(\gamma_t f_{i(t)}'(x_t)).$$

When $t < N$, loop to step $t + 1$.

3. *Termination:* After N steps are executed, output, as approximate solution $\widehat{x}^N$, the best (with the smallest value of f_0) of the points x_t associated with productive steps t; if there were no productive steps, claim (5.15) is infeasible.

Proposition 5.2. *Let $\mathcal{X}$ be bounded. Given integer $N \geq 1$, set $\gamma = \sqrt{2\Omega}/\sqrt{N}$. Then*

  (i) *If (5.15) is feasible, $\widehat{x}^N$ is well defined.*
  (ii) *Whenever $\widehat{x}^N$ is well defined, one has*

$$\max\left[f_0(\widehat{x}^N) - \text{Opt}, f_1(\widehat{x}^N), ..., f_m(\widehat{x}^N)\right] \leq \gamma L = \frac{\sqrt{2\Omega}L}{\sqrt{N}},$$
$$L = \max_{0 \leq i \leq m} \sup_{x \in \mathcal{X}} \|f_i'(x)\|_*. \tag{5.17}$$

Proof. By construction, when $\widehat{x}^N$ is well defined, it is some x_t with productive t, whence $f_i(\widehat{x}^N) \leq \gamma L$ for $1 \leq i \leq m$ by (5.16). It remains to verify that when (5.15) is feasible, $\widehat{x}^N$ is well defined and $f_0(\widehat{x}^N) \leq \text{Opt} + \gamma L$. Assume that it is not the case, whence at every productive step t (if any) we have $f_0(x_t) - \text{Opt} > \gamma \|f_0'(x_t)\|_*$. Let x_* be an optimal solution to (5.15). Exactly the same reasoning as in the proof of Proposition 5.1 yields the following analogy of (5.7) (with $u = x_*$):

$$\sum\nolimits_{t=1}^{N} \gamma_t \langle f'_{i(t)}(x_t), x_t - x_* \rangle \leq \Omega + \frac{1}{2} \sum\nolimits_{t=1}^{N} \gamma_t^2 \|f'_{i(t)}(x_t)\|_*^2 = 2\Omega. \quad (5.18)$$

When t is nonproductive, we have $\gamma_t \langle f'_{i(t)}(x_t), x_t - x_* \rangle \geq \gamma_t f_{i(t)}(x_t) > \gamma^2$, the concluding inequality being given by the definition of $i(t)$ and γ_t. When t is productive, we have $\gamma_t \langle f'_{i(t)}(x_t), x_t - x_* \rangle = \gamma_t \langle f'_0(x_t), x_t - x_* \rangle \geq \gamma_t(f_0(x_t) - \mathrm{Opt}) > \gamma^2$, the concluding inequality being given by the definition of γ_t and our assumption that $f_0(x_t) - \mathrm{Opt} > \gamma \|f'_0(x_t)\|_*$ at all productive steps t. The bottom line is that the left-hand side in (5.18) is $> N\gamma^2 = 2\Omega$, which contradicts (5.18). $\square$

5.4   Minimizing Strongly Convex Functions

The MD algorithm can be modified to obtain the rate $O(1/t)$ in the case where the objective f in (5.2) is *strongly convex*. The strong convexity of f *with modulus* $\kappa > 0$ means that

$$\forall (x, x' \in \mathcal{X}) \quad \langle f'(x) - f'(x'), \, x - x' \rangle \geq \kappa \|x - x'\|^2. \quad (5.19)$$

Further, let ω be the d.-g.f. *for the entire E* (not just for $\mathcal{X}$, which may be unbounded in this case), compatible with $\|\cdot\|$. W.l.o.g. let $0 = \mathrm{argmin}_E \, \omega$, and let

$$\Omega = \max_{\|u\| \leq 1} \omega(u) - \omega(0)$$

be the variation of ω on the unit ball of $\|\cdot\|$. Now, let $\omega^{R,z}(u) = \omega\left(\frac{u-z}{R}\right)$ and $V_x^{R,z}(u) = \omega^{R,z}(u) - \omega^{R,z}(x) - \langle (\omega^{R,z}(x))', \, u - x \rangle$. Given $z \in \mathcal{X}$ and $R > 0$ we define the prox-mapping

$$\mathrm{Prox}_x^{R,z}(\xi) = \mathrm{argmin}_{u \in \mathcal{X}}[\langle \xi, u \rangle + V_x^{R,z}(u)]$$

and the recurrence (cf. (5.6))

$$\begin{aligned} x_{t+1} &= \mathrm{Prox}_{x_t}^{R,z}(\gamma_t f'(x_t)), \quad t = 1, 2, \dots \\ x^t(R, z) &= \left[\sum\nolimits_{\tau=1}^{t} \gamma_\tau\right]^{-1} \sum\nolimits_{\tau=1}^{t} \gamma_\tau x_\tau. \end{aligned} \quad (5.20)$$

We start with the following analogue of Proposition 5.1.

Proposition 5.3. *Let f be strongly convex on $\mathcal{X}$ with modulus $\kappa > 0$ and Lipschitz continuous on $\mathcal{X}$ with $L := \sup_{x \in \mathcal{X}} \|f'(x)\|_* < \infty$. Given $R > 0$, $t \geq 1$, suppose that $\|x_1 - x_*\| \leq R$, where x_* is the minimizer of f on $\mathcal{X}$,* *and let the stepsizes γ_τ satisfy*

$$\gamma_\tau = \frac{\sqrt{2\Omega}}{RL\sqrt{t}}, \ 1 \le \tau \le t. \tag{5.21}$$

Then, after t iterations (5.20) one has

$$f(x^t(R, x_1)) - \text{Opt} \ \le \ \frac{1}{t}\sum_{\tau=1}^{t}\langle f'(x_\tau), x_\tau - x_* \rangle \le \frac{LR\sqrt{2\Omega}}{\sqrt{t}}, \tag{5.22}$$

$$\|x^t(R, x_1) - x_*\|^2 \ \le \ \frac{1}{t\kappa}\sum_{\tau=1}^{t}\langle f'(x_\tau), x_\tau - x_* \rangle \le \frac{LR\sqrt{2\Omega}}{\kappa\sqrt{t}}. \tag{5.23}$$

Proof. Observe that the modulus of strong convexity of the function $\omega^{R,x_1}(\cdot)$ w.r.t. the norm $\|\cdot\|_R = \|\cdot\|/R$ is 1, and the conjugate of the latter norm is $R\|\cdot\|_*$. Following the steps of the proof of Proposition 5.1, with $\|\cdot\|_R$ and $\omega^{R,x_1}(\cdot)$ in the roles of $\|\cdot\|$, respectively, we come to the analogue of (5.7) as follows:

$$\forall u \in \mathcal{X}: \sum_{\tau=1}^{t}\gamma_\tau\langle f'(x_\tau), x_\tau - u \rangle \le V_{x_1}^{R,x_1}(u) + \frac{R^2L^2}{2}\sum_{\tau=1}^{t}\gamma_\tau^2 \le \Omega + \frac{R^2L^2}{2}\sum_{\tau=1}^{t}\gamma_\tau^2.$$

Setting $u = x_*$ (so that $V^{R,x_1}(x_*) \le \Omega$ due to $\|x_1 - x_*\| \le R$), and substituting the value (5.21) of γ_τ, we come to (5.22). Further, from the strong convexity of f it follows that $\langle f'(x_\tau), x_\tau - x_* \rangle \ge \kappa\|x_\tau - x_*\|^2$, which combines with the definition of $x^t(R, x_1)$ to imply the first inequality in (5.23) (recall that γ_τ is independent of τ, so that $x^t(R, x_1) = \frac{1}{t}\sum_{\tau=1}^{t}x_\tau$). The second inequality in (5.23) follows from (5.22). □

Proposition 5.21 states that the smaller R is (i.e., the closer the initial guess x_1 is to x_*), the better the accuracy of the approximate solution $x^t(R, x_1)$ will be in terms of f and in terms of the distance to x_*. When the upper bound on this distance, as given by (5.22), becomes small, we can restart the MD using $x^t(\cdot)$ as the improved initial point, compute a new approximate solution, and so on. The algorithm below is a simple implementation of this idea.

Suppose that $x_1 \in \mathcal{X}$ and $R_0 \ge \|x_* - x_1\|$ are given. The algorithm is as follows:

1. *Initialization:* Set $y_0 = x_1$.
2. *Stage $k = 1, 2, \ldots$:* Set $N_k = \text{Ceil}(2^{k+2}\frac{L^2\Omega}{\kappa^2 R_0^2})$, where $\text{Ceil}(t)$ is the smallest integer $\ge t$, and compute $y_k = x^{N_k}(R_{k-1}, y_{k-1})$ according to (5.20), with $\gamma_t = \gamma^k := \frac{\sqrt{2\Omega}}{LR_{k-1}\sqrt{N_k}}$, $1 \le t \le N_k$. Set $R_k^2 = 2^{-k}R_0^2$ and pass to stage $k + 1$.

For the search points $x_1, ..., x_{N_k}$ of the kth stage of the method, we define

$$\delta_k = \frac{1}{N_k} \sum_{\tau=1}^{N_k} \langle f'(x_\tau), x_\tau - x_* \rangle.$$

Let k_* be the smallest integer such that $k \geq 1$ and $2^{k+2} \frac{L^2 \Omega}{\kappa^2 R_0^2} > k$, and let $M_k = \sum_{j=1}^{k} N_j$, $k = 1, 2, ...$. M_k is the total number of prox-steps carried out at the first k stages.

Proposition 5.4. *Setting $y_0 = x_1$, the points y_k, $k = 0, 1, ...$, generated by the above algorithm satisfy the following relations:*

$$\|y_k - x_*\|^2 \leq R_k^2 = 2^{-k} R_0^2, \tag{I_k}$$

$k = 0, 1, ...,$

$$f(y_k) - \text{Opt} \leq \delta_k \leq \kappa R_k^2 = \kappa 2^{-k} R_0^2, \tag{J_k}$$

$k = 1, 2, ...$ *As a result,*
(i) *When $1 \leq k < k_*$, one has $M_k \leq 5k$ and*

$$f(y_k) - \text{Opt} \leq \kappa 2^{-k} R_0^2; \tag{5.24}$$

(ii) *When $k \geq k_*$, one has*

$$f(y_k) - \text{Opt} \leq \frac{16 L^2 \Omega}{\kappa M_k}. \tag{5.25}$$

The proposition says that when the approximate solution y_k is far from x_*, the method converges linearly; when approaching x_*, it slows down and switches to the rate $O(1/t)$.

Proof. We prove (I_k), (J_k) by induction in k. (I_0) is valid due to $y_0 = x_1$ and the origin of R_0. Assume that for some $m \geq 1$ relations (I_k) and (J_k) are valid for $1 \leq k \leq m-1$, and prove that then (I_m), (J_m) are valid as well. Applying Proposition 5.3 with $R = R_{m-1}$, $x_1 = y_{m-1}$ (so that $\|x_* - x_1\| \leq R$ by (I_{m-1})) and $t = N_m$, we get

$$(a): f(y_m) - \text{Opt} \leq \delta_m \leq \frac{L R_{m-1} \sqrt{2\Omega}}{\sqrt{N_m}}, \quad (b): \|y_m - x_*\|^2 \leq L R_{m-1} \frac{\sqrt{2\Omega}}{\kappa \sqrt{N_m}}.$$

Since $R_{m-1}^2 = 2^{1-m} R_0^2$ by (I_{m-1}) and $N_m \geq 2^{m+2} \frac{L^2 \Omega}{\kappa^2 R_0^2}$, (b) implies (I_m) and (a) implies (J_m). Induction is completed.

Now prove that $M_k \leq 5k$ for $1 \leq k < k_*$. For such a k and for $1 \leq j \leq k$ we have $N_j = 1$ when $2^{j+2} \frac{L^2 \Omega}{\kappa^2 R_0^2} < 1$; let it be so for $j < j_*$; and $N_j \leq 2^{j+3} \frac{L^2 \Omega}{\kappa^2 R_0^2}$ for $j_* \leq j \leq k$. It follows that when $j_* > k$, we have $M_k = k$. When $j_* \leq k$, we have $M := \sum_{j=j_*}^{k} N_j \leq 2^{k+4} \frac{L^2\Omega}{\kappa^2 R_0^2} \leq 4k$ (the concluding inequality is due to $k < k_*$), whence $M_k = j_* - 1 + M \leq 5k$, as claimed. Invoking (J_k), we arrive at (i).

To prove (ii), let $k \geq k_*$, whence $N_k \geq k + 1$. We have

$$2^{k+3} \frac{L^2\Omega}{\kappa^2 R_0^2} > \sum_{j=1}^{k} 2^{j+2} \frac{L^2\Omega}{\kappa^2 R_0^2} \geq \sum_{j=1}^{k} (N_j - 1) = M_k - k \geq M_k/2,$$

where the concluding $\geq$ stems from the fact that $N_k \geq k + 1$, and therefore $M_k \geq \sum_{j=1}^{k-1} N_j + N_k \geq (k-1) + (k+1) = 2k$. Thus $M_k \leq 2^{k+4} \frac{L^2\Omega}{\kappa^2 R_0^2}$, that is, $2^{-k} \leq \frac{16L^2\Omega}{M_k \kappa^2 R_0^2}$, and the right-hand side of (J_k) is $\leq \frac{16L^2\Omega}{M_k \kappa}$. $\qquad\square$

5.5   Mirror Descent Stochastic Approximation

The MD algorithm can be extended to the case when the objective f in (5.2) is given by the *stochastic oracle*—a routine which at tth call, the query point being $x_t \in \mathcal{X}$, returns a vector $G(x_t, \xi_t)$, where $\xi_1, \xi_2, \ldots$ are independent, identically distributed oracle noises. We assume that for all $x \in \mathcal{X}$ it holds that

$$\mathbf{E}\left\{\|G(x,\xi)\|_*^2\right\} \leq L^2 < \infty \ \& \ \|g(x) - f'(x)\|_* \leq \mu, \ g(x) = \mathbf{E}\{G(x,\xi)\}. \quad (5.26)$$

In (5.6), replacing the subgradients $f'(x_t)$ with their stochastic estimates $G(x_t, \xi_t)$, we arrive at *robust mirror descent stochastic approximation* (RMDSA). The convergence properties of this procedure are presented in the following counterpart of Proposition 5.1:

Proposition 5.5. *Let $\mathcal{X}$ be bounded. Given an integer $N \geq 1$, consider N-step RMDSA with the stepsizes*

$$\gamma_t = \sqrt{2\Omega}/[L\sqrt{N}], \ 1 \leq t \leq N. \quad (5.27)$$

Then

$$\mathbf{E}\left\{f(x^N) - \mathrm{Opt}\right\} \leq \sqrt{2\Omega}L/\sqrt{N} + 2\sqrt{2\Omega}\mu. \quad (5.28)$$

Proof. Let $\xi^t = [\xi_1; \ldots; \xi_t]$, so that x_t is a deterministic function of ξ^{t-1}. Exactly the same reasoning as in the proof of Proposition 5.1 results in the following analogy of (5.7):

$$\sum_{\tau=1}^{N} \gamma_\tau \langle G(x_\tau, \xi_\tau), x_\tau - x_* \rangle \leq \Omega + \tfrac{1}{2} \sum_{\tau=1}^{N} \gamma_\tau^2 \|G(x_\tau, \xi_\tau)\|_*^2. \quad (5.29)$$

Observe that x_τ is a deterministic function of ξ^{t-1}, so that

$$\mathbf{E}_{\xi_\tau}\{\langle G(x_\tau, \xi_\tau), x_\tau - x_* \rangle\} = \langle g(x_\tau), x_\tau - x_* \rangle \geq \langle f'(x_\tau), x_\tau - x_* \rangle - \mu D,$$

where $D = \max_{x, x' \in \mathcal{X}} \|x - x'\|$ is the $\|\cdot\|$-diameter of $\mathcal{X}$. Now, taking expectations of both sides of (5.29), we get

$$\mathbf{E}\left\{\sum\nolimits_{\tau=1}^{N} \gamma_\tau \langle f'(x_\tau), x_\tau - x_* \rangle \right\} \leq \Omega + \frac{L^2}{2}\sum\nolimits_{\tau=1}^{N} \gamma_\tau^2 + \mu D \sum\nolimits_{\tau=1}^{N} \gamma_\tau.$$

In the same way as in the proof of Proposition 5.1 we conclude that the left-hand side in this inequality is $\geq [\sum_{\tau=1}^{N} \gamma_\tau]\mathbf{E}\{f(x^N) - \text{Opt}\}$, so that

$$\mathbf{E}\{f(x^N) - \text{Opt}\} \leq \frac{\Omega + \frac{L^2}{2}\sum_{\tau=1}^{N} \gamma_\tau^2}{\sum_{\tau=1}^{N} \gamma_\tau} + \mu D. \tag{5.30}$$

Observe that when $x \in \mathcal{X}$, we have $\omega(x) - \omega(x_1) - \langle \omega'(x_1), x - x_1 \rangle \geq \frac{1}{2}\|x - x_1\|^2$ by the strong convexity of ω, and $\omega(x) - \omega(x_1) - \langle \omega'(x_1), x - x_1 \rangle \leq \omega(x) - \omega(x_1) \leq \Omega$ (since $x_1 = \text{argmin}_\mathcal{X}\, \omega$, and thus $\langle \omega'(x_1), x - x_1 \rangle \geq 0$). Thus, $\|x - x_1\| \leq \sqrt{2\Omega}$ for every $x \in \mathcal{X}$, whence $D := \max_{x, x' \in \mathcal{X}} \|x - x'\| \leq 2\sqrt{2\Omega}$. This relation combines with (5.30) and (5.27) to imply (5.28). $\qquad\square$

5.6 Mirror Descent for Convex-Concave Saddle-Point Problems

Now we shall demonstrate that the MD scheme can be naturally extended from problems of convex minimization to the *convex-concave saddle-point problems*.

5.6.1 Preliminaries

Convex-concave Saddle-Point Problem. A convex-concave saddle-point (c.-c.s.p.) problem reads

$$\text{SadVal} = \inf_{x \in \mathcal{X}} \sup_{y \in \mathcal{Y}} \phi(x, y), \tag{5.31}$$

where $\mathcal{X} \subset E_x$, $\mathcal{Y} \subset E_y$ are nonempty closed convex sets in the respective Euclidean spaces E_x and E_y. The *cost function* $\phi(x, y)$ is continuous on $\mathcal{Z} = \mathcal{X} \times \mathcal{Y} \in E = E_x \times E_y$ and convex in the variable $x \in \mathcal{X}$ and concave in the variable $y \in \mathcal{Y}$; the quantity SadVal is called the *saddle-point value* of ϕ on $\mathcal{Z}$. By definition, (precise) solutions to (5.31) are *saddle points* of ϕ on $\mathcal{Z}$, that is, points $(x_*, y_*) \in \mathcal{Z}$ such that $\phi(x, y_*) \geq \phi(x_*, y_*) \geq \phi(x_*, y)$ for all $(x, y) \in \mathcal{Z}$. The data of problem (5.31) give rise to a *primal-dual pair of* convex optimization problems

$$\mathrm{Opt}(P) \;=\; \min_{x \in \mathcal{X}} \overline{\phi}(x), \quad \overline{\phi}(x) = \sup_{y \in \mathcal{Y}} \phi(x,y) \qquad (P)$$

$$\mathrm{Opt}(D) \;=\; \max_{y \in \mathcal{Y}} \underline{\phi}(y), \quad \underline{\phi}(y) = \inf_{x \in \mathcal{X}} \phi(x,y). \qquad (D)$$

ϕ possesses saddle-points on $\mathcal{Z}$ if and only if problems (P) and (D) are solvable with equal optimal values. Whenever saddle-points exist, they are exactly the pairs (x_*, y_*) comprising optimal solutions x_*, y_* to the respective problems (P) and (D), and for every such pair (x_*, y_*) we have

$$\phi(x_*, y_*) = \overline{\phi}(x_*) = \mathrm{Opt}(P) = \mathrm{SadVal} := \inf_{x \in \mathcal{X}} \sup_{y \in \mathcal{Y}} \phi(x,y)$$

$$= \sup_{y \in \mathcal{Y}} \inf_{x \in \mathcal{X}} \phi(x,y) = \mathrm{Opt}(D) = \underline{\phi}(y_*).$$

From now on, we assume that (5.31) is solvable.

Remark 5.1. *With our basic assumptions on ϕ (continuity and convexity-concavity on $\mathcal{X} \times \mathcal{Y}$) and on $\mathcal{X}, \mathcal{Y}$ (nonemptiness, convexity and closedness), (5.31) definitely is solvable either if $\mathcal{X}$ and $\mathcal{Y}$ are bounded, or if both $\mathcal{X}$ and all level sets $\{y \in \mathcal{Y} : \underline{\phi}(y) \geq a\}$, $a \in \mathbb{R}$, of $\underline{\phi}$ are bounded; these are the only situations we are about to consider in this chapter and in Chapter 6.*

Saddle-Point Accuracy Measure. A natural way to quantify the accuracy of a candidate solution $z = (x, y) \in \mathcal{Z}$ to the c.-c.s.p. problem (5.31) is given by the *gap*

$$\begin{aligned}
\epsilon_{\mathrm{sad}}(z) &= \sup_{\eta \in \mathcal{Y}} \phi(x, \eta) - \inf_{\xi \in \mathcal{X}} \phi(\xi, y) = \overline{\phi}(x) - \underline{\phi}(y) \\
&= \left[\overline{\phi}(x) - \mathrm{Opt}(P)\right] + \left[\mathrm{Opt}(D) - \underline{\phi}(y)\right]
\end{aligned} \qquad (5.32)$$

where the concluding equality is given by the fact that, by our standing assumption, ϕ has a saddle point and thus $\mathrm{Opt}(P) = \mathrm{Opt}(D)$. We see that $\epsilon_{\mathrm{sad}}(x, y)$ is the sum of nonoptimalities, in terms of the respective objectives: of x as an approximate solution to (P) and of y as an approximate solution to (D).

Monotone Operator Associated with (5.31). Let $\partial_x \phi(x, y)$ be the set of all subgradients w.r.t. $\mathcal{X}$ of (the convex function) $\phi(\cdot, y)$, taken at a point $x \in \mathcal{X}$, and let $\partial_y[-\phi(x, y)]$ be the set of all subgradients w.r.t. $\mathcal{Y}$ (of the convex function) $-\phi(x, \cdot)$, taken at a point $y \in \mathcal{Y}$. We can associate with ϕ the point-to-set operator

$$\Phi(x, y) = \{\Phi_x(x, y) = \partial_x \phi(x, y)\} \times \{\Phi_y(x, y) = \partial_y[-\phi(x, y)]\}.$$

The domain Dom $\Phi := \{(x,y) : \Phi(x,y) \neq \emptyset\}$ of this operator comprises all pairs $(x,y) \in \mathcal{Z}$ for which the corresponding subdifferentials are nonempty; it definitely contains the relative interior rint $\mathcal{Z} = \text{rint } \mathcal{X} \times \text{rint } \mathcal{Y}$ of $\mathcal{Z}$, and the values of Φ in its domain are direct products of nonempty closed convex sets in E_x and E_y. It is well known (and easily seen) that Φ is monotone:

$$\forall (z, z' \in \text{Dom } \Phi, F \in \Phi(z), F' \in \Phi(z')) : \langle F - F', z - z' \rangle \geq 0,$$

and the saddle points of ϕ are exactly the points z_* such that $0 \in \Phi(z_*)$. An equivalent characterization of saddle points, more convenient in our context, is as follows: z_* is a saddle point of ϕ if and only if for some (and then for every) *selection* $F(\cdot)$ of Φ (i.e., a vector field $F(z) : \text{rint } \mathcal{Z} \to E$ such that $F(z) \in \Phi(z)$ for every $z \in \text{rint } \mathcal{Z}$) one has

$$\langle F(z), z - z_* \rangle \geq 0 \, \forall z \in \text{rint } \mathcal{Z}. \tag{5.33}$$

5.6.2   Saddle-Point Mirror Descent

Here we assume that $\mathcal{Z}$ is bounded and ϕ is Lipschitz continuous on $\mathcal{Z}$ (whence, in particular, the domain of the associated monotone operator Φ is the entire $\mathcal{Z}$).

The setup of the MP algorithm involves a norm $\|\cdot\|$ on the embedding space $E = E_x \times E_y$ of $\mathcal{Z}$ and a d.-g.f. $\omega(\cdot)$ for $\mathcal{Z}$ compatible with this norm. For $z \in \mathcal{Z}^o$, $u \in \mathcal{Z}$ let (cf. (5.4))

$$V_z(u) = \omega(u) - \omega(z) - \langle \omega'(z), u - z \rangle,$$

and let $z_c = \text{argmin}_{u \in \mathcal{Z}} \omega(u)$. We assume that given $z \in \mathcal{Z}^o$ and $\xi \in E$, it is easy to compute the prox-mapping

$$\text{Prox}_z(\xi) = \underset{u \in \mathcal{Z}}{\text{argmin}} \left[\langle \xi, u \rangle + V_z(u) \right] \left(= \underset{u \in \mathcal{Z}}{\text{argmin}} \left[\langle \xi - \omega'(z), u \rangle + \omega(u) \right] \right).$$

We denote, by $\Omega = \max_{u \in \mathcal{Z}} V_{z_c}(u) \leq \max_{\mathcal{Z}} \omega(\cdot) - \min_{\mathcal{Z}} \omega(\cdot)$, the $\omega(\cdot)$-diameter of $\mathcal{Z}$ (cf. Section 5.2.2).

Let a first-order oracle for ϕ be available, so that for every $z = (x,y) \in \mathcal{Z}$ we can compute a vector $F(z) \in \Phi(z = (x,y)) := \{\partial_x \phi(x,y)\} \times \{\partial_y [-\phi(x,y)]\}$. The saddle-point MD algorithm is given by the recurrence

$$\begin{aligned} (a) : \quad & z_1 = z_c, \\ (b) : \quad & z_{\tau+1} = \text{Prox}_{z_\tau}(\gamma_\tau F(z_\tau)), \\ (c) : \quad & z^\tau = [\textstyle\sum_{s=1}^{\tau} \gamma_s]^{-1} \sum_{s=1}^{\tau} \gamma_s w_s, \end{aligned} \tag{5.34}$$

where $\gamma_\tau > 0$ are the stepsizes. Note that $z_\tau, w_\tau \in \mathcal{Z}^o$, whence $z^t \in \mathcal{Z}$.

The convergence properties of the algorithm are given by the following.

Proposition 5.6. *Suppose that $F(\cdot)$ is bounded on $\mathcal{Z}$, and L is such that $\|F(z)\|_* \leq L$ for all $z \in \mathcal{Z}$.*

(i) For every $t \geq 1$ it holds that

$$\epsilon_{\text{sad}}(z^t) \leq \left[\sum_{\tau=1}^{t}\gamma_\tau\right]^{-1}\left[\Omega + \frac{L^2}{2}\sum_{\tau=1}^{t}\gamma_\tau^2\right]. \tag{5.35}$$

(ii) As a consequence, the N-step MD algorithm with constant stepsizes $\gamma_\tau = \gamma/L\sqrt{N}$, $\tau = 1, ..., N$ satisfies

$$\epsilon_{\text{sad}}(z^N) \leq \frac{L}{\sqrt{N}}\left[\frac{\Omega}{\gamma} + \frac{L\gamma}{2}\right].$$

In particular, the N-step MD. algorithm with constant stepsizes $\gamma_\tau = L^{-1}\sqrt{\frac{2\Omega}{N}}$, $\tau = 1, ..., N$ satisfies

$$\epsilon_{\text{sad}}(z^N) \leq L\sqrt{\frac{2\Omega}{N}}.$$

Proof. By the definition $z_{\tau+1} = \text{Prox}_{z_\tau}(\gamma_\tau F(z_\tau))$ we get

$$\forall u \in \mathcal{Z}, \ \gamma_\tau\langle F(z_\tau), z_\tau - u\rangle \leq V_{z_\tau}(u) - V_{z_{\tau+1}}(u) + \gamma_\tau^2\|F(z_\tau)\|_*^2/2.$$

(It suffices to repeat the derivation of (5.13) in the proof of Proposition 5.1 with $f'(x_\tau)$, x_τ, and $x_{\tau+1}$ substituted, respectively, with $F(z_\tau)$, z_τ, and $z_{\tau+1}$.) When summing for $i = 1, ..., t$ we get, for all $u \in \mathcal{Z}$:

$$\sum_{\tau=1}^{t}\gamma_\tau\langle F(z_\tau), z_\tau - u\rangle \leq V_{z_1}(u) + \sum_{\tau=1}^{t}\gamma_\tau^2\|F(z_\tau)\|_*^2/2 \leq \Omega + \frac{L^2}{2}\sum_{\tau=1}^{t}\gamma_\tau^2 \tag{5.36}$$

Let $z_\tau = (x_\tau, y_\tau)$, $z^t = (x^t, y^t)$, and $\lambda_\tau = \left[\sum_{s=1}^{t}\gamma_s\right]^{-1}\gamma_\tau$. Note that $\sum_{s=1}^{t}\lambda_s = 1$, and for

$$\sum_{\tau=1}^{t}\lambda_\tau\langle F(z_\tau), z_\tau - u\rangle = \sum_{\tau=1}^{t}\lambda_\tau\left[\langle\nabla_x\phi(x_\tau, y_\tau), x_\tau - x\rangle + \langle\nabla_y\phi(x_\tau, y_\tau), y - y_\tau\rangle\right]$$

we have

$$\begin{aligned}
&\sum_{\tau=1}^{t}\lambda_\tau\left[\langle\nabla_x\phi(x_\tau, y_\tau), x_\tau - x\rangle + \langle\nabla_y\phi(x_\tau, y_\tau), y - y_\tau\rangle\right]\\
&\geq \sum_{\tau=1}^{t}\lambda_\tau\left[[\phi(x_\tau, y_\tau) - \phi(x, y_\tau)] + [\phi(x_\tau, y) - \phi(x_\tau, y_\tau)]\right] \quad (a)\\
&= \sum_{\tau=1}^{t}\lambda_\tau[\phi(x_\tau, y) - \phi(x, y_\tau)]\\
&\geq \phi(\sum_{\tau=1}^{t}\lambda_\tau x_\tau, y) - \phi(x, \sum_{\tau=1}^{t}\lambda_\tau y_\tau) = \phi(x^t, y) - \phi(x, y^t) \quad (b)
\end{aligned} \tag{5.37}$$

(inequalities in (a) and (b) are due to the convexity-concavity of ϕ). Thus

(5.36) results in

$$\phi(x^t, y) - \phi(x, y^t) \le \frac{\Omega + \frac{L^2}{2} \sum_{\tau=1}^{t} \gamma_\tau^2}{\sum_{\tau=1}^{t} \gamma_\tau} \quad \forall (x, y) \in \mathcal{Z}.$$

Taking the supremum in $(x, y) \in \mathcal{Z}$, we arrive at (5.35). $\square$

5.7 Setting up a Mirror Descent Method

An advantage of the mirror descent scheme is that its degrees of freedom (the norm $\|\cdot\|$ and the d.-g.f. $\omega(\cdot)$) allow one to adjust the method, to some extent, to the geometry of the problem under consideration. This is the issue we are focusing on in this section. For the sake of definiteness, we restrict ourselves to the minimization problem (5.2); the saddle-point case (5.31) is completely similar, with $\mathcal{Z}$ in the role of $\mathcal{X}$.

5.7.1 Building blocks

The basic MD setups are as follows:

1. *Euclidean setup:* $\|\cdot\| = \|\cdot\|_2$, $\omega(x) = \frac{1}{2} x^T x$.

2. *ℓ_1-setup:* For this setup, $E = \mathbb{R}^n$, $n > 1$, and $\|\cdot\| = \|\cdot\|_1$. As for $\omega(\cdot)$, there could be several choices, depending on what $\mathcal{X}$ is:

   (a) When $\mathcal{X}$ is unbounded, seemingly the only good choice is $\omega(x) = C \ln(n) \|x\|_{p(n)}^2$ with $p(n) = 1 + \frac{1}{2\ln(n)}$, where an *absolute constant* C is chosen in a way which ensures (5.3) (one can take $C = e$).

   (b) When $\mathcal{X}$ is bounded, assuming w.l.o.g. that $\mathcal{X} \subset B^{n,1} := \{x \in \mathbb{R}^n : \|x\|_1 \le 1\}$, one can set $\omega(x) = C \ln(n) \sum_{i=1}^{n} |x_i|^{p(n)}$ with the same as above value of $p(n)$ and $C = 2e$.

   (c) When $\mathcal{X}$ is a part of the simplex $S_n^+ = \{x \in \mathbb{R}_+^n : \sum_{i=1}^{n} x_i \le 1\}$ (or the flat simplex $S_n = \{x \in \mathbb{R}_+^n : \sum_{i=1}^{n} x_i = 1\}$) intersecting int $\mathbb{R}_+^n$, a good choice of $\omega(x)$ is the entropy

$$\omega(x) = \operatorname{Ent}(x) := \sum_{i=1}^{n} x_i \ln(x_i). \tag{5.38}$$

3. *Matrix setup:* This is the matrix analogy of the ℓ_1-setup. Here the embedding space E of $\mathcal{X}$ is the space $\mathbf{S}^\nu$ of block-diagonal symmetric matrices with fixed block-diagonal structure $\nu = [\nu_1; ...; \nu_k]$ (k diagonal blocks of row sizes $\nu_1, ..., \nu_k$). $\mathbf{S}^\nu$ is equipped with the Frobenius inner product $\langle X, Y \rangle = \operatorname{Tr}(XY)$ and the trace norm $|X|_1 = \|\lambda(X)\|_1$, where $\lambda(X)$ is the vector of eigenvalues (taken with their multiplicities in the

nonascending order) of a symmetric matrix X. The d.-g.f.s are the matrix analogies of those for the ℓ_1-setup. Specifically,

(a) When $\mathcal{X}$ is unbounded, we set $\omega(X) = C \ln(|\nu|) \|\lambda(X)\|_{p(|\nu|)}^2$, where $|\nu| = \sum_{\ell=1}^{k} \nu_\ell$ is the total row size of matrices from $\mathbf{S}^\nu$, and C is an appropriate absolute constant which ensures (5.3) (one can take $C = 2e$).

(b) When $\mathcal{X}$ is bounded, assuming w.l.o.g. that $\mathcal{X} \subset B^{\nu,1} = \{X \in \mathbf{S}^\nu : |X|_1 \leq 1\}$, we can take $\omega(X) = 4e \ln(|\nu|) \sum_{i=1}^{|\nu|} |\lambda_i(X)|^{p(|\nu|)}$.

(c) When $\mathcal{X}$ is a part of the *spectahedron* $\Sigma_\nu^+ = \{X \in \mathbf{S}^\nu : X \succeq 0, \mathrm{Tr}(X) \leq 1\}$ (or the flat spectahedron $\Sigma_\nu = \{X \in \mathbf{S}^\nu : X \succeq 0, \mathrm{Tr}(X) = 1\}$) intersecting the interior $\{X \succ 0\}$ of the positive semidefinite cone $\mathbf{S}_+^\nu = \{X \in \mathbf{S}^\nu : X \succeq 0\}$, one can take $\omega(X)$ as the matrix entropy: $\omega(X) = 2\mathrm{Ent}(\lambda(X)) = 2\sum_{i=1}^{|\nu|} \lambda_i(X) \ln(\lambda_i(X))$.

Note that the ℓ_1-setup can be viewed as a particular case of the matrix setup, corresponding to the case when the block-diagonal matrices in question are diagonal, and we identify a diagonal matrix with the vector of its diagonal entries.

With the outlined setups, the simplicity assumption holds, provided that $\mathcal{X}$ is simple enough. Specifically:

■ Within the Euclidean setup, $\mathrm{Prox}_x(\xi)$ is the metric projection of the vector $x - \xi$ onto $\mathcal{X}$ (that is, the point of $\mathcal{X}$ which is the closest to $x - \xi$ in ℓ_2-norm). Examples of sets $\mathcal{X} \subset \mathbb{R}^n$ for which metric projection is easy include $\|\cdot\|_p$-balls and intersections of $\|\cdot\|_p$-balls centered at the origin with the nonnegative orthant $\mathbb{R}_+^n$.

■ Within the ℓ_1-setup, computing the prox-mapping is reasonably easy
—in the case of 2a, when $\mathcal{X}$ is the entire $\mathbb{R}^n$ or $\mathbb{R}_+^n$,
—in the case of 2b, when $\mathcal{X}$ is $B^{n,1}$ or $B^{n,1} \cap \mathbb{R}_+^n$,
—in the case of 2c, when $\mathcal{X}$ is the entire S_n^+ or S_n.
With the indicated sets $\mathcal{X}$, in the cases of 2a and 2b computing the prox-mapping requires solving auxiliary one- or two-dimensional convex problems, which can be done within machine accuracy by, e.g., the ellipsoid algorithm in $O(n)$ operations (cf. Nemirovsky and Yudin, 1983, Chapter 2). In the case of 2c, the prox-mappings are given by the explicit formulas

$$\mathcal{X} = \mathrm{S}_n^+ \Rightarrow \mathrm{Prox}_x(\xi) = \begin{cases} [x_1 e^{\xi_1 - 1}; \ldots; x_n e^{\xi_n - 1}], & \sum_i e^{\eta_i - 1} \leq 1 \\ [\sum_i x_i e^{\xi_i}]^{-1} [x_1 e^{\eta_1}; \ldots; x_n e^{\eta_n}], & \text{otherwise} \end{cases}$$
$$\mathcal{X} = \mathrm{S}_n \Rightarrow \mathrm{Prox}_x(\xi) = [\sum_i x_i e^{\xi_i}]^{-1} [x_1 e^{\eta_1}; \ldots; x_n e^{\eta_n}].$$

$$(5.39)$$

▪ Within the matrix setup, computing the prox-mapping is relatively easy
—in the case of 3a, when $\mathcal{X}$ is the entire $\mathbf{S}^\nu$ or the positive semidefinite cone
$\mathbf{S}^\nu_+ = \{X \in \mathbf{S}^\nu : X \succeq 0\}$,
—in the case of 3b, when $\mathcal{X}$ is the entire $B^{\nu,1}$ or the intersection of $B^{\nu,1}$ with
$\mathbf{S}^\nu_+$,
—in the case of 3c, when $\mathcal{X}$ is the entire spectahedron Σ^+_ν or Σ_ν.
Indeed, in the cases, outlined above, computing $W = \mathrm{Prox}_X(\Xi)$ reduces to
computing the eigenvalue decomposition of the matrix X (which allows one
to get $\omega'(X)$), and subsequent eigenvalue decomposition of the matrix $H =$
$\Xi - \omega'(X)$: $H = U \mathrm{Diag}\{h\}U^T$ (here $\mathrm{Diag}(A)$ stands for the diagonal matrix
with the same diagonal as A). It is easily seen that in the cases in question,
$$W = U\,\mathrm{Diag}\{w\}U^T, \quad w = \underset{z:\,\mathrm{Diag}\{z\}\in\mathcal{X}}{\mathrm{argmin}}\ \{\langle\mathrm{Diag}\{h\},\mathrm{Diag}\{z\}\rangle + \omega(\mathrm{Diag}\{z\})\},$$
and the latter problem is exactly the one arising in the ℓ_1-setup.

Illustration: Euclidean setup vs. ℓ_1-setup. To illustrate the ability of the
MD scheme to adjust, to some extent, the method to the problem's geometry,
consider problem (5.2) when $\mathcal{X}$ is the unit $\|\cdot\|_p$-ball in $\mathbb{R}^n$, where $p = 1$ or
$p = 2$, and compare the respective performances of the Euclidean and the
ℓ_1-setups. (To make optimization over the unit Euclidean ball $B^{n,2}$ available
for the ℓ_1-setup, we pass from $\min_{\|x\|_2 \leq 1} f(x)$ to the equivalent problem
$\min_{\|u\|_2 \leq n^{-1/2}} f(n^{1/2}u)$ and use the setup from Section 5.7.1, item 2b.) The ratio
of the corresponding efficiency estimates (the right-hand sides in (5.11))
within an absolute constant factor is

$$\Theta := \frac{\mathrm{EffEst(Eucl)}}{\mathrm{EffEst}(\ell_1)} = \underbrace{\frac{1}{n^{1-1/p}\sqrt{\ln(n)}}}_{A} \cdot \underbrace{\frac{\sup_{x\in\mathcal{X}}\|f'(x)\|_2}{\sup_{x\in\mathcal{X}}\|f'(x)\|_{1\infty}}}_{B}.$$

Note that $\Theta \ll 1$ means that the MD with the Euclidean setup significantly
outperforms the MD with the ℓ_1-setup, while $\Theta \gg 1$ means exactly the
opposite. Now, A is ≤ 1 and thus is always in favor of the Euclidean setup,
and is as small as $1/\sqrt{n\ln(n)}$ when $\mathcal{X}$ is the Euclidean ball ($p = 2$). The
factor B is in favor of the ℓ_1-setup—it is ≥ 1 and $\leq \sqrt{n}$, and can well be of
the order of $\sqrt{n}$ (look what happens when all entries in $f'(x)$ are of the same
order of magnitude). Which one of the factors overweights depends on f;
however, a reasonable choice can be made independently of the fine structure
of f. Specifically, when $\mathcal{X}$ is the Euclidean ball, the factor $A = 1/\sqrt{n\ln n}$ is so
small that the product AB definitely is ≤ 1, that is, the situation is in favor
of the Euclidean setup. In contrast to this, when $\mathcal{X}$ is the ℓ_1-ball ($p = 1$),
A is nearly constant—just $O(1/\sqrt{\ln(n)})$, since B can be as large as $\sqrt{n}$,
the situation is definitely in favor of the ℓ_1-setup—it can be outperformed

by the Euclidean setup only marginally (by the factor $\leq \sqrt{\ln n}$), and it has a reasonable chance to outperform its adversary quite significantly, by the factor $O(\sqrt{n/\ln(n)})$. Thus, there are all reasons to select the Euclidean setup when $p = 2$ and the ℓ_1-setup when $p = 1$.[6]

5.7.2   Favorable Geometry Case

Consider the case when the domain $\mathcal{X}$ of (5.2) is bounded and, moreover, is a *subset* of the direct product $\mathcal{X}^+$ of standard blocks:

$$\mathcal{X}^+ = \mathcal{X}_1 \times ... \times \mathcal{X}_K \in E_1 \times ... \times E_K, \tag{5.40}$$

where for every $\ell = 1, ..., K$ the pair $(\mathcal{X}_\ell, E_\ell \supset \mathcal{X}_\ell)$ is

- either a *ball block*, that is, $E_\ell = \mathbb{R}^{n_\ell}$ and $\mathcal{X}_\ell$ is either the unit Euclidean ball $B^{n_\ell,2} = \{x \in \mathbb{R}^{n_\ell} : \|x\|_2 \leq 1\}$ in E_ℓ, or the intersection of this ball with $\mathbb{R}^{n_\ell}_+$;

- or a *spectahedron block*, that is, $E_\ell = \mathbf{S}^{\nu^\ell}$ is the space of block-diagonal symmetric matrices with block-diagonal structure ν^ℓ, and $\mathcal{X}_\ell$ is either the unit trace-norm ball $\{X \in \mathbf{S}^{\nu^\ell} : |X|_1 \leq 1\}$, or the intersection of this ball with $\mathbf{S}^{\nu^\ell}_+$, or the spectahedron $\Sigma^+_{\nu^\ell} = \{X \in \mathbf{S}^{\nu^\ell}_+ : \mathrm{Tr}(X) \leq 1\}$, or the flat spectahedron $\Sigma_{\nu^\ell} = \{X \in \mathbf{S}^{\nu^\ell}_+ : \mathrm{Tr}(X) = 1\}$.

Note that according to our convention of identifying vectors with diagonals of diagonal matrices, we allow for some of $\mathcal{X}_\ell$ to be the unit ℓ_1-balls, or their nonnegative parts, or simplexes—they are nothing but spectahedron blocks with purely diagonal structure ν^ℓ.

We equip the embedding spaces E_ℓ of blocks with the natural inner products (the standard inner products when $E_\ell = \mathbb{R}^{n_\ell}$ and the Frobenius inner product when $E_\ell = \mathbf{S}^{\nu^\ell}$) and norms $\|\cdot\|_{(\ell)}$ (the standard Euclidean norm when $E_\ell = \mathbb{R}^{n_\ell}$ and the trace-norm when $E_\ell = \mathbf{S}^{\nu^\ell}$), and the standard

---

6. In fact, with this recommendation we get *theoretically unimprovable*, in terms of the information-based complexity theory, methods for large-scale nonsmooth convex optimization on Euclidean and ℓ_1-balls (for details, see Nemirovsky and Yudin, 1983; Ben-Tal et al., 2001). Numerical experiments reported in Ben-Tal et al. (2001) and Nemirovski et al. (2009) seem to fully support the advantages of the ℓ_1-setup when minimizing over large-scale simplexes.

blocks $\mathcal{X}_\ell$ with d.-g.f.'s

$$
\omega_\ell(x^\ell) = \begin{cases} \frac{1}{2}[x^\ell]^T x^\ell, & \mathcal{X}_\ell \text{ is a ball block} \\[2mm] 4e\ln(|\nu^\ell|)\sum_i |\lambda_i(X^\ell)|^{p(|\nu^\ell|)}, & \begin{array}{l} \mathcal{X}_\ell \text{ is the unit } |\cdot|_1 \text{ ball } B^{\nu^\ell,1} \text{ in} \\ E_\ell = \mathbf{S}^{\nu^\ell}, \text{ or } B^{\nu^\ell,1} \cap \mathbf{S}^{\nu^\ell}_+ \end{array} \\[2mm] 2\text{Ent}(\lambda(X^\ell)), & \begin{array}{l} \mathcal{X}_\ell \text{ is the spectahedron } (\Sigma^+_{\nu^\ell} \text{ or} \\ \Sigma_{\nu^\ell}) \text{ in } E_\ell = \mathbf{S}^{\nu^\ell} \end{array} \end{cases}
$$

$$(5.41)$$

(cf. Section 5.7.1). Finally, the embedding space $E = E_1 \times \dots \times E_K$ of $\mathcal{X}^+$ (and thus of $\mathcal{X} \subset \mathcal{X}^+$) is equipped with the direct product type Euclidean structure induced by the inner products on $E_1, \dots, E_K$ and with the norm

$$
\|(x^1, \dots, x^K)\| = \sqrt{\sum_{\ell=1}^K \alpha_\ell \|x^\ell\|^2_{(\ell)}} \tag{5.42}
$$

where $\alpha_\ell > 0$ are construction parameters. $\mathcal{X}^+$ is equipped with the d.-g.f.

$$
\omega(x^1, \dots, x^K) = \sum_{\ell=1}^K \alpha_\ell \omega_\ell(x^\ell) \tag{5.43}
$$

which, it is easy to see, is compatible with the norm $\|\cdot\|$.

Assuming from now on that $\mathcal{X}$ intersects the relative interior rint $\mathcal{X}^+$, the restriction of $\omega(\cdot)$ onto $\mathcal{X}$ is a d.-g.f. for $\mathcal{X}$ compatible with the norm $\|\cdot\|$ on the space E embedding $\mathcal{X}$, and we can solve (5.2) by the MD algorithm associated with $\|\cdot\|$ and $\omega(\cdot)$. Let us optimize the efficiency estimate of this algorithm over the parameters α_ℓ of our construction. For the sake of definiteness, consider the case where f is represented by a deterministic first-order oracle (the tuning of the MD setup in the case of the stochastic oracle is being completely similar). To this end, assume that we have at our disposal upper bounds $L_\ell < \infty$, $1 \le \ell \le K$, on the quantities $\|f'_{x^\ell}(x^1, \dots, x^K)\|_{(\ell),*}$, $x = (x^1, \dots, x^K) \in \mathcal{X}$. Here $f'_{x^\ell}(x)$ is the projection of $f'(x)$ onto E_ℓ and $\|\cdot\|_{(\ell),*}$ is the norm on E_ℓ conjugate to $\|\cdot\|_{(\ell)}$ (that is, $\|\cdot\|_{(\ell),*}$ is the standard Euclidean norm $\|\cdot\|_2$ on E_ℓ when $E_\ell = \mathbb{R}^{n_\ell}$, and $\|\cdot\|_{(\ell),*}$ is the standard matrix norm (maximal singular value) when $E_\ell = \mathbf{S}^{\nu^\ell}$). The norm $\|\cdot\|_*$ conjugate to the norm $\|\cdot\|$ on E is

$$
\begin{aligned} \|(\xi^1, \dots, \xi^K)\|_* &= \sqrt{\sum_{\ell=1}^K \alpha_\ell^{-1} \|\xi^\ell\|^2_{(\ell),*}} \\ \Rightarrow (\forall x \in \mathcal{X}): \|f'(x)\|_* &\le L := \sqrt{\sum_{\ell=1}^K \alpha_\ell^{-1} L_\ell^2}. \end{aligned} \tag{5.44}
$$

The quantity we need to minimize in order to get as efficient an MD method as possible within our framework is $\sqrt{\Omega}L$ (see, e.g., (5.11)). We clearly have $\Omega \le \Omega[\mathcal{X}^+] \le \sum_{\ell=1}^K \alpha_\ell \Omega_\ell[\mathcal{X}_\ell]$, where $\Omega_\ell[\mathcal{X}_\ell]$ is the variation (maximum minus minimum) of ω_ℓ on $\mathcal{X}_\ell$. These variations are upper-bounded by the quantities

$$\Omega_\ell = \begin{cases} \frac{1}{2} & \text{for ball blocks } \mathcal{X}_\ell \\ 4\mathrm{e}\ln(|\nu^\ell|) & \text{for spectahedron blocks } \mathcal{X}_\ell \end{cases}. \tag{5.45}$$

Assuming that we have K_b ball blocks $\mathcal{X}_1, ..., \mathcal{X}_{K_b}$ and K_s spectahedron blocks $\mathcal{X}_{K_b+1}, ..., \mathcal{X}_{K=K_b+K_s}$, we get

$$\Omega L \leq \Omega[\mathcal{X}^+]L \leq \left[\frac{1}{2} \sum\nolimits_{\ell=1}^{K_b} \alpha_\ell + 4\mathrm{e} \sum\nolimits_{\ell=K_b+1}^{K_b+K_s} \alpha_\ell \ln(|\nu^\ell|) \right] \sqrt{\sum\nolimits_{\ell=1}^{K} \alpha_\ell^{-1} L_\ell^2}.$$

When optimizing the right-hand side bound in $\alpha_1, ..., \alpha_L$, we get

$$\alpha_\ell = \frac{L_\ell}{\sqrt{\Omega_\ell} \sum_{i=1}^{K} L_i \sqrt{\Omega_i}}, \ \Omega[\mathcal{X}^+] = 1, \ L = \mathcal{L} := \sum\nolimits_{\ell=1}^{K} L_\ell \sqrt{\Omega_\ell}. \tag{5.46}$$

The efficiency estimate (5.11) associated with our optimized setup reads as follows

$$\overline{f}_N - \mathrm{Opt} \leq O(1)\mathcal{L}N^{-1/2}$$
$$= O(1)[\max_{1 \leq \ell \leq K} L_\ell] \left[K_b + \sum\nolimits_{\ell=K_b+1}^{K_b+K_s} \sqrt{\ln(|\nu^\ell|)} \right] N^{-1/2}. \tag{5.47}$$

If we consider $\max_{1 \leq \ell \leq K} L_\ell$, K_b, *and* K_s *as given constants, the rate of convergence of the MD algorithm is* $O(1/\sqrt{N})$, N *being the number of steps, with the factor hidden in* $O(\cdot)$ *completely independent of the dimensions of the ball blocks and nearly independent of the sizes of the spectahedron blocks. In other words, when the total number* K *of standard blocks in* $\mathcal{X}^+$ *is* $O(1)$, *the MD algorithm exhibits a nearly dimension-independent* $O(N^{-1/2})$ *rate of convergence, which is good news when solving large-scale* problems. Needless to say, the rate of convergence is not the only entity of interest; what matters is the arithmetic cost of an iteration. The latter, modulo the computational effort for obtaining the first-order information on f, is dominated by the computational complexity of the prox-mapping. This complexity—let us denote it $\mathcal{C}$—depends on exactly what $\mathcal{X}$ is. As it was explained in Section 5.7.1, *in the case of* $\mathcal{X} = \mathcal{X}^+$, $\mathcal{C}$ *is* $O(\sum_{\ell=1}^{K_b} \dim \mathcal{X}_\ell)$ plus the complexity of the eigenvalue decomposition of a matrix from $\mathbf{S}^{\nu^1} \times ... \times \mathbf{S}^{\nu^{K_s}}$. *In particular, when all spectahedron blocks are ℓ_1 balls and simplexes, $\mathcal{C}$ is just linear in the dimension of $\mathcal{X}^+$. Further, when $\mathcal{X}$ is cut off $\mathcal{X}^+$ by $O(1)$ linear inequalities, $\mathcal{C}$ is essentially the same as when $\mathcal{X} = \mathcal{X}_+$*. Indeed, here computing the prox-mapping for $\mathcal{X}$ reduces to solving the problem

$$\min_{z \in \mathcal{X}^+} \left\{ \langle a, z \rangle + \omega(z) : z \in \mathcal{X}^+, Az \leq b \right\}, \ \dim b = k = O(1),$$

or, which is the same, by duality, to solving the problem

$$\max_{\lambda \in \mathbb{R}_+^k} f_*(\lambda), \ f_*(\lambda) = \left[-b^T \lambda + \min_{z \in \mathcal{X}^+} \left[\langle a + A^T \lambda, z \rangle + \omega(z) \right] \right].$$

We are in the situation of $O(1)$ λ-variables, and thus the latter problem can be solved to machine precision in $O(1)$ steps of a simple first-order algorithm like the ellipsoid method. The first order information for f_* required by this method costs the computation of a single prox-mapping for $\mathcal{X}^+$, so that computing the prox-mapping for $\mathcal{X}_+$ is, for all practical purposes, more costly by just an absolute constant factor than computing this mapping for $\mathcal{X}^+$.

When $\mathcal{X}$ is a sophisticated subset of $\mathcal{X}^+$, computing the prox-mapping for $\mathcal{X}$ may become more involved, and the outlined setup could become difficult to implement. One of the potential remedies is to rewrite the problem (5.2) in the form of (5.15) with $\mathcal{X}$ extended to $\mathcal{X}^+$, with f in the role of f_0 and the constraints which cut $\mathcal{X}$ off $\mathcal{X}^+$ in the role of the functional constraints $f_1(x) \leq 0, ..., f_m(x) \leq 0$ of (5.15).

5.8 Notes and Remarks

1. The research of the second author was partly supported by ONR grant N000140811104 and NSF grants DMI-0619977 and DMS-0914785.

2. The very first mirror descent method, *subgradient descent*, originates from Shor (1967) and Polyak (1967); SD is merely the MD algorithm with Euclidean setup: $x_{t+1} = \text{argmin}_{u \in \mathcal{X}} \|(x_t - \gamma_t f'(x_t)) - u\|_2$. Non-Euclidean extensions (i.e., the general MD scheme) originated with Nemirovskii (1979) and Nemirovsky and Yudin (1983); the form of this scheme used in our presentation is due to Beck and Teboulle (2003). An ingenious version of the method, which also allows one to recover dual solutions is proposed by Nesterov (2009). The construction presented in Section 5.3 originated with Nemirovsky and Yudin (1983), for a more recent version, see Beck et al. (2010).

3. The practical performance of FOMs of the type we have considered can be improved significantly by passing to their *bundle* versions, explicitly utilizing both the latest and the past first-order information (in MD, only the latest first-order information is used explicitly, while the past information is loosely summarized in the current iterate). The Euclidean bundle methods originate from Lemaréchal (1978) and are the subject of numerous papers (see, e.g., Lemaréchal et al., 1981; Mifflin, 1982; Kiwiel, 1983, 1995, 1997; Schramm and Zowe, 1992; Lemaréchal et al., 1995; Kiwiel et al., 1999, and

references therein). For an MD version of the bundle scheme, see Ben-Tal and Nemirovski (2005).

4. Classical stochastic approximation (the Euclidean setup version of the algorithm from Proposition 5.5 without averaging: $x^t = x_t$) originated with Robbins and Monro (1951) and assumes the objective f to be smooth and strongly convex; there is a huge related literature (see Nevelson and Hasminskii, 1976; Benveniste et al., 1987, and references therein). The averaging of the trajectory which allows one to extend the method to the case of nonsmooth convex minimization and plays the crucial role in FOMs for saddle-point problems and variational inequalities, was introduced, in the Euclidean setup, in Bruck (1977) and Nemirovskii and Yudin (1978). For more results on "classical" and robust stochastic approximation, see, for instance, Nemirovsky and Yudin (1983); Polyak (1991); Polyak and Juditsky (1992); Nemirovski and Rubinstein (2002); Kushner and Yin (2003); Nemirovski et al. (2009) and references therein.

5. The extensions of the MD scheme from convex minimization to convex-concave saddle-point problems and variational inequalities with monotone operators originated from Nemirovskii (1981) and Nemirovsky and Yudin (1983). For a comprehensive presentation, see Ben-Tal and Nemirovski (2005).

5.9 References

A. Beck and M. Teboulle. Mirror descent and nonlinear projected subgradient methods for convex optimization. *Operations Research Letters*, 31(3):167–175, 2003.

A. Beck, A. Ben-Tal, N. Guttmann-Beck, and L. Tetruashvili. The comirror algorithm for solving nonsmooth constrained convex problems. *Operations Research Letters*, 38(6):493–498, 2010.

A. Ben-Tal and A. Nemirovski. Non-Euclidean restricted memory level method for large-scale convex optimization. *Mathematical Programming*, 102(3):407–456, 2005.

A. Ben-Tal, T. Margalit, and A. Nemirovski. The ordered subsets mirror descent optimization method with applications to tomography. *SIAM Journal on Optimization*, 12(1):79–108, 2001.

A. Benveniste, M. Métivier, and P. Priouret. *Algorithmes Adaptatifs et Approximations Stochastiques*. Masson, 1987.

R. Bruck. On weak convergence of an ergodic iteration for the solution of variational inequalities for monotone operators in Hilbert space. *Journal of Mathematical Analalysis and Applications*, 61(1):159–164, 1977.

K. C. Kiwiel. An aggregate subgradient method for nonsmooth convex minimization. *Mathematical Programming*, 27(3):320–341, 1983.

K. C. Kiwiel. Proximal level bundle method for convex nondifferentiable optimization, saddle point problems and variational inequalities. *Mathematical Programming, Series B*, 69(1):89–109, 1995.

K. C. Kiwiel. Proximal minimization methods with generalized Bregman functions. *SIAM Journal on Control and Optimization*, 35(4):1142–1168, 1997.

K. C. Kiwiel, T. Larsson, and P. O. Lindberg. The efficiency of ballstep subgradient level methods for convex optimization. *Mathematics of Operations Research*, 24 (1):237–254, 1999.

H. J. Kushner and G. G. Yin. *Stochastic Approximation and Recursive Algorithms and Applications*, volume 35 of *Stochastic Modelling and Applied Probability*. Springer, 2003.

C. Lemaréchal. Nonsmooth optimization and descent methods. Technical Report 78-4, International Institute for Applied System Analysis, Laxenburg, Austria, 1978.

C. Lemaréchal, J. J. Strodiot, and A. Bihain. On a bundle algorithm for nonsmooth optimization. In O. L. Mangasarian, R. R.Meyer, and S. M. Robinson, editors, *Nonlinear Programming*, volume 4, pages 245–282. Academic Press, 1981.

C. Lemaréchal, A. Nemirovski, and Y. Nesterov. New variants of bundle methods. *Mathematical Programming, Series B*, 69(1-3):111–147, 1995.

R. Mifflin. A modification and an extension of Lemaréchal's algorithm for nonsmooth minimization. In D. C. Sorensen and R. J.-B. Wets, editors, *Nondifferential and Variational Techniques in Optimization*, volume 17 of *Mathematical Programming Study*, pages 77–90. Springer, 1982.

A. Nemirovski and R. Rubinstein. An efficient stochastic approximation algorithm for stochastic saddle point problems. In *Modeling Uncertainty: Examination of Stochastic Theory, Methods, and Applications*, pages 155–184. Kluwer Academic Publishers, 2002.

A. Nemirovski, A. Juditsky, G. Lan, and A. Shapiro. Robust stochastic approximation approach to stochastic programming. *SIAM Journal on Optimization*, 19 (4):1574–1609, 2009.

A. Nemirovskii. Efficient methods for large-scale convex optimization problems. *Ekonomika i Matematicheskie Metody*, 15, 1979. (In Russian).

A. Nemirovskii. Efficient iterative algorithms for variational inequalities with monotone operators. *Ekonomika i Matematicheskie Metody*, 17(2):344–359, 1981. (In Russian).

A. Nemirovskii and D. Yudin. On Cezari's convergence of the steepest descent method for approximating saddle points of convex-concave functions. *Soviet Math. Doklady*, 19(2), 1978.

A. S. Nemirovsky and D. B. Yudin. *Problem Complexity and Method Efficiency in Optimization*. Wiley Intersciences, 1983.

Y. Nesterov. A method for solving a convex programming problem with rate of convergence $o(1/k^2)$. *Soviet Math. Doklady*, 27(2):372–376, 1983.

Y. Nesterov. Smooth minimization of non-smooth functions. *Mathematical Programming, Series A*, 103(1):127–152, 2005.

Y. Nesterov. Primal-dual subgradient methods for convex problems. *Mathematical Programming, Series A*, 120(1):221–259, 2009.

M. B. Nevelson and R. Z. Hasminskii. *Stochastic Approximation and Recursive Estimation*. Translations of Mathematical Monographs. American Mathematical

Society, 1976.

B. T. Polyak. A general method for solving extremal problems. *Soviet Math. Doklady*, 174:33–36, 1967.

B. T. Polyak. New stochastic approximation type procedures. *Automation and Remote Control*, 51:937–946, 1991.

B. T. Polyak and A. B. Juditsky. Acceleration of stochastic approximation by averaging. *SIAM Journal on Control and Optimization*, 30(4):838–855, 1992.

H. Robbins and S. Monro. A stochastic approximation method. *Annals of Mathematical Statistics*, 22(3):400–407, 1951.

H. Schramm and J. Zowe. A version of the bundle idea for minimizing a nonsmooth function: Conceptual idea, convergence analysis, numerical results. *SIAM Journal on Optimization*, 2(1):121–152, 1992.

N. Z. Shor. Generalized gradient descent with application to block programming. *Kibernetika*, 3(3):53–55, 1967. (In Russian).

6 First-Order Methods for Nonsmooth Convex Large-Scale Optimization, II: Utilizing Problem's Structure

Anatoli Juditsky   Anatoli.Juditsky@imag.fr
Laboratoire Jean Kuntzmann , Université J. Fourier
B. P. 53 38041 Grenoble Cedex, France

Arkadi Nemirovski   nemirovs@isye.gatech.edu
School of Industrial and Systems Engineering, Georgia Institute of Technology
765 Ferst Drive NW, Atlanta Georgia 30332, USA

We present several state-of-the-art first-order methods for well-structured large-scale nonsmooth convex programs. In contrast to their black-box-oriented prototypes considered in Chapter 5, the methods in question utilize the problem structure in order to convert the original nonsmooth minimization problem into a saddle-point problem with a smooth convex-concave cost function. This reformulation allows us to accelerate the solution process significantly. As in Chapter 5, our emphasis is on methods which, under favorable circumstances, exhibit a (nearly) dimension-independent convergence rate. Along with investigating the general well-structured situation, we outline possibilities to further accelerate first-order methods by randomization.

6.1   Introduction

The major drawback of the first-order methods (FOMs) considered in Chapter 5 is their slow convergence: as the number of steps t grows, the inaccuracy decreases as slowly as $O(1/\sqrt{t})$. As explained in Chapter 5, Section 5.1, this rate of convergence is unimprovable in the *unstructured* large-scale case;

however, convex problems usually have a lot of structure (otherwise, how could we know that the problem is convex?), and "good" algorithms should utilize this structure rather than be completely black-box-oriented. For example, by utilizing a problem's structure, we usually can represent it as a linear/conic quadratic/semidefinite program (which usually is easy), and thus make the problem amenable to polynomial time interior-point methods for LP/CQP/SDP. Unfortunately, these algorithms, aimed at generating high accuracy solutions, can become prohibitively time-consuming in the large-scale case. A much cheaper way to exploit a problem's structure when looking for medium-accuracy solutions was proposed by Nesterov (2005a); his main observation (although simple in the hindsight, it led to a real breakthrough) is that typical problems of nonsmooth convex minimization can be reformulated (and this is where a problem's structure is used!) as *smooth* (often just bilinear) convex-concave saddle-point problems, and the latter can be solved by appropriate black-box-oriented FOMs with $O(1/t)$ rate of convergence. More often than not, this simple observation allows for dramatic acceleration of the solution process, compared to the case where a problem's structure is ignored while constantly staying within the scope of computationally cheap FOMs.

In Nesterov's seminal paper (Nesterov, 2005a) the saddle-point reformulation of the (convex) problem of interest, $\min_{x \in \mathcal{X}} f(x)$, is used to construct a computationally cheap smooth convex approximation $\widetilde{f}$ of f, which further is minimized, at the rate $O(1/t^2)$, by Nesterov's method for smooth convex minimization (Nesterov, 1983, 2005a). Since the smoothness parameters of $\widetilde{f}$ deteriorate as $\widetilde{f}$ approaches f, the accuracy to which the problem of interest can be solved in t iterations turns out to be $O(1/t)$; from discussion in Section 5.1 (see item (c)), this is the best we can get in the large-scale case when solving a simple-looking problem such as $\min_{\|x\|_2 \leq R} \|Ax - b\|_2$. In what follows, we use as a "workhorse" the mirror prox (MP) saddle-point algorithm of Nemirovski (2004), which converges at the same rate $O(1/t)$ as Nesterov's smoothing, but is different from the latter algorithm. One of the reasons motivating this choice is a transparent structure of the MP algorithm (in this respect, it is just a simple-looking modification of the saddle-point mirror descent algorithm from Chapter 5, Section 5.6). Another reason is that, compared to smoothing, MP is better suited for accelerating by randomization (to be considered in Section 6.5).

The main body of this chapter is organized as follows. In Section 6.2, we present instructive examples of saddle-point reformulations of well-structured nonsmooth convex minimization problems, along with a kind of simple algorithmic calculus of convex functions admitting *bilinear* saddle-point representation. Our major workhorse — the mirror prox algorithm with the rate of convergence $O(1/t)$ for solving smooth convex-concave saddle-point problems — is presented in Section 6.3. In Section 6.4 we consider two special cases where the MP algorithm can be further accelerated. Another acceleration option is considered in Section 6.5, where we focus on bilinear saddle-point problems. We show that in this case, the MP algorithm, under favorable circumstances (e.g., when applied to saddle-point reformulations of ℓ_1 minimization problems $\min_{\|x\|_1 \leq R} \|Ax - b\|_p$, $p \in \{2, \infty\}$), can be accelerated by randomization — by passing from the precise first-order saddle-point oracle, which can be too time-consuming in the large-scale case, to a computationally much cheaper stochastic counterpart of this oracle.

The terminology and notation we use in this chapter follow those introduced in Sections 5.2.2, 5.6.1, and 5.7 of Chapter 5.

6.2   Saddle-Point Reformulations of Convex Minimization Problems

6.2.1   Saddle-Point Representations of Convex Functions

Let $\mathcal{X} \subset E$ be a nonempty closed convex set in Euclidean space E_x, let $f(x) : \mathcal{X} \to \mathbb{R}$ be a convex function, and let $\phi(x, y)$ be a continuous convex-concave function on $\mathcal{Z} = \mathcal{X} \times \mathcal{Y}$, where $\mathcal{Y} \subset E_y$ is a closed convex set, such that

$$\forall x \in \mathcal{X} : f(x) = \overline{\phi}(x) := \sup_{y \in \mathcal{Y}} \phi(x, y). \tag{6.1}$$

In this chapter, we refer to such a pair $\phi, \mathcal{Y}$ as a *saddle-point representation of* f. Given such a representation, we can reduce the problem

$$\min_{x \in \mathcal{X}} f(x) \tag{6.2}$$

of minimizing f over $\mathcal{X}$ (cf. (5.2)) to the convex-concave saddle-point (c.-c.s.p.) problem

$$\mathrm{SadVal} = \inf_{x \in \mathcal{X}} \sup_{y \in \mathcal{Y}} \phi(x, y), \tag{6.3}$$

(cf. (5.31)). Namely, assuming that ϕ has a saddle-point on $\mathcal{X} \times \mathcal{Y}$, (6.2) is solvable and, invoking (5.32), we get, for all $(x, y) \in \mathcal{X} \times \mathcal{Y}$:

$$\begin{aligned} f(x) - \min_{\mathcal{X}} f &= \overline{\phi}(x) - \mathrm{Opt}(P) = \overline{\phi}(x) - \mathrm{SadVal} \\ &\leq \overline{\phi}(x) - \underline{\phi}(y) = \epsilon_{\mathrm{sad}}(x, y). \end{aligned} \tag{6.4}$$

That is, the x-component of an ϵ-solution to (6.3) (i.e., a point $(x, y) \in \mathcal{X} \times \mathcal{Y}$ with $\epsilon_{\text{sad}}(x, y) \leq \epsilon$) is an ϵ-solution to (6.2): $f(x) - \min_{\mathcal{X}} f \leq \epsilon$.

The potential benefits of saddle-point representations stem from the fact that in many important cases a nonsmooth, but well-structured, convex function f admits an explicit saddle-point representation involving smooth function ϕ and simple $\mathcal{Y}$; as a result, the saddle-point reformulation (6.3) of the problem (6.2) associated with f can be much better suited for processing by FOMs than problem (6.2) as it is. Let us consider some examples (where S_n, S_n^+, Σ_ν, Σ_ν^+ are the standard flat and full-dimensional simplexes/spectahedrons, see Chapter 5, Section 5.7.1):

1. $f(x) := \max_{1 \leq \ell \leq L} f_\ell(x) = \max_{y \in S_L} \left[\phi(x, y) := \sum_{\ell=1}^{L} y_\ell f_\ell(x) \right]$; when all f_ℓ are smooth, so is ϕ.

2. $f(x) := \|Ax - b\|_p = \max_{\|y\|_q \leq 1} \left[\phi(x, y) := y^T(Ax - b) \right]$, $q = \frac{p}{p-1}$. With the same $\phi(x, y) = y^T(Ax - b)$, and with the coordinate wise interpretation of $[u]_+ = \max[u, 0]$ for vectors u, we have $f(x) := \|[Ax - b]_+\|_p = \max_{\|y\|_q \leq 1, y \geq 0} \phi(x, y)$ and $f(x) := \min_s \|[Ax - b - sc]_+\|_p = \max_{\|y\|_q \leq 1, y \geq 0, c^T y = 0} \phi(x, y)$. In particular,

(a) Let $\mathcal{A}(\cdot)$ be an affine mapping. The problem

$$\text{Opt} = \min_{\xi \in \Xi} \left[f(\xi) := \|\mathcal{A}(\xi)\|_p \right] \tag{6.5}$$

with $\Xi = \{\xi \in \mathbb{R}^n : \|\xi\|_1 \leq 1\}$ (cf. Lasso and Dantzig selector) reduces to the bilinear saddle-point problem

$$\min_{x \in S_{2n}^+} \max_{\|y\|_q \leq 1} y^T \mathcal{A}(Jx) \quad \left[J = [I, -I], \, q = \frac{p}{p-1} \right] \tag{6.6}$$

on the product of the standard simplex and the unit $\|\cdot\|_q$-ball. When $\Xi = \{\xi \in \mathbb{R}^{m \times n} : \|\xi\|_n \leq 1\}$, with $\|\cdot\|_n$ being the nuclear norm (cf. nuclear norm minimization) representing Ξ as the image of the spectahedron Σ_{m+n}^+ under the linear mapping $x = \begin{bmatrix} u & v \\ v^T & w \end{bmatrix} \mapsto \mathcal{J}x := 2v$, (6.5) reduces to the bilinear saddle-point problem

$$\min_{x \in \Sigma_{m+n}^+} \max_{\|y\|_q \leq 1} y^T \mathcal{A}(\mathcal{J}x); \tag{6.7}$$

(b) the SVM-type problem

$$\min_{\substack{w \in \mathbb{R}^n, \|w\| \leq R, \\ s \in \mathbb{R}}} \left\| \left[\mathbf{1} - \text{Diag}\{\eta\}(M^T w + s\mathbf{1}) \right]_+ \right\|_p, \quad \mathbf{1} = [1; ...; 1]$$

reduces to the bilinear saddle-point problem

$$\min_{\|x\|\le 1} \max_{\substack{\|y\|_q \le 1, \\ y \ge 0, \, \eta^T y = 0}} \left[\phi(x,y) := \sum_j y_j - y^T \operatorname{Diag}\{\eta\} R M^T x \right], \qquad (6.8)$$

where $x = w/R$.

3. Let $\mathcal{A}(x) = A_0 + \sum_{i=1}^{n} x_i A_i$ with $A_0, ..., A_n \in \mathbf{S}^\nu$, and let $S_k(A)$ be the sum of the k largest eigenvalues of a symmetric matrix A. Then $f(x) := S_k(\mathcal{A}(x)) = \max_{y \in \Sigma_\nu, \, y \preceq k^{-1} I} [\phi(x,y) := k \langle y, \mathcal{A}(x) \rangle]$.

In the above examples, except for the first one, ϕ is as simple as it could be — it is just bilinear. The number of examples of this type can easily be increased due to the observation that the family of convex functions f admitting explicit bilinear saddle-point representations (b.s.p.r.'s),

$$f(x) = \max_{y \in \mathcal{Y}} [\langle y, \mathcal{A}x + a \rangle + \langle b, x \rangle + c] \qquad (6.9)$$

with nonempty *compact* convex sets $\mathcal{Y}$ (with unbounded $\mathcal{Y}$, f typically would be poorly defined) admits a simple calculus. Namely, it is closed w.r.t. taking the basic convexity-preserving operations: (a) affine substitution of the argument $x \leftarrow P\xi + p$, (b) multiplication by nonnegative reals, (c) summation, (d) direct summation $\{f_i(x^i)\}_{i=1}^k \mapsto f(x^1, ..., x^k) = \sum_{i=1}^k f_i(x^i)$, and (e) taking the maximum. Here (a) and (b) are evident, and (c) is nearly so: if

$$f_i(x) = \max_{y^i \in \mathcal{Y}_i} \left[\langle y^i, \mathcal{A}_i x + a^i \rangle + \langle b^i, x \rangle + c_i \right], \quad i = 1, ..., k, \qquad (6.10)$$

with nonempty convex compact $\mathcal{Y}_i$, then

$$\sum_{i=1}^k f_i(x) = \max_{y = (y^1, ..., y^k) \in \mathcal{Y}^1 \times ... \times \mathcal{Y}^k} \left[\sum_{i=1}^k \overbrace{[\langle y^i, \mathcal{A}_i x + a^i \rangle + \langle b^i, x \rangle + c_i]}^{\langle y, \mathcal{A}x + a \rangle + \langle b, x \rangle + c} \right].$$

(d) is an immediate consequence of (a) and (c). To verify (e), let f_i be given by (6.10), let E_i be the embedding space of $\mathcal{Y}_i$, and let $\mathcal{U}_i = \{(u^i, \lambda_i) = (\lambda_i y^i, \lambda_i) : y^i \in \mathcal{Y}_i, \lambda_i \ge 0\} \subset E_i^+ = E_i \times \mathbb{R}$. Since $\mathcal{Y}_i$ are convex and compact, the sets $\mathcal{U}_i$ are closed convex cones. Now let

$$\mathcal{U} = \{y = ((u^1, \lambda_1), ..., (u^k, \lambda_k)) \in \mathcal{U}_1 \times ... \times \mathcal{U}_k : \sum_i \lambda_i = 1\}.$$

This set clearly is nonempty, convex, and closed; it is immediately seen that it is bounded as well. We have

$$
\begin{aligned}
\max_{1 \le i \le k} f_i(x) &= \max_{\lambda \ge 0: \sum_i \lambda_i = 1} \sum_{i=1}^k \lambda_i f_i(x) = \max_{\lambda, y^1, \dots, y^k} \left\{ \sum_{i=1}^k \left[\langle \overbrace{\lambda_i y^i}^{u^i}, \mathcal{A}_i x + a^i \rangle \right. \right. \\
&\qquad \left. \left. + \langle \lambda_i b^i, x \rangle + \lambda_i c_i \right] : \lambda \ge 0, \sum_i \lambda_i = 1, y^i \in \mathcal{Y}_i, 1 \le i \le k \right\} \\
&= \max_{u = \{(u^i, \lambda_i) : 1 \le i \le k\} \in \mathcal{U}} \left[\sum_{i=1}^k [\langle u^i, \mathcal{A}_i x + a^i \rangle + \langle \lambda_i b^i, x \rangle + \lambda_i c_i] \right],
\end{aligned}
$$

and we end up with a b.s.p.r. of $\max_i f_i$.

6.3 Mirror-Prox Algorithm

We are about to present the basic MP algorithm for the problem (6.3).

6.3.1 Assumptions and Setup

Here we assume that

A. The closed and convex sets $\mathcal{X}$, $\mathcal{Y}$ are bounded.

B. The convex-concave function $\phi(x, y) : \mathcal{Z} = \mathcal{X} \times \mathcal{Y} \to \mathbb{R}$ possesses a Lipschitz continuous gradient $\nabla \phi(x, y) = (\nabla_x \phi(x, y), \nabla_y \phi(x, y))$.

We set $F(x, y) = (F_x(x, y) := \nabla_x \phi(x, y), F_y(x, y) := -\nabla_y \phi(x, y))$, thus getting a Lipschitz continuous selection for the monotone operator associated with (6.3) (see Section 5.6.1).

The setup for the MP algorithm is given by a norm $\| \cdot \|$ on the embedding space $E = E_x \times E_y$ of $\mathcal{Z}$ and by a d.-g.f. $\omega(\cdot)$ for $\mathcal{Z}$ compatible with this norm (cf. Section 5.2.2). For $z \in \mathcal{Z}^o$ and $w \in \mathcal{Z}$, let

$$
V_z(w) = \omega(w) - \omega(z) - \langle \omega'(z), w - z \rangle, \tag{6.11}
$$

(cf. the definition (5.4)) and let $z_c = \operatorname{argmin}_{w \in \mathcal{Z}} \omega(w)$. Further, we assume that given $z \in \mathcal{Z}^o$ and $\xi \in E$, it is easy to compute the prox-mapping

$$
\operatorname{Prox}_z(\xi) = \operatorname*{argmin}_{w \in \mathcal{Z}} [\langle \xi, w \rangle + V_z(w)] \left(= \operatorname*{argmin}_{w \in \mathcal{Z}} [\langle \xi - \omega'(z), w \rangle + \omega(w)] \right),
$$

and set

$$
\Omega = \max_{w \in \mathcal{Z}} V_{z_c}(w) \le \max_{\mathcal{Z}} \omega(\cdot) - \min_{\mathcal{Z}} \omega(\cdot) \tag{6.12}
$$

(cf. Chapter 5, Section 5.2.2). We also assume that we have at our disposal an upper bound L on the Lipschitz constant of F from the norm $\| \cdot \|$ to the conjugate norm $\| \cdot \|_*$:

$$\forall (z, z' \in \mathcal{Z}) : \|F(z) - F(z')\|_* \leq L\|z - z'\|. \tag{6.13}$$

6.3.2 The Algorithm

The MP algorithm is given by the recurrence

$$
\begin{aligned}
(a): \quad & z_1 = z_c, \\
(b): \quad & w_\tau = \mathrm{Prox}_{z_\tau}(\gamma_\tau F(z_\tau)), \; z_{\tau+1} = \mathrm{Prox}_{z_\tau}(\gamma_\tau F(w_\tau)), \\
(c): \quad & z^\tau = [\textstyle\sum_{s=1}^\tau \gamma_s]^{-1} \sum_{s=1}^\tau \gamma_s w_s,
\end{aligned}
\tag{6.14}
$$

where $\gamma_\tau > 0$ are the stepsizes. Note that $z_\tau, w_\tau \in \mathcal{Z}^o$, whence $z^\tau \in \mathcal{Z}$. Let

$$\delta_\tau = \gamma_\tau \langle F(w_\tau), w_\tau - z_{\tau+1} \rangle - V_{z_\tau}(z_{\tau+1}) \tag{6.15}$$

(cf. (5.4)). The convergence properties of the algorithm are given by the following

Proposition 6.1. *Under assumptions* **A** *and* **B***:*
(i) *For every* $t \geq 1$ *it holds (for notation, see (6.12) and (6.15)) that*

$$\epsilon_{\mathrm{sad}}(z^t) \leq \left[\textstyle\sum_{\tau=1}^t \gamma_\tau \right]^{-1} \left[\Omega + \sum_{\tau=1}^t \delta_\tau \right]. \tag{6.16}$$

(ii) *If the stepsizes satisfy the conditions* $\gamma_\tau \geq L^{-1}$ *and* $\delta_\tau \leq 0$ *for all* τ *(which certainly is so when* $\gamma_\tau \equiv L^{-1}$*), we have*

$$\forall t \geq 1 : \epsilon_{\mathrm{sad}}(z^t) \leq \Omega \left[\textstyle\sum_{\tau=1}^t \gamma_\tau \right]^{-1} \leq \Omega L / t. \tag{6.17}$$

Proof. 1^0. We start with the following basic observation:

Lemma 6.2. *Given* $z \in \mathcal{Z}^o$, $\xi, \eta \in E$, *let* $w = \mathrm{Prox}_z(\xi)$ *and* $z_+ = \mathrm{Prox}_z(\eta)$. *Then for all* $u \in \mathcal{Z}$ *it holds that*

$$
\begin{aligned}
\langle \eta, w - u \rangle &\leq V_z(u) - V_{z_+}(u) + \langle \eta, w - z_+ \rangle - V_z(z_+) && (a) \\
&\leq V_z(u) - V_{z_+}(u) + \langle \eta - \xi, w - z_+ \rangle - V_z(w) - V_w(z_+) && (b) \\
&\leq V_z(u) - V_{z_+}(u) + \left[\tfrac{1}{2}\|\eta - \xi\|_* \|w - z_+\| - \tfrac{1}{2}\|z - w\|^2 - \tfrac{1}{2}\|z_+ - w\|^2 \right] && (c) \\
&\leq V_z(u) - V_{z_+}(u) + \tfrac{1}{2}[\|\eta - \xi\|_*^2 - \|w - z\|^2] && (d)
\end{aligned}
$$

$$\tag{6.18}$$

Proof. By the definition of $z_+ = \mathrm{Prox}_z(\eta)$ we have $\langle \eta - \omega'(z) + \omega'(z_+), u - z_+ \rangle \geq 0$; we obtain (6.18.a) by rearranging terms and taking into account the definition of $V_v(u)$, (cf. the derivation of (5.12)). By the definition of $w = \mathrm{Prox}_z(\xi)$ we have $\langle \xi - \omega'(z) + \omega'(w), z_+ - w \rangle \geq 0$, whence $\langle \eta, w - z_+ \rangle \leq$

$\langle \eta - \xi, w - z_+ \rangle + \langle \omega'(w) - \omega'(z), z_+ - w \rangle$; replacing the third term in the right-hand side of (a) with this upper bound and rearranging terms, we get (b). (c) follows from (b) due to the strong convexity of ω, implying that $V_v(u) \geq \frac{1}{2}\|u - v\|^2$, and (d) is an immediate consequence of (c). $\square$

2^0. Applying Lemma 6.2 to $z = z_\tau$, $\xi = \gamma_\tau F(z_\tau)$ (which results in $w = w_\tau$) and to $\eta = \gamma_\tau F(w_\tau)$ (which results in $z_+ = z_{\tau+1}$), we obtain, due to (6.18.d):

$$(a) \quad \gamma_\tau \langle F(w_\tau), w_\tau - u \rangle \leq V_{z_\tau}(u) - V_{z_{\tau+1}}(u) + \delta_\tau \,\forall u \in \mathcal{Z},$$
$$(b) \quad \delta_\tau \leq \frac{1}{2}\left[\gamma_\tau^2 \|F(w_\tau) - F(z_\tau)\|_*^2 - \|w_\tau - z_\tau\|^2\right] \tag{6.19}$$

Summing (6.19.a) over $\tau = 1, ..., t$, taking into account that $V_{z_1}(u) = V_{z_c}(u) \leq \Omega$ by (6.12) and setting, for a given t, $\lambda_\tau = \gamma_\tau / \sum_{\tau=1}^t \gamma_\tau$, we get $\lambda_\tau \geq 0$, $\sum_{\tau=1}^t \lambda_\tau = 1$, and

$$\forall u \in \mathcal{Z} : \sum_{\tau=1}^t \lambda_\tau \langle F(w_\tau), w_\tau - u \rangle \leq A := \frac{\Omega + \sum_{\tau=1}^t \delta_\tau}{\sum_{\tau=1}^t \gamma_\tau}. \tag{6.20}$$

On the other hand, setting $w_\tau = (x_\tau, y_\tau)$, $z^t = (x^t, y^t)$, $u = (x, y)$, and using (5.37), we have

$$\sum_{\tau=1}^t \lambda_\tau \langle F(w_\tau), w_\tau - u \rangle \geq \phi(x^t, y) - \phi(x, y^t),$$

so that (6.20) results in $\phi(x^t, y) - \phi(x, y^t) \leq A$ for all $(x, y) \in \mathcal{Z}$. Taking the supremum in $(x, y) \in \mathcal{Z}$, we arrive at (6.16); (i) is proved. To prove (ii), note that with $\gamma_t \leq L^{-1}$, (6.19.b) implies that $\delta_\tau \leq 0$, see (6.13). $\square$

6.3.3 Setting up the MP Algorithm

Let us restrict ourselves to the *favorable geometry* case defined completely similarly to Chapter 5, Section 5.7.2, but with $\mathcal{Z}$ in the role of $\mathcal{X}$. Specifically, we assume that $\mathcal{Z} = \mathcal{X} \times \mathcal{Y}$ is a subset of the direct product $\mathcal{Z}^+$ of K standard blocks $\mathcal{Z}_\ell$ (K_b ball blocks and $K_s = K - K_b$ spectahedron blocks) and that $\mathcal{Z}$ intersects rint $\mathcal{Z}^+$. We assume that the representation $\mathcal{Z}^+ = \mathcal{Z}_1 \times ... \times \mathcal{Z}_K$ is coherent with the representation $\mathcal{Z} = \mathcal{X} \times \mathcal{Y}$, meaning that $\mathcal{X}$ is a subset of the direct product of some of the blocks $\mathcal{Z}_\ell$, while $\mathcal{Y}$ is a subset of the direct product of the remaining blocks. We equip the embedding space $E = E_1 \times ... \times E_K$ of $\mathcal{Z} \subset \mathcal{Z}^+$ with the norm $\|\cdot\|$ and a d.-g.f. $\omega(\cdot)$ according to (5.42) and (5.43) (where, for notational consistency, we should replace x^ℓ with z^ℓ and $\mathcal{X}_\ell$ with $\mathcal{Z}_\ell$). Our current goal is to optimize the efficiency estimate of the associated MP algorithm over the coefficients α_ℓ in (5.42), (5.43). To this end assume that we have at our disposal upper bounds

$L_{\mu\nu} = L_{\nu\mu}$ on the partial Lipschitz constants of the (Lipschitz continuous by assumption **B**) vector field $F(z = (x, y)) = (\nabla_x \phi(x, y), -\nabla_y \phi(x, y))$, so that for $1 \leq \mu \leq K$ and all $u, v \in \mathcal{Z}$, we have

$$\|F_\mu(u) - F_\mu(v)\|_{(\mu),*} \leq \sum_{\nu=1}^{K} L_{\mu\nu} \|u^\nu - v^\nu\|_{(\nu)},$$

where the decomposition $F(z = (z^1, ..., z^K)) = (F_1(z), ..., F_K(z))$ is induced by the representation $E = E_1 \times ... \times E_K$.

Let Ω_ℓ be defined by (5.46) with $\mathcal{Z}_\ell$ in the role of $\mathcal{X}_\ell$. The choice

$$\alpha_\ell = \frac{\sum_{\nu=1}^{K} L_{\ell\nu} \sqrt{\Omega_\nu}}{\sqrt{\Omega_\ell} \sum_{\mu,\nu} L_{\mu\nu} \sqrt{\Omega_\mu \Omega_\nu}}$$

(cf. Nemirovski, 2004) results in

$$\Omega \leq 1 \quad \text{and} \quad L \leq \mathcal{L} := \sum_{\mu,\nu} L_{\mu\nu} \sqrt{\Omega_\mu \Omega_\nu},$$

so that the bound (6.17) is

$$\epsilon_{\text{sad}}(z^t) \leq \mathcal{L}/t, \quad \mathcal{L} = \sum_{\mu,\nu} L_{\mu\nu} \sqrt{\Omega_\mu \Omega_\nu}. \tag{6.21}$$

As far as complexity of a step and dependence of the efficiency estimate on a problem's dimension are concerned, the present situation is identical to that of MD (studied in Chapter 5, Section 5.7). In particular, all our considerations in the discussion at the end of Section 5.7.2 remain valid here.

6.3.3.1 Illustration I

As simple and instructive illustrations, consider problems (6.8) and (6.5).
1. Consider problem (6.8), and assume, in full accordance with the SVM origin of the problem, that $\|w\| = \|w\|_r$ with $r \in \{1, 2\}$, $p \in \{2, \infty\}$, and that η is a ± 1 vector which has both positive and negative entries. When $p = 2$, (6.8) is a bilinear saddle-point problem on the product of the unit $\|\cdot\|_r$-ball and a simple part of $\|\cdot\|_2$-ball. Combining (6.21) with what was said in Section 5.7.2, we arrive at the efficiency estimate

$$\epsilon_{\text{sad}}(x^t, y^t) \leq O(1)(\ln(\dim w))^{1-r/2} R\|M\|_{2,r_*} t^{-1}, \quad r_* = r/(r-1),$$

where $\|M\|_{2,2}$ is the spectral norm of M, and $\|M\|_{2,\infty}$ is the maximum of the Euclidean norms of the rows in M. When $p = 1$, the situation becomes worse: (6.8) is now a bilinear saddle-point problem on the product of the unit $\|\cdot\|_r$-ball and a simple subset of the unit box $\{y : \|y\|_\infty \leq 1\}$, or, which is the same, a simple subset of the Euclidean ball of the radius $\rho = \sqrt{\dim \eta}$ centered at the origin. Substituting $y = \rho u$, we end up with a bilinear saddle-point problem on the direct product of the unit $\|\cdot\|_r$ ball and a simple subset of the unit Euclidean ball, the matrix of the bilinear part of the cost function being $\rho R \operatorname{Diag}\{\eta\} M^T$. As a result, we arrive at the dimension-dependent efficiency estimate

$$\epsilon_{\text{sad}}(x^t, y^t) \leq O(1)(\ln(\dim w))^{1-r/2}\sqrt{\dim \eta} R\|M\|_{2,r_*} t^{-1}, \ r_* = r/(r-1).$$

Note that in all cases the computational effort at a step of the MP is dominated by computing $O(1)$ matrix-vector products involving M and M^T.

2. Now consider problem (6.5), and let $p \in \{2, \infty\}$.

2.1. Let us start with the case of $\Xi = \{\xi \in \mathbb{R}^n : \|\xi\|_1 \leq 1\}$, so that $\mathcal{A}(Jx) = A_0 + Ax$, where A is an $m \times 2n$ matrix. Here (6.6) is a bilinear saddle-point problem on the direct product of the standard simplex S_{2n}^+ in $\mathbb{R}^{2n}$ and the unit $\|\cdot\|_q$-ball in $\mathbb{R}^m$. Combining (6.21) with derivations in Section 5.7.2, the efficiency estimate of MP is

$$\epsilon_{\text{sad}}(x^t, y^t) \leq O(1)\sqrt{\ln(n)}(\ln(m))^{\frac{1}{2}-\frac{1}{p}}\left[\max_{1 \leq j \leq \dim x}\|A_j\|_p\right]t^{-1}, \quad (6.22)$$

where A_j are columns of A. The complexity of a step is dominated by the necessity to compute $O(1)$ matrix-vector products involving A and A^T.

2.2. The next case, inspired by K. Scheinberg, is the one where $\Xi = \{(\xi^1, ..., \xi^k) \in \mathbb{R}^{d_1} \times ... \times \mathbb{R}^{d_k} : \sum_j \|\xi^j\|_2 \leq 1\}$, so that problem (6.5) is of the form arising in block Lasso ($p = 2$) or block Dantzig selector ($p = \infty$). Setting $\mathcal{X} = \Xi$, let us equip the embedding space $E_x = \mathbb{R}^{d_1} \times ... \times \mathbb{R}^{d_k}$ of $\mathcal{X}$ with the norm $\|x\|_x = \sum_{i=1}^k \|x^i[x]\|_2$, where $x^i[x] \in \mathbb{R}^{d_i}$ are the blocks of $x \in E_x$. It is easily seen that the function

$$\omega_x(x) = \tfrac{1}{p\gamma}\sum_{i=1}^k \|x^i[x]\|_2^p : \mathcal{X} \to \mathbf{R},$$

$$p = \begin{cases} 2, & k \leq 2 \\ 1 + 1/\ln(k), & k \geq 3 \end{cases}, \ \gamma = \begin{cases} 1, & k = 1 \\ 1/2, & k = 2 \\ 1/(e\ln(k)), & k > 2 \end{cases}$$

is a d.-g.f. for $\mathcal{X}$ compatible with the norm $\|\cdot\|_x$, and that the $\omega_x(\cdot)$ diameter Ω_x of $\mathcal{X}$ does not exceed $O(1)\ln(k+1)$. Note that in the case of $k = 1$ (where $\mathcal{X} = \Xi$ is the unit $\|\cdot\|_2$-ball), $\|\cdot\|_x = \|\cdot\|_2$, $\omega_x(\cdot)$ are exactly as in the Euclidean MD setup, and in the case of $d_1 = ... = d_k = 1$ (where $\mathcal{X} = \Xi$ is the unit ℓ_1 ball), $\|\cdot\|_x = \|\cdot\|_1$ and $\omega_x(\cdot)$ is, basically, the d.-g.f. from item 2b of Section 5.7.1. Applying the results of Section 6.3.3, the efficiency estimate of MP is

$$\epsilon_{\text{sad}}(z^t) \leq O(1)(\ln(k+1))^{\frac{1}{2}}(\ln(m))^{\frac{1}{2}-\frac{1}{p}}\pi(A)t^{-1}, \quad (6.23)$$

where $\pi(A)$ is the norm of the linear mapping $x \mapsto Ax$ induced by the norms $\|x\|_x = \sum_{i=1}^{k} \|x^i[x]\|_2$ and $\|\cdot\|_p$ in the argument and the image spaces.

Note that the prox-mapping is easy to compute in this setup. The only nonevident part of this claim is that it is easy to minimize over X a function of the form $\omega_x(x) + \langle a, x \rangle$ or, which is the same, a function of the form $g(x) = \frac{1}{p} \sum_{i=1}^{k} \|x^i[x]\|_2^p + \langle b, x \rangle$. Here is the verification: setting $\beta_i = \|x^i[b]\|_2$, $1 \le i \le k$, it is easily seen that at the minimizer x_* of $g(\cdot)$ over X the blocks $x^i[x_*]$ are nonpositive multiples of $x^i[b]$; thus, all we need is to find $\sigma_i^* = \|x^i[x_*]\|_2$, $1 \le i \le k$. Clearly, $\sigma^* = [\sigma_1^*; ...; \sigma_k^*]$ is nothing but

$$\operatorname*{argmin}_{\sigma} \left\{ \frac{1}{p} \sum_{i=1}^{k} \sigma_i^p - \sum_{i=1}^{k} \beta_i \sigma_i : \sigma \ge 0, \sum_i \sigma_i \le 1 \right\}.$$

After β_i are computed, the resulting "nearly separable" convex problem clearly can be solved within machine accuracy in $O(k)$ a.o. As a result, the arithmetic cost of a step of MP in our situation is dominated by $O(1)$ computations of matrix-vector products involving A and A^T.

Note that slightly modifying the above d.-g.f. for the unit ball X of $\|\cdot\|_x$, we get a d.-g.f. compatible with $\|\cdot\|_x$ on the entire E_x, namely, the function

$$\widehat{\omega}_x(x) = \frac{k^{(p-1)(2-p)/p}}{2\gamma} \left[\sum_{i=1}^{k} \|x^i[x]\|_2^p \right]^{\frac{2}{p}},$$

with the same as above p, γ.[1] The associated prox-mapping is a "closed form" one. Indeed, here again the blocks $x^i[x_*]$ of the minimizer, over the entire E_x, of $g(x) = \frac{1}{2} \left(\sum_{i=1}^{k} \|x^i[x]\|_2^p \right)^{\frac{2}{p}} - \langle b, x \rangle$ are nonpositive multiples of the blocks $x^i[b]$; thus, all we need is to find $\sigma^* = [\|x^1[x_*]\|_2; ...; x^k[x_*]]$. Setting $\beta = [\|x^1[b]\|_2; ...; \|x^k[b]\|_2]$, we have

$$\sigma^* = \operatorname*{argmin}_{\sigma \in \mathbb{R}^k} \left\{ \frac{1}{2} \|\sigma\|_p^2 - \langle \beta, \sigma \rangle : \sigma \ge 0 \right\} = \nabla \left(\frac{1}{2} \|\beta\|_{\frac{p}{p-1}}^2 \right).$$

2.3. Finally, consider the case when Ξ is the unit nuclear-norm ball, so that $\mathcal{A}(\mathcal{J}x) = a_0 + [\operatorname{Tr}(A_1 x); ...; \operatorname{Tr}(A_k x)]$ with $A_i \in \mathbf{S}^{m+n}$, and (6.7) is a bilinear saddle-point problem on the direct product of the spectahedron Σ_{m+n}^+ and the unit $\|\cdot\|_q$-ball in $\mathbb{R}^k$. Applying the results of Section 6.3.3, the efficiency estimate of MP is

$$\epsilon_{\text{sad}}(x^t, y^t) \le O(1)\sqrt{\ln(m+n)}(\ln(k))^{\frac{1}{2} - \frac{1}{p}} \left[\max_{\|\zeta\|_2 \le 1} \|[\zeta^T A_1 \zeta; ...; \zeta^T A_k \zeta]\|_p \right] t^{-1}.$$

---

1. Note that in the "extreme" cases (a): $k = 1$ and (b): $d_1 = ... = d_k = 1$, our d.-g.f. recovers the Euclidean d.-g.f. and the d.-g.f. similar to the one in item 2a of Section 5.7.1.

The complexity of a step is dominated by $O(1)$ computations of the values of $\mathcal{A}$ and of matrices of the form $\sum_{i=1}^{k} y_i A_i$, plus computing a single eigenvalue decomposition of a matrix from $\mathbf{S}^{m+n}$.

In all cases, the approximate solution (x^t, y^t) to the saddle-point reformulation of (6.5) straightforwardly induces a feasible solution ξ^t to the problem of interest (6.5) such that $f(\xi^t) - \text{Opt} \le \epsilon_{\text{sad}}(x^t, y^t)$.

6.4   Accelerating the Mirror-Prox Algorithm

In what follows, we present two modifications of the MP algorithm.

6.4.1   Splitting

6.4.1.1   *Situation and Assumptions*

Consider the c.-c.s.p. problem (6.3) and assume that both $\mathcal{X}$ and $\mathcal{Y}$ are bounded. Assume also that we are given norms $\| \cdot \|_x$, $\| \cdot \|_y$ on the corresponding embedding spaces E_x, E_y, along with d.-g.f.'s $\omega_x(\cdot)$ for $\mathcal{X}$ and $\omega_y(\cdot)$ for $\mathcal{Y}$ which are compatible with the respective norms.

We already know that if the convex-concave cost function ϕ is smooth (i.e., possesses a Lipschitz continuous gradient), the problem can be solved at the rate $O(1/t)$. We are about to demonstrate that the same holds true when, roughly speaking, ϕ can be represented as a sum of a "simple" part and a smooth parts. Specifically, let us assume the following:

C.1. The monotone operator Φ associated with (6.3) (see Section 5.6.1) admits splitting: we can point out a *Lipschitz continuous on* $\mathcal{Z}$ vector field $G(z) = (G_x(z), G_y(z)) : \mathcal{Z} \to E = E_x \times E_y$, and a point-to-set monotone operator $\mathcal{H}$ with the same domain as Φ such that the sets $\mathcal{H}(z)$, $z \in \text{Dom } \mathcal{H}$, are convex and nonempty, the graph of $\mathcal{H}$ (the set $\{(z, h) : z \in \text{Dom } \mathcal{H}, h \in \mathcal{H}(z)\}$) is closed, and

$$\forall z \in \text{Dom } \mathcal{H} : \mathcal{H}(z) + G(z) \subset \Phi(z). \tag{6.24}$$

C.2. $\mathcal{H}$ is simple, specifically, it is easy to find a weak solution to the variational inequality associated with $\mathcal{Z}$ and a monotone operator of the form $\Psi(x, y) = \alpha \mathcal{H}(x, y) + [\alpha_x \omega_x'(x) + e; \alpha_y \omega_y'(y) + f]$ (where $\alpha, \alpha_x, \alpha_y$ are positive), that is, it is easy to find a point $\hat{z} \in \mathcal{Z}$ satisfying

$$\forall(z \in \text{rint } \mathcal{Z}, F \in \Psi(z)) : \langle F, z - \hat{z} \rangle \ge 0. \tag{6.25}$$

It is easily seen that in the case of **C.1**, (6.25) has a unique solution $\hat{z} = (\hat{x}, \hat{y})$

which belongs to Dom $\Phi \cap Z^o$ and in fact is a strong solution: there exists $\zeta \in \mathcal{H}(\widehat{z})$ such that

$$\forall z \in Z : \langle \alpha\zeta + [\alpha_z \omega_x'(\widehat{x}) + e; \alpha_y \omega_y'(\widehat{y}) + f], z - \widehat{z} \rangle \geq 0. \qquad (6.26)$$

We assume that when solving (6.25), we get both $\widehat{z}$ and ζ.

We intend to demonstrate that under assumptions **C.1** and **C.2** we can solve (6.3) as if there were no $\mathcal{H}$-component at all.

6.4.2 Algorithm MPa

6.4.2.1 *Preliminaries*

Recall that the mapping $G(x, y) = (G_x(x, y), G_y(x, y)) : Z \to E$ defined in **C.1** is Lipschitz continuous. We assume that we have at our disposal nonnegative constants L_{xx}, L_{yy}, L_{xy} such that

$$\forall (z = (x, y) \in Z, z' = (x', y') \in Z) :$$
$$\begin{cases} \|G_x(x', y) - G_x(x, y)\|_{x,*} \leq L_{xx}\|x' - x\|_x, \\ \|G_y(x, y') - G_y(x, y)\|_{y,*} \leq L_{yy}\|y' - y\|_y \\ \|G_x(x, y') - G_x(x, y)\|_{x,*} \leq L_{xy}\|y' - y\|_y, \\ \|G_y(x', y) - G_y(x, y)\|_{y,*} \leq L_{xy}\|x' - x\|_x \end{cases} \qquad (6.27)$$

where $\|\cdot\|_{x,*}$ and $\|\cdot\|_{y,*}$ are the norms conjugate to $\|\cdot\|_x$ and $\|\cdot\|_y$, respectively. We set

$$\begin{aligned} &\Omega_x = \max_\mathcal{X} \omega_x(\cdot) - \min_\mathcal{X} \omega_x(\cdot), \quad \Omega_y = \max_\mathcal{Y} \omega_y(\cdot) - \min_\mathcal{Y} \omega_y(\cdot), \\ &\mathcal{L} = L_{xx}\Omega_x + L_{xy}\Omega_y + 2L_{xy}\sqrt{\Omega_x \Omega_y}, \\ &\alpha = [L_{xx}\Omega_x + L_{xy}\sqrt{\Omega_x \Omega_y}]/\mathcal{L}, \quad \beta = [L_{yy}\Omega_y + L_{xy}\sqrt{\Omega_x \Omega_y}]/\mathcal{L}, \quad (6.28) \\ &\omega(x, y) = \tfrac{\alpha}{\Omega_x}\omega_x(x) + \tfrac{\beta}{\Omega_y}\omega_y(y) : Z \to \mathbb{R}, \\ &\|(x, y)\| = \sqrt{\tfrac{\alpha}{\Omega_x}\|x\|_x^2 + \tfrac{\beta}{\Omega_y}\|y\|_y^2} \end{aligned}$$

so that the conjugate norm is $\|(x, y)\|_* = \sqrt{\tfrac{\Omega_x}{\alpha}\|x\|_{x,*}^2 + \tfrac{\Omega_y}{\beta}\|y\|_{y,*}^2}$ (cf. Section 6.3.3). Observe that $\omega(\cdot)$ is a d.-g.f. on Z compatible with the norm $\|\cdot\|$. It is easily seen that $\Omega := 1 \geq \max_{z \in Z} \omega(z) - \min_{z \in Z} \omega(z)$ and

$$\forall (z, z' \in Z) : \|G(z) - G(z')\|_* \leq \mathcal{L}\|z - z'\|. \qquad (6.29)$$

6.4.2.2 *Algorithm MPa*

Our new version, MPa, of the MP algorithm is as follows:

1. *Initialization:* Set $z_1 = \text{argmin}_z \, \omega(\cdot)$.

2. *Step* $\tau = 1, 2, ...$: Given $z_\tau \in \mathcal{Z}^o$ and a stepsize $\gamma_\tau > 0$, we find w_τ that satisfies

$$(\forall u \in \text{rint } \mathcal{Z}, \; F \in \mathcal{H}(u)) : \langle \gamma_\tau (F + G(z_\tau)) + \omega'(u) - \omega'(z_\tau), u - w_\tau \rangle \geq 0$$

and find $\zeta_\tau \in \mathcal{H}(w_\tau)$ such that

$$\forall (u \in \mathcal{Z}) : \langle \gamma_\tau (\zeta_\tau + G(z_\tau)) + \omega'(w_\tau) - \omega'(z_\tau), u - w_\tau \rangle \geq 0; \qquad (6.30)$$

by assumption **C.2**, computation of ω_τ and ζ_τ is easy. Next, we compute

$$
\begin{aligned}
z_{\tau+1} &= \text{Prox}_{z_\tau} (\gamma_\tau (\zeta_\tau + G(w_\tau))) \\
&:= \text{argmin}_{z \in \mathcal{Z}} \left[\langle \gamma_\tau (\zeta_\tau + G(w_\tau)), z \rangle + V_{z_\tau}(z) \right],
\end{aligned}
\qquad (6.31)
$$

where $V.(\cdot)$ is defined in (6.11). We set

$$z^\tau = \left[\sum\nolimits_{s=1}^\tau \gamma_s \right]^{-1} \sum\nolimits_{s=1}^\tau \gamma_s w_s$$

and loop to step $\tau + 1$.

Let

$$\delta_\tau = \langle \gamma_\tau (\zeta_\tau + G(w_\tau)), w_\tau - z_{\tau+1} \rangle - V_{z_\tau}(z_{\tau+1})$$

(cf. (6.27)). The convergence properties of the algorithm are given by

Proposition 6.3. *Under assumptions* **C.1** *and* **C.2**, *algorithm MPa ensures that*

(i) *For every* $t \geq 1$ *it holds that*

$$\epsilon_{\text{sad}}(z^t) \leq \left[\sum\nolimits_{\tau=1}^t \gamma_\tau \right]^{-1} \left[1 + \sum\nolimits_{\tau=1}^t \delta_\tau \right]. \qquad (6.32)$$

(ii) *If the stepsizes satisfy the condition* $\gamma_\tau \geq \mathcal{L}^{-1}$, $\delta_\tau \leq 0$ *for all* τ *(which certainly is so when* $\gamma_\tau \equiv \mathcal{L}^{-1}$), *we have*

$$\forall t \geq 1 : \epsilon_{\text{sad}}(z^t) \leq \left[\sum\nolimits_{\tau=1}^t \gamma_\tau \right]^{-1} \leq \mathcal{L}/t. \qquad (6.33)$$

Proof. Relation (6.30) exactly expresses the fact that $w_\tau = \text{Prox}_{z_\tau}(\gamma_\tau(\zeta_\tau + G(z_\tau)))$. With this in mind, Lemma 6.2 implies that

$$
\begin{aligned}
&(a) \quad \gamma_\tau \langle \zeta_\tau + G(w_\tau), w_\tau - u \rangle \leq V_{z_\tau}(u) - V_{z_{\tau+1}}(u) + \delta_\tau \; \forall u \in \mathcal{Z}, \\
&(b) \quad \delta_\tau \leq \tfrac{1}{2} \left[\gamma_\tau^2 \|G(w_\tau) - G(z_\tau)\|_*^2 - \|w_\tau - z_\tau\|^2 \right]
\end{aligned}
\qquad (6.34)
$$

(cf. (6.19)). It remains to repeat word by word the reasoning in items 2^0–3^0 of the proof of Proposition 6.1, keeping in mind (6.29) and the fact that, by the origin of ζ_τ and in view of (6.24), we have $\zeta_\tau + G(w_\tau) \in \Phi(w_\tau)$. $\qquad\square$

6.4.2.3 Illustration II

Consider a problem of the Dantzig selector type

$$\text{Opt} = \min_{\|x\|_1 \leq 1}\|A^T(Ax - b)\|_\infty \quad [A: m \times n, m \leq n] \qquad (6.35)$$

(cf. (6.5)) along with its saddle-point reformulation:

$$\text{Opt} = \min_{\|x\|_1 \leq 1}\max_{\|y\|_1 \leq 1}y^T[Bx - c], \ B = A^TA, \ c = A^Tb. \qquad (6.36)$$

As already mentioned, the efficiency estimate for the basic MP as applied to this problem is $\epsilon_{\text{sad}}(z^t) \leq O(1)\sqrt{\ln(n)}\|B\|_{1,\infty}t^{-1}$, where $\|B\|_{1,\infty}$ is the maximum of magnitudes of entries in B. Now, in typical large-scale compressed sensing applications, columns A_i of A are of nearly unit $\|\cdot\|_2$-norm and are nearly orthogonal: the *mutual incoherence* $\mu(A) = \max_{i \neq j}|A_i^TA_j|/A_i^TA_i$ is $\ll 1$. In other words, the diagonal entries in B are of order 1, and the magnitudes of off-diagonal entries do not exceed $\mu \ll 1$. For example, for a typical randomly selected A, μ is as small as $O(\sqrt{\ln(n)/m})$. Now, the monotone operator associated with (6.36) admits an affine selection $F(x,y) = (B^Ty, c - Bx)$ and can be split as

$$F(x,y) = \overbrace{(Dy, -Dx)}^{\mathcal{H}(x,y)} + \overbrace{(\widehat{B}^Ty, c - \widehat{B}x)}^{G(x,y)},$$

where D is the diagonal matrix with the same diagonal as in B, and $\widehat{B} = B - D$. Now, the domains $\mathcal{X} = \mathcal{Y}$ associated with (6.36) are unit ℓ_1-balls in the respective embedding spaces $E_x = E_y = \mathbb{R}^n$. Equipping $E_x = E_y$ with the norm $\|\cdot\|_1$, and the unit $\|\cdot\|_1$ ball $\mathcal{X} = \mathcal{Y}$ in $\mathbb{R}^n$ with the d.-g.f. presented in item 2b of Chapter 5, Section 5.7.1, we clearly satisfy **C.1** and, on a closest inspection, satisfy **C.2** as well. As a result, we can solve the problem by MPa, the efficiency estimate being $\epsilon_{\text{sad}}(z^t) \leq O(1)\ln(n)\|\widehat{B}\|_{1,\infty}t^{-1}$, which is much better than the estimate $\epsilon_{\text{sad}}(z^t) \leq O(1)\ln(n)\|B\|_{1,\infty}t^{-1}$ for the plain MP (recall that we are dealing with the case of $\mu := \|\widehat{B}\|_{1,\infty} \ll \|B\|_{1,\infty} = O(1)$). To see that **C.2** indeed takes place, note that in our situation, finding a solution $\widehat{z}$ to (6.25) reduces to solving the c.-c.s.p. problem (where $\alpha > 0, \beta > 0, p \in (1,2)$)

$$\min_{\|x\|_1 \leq 1}\max_{\|y\|_1 \leq 1}\left[\alpha\sum_i|x_i|^p - \beta\sum_i|y_i|^p + \sum_i[a_ix_i + b_iy_i + c_ix_iy_i]\right]. \quad (6.37)$$

By duality, this is equivalent to solving the c.-c.s.p. problem

$$\sup_{\mu \geq 0} \inf_{\nu \geq 0} \left[f(\mu, \nu) := \nu - \mu \right.$$
$$\left. + \sum_i \min_{x_i} \max_{y_i} \left[\alpha |x_i|^p + \mu |x_i| - \beta |y_i|^p - \nu |y_i| + a_i x_i + b_i y_i + c_i x_i y_i \right] \right].$$

The function $f(\mu, \nu)$ is convex-concave; computing first-order information on f reduces to solving n simple two-dimensional c.-c.s.p. problems $\min_{x_i} \max_{y_i} [...]$ and, for all practical purposes, costs only $O(n)$ operations. Then we can solve the (two-dimensional) c.-c.s.p. problem $\max_{\mu \geq 0} \min_{\nu \geq 0} f(\mu, \nu)$ by a polynomial-time first-order algorithm, such as the saddle-point version of the Ellipsoid method (see, e.g., Nemirovski et al., 2010). Thus, solving (6.37) within machine accuracy takes just $O(n)$ operations.

6.4.3   The Strongly Concave Case

6.4.3.1   *Situation and Assumptions*

Our current goal is to demonstrate that in the situation of the previous section, assuming that ϕ is strongly concave, we can improve the rate of convergence from $O(1/t)$ to $O(1/t^2)$. Let us consider the c.-c.s.p. problem (6.3) and assume that $\mathcal{X}$ is bounded (while $\mathcal{Y}$ can be unbounded), and that we are given norms $\|\cdot\|_x$, $\|\cdot\|_y$ on the corresponding embedding spaces E_x, E_y. We assume that we are also given a d.-g.f. $\omega_x(\cdot)$, compatible with $\|\cdot\|_x$, for $\mathcal{X}$, and a d.-g.f. $\omega_y(\cdot)$ compatible with $\|\cdot\|_y$, *for the entire E_y* (and not just for $\mathcal{Y}$). W.l.o.g. let $0 = \operatorname{argmin}_{E_y} \omega_y$. We keep assumption **C.1** intact and replace assumption **C.2** with its modification:

C.2′. It is easy to find a solution $\widehat{z}$ to the variational inequality (6.25) associated with $\mathcal{Z}$ and a monotone operator of the form $\Psi(x, y) = \alpha \mathcal{H}(x, y) + [\alpha_x \omega_x'(x) + e; \alpha_y \omega_y'((y - \bar{y})/R) + f]$ (where $\alpha, \alpha_x, \alpha_y, R$ are positive and $\bar{y} \in \mathcal{Y}$).

As above, it is easily seen that $\widehat{z} = (\widehat{x}, \widehat{y})$ is in fact a strong solution to the variational inequality: there exists $\zeta \in \mathcal{H}(\widehat{z})$ such that

$$\langle \alpha \zeta + [\alpha_x \omega_x'(\widehat{x}) + e; \alpha_y \omega_y'((\widehat{y} - \bar{y})/R) + f], u - \widehat{z} \rangle \geq 0 \ \forall u \in \mathcal{Z}. \quad (6.38)$$

We assume, as in the case of **C.2**, that when solving (6.25), we get both $\widehat{z}$ and ζ.

Furthermore, there are two new assumptions:

C.3. The function ϕ is strongly concave with modulus $\kappa > 0$ w.r.t. $\|\cdot\|_y$:

$$\forall \left(\begin{array}{c} x \in \mathcal{X}, y \in \operatorname{rint} \mathcal{Y}, f \in \partial_y[-\phi(x, y)], \\ y' \in \operatorname{rint} \mathcal{Y}, g \in \partial_y[-\phi(x, y')] \end{array} \right) : \langle f - g, y - y' \rangle \geq \kappa \|y - y'\|_y^2.$$

C.4. The E_x-component of $G(x, y)$ is independent of x, that is, $L_{xx} = 0$ (see (6.27)).

Note that **C.4** is automatically satisfied when $G(\cdot) = (\nabla_x \widetilde{\phi}(\cdot), -\nabla_y \widetilde{\phi}(\cdot))$ comes from a bilinear component $\widetilde{\phi}(x, y) = \langle a, x \rangle + \langle b, y \rangle + \langle y, Ax \rangle$ of ϕ.

Observe that since $\mathcal{X}$ is bounded, the function $\underline{\phi}(y) = \min_{x \in \mathcal{X}} \phi(x, y)$ is well defined and continuous on $\mathcal{Y}$; by **C.3**, this function is strongly concave and thus has bounded level sets. By remark 5.1, ϕ possesses saddle points, and since $\underline{\phi}$ is strongly convex, the y-component of a saddle point is the unique maximizer y_* of $\underline{\phi}$ on $\mathcal{Y}$. We set

$$x_c = \mathrm{argmin}_{\mathcal{X}} \omega_x(\cdot), \quad \Omega_x = \max_{\mathcal{X}} \omega_x(\cdot) - \min_{\mathcal{X}} \omega_x(\cdot),$$

$$\Omega_y = \max_{\|y\|_y \leq 1} \omega_y(y) - \min_y \omega_y(y) = \max_{\|y\|_y \leq 1} \omega_y(y) - \omega_y(0).$$

6.4.3.2 Algorithm MPb

The idea we intend to implement is the same one we used in Section 5.4 when designing MD for strongly convex optimization: all other things being equal, the efficiency estimate (5.28) is the better, the smaller the domain $\mathcal{Z}$ (cf. the factor Ω in (6.17)). On the other hand, when applying MP to a saddle-point problem with $\phi(x, y)$ which is strongly concave in y, we ensure a qualified rate of convergence of y^t to y_*, and thus eventually could replace the original domain $\mathcal{Z}$ with a smaller one by reducing the y-component. When it happens, we can run MP on this smaller domain, thus accelerating the solution process. This, roughly speaking, is what is going on in the algorithm MPb we are about to present.

Building Blocks. Let $R > 0$, $\bar{y} \in \mathcal{Y}$ and $\bar{z} = (x_c, \bar{y}) \in \mathcal{Z}$, so that $\bar{z} \in \mathcal{Z}$. Define the following entities:

$$\mathcal{Z}_R = \{(x; y) \in \mathcal{Z} : \|y - \bar{y}\|_y \leq R\},$$
$$\mathcal{L}_R = 2 L_{xy} \sqrt{\Omega_x \Omega_y} R + L_{yy} \Omega_y R^2,$$
$$\alpha = [L_{xy} \sqrt{\Omega_x \Omega_y} R]/\mathcal{L}_R, \quad \beta = [L_{xy} \sqrt{\Omega_x \Omega_y} R + L_{yy} \Omega_y R^2]/\mathcal{L}_R, \quad (6.39)$$
$$\omega^{R, \bar{y}}(x, y) = \frac{\alpha}{\Omega_x} \omega_x(x) + \frac{\beta}{\Omega_y} \omega_y([y - \bar{y}]/R),$$
$$\|(x, y)\| = \sqrt{\frac{\alpha}{\Omega_x} \|x\|_x^2 + \frac{\beta}{\Omega_y R^2} \|y\|_y^2}$$

with $\|(\xi, \eta)\|_* = \sqrt{\frac{\Omega_x}{\alpha} \|\xi\|_{x,*}^2 + \frac{\Omega_y R^2}{\beta} \|\eta\|_{y,*}^2}$. It is easily seen that $\omega^{R, \bar{y}}$ is a d.-g.f. for $\mathcal{Z}$ compatible with the norm $\| \cdot \|$, $\bar{z} = \mathrm{argmin}_{\mathcal{Z}} \omega^{R, \bar{y}}(\cdot)$, and

$$
\begin{aligned}
&(a) \quad \max_{\mathcal{Z}_R} \omega^{R, \bar{y}}(\cdot) - \min_{\mathcal{Z}_R} \omega^{R, \bar{y}}(\cdot) \leq 1, \\
&(b) \quad \forall (z, z' \in \mathcal{Z}) : \|G(z) - G(z')\|_* \leq \mathcal{L}_R \|z - z'\|.
\end{aligned}
\quad (6.40)
$$

For $u \in \mathcal{Z}$ and $z \in \mathcal{Z}^o$ we set $V_z^{R,\bar{y}}(u) = \omega^{R,\bar{y}}(u) - \omega^{R,\bar{y}}(z) - \langle (\omega^{R,\bar{y}}(z))', u - z \rangle$ and define the prox-mapping

$$\mathrm{Prox}_z^{R,\bar{y}}(\xi) = \mathrm{argmin}_{u \in \mathcal{Z}}[\langle \xi, u \rangle + V_z^{R,\bar{y}}(u)].$$

Let $z_1 = \bar{z}$ and $\gamma_t > 0$, $t = 1, 2, \dots$. Consider the following recurrence $\mathcal{B}$ (cf. Section 6.4.1):

(a) Given $z_t \in \mathcal{Z}^o$, we form the monotone operator $\Psi(z) = \gamma_t \mathcal{H}(z) + (\omega^{R,\bar{y}})'(z) - (\omega^{R,\bar{y}})'(z_t) + \gamma_t G(z_t)$ and solve the variational inequality (6.25) associated with $\mathcal{Z}$ and this operator; let the solution be denoted by w_t. Since the operator Ψ is of the form considered in **C.2'**, as a by-product of our computation we get a vector ζ_t such that $\forall u \in \mathcal{Z}$:

$$\zeta_t \in \mathcal{H}(w_t) \ \& \ \langle \gamma_t[\zeta_t + G(z_t)] + (\omega^{R,\bar{y}})'(w_t) - (\omega^{R,\bar{y}})'(z_t), u - w_t \rangle \geq 0 \quad (6.41)$$

(cf. (6.38)).

(b) Compute $z_{t+1} = \mathrm{Prox}_{z_t}^{R,\bar{y}}(\gamma_t(\zeta_t + G(w_t)))$ and

$$z^t(R, \bar{y}) \equiv (x^t(R, \bar{y}), y^t(R, \bar{y})) = \left[\sum\nolimits_{\tau=1}^{t} \gamma_\tau\right]^{-1} \sum\nolimits_{\tau=1}^{t} \gamma_\tau w_\tau.$$

Let

$$F_t = \zeta_t + G(w_t), \quad \delta_t = \langle \gamma_t F_t, w_t - z_{t+1} \rangle - V_{z_t}^{R,\bar{y}}(z_{t+1}).$$

Proposition 6.4. *Let assumptions* **C.1** *and* **C.2'**-**C.4** *hold. Let the stepsizes satisfy the conditions $\gamma_\tau \geq \mathcal{L}_R^{-1}$ and $\delta_\tau \leq 0$ for all τ (which certainly is so when $\gamma_\tau = \mathcal{L}_R^{-1}$ for all τ).*

(i) Assume that $\|\bar{y} - y_*\|_y \leq R$. Then for $x^t = x^t(R, \bar{y})$, $y^t = y^t(R, \bar{y})$ it holds that

$$
\begin{aligned}
(a) \quad \widetilde{\phi}_R(x^t) - \underline{\phi}(y^t) &\leq \left[\sum\nolimits_{\tau=1}^{t} \gamma_\tau\right]^{-1} \sum\nolimits_{\tau=1}^{t} \gamma_\tau \langle F_\tau, w_\tau - z_* \rangle \\
&\leq \left[\sum\nolimits_{\tau=1}^{t} \gamma_\tau\right]^{-1} \leq \frac{\mathcal{L}_R}{t}, \quad (6.42) \\
(b) \quad \|y^t - y_*\|_y^2 &\leq \frac{2}{\kappa}[\widetilde{\phi}_R(x^t) - \underline{\phi}(y^t)] \leq \frac{2\mathcal{L}_R}{\kappa t},
\end{aligned}
$$

where $\widetilde{\phi}_R(x) = \max_{y \in \mathcal{Y}: \|y - \bar{y}\|_y \leq R} \phi(x, y)$.

(ii) Further, if $\|\bar{y} - y_*\|_y \leq R/2$ and $t > \frac{8\mathcal{L}_R}{\kappa R^2}$, then $\widetilde{\phi}_R(x^t) = \overline{\phi}(x^t) :=$ $\max_{y \in \mathcal{Y}} \phi(x^t, y)$, and therefore

$$\epsilon_{\mathrm{sad}}(x^t, y^t) := \overline{\phi}(x^t) - \underline{\phi}(y^t) \leq \frac{\mathcal{L}_R}{t}. \quad (6.43)$$

Proof. (i): Exactly the same argument as in the proof of Proposition 6.3, with (6.40.b) in the role of (6.29), shows that

$$\forall u \in \mathcal{Z} : \sum_{\tau=1}^{t} \gamma_\tau \langle F_\tau, z_\tau - u \rangle \le V_{z_1}^{R,\bar{y}}(u) + \sum_{\tau=1}^{t} \delta_\tau$$

and that $\delta_\tau \le 0$, provided $\gamma_\tau = \mathcal{L}_R^{-1}$. Thus, under the premise of Proposition 6.4 we have

$$\sum_{\tau=1}^{t} \gamma_\tau \langle F_\tau, z_\tau - u \rangle \le V_{z_1}^{R,\bar{y}}(u) \ \forall u \in \mathcal{Z}.$$

When $u = (x, y) \in \mathcal{Z}_R$, the right-hand side of this inequality is ≤ 1 by (6.40.a) and due to $z_1 = \bar{z}$. Using the same argument as in item 2^0 of the proof of Proposition 6.1, we conclude that the left-hand side in the inequality is $\ge \left[\sum_{\tau=1}^{t} \gamma_\tau \right] \left[\phi(x^t, y) - \phi(x, y^t) \right]$. Thus,

$$\forall u \in \mathcal{Z}_R : \phi(x^t, y) - \phi(x, y^t) \le \left[\sum_{\tau=1}^{t} \gamma_\tau \right]^{-1} \sum_{\tau=1}^{t} \gamma_\tau \langle F_\tau, z_\tau - u \rangle.$$

Taking the supremum of the left hand side of this inequality over $u \in \mathcal{Z}_R$ and noting that $\gamma_\tau \ge \mathcal{L}_R^{-1}$, we arrive at (6.42.a). Further, $\|\bar{y} - y_*\| \le R$, whence $\widetilde{\phi}_R(x^t) \ge \phi(x^t, y_*) \ge \underline{\phi}(y_*)$. Since y_* is the maximizer of the strongly concave, modulus κ w.r.t. $\| \cdot \|_y$, function $\underline{\phi}(\cdot)$ over $\mathcal{Y}$, we have

$$\|y^t - y_*\|_y^2 \le \frac{2}{\kappa} [\underline{\phi}(y_*) - \underline{\phi}(y^t)] \le \frac{2}{\kappa} [\widetilde{\phi}_R(x^t) - \underline{\phi}(y^t)],$$

which is the first inequality in (6.42.b); the second inequality in (6.42.b) is given by (6.42.a). (i) is proved.

(ii): All we need to derive (ii) from (i) is to prove that under the premise of (ii), the quantities $\overline{\phi}(x^t) := \max_{y \in \mathcal{Y}} \phi(x^t, y)$ and $\widetilde{\phi}_R(x^t) := \max_{y \in \mathcal{Y}, \|y - \bar{y}\|_y \le R} \phi(x^t, y)$ are equal to each other. Assume that this is not the case, and let us lead this assumption to a contradiction. Looking at the definitions of $\overline{\phi}$ and $\widetilde{\phi}_R$, we see that in the case in question the maximizer $\widetilde{y}$ of $\phi(x^t, y)$ over $\mathcal{Y}_R = \{y :\in \mathcal{Y} : \|y - \bar{y}\|_y \le R\}$ satisfies $\|\bar{y} - \widetilde{y}\|_y = R$. Since $\|\bar{y} - y_*\|_y \le R/2$, it follows that $\|y_* - \widetilde{y}\|_y \ge R/2$. Because $y_* \in \mathcal{Y}_R$, $\widetilde{y} = \mathrm{argmax}_{y \in \mathcal{Y}_R} \phi(x^t, y)$ and $\phi(x^t, y)$ is strongly concave, modulus κ w.r.t. $\| \cdot \|_y$, we get $\phi(x^t, y_*) \le \phi(x^t, \widetilde{y}) - \frac{\kappa}{2} \|y_* - \widetilde{y}\|_y^2 \le \phi(x^t, \widetilde{y}) - \frac{\kappa R^2}{8}$, whence $\widetilde{\phi}_R(x^t) = \phi(x^t, \widetilde{y}) \ge \phi(x^t, y_*) + \frac{\kappa R^2}{8}$. On the other hand, $\phi(x^t, y_*) \ge \underline{\phi}(y_*) \ge \underline{\phi}(y^t)$, and we arrive at $\widetilde{\phi}_R(x^t) - \underline{\phi}(y^t) \ge \frac{\kappa R^2}{8}$. At the same time, (6.42.a) says that $\widetilde{\phi}_R(x^t) - \underline{\phi}(y^t) \le \mathcal{L}_R t^{-1} < \frac{\kappa R^2}{8}$, where the latter inequality is due to $t > \frac{8 \mathcal{L}_R}{\kappa R^2}$. We arrive at the desired contradiction. $\qquad\square$

Algorithm MPb. Let $R_0 > 0$ and $y^0 \in \mathcal{Y}$ such that

$$\|y^0 - y_*\| \le R_0/2 \tag{6.44}$$

are given, and let

$$R_k = 2^{-k/2} R_0,$$
$$N_k = \text{Ceil}\left(16\kappa^{-1}\left[2^{\frac{k+1}{2}} L_{xy}\sqrt{\Omega_x \Omega_y} R_0^{-1} + L_{yy}\Omega_y\right]\right),$$
$$M_k = \sum_{j=1}^{k} N_j, \ k = 1, 2, \dots$$

Execution of MPb is split into *stages* $k = 1, 2, \dots$. At the beginning of stage k, we have at our disposal $y^{k-1} \in \mathcal{Y}$ such that

$$\|y^{k-1} - y_*\|_y \le R_{k-1}/2. \tag{I_{k-1}}$$

At stage k, we compute $(\widehat{x}^k, \widehat{y}^k) = z^{N_k}(R_{k-1}, y^{k-1})$, which takes N_k steps of the recurrence $\mathcal{B}$ (where R is set to R_{k-1} and $\bar{y}$ is set to y^{k-1}). The stepsize policy can be an arbitrary policy satisfying $\gamma_\tau \ge \mathcal{L}_{R_{k-1}}^{-1}$ and $\delta_\tau \le 0$, e.g., $\gamma_\tau \equiv L_{R_{k-1}}^{-1}$; see Proposition 6.4. After $(\widehat{x}^k, \widehat{y}^k)$ is built, we set $y^k = \widehat{y}^k$ and pass to stage $k+1$.

Note that M_k is merely the total number of steps of $\mathcal{B}$ carried out in course of the first k stages of MPb.

The convergence properties of MPb are given by the following statement (which can be derived from Proposition 6.4 in exactly the same way that Proposition 5.4 was derived from Proposition 5.3):

Proposition 6.5. *Let assumptions* **C.1**, **C.2′–C.4** *hold, and let $R_0 > 0$ and $y^0 \in \mathcal{Y}$ satisfy (6.44). Then algorithm MPb maintains relations (I_{k-1}) and*

$$\epsilon_{\text{sad}}(\widehat{x}^k, \widehat{y}^k) \le \kappa 2^{-(k+3)} R_0^2, \tag{J_k}$$

$k = 1, 2, \dots$ *Further, let k_* be the smallest integer k such that $k \ge 1$ and $2^{\frac{k}{2}} \ge kR_0\frac{L_{xy}\sqrt{\Omega_x\Omega_y}}{L_{yy}\Omega_y + \kappa}$. Then*
— for $1 \le k < k_*$, we have $M_k \le O(1)k\frac{L_{yy}\Omega_y+\kappa}{\kappa}$ and $\epsilon_{\text{sad}}(\widehat{x}^k, \widehat{y}^k) \le \kappa 2^{-k}R_0^2$;
— for $k \ge k_*$, we have $M_k \le O(1)N_k$ and $\epsilon_{\text{sad}}(\widehat{x}^k, \widehat{y}^k) \le O(1)\frac{L_{xy}^2\Omega_x\Omega_y}{\kappa M_k^2}$.

Note that MPb behaves in the same way as the MD algorithm for strongly convex objectives (cf. Chapter 5, Section 5.4). Specifically, when the approximate solution y_k is far from the optimal solution y_*, the method converges linearly and switches to the sublinear rate (now it is $O(1/t^2)$) when approaching y_*.

6.4.3.3 *Illustration III*

As an instructive application example for algorithm MPb, consider the convex minimization problem

$$\text{Opt} = \min_{\xi \in \Xi} f(\xi), \quad f(\xi) = f_0(\xi) + \sum_{\ell=1}^{L} \tfrac{1}{2}\text{dist}^2(A_\ell \xi - b_\ell, U_\ell + V_\ell),$$
$$\text{dist}^2(w, W) = \min_{w' \in W} \|w - w'\|_2^2 \tag{6.45}$$

where

• $\Xi \subset E_\xi = \mathbb{R}^{n_\xi}$ is a convex compact set with a nonempty interior, E_ξ is equipped with a norm $\|\cdot\|_\xi$, and Ξ is equipped with a d.-g.f. $\omega_\xi(\xi)$ compatible with $\|\cdot\|_\xi$;

• $f_0(\xi) : \Xi \to \mathbb{R}$ is a simple continuous convex function, "simple" meaning that it is easy to solve auxiliary problems

$$\min_{\xi \in \Xi} \left\{ \alpha f_0(\xi) + a^T \xi + \beta \omega_\xi(\xi) \right\} \qquad [\alpha, \beta > 0]$$

• $U_\ell \subset \mathbb{R}^{m_\ell}$ are convex compact sets such that computing metric projection $\text{Proj}_{U_\ell}(u) = \text{argmin}_{u' \in U_\ell} \|u - u'\|_2$ onto U_ℓ is easy;

• $V_\ell \subset \mathbb{R}^{m_\ell}$ are polytopes given as $V_\ell = \text{Conv}\{v_{\ell,1}, ..., v_{\ell,n_\ell}\}$.

On a close inspection, problem (6.45) admits a saddle-point reformulation. Specifically, recalling that $S_k = \{x \in \mathbb{R}_+^k : \sum_i x_i = 1\}$ and setting

$$\mathcal{X} = \{x = [\xi; x^1; ...; x^L] \in \Xi \times S_{n_1} \times ... \times S_{n_L}\} \subset E_x = \mathbb{R}^{n_\xi + n_1 + ... + n_L},$$
$$\mathcal{Y} = E_y := \mathbb{R}_{y^1}^{m_1} \times ... \times \mathbb{R}_{y^L}^{m_L},$$
$$g(y = (y^1, ..., y^L)) = \sum_\ell g_\ell(y^\ell), \quad g_\ell(y^\ell) = \frac{1}{2}[y^\ell]^T y^\ell + \max_{u \in U_\ell} u_\ell^T y^\ell,$$
$$B_\ell = [v_{\ell,1}, ..., v_{\ell,n_\ell}],$$
$$A[\xi; x^1; ...; x^L] - b = [A_1 \xi - B_1 x^1; ...; A_L \xi - B_\ell x^L] - [b_1; ...; b_L],$$
$$\phi(x, y) = f_0(\xi) + y^T[Ax - b] - g(y),$$

we get a continuous convex-concave function ϕ on $\mathcal{X} \times \mathcal{Y}$ such that

$$f(\xi) = \min_{\eta = (x^1, ..., x^L):(\xi, \eta) \in \mathcal{X}} \max_{y \in \mathcal{Y}} \phi((\xi, \eta), y),$$

so that if a point $(x = [\xi; x^1; ...; x^L], y = [y^1; ...; y^L]) \in \mathcal{X} \times \mathcal{Y}$ is an ϵ-solution to the c.-c.s.p. problem $\inf_{x \in \mathcal{X}} \sup_y \phi(x, y)$, ξ is an ϵ-solution to the problem of interest (6.45):

$$\epsilon_{\text{sad}}(x, y) \leq \epsilon \Rightarrow f(\xi) - \text{Opt} \leq \epsilon.$$

Now we apply algorithm MPb to the saddle-point problem $\inf_{x \in \mathcal{X}} \sup_{y \in \mathcal{Y}} \phi(x, y)$.

The required setup is as follows:

1. Given positive $\alpha, \alpha_1, ..., \alpha_L$ (parameters of the construction), we equip the embedding space E_x of $\mathcal{X}$ with the norm

$$\|[\xi; x^1; ...; x^L]\|_x = \sqrt{\alpha\|\xi\|^2 + \sum\nolimits_{\ell=1}^{L} \alpha_\ell \|x^\ell\|_1^2},$$

and $\mathcal{X}$ itself with the d.-g.f.

$$\omega_x([\xi; x^1; ...; x^L]) = \alpha\omega_\xi(\xi) + \sum\nolimits_{\ell=1}^{L} \alpha_\ell \text{Ent}(x^\ell), \ \text{Ent}(u) = \sum\nolimits_{i=1}^{\dim u} u_i \ln u_i,$$

which, it can immediately be seen, is compatible with $\|\cdot\|_x$.

2. We equip $\mathcal{Y} = E_y = \mathbb{R}_y^{m_1 + ... + m_L}$ with the standard Euclidean norm $\|y\|_2$ and the d.-g.f. $\omega_y(y) = \frac{1}{2}y^T y$.

3. The monotone operator Φ associated with (ϕ, z) is

$$\Phi(x, y) = \{\partial_x[\phi(x,y) + \chi_\mathcal{X}(x)]\} \times \{\partial_y[-\phi(x,y)]\}, \ \chi_Q(u) = \begin{cases} 0, & u \in Q \\ +\infty, & u \notin Q \end{cases}.$$

We define its splitting, required by **C.1**, as

$$\mathcal{H}(x,y) = \{\{\partial_\xi[f_0(\xi) + \chi_\Xi(\xi)\} \times \{0\}... \times \{0\}\} \times \{\partial_y[\sum\nolimits_{\ell=1}^{L} g_\ell(y^\ell)]\},$$
$$G(x,y) = (\nabla_x[y^T[Ax - b]] = A^T y, -\nabla_y[y^T[Ax - b]] = b - Ax).$$

With this setup, we satisfy **C.1** and **C.3**-**C.4** (**C.3** is satisfied with $\kappa = 1$). Let us verify that **C.2'** is satisfied as well. Indeed, in our current situation, finding a solution $\hat{z}$ to (6.25) means solving the pair of convex optimization problems

$$\begin{aligned} (a) \quad & \min_{[\xi; x^1; ...; x^L] \in \mathcal{X}} \left[pa\omega_\xi(\xi) + qf_0(\xi) + e^T\xi \right. \\ & \left. \qquad\qquad + \sum\nolimits_{\ell=1}^{L} \left[pa_\ell \text{Ent}(x^\ell) + e_\ell^T x^\ell \right] \right] \\ (b) \quad & \min_{y=[y^1; ...; y^L]} \sum\nolimits_{\ell=1}^{L} \left[\frac{r}{2}[y^\ell]^T y^\ell + sg_\ell(y^\ell) + f_\ell^T y^\ell \right] \end{aligned} \tag{6.46}$$

where p, q, r, and s are positive. Due to the direct product structure of $\mathcal{X}$, (6.46.a) decomposes into the uncoupled problems $\min_{\xi \in \Xi}[pa\omega_\xi(\xi) + f_0(\xi) + e^T\xi]$ and $\min_{x^\ell \in S_\ell}[pa_\ell\text{Ent}(x^\ell) + e_\ell^T x^\ell]$. We have explicitly assumed that the first of these problems is easy; the remaining ones admit closed form solutions (cf. (5.39)). (6.46.b) also is easy: a simple computation yields $y^\ell = -\frac{1}{r+s}[s\text{Proj}_{U_\ell}(-s^{-1}f_\ell) + f_\ell]$, and it was assumed that it is easy to project onto U_ℓ.

The bottom line is that *we can solve* (6.45) *by algorithm MPb, the resulting efficiency estimate being*

$$f(\widehat{\xi}^k) - \text{Opt} \leq O(1)\frac{L_{xy}^2 \Omega_x}{M_k^2}, k \geq k_* = O(1)\ln(R_0 L_{xy}\sqrt{\Omega_x} + 2)$$

(see Proposition 6.5 and take into account that we are in the situation of $\kappa = 1, \Omega_y = \frac{1}{2}, L_{yy} = 0$). We can further use the parameters $\alpha, \alpha_1, ..., \alpha_L$ to optimize the quantity $L_{xy}^2 \Omega_x$. A rough optimization leads to the following: let μ_ℓ be the norm of the linear mapping $\xi \rightarrow A_\ell \xi$ induced by the norms $\|\cdot\|_\xi, \|\cdot\|_2$ in the argument and the image spaces, respectively, and let $\nu_\ell = \max_{1 \leq j \leq n_\ell}\|v_{\ell,j}\|_2$. Choosing

$$\alpha = \sum\nolimits_{\ell=1}^{L}\mu_\ell^2, \ \alpha_\ell = \nu_\ell^2, \ 1 \leq \ell \leq L$$

results in $L_{xy}^2 \Omega_x \leq O(1)\left[\Omega_\xi \sum_\ell \mu_\ell^2 + \sum_\ell \nu_\ell^2 \ln(n_\ell + 1)\right]$, $\Omega_\xi = \max_\Xi \omega_\xi(\cdot) - \min_\Xi \omega_\xi(\cdot)$.

6.5   Accelerating First-Order Methods by Randomization

We have seen in Section 6.2.1 that many important well-structured convex minimization programs reduce to just *bilinear* saddle-point problems

$$\text{SadVal} = \min_{x \in \mathcal{X} \subset E_x} \max_{y \in \mathcal{Y} \subset E_y} \ [\phi(x, y) := \langle a, x \rangle + \langle y, Ax - b \rangle], \tag{6.47}$$

the corresponding monotone operator admitting an *affine selection*

$$\begin{aligned}F(z = (x, y)) &= (a + A^T y, b - Ax) = (a, b) + \mathcal{F}z,\\ \mathcal{F}(x, y) &= (A^T y, -Ax).\end{aligned} \tag{6.48}$$

Computing the value of F requires two matrix-vector multiplications involving A and A^T. When $\mathcal{X}, \mathcal{Y}$ are simple and the problem is large-scale with dense A (which is the case in many machine learning and signal processing applications), these matrix-vector multiplications dominate the computational cost of an iteration of an FOM; as the sizes of A grow, these multiplications can become prohibitively time consuming. The idea of what follows is that *matrix-vector multiplications is easy to randomize, and this randomization, under favorable circumstances, allows for dramatic acceleration of FOMs in the extremely large-scale case.*

6.5.1 Randomizing Matrix-Vector Multiplications

Let $u \in \mathbf{R}^n$. Computing the image of u under a linear mapping $u \mapsto Bu = \sum_{j=1}^n u_j b_j : \mathbf{R}^n \to E$ are easy to randomize: treat the vector $[|u_1|; ...; |u_n|]/\|u\|_1$ as a probability distribution on the set $\{b_1, ..., b_n\}$, draw from this distribution a sample b_j and set $\xi_u = \|u\|_1 \text{sign}(u_j) b_j$, thus getting an *unbiased* ($\mathbf{E}\{\xi_u\} = Bu$) random estimate of Bu. When b_j are represented by readily available arrays, the arithmetic cost of sampling from the distribution P_u of ξ_u, modulo the setup cost $O(n)$ a.o. of computing the cumulative distribution $\{\|u\|_1^{-1} \sum_{i=1}^j |u_i|\}_{j=1}^n$ is just $O(\ln(n))$ a.o. to generate j plus $O(\dim E)$ a.o. to compute $\|u\|_1 \text{sign}(u_j) b_j$. Thus, the total cost of getting a single realization of ξ_u is $O(n) + \dim E$. For large n and $\dim E$ this is much less than the cost $O(n \dim E)$, assuming b_j are dense, of a straightforward precise computation of Bu.

We can generate a number k of independent samples $\xi^\ell \sim P_u$, $\ell = 1, ..., k$, and take, as an unbiased estimate of Bu, the average $\xi = \frac{1}{k} \sum_{\ell=1}^k \xi^\ell$, thus reducing the estimate's variability; with this approach, the setup cost is paid only once.

6.5.2 Randomized Algorithm for Solving Bilinear Saddle-Point Problem

We are about to present a randomized version MPr of the mirror-prox algorithm for solving the bilinear saddle-point problem (6.47).

6.5.2.1 *Assumptions and Setup*

1. As usual, we assume that $\mathcal{X}$ and $\mathcal{Y}$ are nonempty compact convex subsets of Euclidean spaces E_x, E_y; these spaces are equipped with the respective norms $\| \cdot \|_x$, $\| \cdot \|_y$, while $\mathcal{X}$, $\mathcal{Y}$ are equipped with d.-g.f.'s $\omega_x(\cdot)$, $\omega_y(\cdot)$ compatible with $\| \cdot \|_x$, resp., $\| \cdot \|_y$, and define Ω_x, Ω_y according to (6.28). Further, we define $\|A\|_{x,y}$ as the norm of the linear mapping $x \mapsto Ax : E_x \to E_y$ induced by the norms $\| \cdot \|_x$, $\| \cdot \|_y$ on the argument and the image spaces.

2. We assume that every point $u \in \mathcal{X}$ is associated with a probability distribution Π_u supported on $\mathcal{X}$ such that $\mathbf{E}_{\xi \sim \Pi_u}\{\xi\} = u$, for all $u \in \mathcal{X}$. Similarly, we assume that every point $v \in \mathcal{Y}$ is associated with a probability distribution P_v on E_y with a bounded support and such that $\mathbf{E}_{\eta \sim P_v}\{\eta\} = v$ for all $v \in \mathcal{Y}$. We refer to the case when P_v, for every $v \in \mathcal{Y}$, is supported on $\mathcal{Y}$, as the *inside* case, as opposed to the *general* case, where support of P_v, $v \in \mathcal{Y}$, does not necessarily belong to $\mathcal{Y}$. We will use Π_x, P_y to randomize matrix-vector multiplications. Specifically, given two positive integers k_x, k_y (parameters of our construction), and given $u \in \mathcal{X}$, we build a randomized estimate of Au as $A\xi_u$, where $\xi_u = \frac{1}{k_x} \sum_{i=1}^{k_x} \xi_i$ and ξ_i are sampled, independently of each other, from Π_u. Similarly, given $v \in \mathcal{Y}$, we estimate $A^T v$ by $A^T \eta_v$, where $\eta_v = \frac{1}{k_y} \sum_{i=1}^{k_y} \eta_i$, with η_i sampled independently of each other from P_v. Note that $\xi_v \in \mathcal{X}$, and in the inside case $\eta_u \in \mathcal{Y}$. Of course, a randomized estimation of Au, $A^T v$ makes sense only when computing $A\xi$, $\xi \in \mathrm{supp}(\Pi_u)$, $A^T \eta$, $\eta \in \mathrm{supp}(P_v)$ is much easier than computing Au, $A^T v$ for a general type u and v.

We introduce the quantities

$$\sigma_x^2 = \sup_{u \in \mathcal{X}} \mathbf{E}\{\|A[\xi_u - u]\|_{y,*}^2\}, \quad \sigma_y^2 = \sup_{v \in \mathcal{Y}} \mathbf{E}\{A^T[\eta_v - v]\|_{x,*}^2\},$$
$$\Theta = 2\left[\Omega_x \sigma_y^2 + \Omega_y \sigma_x^2\right]. \tag{6.49}$$

where ξ_u, η_v are the random vectors just defined, and, as always, $\|\cdot\|_{x,*}$, $\|\cdot\|_{y,*}$ are the norms conjugate to $\|\cdot\|_x$ and $\|\cdot\|_y$.

3. The setup for the algorithm MPr is given by the norm $\|\cdot\|$ on $E = E_x \times E_y \supset \mathcal{Z} = \mathcal{X} \times \mathcal{Y}$, and the compatible with this norm d.-g.f. $\omega(\cdot)$ for $\mathcal{Z}$ which are given by

$$\|(x,y)\| = \sqrt{\frac{1}{2\Omega_x}\|x\|_x^2 + \frac{1}{2\Omega_y}\|y\|_y^2}, \quad \omega(x,y) = \frac{1}{2\Omega_x}\omega_x(x) + \frac{1}{2\Omega_y}\omega_y(y),$$

so that

$$\|(\xi,\eta)\|_* = \sqrt{2\Omega_x\|\xi\|_{x,*}^2 + 2\Omega_y\|\eta\|_{y,*}^2}. \tag{6.50}$$

For $z \in \mathcal{Z}^o$, $w \in \mathcal{Z}$ let (cf. the definition (5.4))

$$V_z(w) = \omega(w) - \omega(z) - \langle \omega'(z), w - z \rangle,$$

and let $z_c = \mathrm{argmin}_{w \in \mathcal{Z}} \omega(w)$. Further, we assume that given $z \in \mathcal{Z}^o$ and $\xi \in E$, it is easy to compute the prox-mapping

$$\mathrm{Prox}_z(\xi) = \underset{w \in \mathcal{Z}}{\mathrm{argmin}}\left[\langle \xi, w \rangle + V_z(w)\right] \left(= \underset{w \in \mathcal{Z}}{\mathrm{argmin}}\left[\langle \xi - \omega'(z), w \rangle + \omega(w)\right]\right).$$

It can immediately be seen that

$$\Omega[\mathcal{Z}] = \max_{\mathcal{Z}} \omega(\cdot) - \min_{\mathcal{Z}} \omega(\cdot) = 1. \tag{6.51}$$

and the affine monotone operator $F(z)$ given by (6.48) satisfies the relation

$$\forall z, z' : \|F(z) - F(z')\|_* \leq \mathcal{L}\|z - z'\|, \quad \mathcal{L} = 2\|A\|_{x,y}\sqrt{\Omega_x\Omega_y}. \tag{6.52}$$

6.5.2.2   Algorithm

For simplicity, we present here the version of MPr where the number of steps, N, is fixed in advance. Given N, we set

$$\gamma = \min\left[\frac{1}{\sqrt{3}\mathcal{L}}, \frac{1}{\sqrt{3\Theta N}}\right] \tag{6.53}$$

and run N steps of the following randomized recurrence:

1. *Initialization:* We set $z_1 = \mathrm{argmin}_z\, \omega(\cdot)$.

2. *Step $t = 1, 2, ..., N$:* Given $z_t = (x_t, y_t) \in Z^o$, we generate ξ_{x_t}, η_{y_t} as explained above, set $\zeta_t = (\xi_{x_t}, \eta_{y_t})$, and compute $F(\zeta_t) = (a + A^T\eta_{y_t}, b - A\xi_{x_t})$ and

$$w_t = (\widehat{x}_t, \widehat{y}_t) = \mathrm{Prox}_{z_t}(\gamma F(\zeta_t)).$$

We next generate $\xi_{\widehat{x}_t}$, $\eta_{\widehat{y}_t}$ as explained above, set $\widehat{\zeta}_t = (\xi_{\widehat{x}_t}, \eta_{\widehat{y}_t})$, and compute $F(\widehat{\zeta}_t) = (a + A^T\eta_{\widehat{y}_t}, b - A\xi_{\widehat{x}_t})$ and $z_{t+1} = \mathrm{Prox}_{z_t}(\gamma F(\widehat{\zeta}_t))$.

3. *Termination $t = N$:* we output

$$z^N = (x^N, y^N) = \frac{1}{N}\sum_{t=1}^N (\xi_{\widehat{x}_t}, \eta_{\widehat{y}_t}), \ \text{ and } \ F(z^N) = \frac{1}{N}\sum_{t=1}^N F(\widehat{\zeta}_t)$$

(recall that $F(\cdot)$ is affine).

The efficiency estimate of algorithm MPr is given by the following

Proposition 6.6. *For every N, the random approximate solution $z^N = (x^N, y^N)$ generated by algorithm MPr possesses the following properties:*
(i) *In the inside case, $z^N \in Z$ and*

$$\mathbf{E}\{\epsilon_{\mathrm{sad}}(z^N)\} \le \epsilon_N := \max\left[\frac{2\sqrt{3\Theta}}{\sqrt{N}}, \frac{4\sqrt{3}\|A\|_{x,y}\sqrt{\Omega_x\Omega_y}}{N}\right]; \tag{6.54}$$

(ii) *In the general case, $x^N \in X$ and $\mathbf{E}\{\overline{\phi}(x^N)\} - \min_X \overline{\phi} \le \epsilon_N$.*

Observe that in the general case we do not control the error $\epsilon_{\mathrm{sad}}(z^N)$ of the saddle-point solution. Yet the bound (ii) of Proposition 6.6 allows us to control the accuracy $f(x^N) - \min_X f$ of the solution x^N when the saddle-point problem is used to minimize the convex function $f = \overline{\phi}$ (cf. (6.4)).

Proof. Setting $F_t = F(\zeta_t), \widehat{F}_t = F(\widehat{\zeta}_t), F_t^* = F(z_t), \widehat{F}_t^* = F(w_t), V_z(u) =$

$\omega(u) - \omega(z) - \langle \omega'(z), u - z \rangle$ and invoking Lemma 6.2, we get

$$\forall u \in \mathcal{Z} : \gamma \langle \widehat{F}_t, w_t - u \rangle \leq V_{z_t}(u) - V_{z_{t+1}}(u) + \Delta_t,$$
$$\Delta_t = \tfrac{1}{2} \left[\gamma^2 \| F_t - \widehat{F}_t \|_*^2 - \| z_t - w_t \|^2 \right],$$

whence, taking into account that $V_{z_1}(u) \leq \Omega[\mathcal{Z}] = 1$ (see (6.51)) and that $V_{z_{N+1}}(u) \geq 0$,

$$\forall u = (x, y) \in \mathcal{Z} : \gamma \sum_{t=1}^{N} \langle \widehat{F}_t, \widehat{\zeta}_t - u \rangle \leq 1 + \overbrace{\sum_{t=1}^{N} \Delta_t}^{\alpha_N} + \overbrace{\gamma \sum_{t=1}^{N} \langle \widehat{F}_t, \widehat{\zeta}_t - w_t \rangle}^{\beta_N}.$$

Substituting the values of $\widehat{F}_t$ and taking expectations, the latter inequality (where the right-hand side is independent of u) implies that

$$\mathbf{E} \left\{ \max_{(x,y) \in \mathcal{Z}} \gamma N \left[\phi(x^N, y) - \phi(x, y^N) \right] \right\} \leq 1 + \mathbf{E}\{\alpha_N\} + \mathbf{E}\{\beta_N\}, \quad (6.55)$$

$$\beta_N = \gamma \sum_{t=1}^{N} \left[\langle a, \xi_{\widehat{x}_t} - \widehat{x}_t \rangle + \langle b, \eta_{\widehat{y}_t} - \widehat{y}_t \rangle + \langle A \xi_{\widehat{x}_t}, \widehat{y}_t \rangle - \langle A^T \eta_{\widehat{x}_t}, \widehat{x}_t \rangle \right].$$

Now let $\mathbf{E}_{w_t}\{\cdot\}$ stand for the expectation conditional to the history of the solution process up to the moment when w_t is generated. We have $\mathbf{E}_{w_t}\{\xi_{\widehat{x}_t}\} = \widehat{x}_t$ and $\mathbf{E}_{w_t}\{\eta_{\widehat{y}_t}\} = \widehat{y}_t$, so that $\mathbf{E}\{\beta_N\} = 0$. Further, we have

$$\Delta_t \leq \frac{1}{2} \left[3\gamma^2 \left[\| \widehat{F}_t^* - F_t^* \|_*^2 + \| \widehat{F}_t^* - \widehat{F}_t \|_*^2 + \| F_t^* - F_t \|_*^2 \right] - \| z_t - w_t \|^2 \right]$$

and, recalling the origin of Fs, $\| F_t^* - \widehat{F}_t^* \|_* \leq \mathcal{L} \| z_t - w_t \|$ by (6.52). Since $3\gamma^2 \leq \mathcal{L}^2$ by (6.53), we get

$$\mathbf{E}\{\Delta_t\} \leq \frac{3\gamma^2}{2} \mathbf{E}\{\| \widehat{F}_t^* - \widehat{F}_t \|_*^2 + \| F_t^* - F_t \|_*^2\} \leq 3\gamma^2 \Theta,$$

where the concluding inequality is due to the definitions of Θ and of the norm $\| \cdot \|_*$ (see (6.49) and (6.50), respectively). Thus, (6.55) implies that

$$\mathbf{E} \left\{ \max_{(x,y) \in \mathcal{Z}} \left[\phi(x^N, y) - \phi(x, y^N) \right] \right\} \leq 1/(N\gamma) + 3\Theta\gamma \leq \epsilon_N, \quad (6.56)$$

due to the definition of ϵ_N. Now, in the inside case we clearly have $(x^N, y^N) \in \mathcal{Z}$, and therefore (6.56) implies (6.54). In the general case we have $x^N \in \mathcal{X}$. In addition, let x_* be the x-component of a saddle point of ϕ on $\mathcal{Z}$. Replacing in the left-hand side of (6.56) maximization over all pairs (x, y) from $\mathcal{Z}$ with maximization only over the pair (x_*, y) with $y \in \mathcal{Y}$ (which can only decrease the left-hand side), we get from (6.56) that

$$\mathbf{E}\{\overline{\phi}(x^N)\} \leq \epsilon_N + \mathbf{E}\{\phi(x_*, y^N)\} = \epsilon_N + \phi\big(x_*, \mathbf{E}\{y^N\}\big). \tag{6.57}$$

Observe that $\mathbf{E}_{w_t}\{\eta_{\widehat{y}_t}\} = \widehat{y}_t \in \mathcal{Y}$. We conclude that

$$\mathbf{E}\{y^N\} = \mathbf{E}\left\{ \frac{1}{N} \sum_{t=1}^{N} \eta_{\widehat{y}_t} \right\} = \mathbf{E}\left\{ \frac{1}{N} \sum_{t=1}^{N} \widehat{y}_t \right\} \in \mathcal{Y}.$$

Thus, the right-hand side in (6.57) is $\leq \epsilon_N + \text{SadVal}$, and (ii) follows. $\qquad \square$

Remark 6.1. *We stress here that MPr, along with the approximate solution (x^N, y^N), returns the value $F(x^N, y^N)$. This allows for easy computation, not requiring matrix-vector multiplications, of $\overline{\phi}(x^N)$ and $\underline{\phi}(y^N)$.*

6.5.2.3   *Illustration IV: ℓ_1-Minimization*

Consider problem (6.5) with $\Xi = \{\xi \in \mathbb{R}^n : \|\xi\|_1 \leq 1\}$. Representing Ξ as the image of the standard simplex $S_{2n} = \{x \in \mathbb{R}_+^{2n} : \sum_i x_i = 1\}$ under the mapping $x \mapsto J_n x$, $J_n = [I_n, -I_n]$, the problem reads

$$\text{Opt} = \min_{x \in S_{2n}} \|Ax - b\|_p \quad [A \in \mathbb{R}^{m \times 2n}]. \tag{6.58}$$

We consider two cases: $p = \infty$ (*uniform fit, as in the Dantzig selector*) and $p = 2$ (*ℓ_2-fit, as in Lasso*).

Uniform Fit. Here (6.58) can be converted into the bilinear saddle-point problem

$$\text{Opt} = \min_{x \in S_{2n}} \max_{y \in S_{2m}} \left[\phi(x, y) := y^T J_m^T [Ax - b] \right]. \tag{6.59}$$

Setting $\|\cdot\|_x = \|\cdot\|_1$, $\omega_x(x) = \text{Ent}(x)$, $\|\cdot\|_y = \|\cdot\|_1$, $\omega_y(y) = \text{Ent}(y)$, let us specify Π_u, $u \in S_{2n}$, and P_v, $v \in S_{2m}$, according to the recipe from Section 6.5.1, that is, the random vector $\xi_u \sim \Pi_u$ with probability u_i is the ith basic orth, $i = 1, ..., m$, and similarly for $\eta_v \sim P_v$. This is the inside case, and when we set $\|A\|_{1,\infty} = \max_{i,j} |A_{ij}|$, we get $\sigma_x^2 = O(1) \frac{\|A\|_{1,\infty}^2 \ln(2m)}{k_x + \ln(2m)}$, $\sigma_y^2 =$

$O(1)\frac{\|A\|_{1,\infty}^2 \ln(2n)}{k_y + \ln(2n)}$,[2] and

$$\Omega_x = \ln(2n),\ \Omega_y = \ln(2m),\ \mathcal{L} = 2\|A\|_{1,\infty}\sqrt{\ln(2n)\ln(2m)},$$
$$\Theta \leq O(1)\|A\|_{1,\infty}^2 \left[\frac{\ln^2(2m)}{k_x + \ln(2m)} + \frac{\ln^2(2n)}{k_y + \ln(2n)}\right]$$

In this setting Proposition 6.6 reads:

Corollary 6.7. *For all positive integers* k_x, k_y, N *one can find a random feasible solution* (x^N, y^N) *to (6.59) along with the quantities* $\overline{\phi}(x^N) = \|Ax^N - b\|_\infty \geq \text{Opt}$ *and a lower bound* $\underline{\phi}(y^N)$ *on* Opt *such that*

$$\text{Prob}\left\{\overline{\phi}(x^N) - \underline{\phi}(y^N) \leq O(1)\frac{\|A\|_{1,\infty}\ln(2mn)}{\sqrt{N}\sqrt{\min[N, k_x + \ln(2m), k_y + \ln(2n)]}}\right\} \geq \frac{1}{2} \tag{6.60}$$

in N *steps, the computational effort per step dominated by the necessity to extract* $2k_x$ *columns and* $2k_y$ *rows from* A *, given their indexes.*

Note that our computation yields, along with (x^N, y^N), the quantities $\overline{\phi}(x^N)$ and $\underline{\phi}(y^N)$. Thus, when repeating the computation ℓ times and choosing the best among the resulting x- and y-components of the solutions we make the probability of the left-hand side event in (6.60) as large as $1 - 2^{-\ell}$. For example, with $k_x = k_y = 1$, assuming $\delta = \epsilon/\|A\|_{1,\infty} \leq 1$, finding an ϵ-solution to (6.59) with reliability $\geq 1 - \beta$ costs $O(1)\ln^2(2mn)\ln(1/\beta)\delta^{-2}$ steps of the outlined type, that is, $O(1)(m+n)\ln^2(2mn)\ln(1/\beta)\delta^{-2}$ a.o. For comparison, when δ stays fixed and m, n are large, the lowest known (so far) cost of finding an ϵ-solution to problem (6.58) with unform fit is $O(1)\sqrt{\ln(m)\ln(n)}\delta^{-1}$ steps, with the effort per step dominated by the necessity to compute $O(1)$ matrix-vector multiplications involving A and A^T (this cost is achieved by Nesterov's smoothing or with MP; see (6.22)). When A is a general-type dense $m \times n$ matrix, the cost of the deterministic computation is $O(1)mn\sqrt{\ln(m)\ln(n)}\delta^{-1}$. We see that for fixed relative accuracy δ and large m, n, randomization does accelerate the solution process, the gain growing with m, n.

---

2. The bound for σ_x^2 and σ_y^2 is readily given by the following fact (see, e.g., Juditsky and Nemirovski, 2008): when $\xi_1, ..., \xi_k \in \mathbf{R}^n$ are independent zero mean random vectors with $\mathbf{E}\{\|\xi_i\|_\infty^2\} \leq 1$ for all i, one has $\mathbf{E}\{\|\frac{1}{k}\sum_{i=1}^k \xi_i\|_\infty^2\} \leq O(1)\min[1, \ln(n)/k]$; this inequality remains true when $\mathbf{R}^n$ is replaced with $\mathbf{S}^n$, and $\|\cdot\|_\infty$ is replaced with the standard matrix norm (largest singular value).

ℓ_2-**fit.** Here (6.58) can be converted into the bilinear saddle-point problem

$$\text{Opt} = \min_{x \in S_{2n}} \max_{\|y\|_2 \leq 1} \left[\phi(x,y) := y^T [Ax - b] \right]. \qquad (6.61)$$

In this case we keep $\| \cdot \|_x = \| \cdot \|_1$, $\omega_x(x) = \text{Ent}(x)$, and set $\| \cdot \|_y = \| \cdot \|_2$, $\omega_y(y) = \frac{1}{2} y^T y$. We specify Π_u, $u \in S_{2n}$, exactly as in the case of uniform fit, and define P_v, $v \in \mathcal{Y} = \{y \in \mathbb{R}^m : \|y\|_2 \leq 1\}$ as follows: $\eta_v \sim P_v$ takes values $\text{sign}(u_i)\|u\|_1 e_i$, e_i being basic orths, with probabilities $|u_i|/\|u\|_1$. Note that we are not in the inside case anymore. Setting $\|A\|_{1,2} = \max_{1 \leq j \leq 2n} \|A_j\|_2$, A_j being the columns of A, we get

$$\Omega_x = \ln(2n), \Omega_y = \tfrac{1}{2}, \mathcal{L} = \|A\|_{1,2}\sqrt{2\ln(2n)},$$
$$\Theta \leq O(1)\left[\tfrac{1}{k_x}\|A\|_{1,2}^2 + \tfrac{\ln^2(2n)}{k_y+\ln(2n)}[\|A\|_{1,2} + \sqrt{m}\|A\|_{1,\infty}]^2 \right].$$

Now Proposition 6.6 reads:

Corollary 6.8. *For all positive integers k_x, k_y, N, one can find a random feasible solution x^N to (6.58) (where $p = 2$), along with the vector Ax^N, such that*

$$\text{Prob}\Big\{ \|Ax^N - b\|_2 \leq \text{Opt}$$
$$+ O(1)\frac{\|A\|_{1,2}\sqrt{\ln(2n)}}{\sqrt{N}}\sqrt{\tfrac{1}{N} + \tfrac{1}{k_x\ln(2n)} + \tfrac{\ln(2n)\Gamma^2(A)}{k_y+\ln(2n)}} \Big\} \geq \tfrac{1}{2}, \quad (6.62)$$
$$\Gamma(A) = \sqrt{m}\|A\|_{1,\infty}/\|A\|_{1,2}$$

in N steps, the computational effort per step dominated by the necessity to extract $2k_x$ columns and $2k_y$ rows from A, given their indexes.

Here again, repeating the computation ℓ times and choosing the best among the resulting solutions to (6.58), we make the probability of the left-hand side event in (6.62) as large as $1-2^{-\ell}$. For instance, with $k_x = k_y = 1$, assuming $\delta := \epsilon/\|A\|_{1,2} \leq 1$, finding an ϵ-solution to (6.58) with reliability $\geq 1-\beta$ costs $O(1)\ln(2n)\ln(1/\beta)\Gamma^2(A)\delta^{-2}$ steps of the outlined type, that is, $O(1)(m+n)\ln(2n)\ln(1/\beta)\Gamma^2(A)\delta^{-2}$ a.o. Assuming that a precise multiplication of a vector by A takes $O(mn)$ a.o., the best known (so far) deterministic counterpart of the above complexity bound is $O(1)mn\sqrt{\ln(2n)}\delta^{-1}$ a.o. (cf. (6.22)). Now the advantages of randomization when δ is fixed and m, n are large are not as evident as in the case of uniform fit, since the complexity bound for the randomized computation contains an extra factor $\Gamma^2(A)$ which may be as large as $O(m)$. Fortunately, we may "nearly kill" $\Gamma(A)$ by randomized preprocessing of the form $[A, b] \mapsto [\bar{A}, \bar{b}] = [UDA, UDb]$, where U is a deterministic orthogonal matrix with entries of order $O(1/\sqrt{m})$, and D is a random diagonal matrix with i.i.d. diagonal entries taking values ± 1 with equal probabilities. This preprocessing converts (6.58) into an equivalent problem, and it is easily seen that for every $\beta \ll 1$, for the transformed matrix $\bar{A}$ with probability $\geq 1 - \beta$ it holds that $\Gamma(\bar{A}) \leq O(1)\sqrt{\ln(mn\beta^{-1})}$. This implies that, modulo preprocessing's cost, the complexity estimate of the randomized computation reduces to $O(1)(m+n)\ln(n)\ln(mn/\beta)\ln(1/\beta)\delta^{-2}$. Choosing U as a cosine transform or Hadamard matrix, so that the cost of computing Uu is $O(m\ln(m))$ a.o., the cost of preprocessing does not exceed $O(mn\ln(m))$, which, for small δ, is a small fraction of the cost of deterministic computation. Thus, there is a meaningful range of values of δ, m, n where randomization is highly profitable. It should be added that in some applications (e.g., in compressed sensing) typical values of $\Gamma(A)$ are quite moderate, and thus no preprocessing is needed.

6.6 Notes and Remarks

1. The research of the second author was partly supported by ONR grant N000140811104, BSF grant 2008302, and NSF grants DMI-0619977 and DMS-0914785.

2. The mirror-prox algorithm was proposed by Nemirovski (2004); its modification able to handle the stochastic case, where the precise values of the monotone operator associated with (6.3) are replaced by unbiased random estimates of these values (cf. Chapter 5, Section 5.5) is developed by Juditsky et al. (2008). The MP combines two basic ideas: (a) averaging of the search trajectory to get approximate solutions (this idea goes back to Bruck (1977) and Nemirovskii and Yudin (1978)) and (b) exploiting *extragradient* steps: instead of the usual gradient-type update $z \mapsto z_+ = \mathrm{Prox}_z(\gamma F(z))$ used in the saddle-point MP (Section 5.6), the update $z \mapsto w = \mathrm{Prox}_z(\gamma F(z)) \mapsto z_+ = \mathrm{Prox}_z(\gamma F(w))$ is used. This construction goes back to Korpelevich (1983, 1976), see also Noor (2003) and references therein. Note that a different implementation of the same ideas is provided by Nesterov (2007b) in his dual extrapolation algorithm.

3. The material in Sections 6.4.1 and 6.4.3 is new; this being said, problem settings and complexity results considered in these sections (but not the associated algorithms) are pretty close, although not fully identical, to those covered by the excessive gap technique of Nesterov (2005b). For example, the situation considered in illustration III can be treated equally well via Nesterov's technique, which perhaps is not the case for illustration II. It should be added that splitting like the one in Section 6.4.1, in a slightly more general context of variational inequalities with monotone

operators, was considered by Tseng (2000), although without averaging and thus without any efficiency estimate. These missing elements were added in the recent papers of Monteiro and Svaiter (2010b,a) which in this respect can be viewed as independently developed Euclidean case version of Section 6.4.1. For other schemes of accelerating FOMs via exploiting a problem's structure, see Nesterov (2007a), Beck and Teboulle (2009), Tseng (2008), Goldfarb and Scheinberg (2010), and references therein.

4. The material of Section 6.5.1 originated with Juditsky et al. (2010), where one can find various versions of MPr and (rather encouraging) results of preliminary numerical experiments. Note that the "cheap randomized matrix-vector multiplication" outlined in Section 6.5.1 admits extensions which can be useful when solving semidefinite programs (see Juditsky et al., 2010, Section 2.1.4).

Obviously, the idea of improving the numerical complexity of optimization algorithms by utilizing random subsampling of problem data is not new. For instance, such techniques have been applied to support vector machine classification in Kumar et al. (2008), and to solving certain semidefinite programs in Arora and Kale (2007) and d'Aspremont (2009). Furthermore, as we have already mentioned, both MD and MP admit modifications (see Nemirovski et al., 2009; Juditsky et al., 2008) capable to handle c.-c.s.p. problems (not necessarily bilinear) in the situation where instead of the precise values of the associated monotone operator, unbiased random estimates of these values are used. A common drawback of these modifications is that while we have at our disposal explicit nonasymptotical upper bounds on the *expected* inaccuracy of random approximate solutions z^N (which, as in the basic MP, are averages of the search points w_t) generated by the algorithm, we do *not* know what the actual quality of z^N is. In the case of a bilinear problem (6.47) and with the randomized estimates of $F(w_t)$ defined as $F(\widehat{\zeta}_t)$, we get a new option: to define z^N as the average of the points $\widehat{\zeta}_t$. As a result, we do know $F(z^N)$ and thus can easily assess the quality of z^N (n.b. remark 6.1). To the best of our knowledge, this option has been realized (implicitly) only once, namely, in the randomized sublinear-time matrix game algorithm of Grigoriadis and Khachiyan (1995) (that ad hoc algorithm is close, although not identical, to MPr as applied to problem (6.59), which is equivalent to a matrix game).

On the other hand, the possibility to assess, in a computationally cheap fashion, the quality of an approximate solution to (6.47) is crucial when solving *parametric* bilinear saddle-point problems. Specifically, many important applications reduce to problems of the form

$$\max\left\{\rho : \mathrm{SadVal}(\rho) := \min_{x\in\mathcal{X}}\max_{y\in\mathcal{Y}}\phi_\rho(x,y) \le 0\right\}, \tag{6.63}$$

where $\phi_\rho(x,y)$ is a bilinear function affinely depending on ρ. For example, the ℓ_1-minimization problem as it arises in sparsity-oriented signal processing is $\mathrm{Opt} = \min_\xi\{\|\xi\|_1 : \|A\xi - b\|_p \le \delta\}$, which is nothing but

$$\frac{1}{\mathrm{Opt}} = \max\left\{\rho : \mathrm{SadVal}(\rho) := \min_{\|x\|_1\le 1}\max_{\|y\|_{p/(p-1)}\le 1}\left[y^T[Ax - \rho b] - \rho\delta\right] \le 0\right\}.$$

From the complexity viewpoint, the best known (to us) way to process (6.63) is to solve the master problem $\max\{\rho : \mathrm{SadVal}(\rho) \le 0\}$ by an appropriate first-order root-finding routine, the (approximate) first-order information on $\mathrm{SadVal}(\cdot)$ being provided by a first-order saddle-point algorithm. The ability of the MPr algorithm to provide accurate bounds of the value $\mathrm{SadVal}(\cdot)$ of the inner saddle-point problems makes it the method of choice when solving extremely large parametric saddle-point problems (6.63). For more details on this subject, see Juditsky et al. (2010).

6.7   References

S. Arora and S. Kale. A combinatorial primal-dual approach to semidefinite programs. In: *Proceedings of the 39th Annual ACM Symposium on the Theory of Computations*, pages 227–236, 2007.

A. Beck and M. Teboulle. A fast iterative shrinkage-thresholding algorithm for linear inverse problems. *SIAM Journal on Imaging Sciences*, 2(1):183–202, 2009.

R. Bruck. On weak convergence of an ergodic iteration for the solution of variational inequalities with monotone operators in Hilbert space. *Journal of Mathematical Analysis and Applications*, 61(1):15–164, 1977.

A. d'Aspremont. Subsampling algorithms for semidefinite programming. Technical report, arXiv:0803.1990v5, http://arxiv.org/abs/0803.1990, November 2009.

D. Goldfarb and K. Scheinberg. Fast first order method for separable convex optimization with line search. Technical report, Department of Industrial Engineering and Operations Research, Columbia University, 2010.

M. D. Grigoriadis and L. G. Khachiyan. A sublinear-time randomized approximation algorithm for matrix games. *Operations Research Letters*, 18(2):53–58, 1995.

A. Juditsky and A. Nemirovski. Large deviations of vector-valued martingales in 2-smooth normed spaces. Technical report, HAL: hal-00318071, http://hal.archives-ouvertes.fr/hal-00318071/, 2008.

A. Juditsky, A. Nemirovski, and C. Tauvel. Solving variational inequalities with stochastic mirror prox algorithm. Technical report, HAL: hal-00318043, http://hal.archives-ouvertes.fr/hal-00318043/, 2008.

A. Juditsky, F. K. Karzan, and A. Nemirovski. ℓ_1-minimization via randomized first order algorithms. Technical report, Optimization Online, http://www.optimization-online.org/DB_FILE/2010/05/2618.pdf, 2010.

G. M. Korpelevich. The extragradient method for finding saddle points and other problems. *Ekonomika i Matematicheskie Metody*, 12:747–756, 1976. (in Russian).

G. M. Korpelevich. Extrapolation gradient methods and relation to modified lagrangeans. *Ekonomika i Matematicheskie Metody*, 19:694–703, 1983. (in Russian).

K. Kumar, C. Bhattacharya, and R. Hariharan. A randomized algorithm for large scale support vector learning. In J. Platt, D. Koller, Y. Singer, and S. Roweis, editors, *Advances in Neural Information Processing Systems*, volume 20. MIT Press, 2008.

R. D. C. Monteiro and B. F. Svaiter. Complexity of vairants of Tseng's modified F-B splitting and Korpelevich's methods for generalized variational inequalities with applications to saddle point and convex optimization problems. Technical report, Optimization Online, http://www.optimization-online.org/DB_HTML/2010/07/2675.html, 2010a.

R. D. C. Monteiro and B. F. Svaiter. On the complexity of the hybrid proximal extragradient method for the iterates and the ergodic mean. *SIAM Journal on Optimization*, 20:2755–2787, 2010b.

A. Nemirovski. Prox-method with rate of convergence $o(1/t)$ for variational inequalities with lipschitz continuous monotone operators and smooth convex-concave saddle-point problems. *SIAM Journal on Optimization*, 15:229–251, 2004.

A. Nemirovski, A. Juditsky, G. Lan, and A. Shapiro. Robust stochastic approximation approach to stochastic programming. *SIAM Journal on Optimization*, 19 (4):1574–1609, 2009.

A. Nemirovski, S. Onn, and U. Rothblum. Accuracy certificates for computational problems with convex structure. *Mathematics of Operations Research*, 35:52–78, 2010.

A. Nemirovskii and D. Yudin. On Cezari's convergence of the steepest descent method for approximating saddle points of convex-concave functions. *Soviet Math. Doklady*, 19(2), 1978.

Y. Nesterov. A method for solving a convex programming problem with rate of convergence $o(1/k^2)$. *Soviet Math. Doklady*, 27(2):372–376, 1983.

Y. Nesterov. Smooth minimization of nonsmooth functions. *Mathematical Programming, Series A*, 103:127–152, 2005a.

Y. Nesterov. Excessive gap technique in nonsmooth convex minimization. *SIAM Journal on Optimization*, 16(1):235–239, 2005b.

Y. Nesterov. Gradient methods for minimizing composite objective function. Technical Report 2007/76, Center for Operations Rersearch and Econometrics, Catholic University of Louvain, http://www.uclouvain.be/cps/ucl/doc/core/documents/coredp2007_76.pdf, 2007a.

Y. Nesterov. Dual extrapolation and its application for solving variational inequalities and related problems. *Mathematical Programming, Series A*, 109(2–3): 319–344, 2007b.

M. A. Noor. New extragradient-type methods for general variational inequalities.

Journal of Mathematical Analysis and Applications, 277:379–394, 2003.

P. Tseng. A modified forward-backward splitting method for maximal monotone mappings. *SIAM Journal on Control and Optimization*, 38(2):431–446, 2000.

P. Tseng. On accelerated proximal gradient methods for convex-concave optimization. Technical report,
http://www.math.washington.edu/~tseng/papers/apgm.pdf, 2008.

7   Cutting-Plane Methods in Machine Learning

Vojtěch Franc   xfrancv@cmp.felk.cvut.cz
Czech Technical University in Prague
Technická 2, 166 27 Prague 6
Czech Republic

Sören Sonnenburg   Soeren.Sonnenburg@tu-berlin.de
Berlin Institute of Technology
Franklinstr. 28/29
10587 Berlin, Germany

Tomáš Werner   werner@cmp.felk.cvut.cz
Czech Technical University in Prague
Technická 2, 166 27 Prague 6
Czech Republic

Cutting-plane methods are optimization techniques that incrementally construct an approximation of a feasible set or an objective function by linear inequalities called cutting planes. Numerous variants of this basic idea are among standard tools used in convex nonsmooth optimization and integer linear programing. Recently, cutting-plane methods have seen growing interest in the field of machine learning. In this chapter, we describe the basic theory behind these methods and show three of their successful applications to solving machine learning problems: regularized risk minimization, multiple kernel learning, and MAP inference in graphical models.

Many problems in machine learning are elegantly translated into convex optimization problems, which, however, are sometimes difficult to solve efficiently with off-the-shelf solvers. This difficulty can stem from complexity of either the feasible set or the objective function. Often, these can be accessed only indirectly via an oracle. To access a feasible set, the oracle either asserts that a given query point lies in the set or finds a hyperplane that separates the point from the set. To access an objective function, the oracle returns the value and a subgradient of the function at the query point. Cutting-plane methods solve the optimization problem by approximating the feasible set or the objective function by a bundle of linear inequalities, called cutting planes. The approximation is iteratively refined by adding new cutting planes computed from the responses of the oracle.

Cutting-plane methods have been extensively studied in the literature. We refer to Boyd and Vandenberge (2008) for an introductory yet comprehensive overview. For the sake of self-consistency, we review the basic theory in Section 7.1. Then, in three separate sections, we describe their successful applications to three machine learning problems.

The first application, Section 7.2, is on learning linear predictors from data based on *regularized risk minimization* (RRM). RRM often leads to a convex but nonsmooth task, which cannot be efficiently solved by general-purpose algorithms, especially for large-scale data. Prominent examples of RRM are support vector machines, logistic regression, and structured output learning. We review a generic risk minimization algorithm proposed by Teo et al. (2007, 2010), inspired by a variant of cutting-plane methods known as proximal bundle methods. We also discuss the accelerated version (Franc and Sonnenburg, 2008, 2010; Teo et al., 2010), which is among the fastest solvers for large-scale learning.

The second application, Section 7.3, is *multiple kernel learning* (MKL). Although classical kernel-based learning algorithms use a single kernel, it is sometimes desirable to use multiple kernels (Lanckriet et al., 2004a). Here, we focus on the convex formulation of the MKL problem for classification as first stated in Zien and Ong (2007) and Rakotomamonjy et al. (2007). We show how this problem can be efficiently solved by a cutting-plane algorithm recycling standard SVM implementations. The resulting MKL solver is equivalent to the column generation approach applied to the semi-infinite programming formulation of the MKL problem proposed by Sonnenburg et al. (2006a).

The third application, Section 7.4, is *maximum a posteriori (MAP) inference in graphical models*. It leads to a combinatorial optimization problem which can be formulated as a linear optimization over the marginal polytope (Wainwright and Jordan, 2008). Cutting-plane methods iteratively construct a sequence of progressively tighter outer bounds of the marginal polytope that corresponds to a sequence of LP relaxations. We revisit the approach by Werner (2008a, 2010), in which a dual cutting-plane method is a straightforward extension of a simple message-passing algorithm. It is a generalization of the dual LP relaxation approach of Shlezinger (1976) and of the max-sum diffusion algorithm by Kovalevsky and Koval.

7.1   Introduction to Cutting-plane Methods

Suppose we want to solve the optimization problem

$$\min\{\, f(\boldsymbol{x}) \mid \boldsymbol{x} \in X \,\}\,, \tag{7.1}$$

where $X \subseteq \mathbb{R}^n$ is a convex set, $f \colon \mathbb{R}^n \to \mathbb{R}$ is a convex function, and we assume that the minimum exists. Set X can be accessed only via the *separation oracle* (or *separation algorithm*). Given $\hat{\boldsymbol{x}} \in \mathbb{R}^n$, the separation oracle either asserts that $\hat{\boldsymbol{x}} \in X$ or returns a hyperplane $\langle \boldsymbol{a}, \boldsymbol{x} \rangle \leq b$ (called a *cutting plane*) that separates $\hat{\boldsymbol{x}}$ from X, that is, $\langle \boldsymbol{a}, \hat{\boldsymbol{x}} \rangle > b$ and $\langle \boldsymbol{a}, \boldsymbol{x} \rangle \leq b$ for all $\boldsymbol{x} \in X$. Figure 7.1(a) illustrates this.

The *cutting-plane algorithm* (algorithm 7.1) solves (7.1) by constructing progressively tighter convex polyhedrons X_t containing the true feasible set X, by cutting off infeasible parts of an initial polyhedron X_0. It stops when $\boldsymbol{x}_t \in X$ (possibly up to some tolerance).

The trick behind the method is not to approximate X well by a convex polyhedron, but to do so *only near the optimum*. This is best seen if X is already a convex polyhedron described by a set of linear inequalities. At optimum, only some of the inequalities are active. We could in fact remove all the inactive inequalities without affecting the problem. Of course, we do not know which ones to remove until we know the optimum. The cutting-plane algorithm imposes more than the minimal set of inequalities, but possibly many fewer than the whole original description of X.

Algorithm 7.1 Cutting-plane algorithm

1:   **Initialization:** $t \leftarrow 0$, $X_0 \supseteq X$
2:   **loop**
3:       Let $\boldsymbol{x}_t \in \operatorname{argmin}_{\boldsymbol{x} \in X_t} f(\boldsymbol{x})$
4:       If $\boldsymbol{x}_t \in X$, then stop, else find a cutting plane $\langle \boldsymbol{a}, \boldsymbol{x} \rangle \leq b$ separating $\boldsymbol{x}_t$ from X
5:       $X_{t+1} \leftarrow X_t \cap \{\, \boldsymbol{x} \mid \langle \boldsymbol{a}, \boldsymbol{x} \rangle \leq b \,\}$
6:       $t \leftarrow t + 1$
7:   **end loop**

This basic idea has many incarnations. Next we describe three of them, which have been used in the three machine learning applications presented in this chapter. Section 7.1.1 describes a cutting-plane method suited for minimization of nonsmooth convex functions. An improved variant thereof, called the *bundle method*, is described in Section 7.1.2. Finally, Section 7.1.3

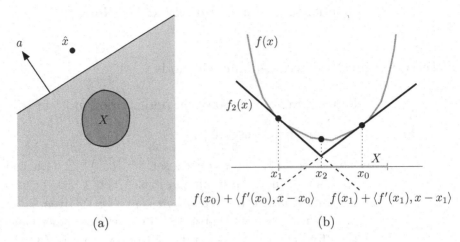

$$f(x_0) + \langle f'(x_0), x - x_0\rangle \qquad f(x_1) + \langle f'(x_1), x - x_1\rangle$$

(a)   (b)

Figure 7.1: (a) illustrates the cutting plane $\langle \boldsymbol{a}, \boldsymbol{x}\rangle \leq b$ cutting off the query point $\hat{x}$ from the light gray half-space $\{\boldsymbol{x} \mid \langle \boldsymbol{a}, \boldsymbol{x}\rangle \leq b\}$, which contains the feasible set X (dark gray). (b) shows a feasible set X (gray interval) and a function $f(\boldsymbol{x})$ which is approximated by a cutting-plane model $f_2(\boldsymbol{x}) = \max\{f(\boldsymbol{x}_0) + \langle f'(\boldsymbol{x}_0), \boldsymbol{x} - \boldsymbol{x}_0\rangle, f(\boldsymbol{x}_1) + \langle f'(\boldsymbol{x}_1), \boldsymbol{x} - \boldsymbol{x}_1\rangle\}$. Starting from $\boldsymbol{x}_0$, the cutting-plane algorithm generates points $\boldsymbol{x}_1$ and $\boldsymbol{x}_2 = \operatorname{argmin}_{\boldsymbol{x} \in X} f_2(\boldsymbol{x})$.

describes application of cutting-plane methods to solving combinatorial optimization problems.

7.1.1   Nonsmooth Optimization

When f is a complicated nonsmooth function while the set X is simple, we want to avoid explicit minimization of f in the algorithm. This can be done by writing (7.1) in epigraph form as

$$\min\{y \mid (\boldsymbol{x}, y) \in Z\} \quad \text{where} \quad Z = \{(\boldsymbol{x}, y) \in X \times \mathbb{R} \mid f(\boldsymbol{x}) \leq y\}. \quad (7.2)$$

In this case, cutting planes can be generated by means of subgradients. Recall that $f'(\hat{\boldsymbol{x}}) \in \mathbb{R}^n$ is a subgradient of f at $\hat{\boldsymbol{x}}$ if

$$f(\boldsymbol{x}) \geq f(\hat{\boldsymbol{x}}) + \langle f'(\hat{\boldsymbol{x}}), \boldsymbol{x} - \hat{\boldsymbol{x}}\rangle, \qquad \boldsymbol{x} \in X. \quad (7.3)$$

Thus, the right-hand side is a linear underestimator of f. Assume that $\hat{\boldsymbol{x}} \in X$. Then, the separation algorithm for the set Z can be constructed as follows. If $f(\hat{\boldsymbol{x}}) \leq \hat{y}$, then $(\hat{\boldsymbol{x}}, \hat{y}) \in Z$. If $f(\hat{\boldsymbol{x}}) > \hat{y}$, then the inequality

$$y \geq f(\hat{\boldsymbol{x}}) + \langle f'(\hat{\boldsymbol{x}}), \boldsymbol{x} - \hat{\boldsymbol{x}}\rangle \quad (7.4)$$

defines a cutting plane separating $(\hat{\boldsymbol{x}}, \hat{y})$ from Z.

This leads to the algorithm proposed independently by Cheney and Goldstein (1959) and Kelley (1960). Starting with $\boldsymbol{x}_0 \in X$, it computes the next iterate $\boldsymbol{x}_t$ by solving

$$(\boldsymbol{x}_t, y_t) \in \underset{(\boldsymbol{x},y)\in Z_t}{\operatorname{argmin}} y \qquad \text{where}$$
$$Z_t = \left\{ (\boldsymbol{x}, y) \in X \times \mathbb{R} \mid y \geq f(\boldsymbol{x}_i) + \langle f'(\boldsymbol{x}_i),\, \boldsymbol{x} - \boldsymbol{x}_i \rangle, i = 0, \ldots, t-1 \right\}. \tag{7.5}$$

Here, Z_t is a polyhedral outer bound of Z defined by X and the cutting planes from previous iterates $\{\boldsymbol{x}_0, \ldots, \boldsymbol{x}_{t-1}\}$. Problem (7.5) simplifies to

$$\boldsymbol{x}_t \in \underset{\boldsymbol{x}\in X}{\operatorname{argmin}} f_t(\boldsymbol{x}) \quad \text{where} \quad f_t(\boldsymbol{x}) = \max_{i=0,\ldots,t-1} \left[f(\boldsymbol{x}_i) + \langle f'(\boldsymbol{x}_i),\, \boldsymbol{x} - \boldsymbol{x}_i \rangle \right]. \tag{7.6}$$

Here, f_t is a *cutting-plane model* of f (see Figure 7.1(b)). Note that $(\boldsymbol{x}_t, f_t(\boldsymbol{x}_t))$ solves (7.5). By (7.3) and (7.6), $f(\boldsymbol{x}_i) = f_t(\boldsymbol{x}_i)$ for $i = 0, \ldots, t-1$ and $f(\boldsymbol{x}) \geq f_t(\boldsymbol{x})$ for $\boldsymbol{x} \in X$, that is, f_t is an underestimator of f which touches f at the points $\{\boldsymbol{x}_0, \ldots, \boldsymbol{x}_{t-1}\}$. By solving (7.6) we get not only an estimate $\boldsymbol{x}_t$ of the optimal point $\boldsymbol{x}^*$ but also a lower bound $f_t(\boldsymbol{x}_t)$ on the optimal value $f(\boldsymbol{x}^*)$. It is natural to terminate when $f(\boldsymbol{x}_t) - f_t(\boldsymbol{x}_t) \leq \varepsilon$, which guarantees that $f(\boldsymbol{x}_t) \leq f(\boldsymbol{x}^*) + \varepsilon$. The method is summarized in algorithm 7.2.

Algorithm 7.2 Cutting-plane algorithm in epigraph form

1: **Initialization:** $t \leftarrow 0$, $\boldsymbol{x}_0 \in X$, $\varepsilon > 0$
2: **repeat**
3:      $t \leftarrow t + 1$
4:      Compute $f(\boldsymbol{x}_{t-1})$ and $f'(\boldsymbol{x}_{t-1})$
5:      Update the cutting-plane model $f_t(\boldsymbol{x}) \leftarrow \max_{i=0,\ldots,t-1} \left[f(\boldsymbol{x}_i) + \langle f'(\boldsymbol{x}_i),\, \boldsymbol{x} - \boldsymbol{x}_i \rangle \right]$
6:      Let $\boldsymbol{x}_t \in \operatorname{argmin}_{\boldsymbol{x}\in X} f_t(\boldsymbol{x})$
7: **until** $f(\boldsymbol{x}_t) - f_t(\boldsymbol{x}_t) \leq \varepsilon$

In Section 7.3, this algorithm is applied to *multiple kernel learning*. This requires solving the problem

$$\min\{\, f(\boldsymbol{x}) \mid \boldsymbol{x} \in X \,\} \qquad \text{where} \qquad f(\boldsymbol{x}) = \max\{\, g(\boldsymbol{\alpha}, \boldsymbol{x}) \mid \boldsymbol{\alpha} \in A \,\}. \tag{7.7}$$

X is a simplex, and function g is linear in $\boldsymbol{x}$ and quadratic negative semi-definite in $\boldsymbol{\alpha}$. In this case, the subgradient $f'(\boldsymbol{x})$ equals the gradient $\nabla_{\boldsymbol{x}} g(\hat{\boldsymbol{\alpha}}, \boldsymbol{x})$ where $\hat{\boldsymbol{\alpha}}$ is obtained by solving a convex quadratic program $\hat{\boldsymbol{\alpha}} \in \operatorname{argmax}_{\boldsymbol{\alpha}\in A} g(\boldsymbol{\alpha}, \boldsymbol{x})$.

7.1.2   Bundle Methods

Algorithm 7.2 may converge slowly (Nemirovskij and Yudin, 1983) because subsequent solutions can be very distant, exhibiting a zig-zag behavior. Thus many cutting planes do not actually contribute to the approximation of f around the optimum $\boldsymbol{x}^*$. Bundle methods (Kiwiel, 1983; Lemaréchal et al., 1995) try to reduce this behavior by adding a stabilization term to (7.6). The *proximal bundle methods* compute the new iterate as

$$\boldsymbol{x}_t \in \operatorname*{argmin}_{\boldsymbol{x} \in X} \{\, \nu_t \|\boldsymbol{x} - \boldsymbol{x}_t^+\|_2^2 + f_t(\boldsymbol{x}) \,\} \,,$$

where $\boldsymbol{x}_t^+$ is a current prox-center selected from $\{\boldsymbol{x}_0, \dots, \boldsymbol{x}_{t-1}\}$ and ν_t is a current stabilization parameter. The added quadratic term ensures that the subsequent solutions are within a ball centered at $\boldsymbol{x}_t^+$ whose radius depends on ν_t. If $f(\boldsymbol{x}_t)$ sufficiently decreases the objective, the *decrease step* is performed by moving the prox-center as $\boldsymbol{x}_{t+1}^+ := \boldsymbol{x}_t$. Otherwise, the *null step* is performed, $\boldsymbol{x}_{t+1}^+ := \boldsymbol{x}_t^+$. If there is an efficient line-search algorithm, the decrease step computes the new prox-center $\boldsymbol{x}_{t+1}^+$ by minimizing f along the line starting at $\boldsymbol{x}_t^+$ and passing through $\boldsymbol{x}_t$. Though bundle methods may improve the convergence significantly, they require two parameters: the stabilization parameter ν_t and the minimal decrease in the objective which defines the null step. Despite significantly influencing the convergence, there is no versatile method for choosing these parameters optimally.

In Section 7.2, a variant of this method is applied to *regularized risk minimization* which requires minimizing $f(\boldsymbol{x}) = g(\boldsymbol{x}) + h(\boldsymbol{x})$ over $\mathbb{R}^n$ where g is a simple (typically differentiable) function and h is a complicated nonsmooth function. In this case, the difficulties with setting two parameters are avoided because g naturally plays the role of the stabilization term.

7.1.3   Combinatorial Optimization

A typical combinatorial optimization problem can be formulated as

$$\min\{\, \langle \boldsymbol{c}, \boldsymbol{x} \rangle \mid \boldsymbol{x} \in C \,\}, \tag{7.8}$$

where $C \subseteq \mathbb{Z}^n$ (often just $C \subseteq \{0,1\}^n$) is a finite set of feasible configurations, and $\boldsymbol{c} \in \mathbb{R}^n$ is a cost vector. Usually C is combinatorially large but highly structured. Consider the problem

$$\min\{\, \langle \boldsymbol{c}, \boldsymbol{x} \rangle \mid \boldsymbol{x} \in X \,\} \quad \text{where} \quad X = \operatorname{conv} C \,. \tag{7.9}$$

Clearly, X is a polytope (bounded convex polyhedron) with integral vertices. Hence, (7.9) is a linear program. Since a solution of a linear program is always attained at a vertex, problems (7.8) and (7.9) have the same optimal value. The set X is called the *integral hull* of problem (7.8).

Integral hulls of hard problems are complex. If problem (7.8) is not polynomially solvable, then inevitably the number of facets of X is not polynomial. Therefore (7.9) cannot be solved explicitly. This is where algorithm 7.1 is used. The initial polyhedron $X_0 \supseteq X$ is described by a tractable number of linear inequalities, and usually it is already a good approximation of X, often, but not necessarily, we also have $X_0 \cap \mathbb{Z}^n = C$. The cutting-plane algorithm then constructs a sequence of gradually tighter LP relaxations of (7.8).

A fundamental result states that a linear optimization problem and the corresponding separation problem are polynomial-time equivalent (Grötschel et al., 1981). Therefore, for an intractable problem (7.8) there is no hope of finding a polynomial algorithm to separate an arbitrary point from X. However, a polynomial separation algorithm may exist for a *subclass* (even intractably large) of linear inequalities describing X.

After this approach was first proposed by Dantzig et al. (1954) for the travelling salesman problem, it became a breakthrough in tackling hard combinatorial optimization problems. Since then, much effort has been devoted to finding good initial LP relaxations X_0 for many such problems, subclasses of inequalities describing integral hulls for these problems, and polynomial separation algorithms for these subclasses. This is the subject of *polyhedral combinatorics* (e.g., Schrijver, 2003).

In Section 7.4, we focus on the NP-hard combinatorial optimization problem arising in MAP inference in graphical models. This problem, in its full generality, has not been properly addressed by the optimization community. We show how its LP relaxation can be incrementally tightened during a message-passing algorithm. Because message-passing algorithms are dual, this can be understood as a *dual* cutting-plane algorithm: it does not add constraints in the primal, but does add variables in the dual. The sequence of approximations of the integral hull X (the marginal polytope) can be seen as arising from lifting and projection.

7.2 Regularized Risk Minimization

Learning predictors from data is a standard machine learning problem. A wide range of such problems are special instances of *regularized risk minimization*. In this case, learning is often formulated as an unconstrained minimization of a convex function:

$$\boldsymbol{w}^* \in \operatorname*{argmin}_{\boldsymbol{w} \in \mathbb{R}^n} F(\boldsymbol{w}) \qquad \text{where} \qquad F(\boldsymbol{w}) = \lambda \Omega(\boldsymbol{w}) + R(\boldsymbol{w}) \, . \qquad (7.10)$$

The objective $F \colon \mathbb{R}^n \to \mathbb{R}$, called *regularized risk*, is composed of a regularization term $\Omega \colon \mathbb{R}^n \to \mathbb{R}$ and an empirical risk $R \colon \mathbb{R}^n \to \mathbb{R}$, both of which are convex functions. The number $\lambda \in \mathbb{R}_+$ is a predefined regularization constant, and $\boldsymbol{w} \in \mathbb{R}^n$ is a parameter vector to be learned. The regularization term Ω is typically a simple, cheap-to-compute function used to constrain the space of solutions in order to improve generalization. The empirical risk R evaluates how well the parameter $\boldsymbol{w}$ explains the training examples. Evaluation of R is often computationally expensive.

Example 7.1. *Given a set of training examples $\{(\boldsymbol{x}_1, y_1), \ldots, (\boldsymbol{x}_m, y_m)\} \in (\mathbb{R}^n \times \{+1, -1\})^m$, the goal is to learn a parameter vector $\boldsymbol{w} \in \mathbb{R}^n$ of a linear classifier $h \colon \mathbb{R}^n \to \{-1, +1\}$ which returns $h(\boldsymbol{x}) = +1$ if $\langle \boldsymbol{x}, \boldsymbol{w} \rangle \geq 0$ and $h(\boldsymbol{x}) = -1$ otherwise. Linear support vector machines (Cortes and Vapnik, 1995) without bias learn the parameter vector $\boldsymbol{w}$ by solving (7.10) with the regularization term $\Omega(\boldsymbol{w}) = \frac{1}{2}\|\boldsymbol{w}\|_2^2$ and the empirical risk $R(\boldsymbol{w}) = \frac{1}{m} \sum_{i=1}^m \max\{0, 1 - y_i \langle \boldsymbol{x}_i, \boldsymbol{w} \rangle\}$, which in this case is a convex upper bound on the number of mistakes the classifier $h(\boldsymbol{x})$ makes on the training examples.*

There is a long list of learning algorithms which at their core are solvers of a special instance of (7.10), see, e.g., Schölkopf and Smola (2002). If F is differentiable, (7.10) is solved by algorithms for a smooth optimization. If F is nonsmooth, (7.10) is typically transformed to an equivalent problem solvable by off-the-shelf methods. For example, learning of the linear SVM classifier in example 7.1 can be equivalently expressed as a quadratic program. Because off-the-shelf solvers are often not efficient enough in practice, a huge effort has been put into development of specialized algorithms tailored to particular instances of (7.10).

Teo et al. (2007, 2010) proposed a generic algorithm to solve (7.10) which is a modification of the proximal bundle methods. The algorithm, called the *bundle method for risk minimization* (BMRM), exploits the specific structure of the objective F in (7.10). In particular, only the risk term R is approximated by the cutting-plane model, while the regularization term Ω is used without any change to stabilize the optimization. In contrast, standard bundle methods introduce the stabilization term artificially. The resulting BMRM is highly modular and was proved to converge to an ε-precise solution in $\mathcal{O}(\frac{1}{\varepsilon})$ iterations. In addition, if an efficient line-search algorithm is available, BMRM can be drastically accelerated with a technique proposed by Franc and Sonnenburg (2008, 2010), and Teo et al. (2010). The acceler-

---

Algorithm 7.3 Bundle Method for Regularized Risk Minimization (BMRM)

1: **input & initialization:** $\varepsilon > 0$, $\boldsymbol{w}_0 \in \mathbb{R}^n$, $t \leftarrow 0$
2: **repeat**
3:      $t \leftarrow t + 1$
4:      Compute $R(\boldsymbol{w}_{t-1})$ and $R'(\boldsymbol{w}_{t-1})$
5:      Update the model $R_t(\boldsymbol{w}) \leftarrow \max_{i=0,\ldots,t-1} R(\boldsymbol{w}_i) + \langle R'(\boldsymbol{w}_i), \boldsymbol{w} - \boldsymbol{w}_i \rangle$
6:      Solve the reduced problem $\boldsymbol{w}_t \leftarrow \operatorname{argmin}_{\boldsymbol{w}} F_t(\boldsymbol{w})$ where $F_t(\boldsymbol{w}) = \lambda\Omega(\boldsymbol{w}) + R_t(\boldsymbol{w})$
7: **until** $F(\boldsymbol{w}_t) - F_t(\boldsymbol{w}_t) \leq \varepsilon$

---

ated BMRM has been shown to be highly competitive with state-of-the-art solvers tailored to particular instances of (7.10).

In the next two sections, we describe the BMRM algorithm and its version accelerated by line-search.

7.2.1 Bundle Method for Regularized Risk Minimization

Following optimization terminology, we will call (7.10) the *master problem*. Using the approach of Teo et al. (2007), one can approximate the master problem (7.10) by its *reduced problem*

$$\boldsymbol{w}_t \in \operatorname*{argmin}_{\boldsymbol{w} \in \mathbb{R}^n} F_t(\boldsymbol{w}) \qquad \text{where} \qquad F_t(\boldsymbol{w}) = \lambda\Omega(\boldsymbol{w}) + R_t(\boldsymbol{w}) . \qquad (7.11)$$

The reduced problem (7.11) is obtained from the master problem (7.10) by substituting the cutting-plane model R_t for the empirical risk R while the regularization term Ω remains unchanged. The cutting-plane model reads

$$R_t(\boldsymbol{w}) = \max_{i=0,\ldots,t-1} \left[R(\boldsymbol{w}_i) + \langle R'(\boldsymbol{w}_i), \boldsymbol{w} - \boldsymbol{w}_i \rangle \right], \qquad (7.12)$$

where $R'(\boldsymbol{w}) \in \mathbb{R}^n$ is a subgradient of R at point $\boldsymbol{w}$. Since $R(\boldsymbol{w}) \geq R_t(\boldsymbol{w})$, $\forall \boldsymbol{w} \in \mathbb{R}^n$, the reduced problem's objective F_t is an underestimator of the master objective F. Starting from $\boldsymbol{w}_0 \in \mathbb{R}^n$, the BMRM of Teo et al. (2007) (Algorithm 7.3) computes a new iterate $\boldsymbol{w}_t$ by solving the reduced problem (7.11). In each iteration t, the cutting-plane model (7.12) is updated by a new cutting plane computed at the intermediate solution $\boldsymbol{w}_t$, leading to a progressively tighter approximation of F. The algorithm halts if the gap between the upper bound $F(\boldsymbol{w}_t)$ and the lower bound $F_t(\boldsymbol{w}_t)$ falls below a desired ε, meaning that $F(\boldsymbol{w}_t) \leq F(\boldsymbol{w}^*) + \varepsilon$.

In practice, the number of cutting planes t required before the algorithm converges is typically much lower than the dimension n of the parameter vector $\boldsymbol{w} \in \mathbb{R}^n$. Thus, it is beneficial to solve the reduced problem (7.11) in its dual formulation. Let $A = [\boldsymbol{a}_0, \ldots, \boldsymbol{a}_{t-1}] \in \mathbb{R}^{n \times t}$ be a matrix whose columns are the subgradients $\boldsymbol{a}_i = R'(\boldsymbol{w}_i)$, and let $\boldsymbol{b} = [b_0, \ldots, b_{t-1}] \in \mathbb{R}^t$ be a column vector whose components equal $b_i = R(\boldsymbol{w}_i) - \langle R'(\boldsymbol{w}_i), \boldsymbol{w}_i \rangle$. Then the reduced problem (7.11) can be equivalently expressed as

$$\boldsymbol{w}_t \in \operatorname*{argmin}_{\boldsymbol{w} \in \mathbb{R}^n, \xi \in \mathbb{R}} \left[\lambda \Omega(\boldsymbol{w}) + \xi \right] \quad \text{s.t.} \quad \xi \geq \langle \boldsymbol{w}, \boldsymbol{a}_i \rangle + b_i, \ i = 0, \ldots, t-1. \quad (7.13)$$

The Lagrange dual of (7.13) reads (Teo et al., 2010, theorem 2)

$$\boldsymbol{\alpha}_t \in \operatorname*{argmin}_{\boldsymbol{\alpha} \in \mathbb{R}^t} \left[- \lambda \Omega^*(-\lambda^{-1} A \boldsymbol{\alpha}) + \langle \boldsymbol{\alpha}, \boldsymbol{b} \rangle \right] \quad \text{s.t.} \quad \|\boldsymbol{\alpha}\|_1 = 1, \boldsymbol{\alpha} \geq 0, \quad (7.14)$$

where $\Omega^* \colon \mathbb{R}^n \to \mathbb{R}^t$ denotes the Fenchel dual of Ω defined as

$$\Omega^*(\boldsymbol{\mu}) = \sup \left\{ \langle \boldsymbol{w}, \boldsymbol{\mu} \rangle - \Omega(\boldsymbol{w}) \mid \boldsymbol{w} \in \mathbb{R}^n \right\}.$$

Having the dual solution $\boldsymbol{\alpha}_t$, the primal solution can be computed by solving $\boldsymbol{w}_t \in \operatorname{argmax}_{\boldsymbol{w} \in \mathbb{R}^n} \left[\langle \boldsymbol{w}, -\lambda^{-1} A \boldsymbol{\alpha}_t \rangle - \Omega(\boldsymbol{w}) \right]$, which for differentiable Ω simplifies to $\boldsymbol{w}_t = \nabla_{\boldsymbol{\mu}} \Omega^*(-\lambda^{-1} A \boldsymbol{\alpha}_t)$.

Example 7.2. *For the quadratic regularizer* $\Omega(\boldsymbol{w}) = \frac{1}{2} \|\boldsymbol{w}\|_2^2$ *the Fenchel dual reads* $\Omega^*(\boldsymbol{\mu}) = \frac{1}{2} \|\boldsymbol{\mu}\|_2^2$. *The dual reduced problem (7.14) boils down to the quadratic program*

$$\boldsymbol{\alpha}_t \in \operatorname*{argmin}_{\boldsymbol{\alpha} \in \mathbb{R}^t} \left[- \frac{1}{2\lambda} \boldsymbol{\alpha}^T A^T A \boldsymbol{\alpha} + \boldsymbol{\alpha}^T \boldsymbol{b} \right] \quad \text{s.t.} \quad \|\boldsymbol{\alpha}\|_1 = 1, \boldsymbol{\alpha} \geq 0,$$

and the primal solution can be computed analytically by $\boldsymbol{w}_t = -\lambda^{-1} A \boldsymbol{\alpha}_t$.

The convergence of Algorithm 7.3 in a finite number of iterations is guaranteed by the following theorem.

Theorem 7.1. *(Teo et al., 2010, theorem 5). Assume that (i)* $F(\boldsymbol{w}) \geq 0$, $\forall \boldsymbol{w} \in \mathbb{R}^n$, *(ii)* $\max_{\boldsymbol{g} \in \partial R(\boldsymbol{w})} \|\boldsymbol{g}\|_2 \leq G$ *for all* $\boldsymbol{w} \in \{\boldsymbol{w}_0, \ldots, \boldsymbol{w}_{t-1}\}$ *where* $\partial R(\boldsymbol{w})$ *denotes the subdifferential of R at point* $\boldsymbol{w}$, *and (iii)* Ω^* *is twice differentiable and has bounded curvature, that is,* $\|\partial^2 \Omega^*(\boldsymbol{\mu})\| \leq H^*$ *for all* $\boldsymbol{\mu} \in \{ \boldsymbol{\mu}' \in \mathbb{R}^t \mid \boldsymbol{\mu}' = \lambda^{-1} A \boldsymbol{\alpha}, \|\boldsymbol{\alpha}\|_1 = 1, \boldsymbol{\alpha} \geq 0 \}$ *where* $\partial^2 \Omega^*(\boldsymbol{\mu})$ *is the Hessian of Ω^* at point* $\boldsymbol{\mu}$. *Then Algorithm 7.3 terminates after at most*

$$T \leq \log_2 \frac{\lambda F(0)}{G^2 H^*} + \frac{8 G^2 H^*}{\lambda \varepsilon} - 1$$

iterations for any $\varepsilon < 4 G^2 H^* \lambda^{-1}$.

Furthermore, for a twice differentiable F with bounded curvature, Algorithm 7.3 requires only $\mathcal{O}(\log \frac{1}{\varepsilon})$ iterations instead of $\mathcal{O}(\frac{1}{\varepsilon})$ (Teo et al., 2010, theorem 5). The most constraining assumption of theorem 7.1 is that it requires Ω^* to be twice differentiable. This assumption holds, for instance, for the quadratic $\Omega(\boldsymbol{w}) = \frac{1}{2} \|\boldsymbol{w}\|_2^2$ and the negative entropy

$\Omega(\boldsymbol{w}) = \sum_{i=1}^{n} w_i \log w_i$ regularizers. Unfortunately, the theorem does not apply to the ℓ_1-norm regularizer $\Omega(\boldsymbol{w}) = \|\boldsymbol{w}\|_1$ that is often used to enforce sparse solutions.

7.2.2 BMRM Algorithm Accelerated by Line-search

BMRM can be drastically accelerated whenever an efficient line-search algorithm for the master objective F is available. An accelerated BMRM for solving linear SVM problems (c.f. Example 7.1) was first proposed by Franc and Sonnenburg (2008). Franc and Sonnenburg (2010) generalized the method for solving (7.10) with an arbitrary risk R and a quadratic regularizer $\Omega(\boldsymbol{w}) = \frac{1}{2}\|\boldsymbol{w}\|_2^2$. Finally, Teo et al. (2010) proposed a fully general version imposing no restrictions on Ω and R. BMRM accelerated by the line-search, in Teo et al. (2010) called LS-BMRM, is described by Algorithm 7.4.

Algorithm 7.4 BMRM accelerated by line-search (LS-BMRM)

1: **input & initialization:** $\varepsilon \geq 0$, $\theta \in (0,1]$, $\boldsymbol{w}_0^b$, $\boldsymbol{w}_0^c \leftarrow \boldsymbol{w}_0^b$, $t \leftarrow 0$
2: **repeat**
3:      $t \leftarrow t + 1$
4:      Compute $R(\boldsymbol{w}_{t-1}^c)$ and $R'(\boldsymbol{w}_{t-1}^c)$
5:      Update the model $R_t(\boldsymbol{w}) \leftarrow \max_{i=1,\ldots,t-1} R(\boldsymbol{w}_i^c) + \langle R'(\boldsymbol{w}_i^c), \boldsymbol{w} - \boldsymbol{w}_i^c \rangle$
6:      $\boldsymbol{w}_t \leftarrow \operatorname{argmin}_{\boldsymbol{w}} F_t(\boldsymbol{w})$ where $F_t(\boldsymbol{w}) = \lambda\Omega(\boldsymbol{w}) + R_t(\boldsymbol{w})$
7:      Line-search: $k_t \leftarrow \operatorname{argmin}_{k \geq 0} F(\boldsymbol{w}_t^b + k(\boldsymbol{w}_t - \boldsymbol{w}_{t-1}^b))$
8:      $\boldsymbol{w}_t^b \leftarrow \boldsymbol{w}_{t-1}^b + k_t(\boldsymbol{w}_t - \boldsymbol{w}_{t-1}^b)$
9:      $\boldsymbol{w}_t^c \leftarrow (1-\theta)\boldsymbol{w}_{t-1}^b + \theta\boldsymbol{w}_t$
10: **until** $F(\boldsymbol{w}_t^b) - F_t(\boldsymbol{w}_t) \leq \varepsilon$

Unlike BMRM, LS-BMRM simultaneously optimizes the master and reduced problems' objectives F and F_t, respectively. In addition, LS-BMRM selects cutting planes that are close to the best-so-far solution which has a stabilization effect. Moreover, such cutting planes have a higher chance of actively contributing to the approximation of the master objective F around the optimum $\boldsymbol{w}^*$. In particular, there are three main changes compared to BMRM:

1. LS-BMRM maintains the best-so-far solution $\boldsymbol{w}_t^b$ obtained during the first t iterations, that is, $F(\boldsymbol{w}_0^b), \ldots, F(\boldsymbol{w}_t^b)$ is a monotonically decreasing sequence.

2. The new best-so-far solution $\boldsymbol{w}_t^b$ is found by searching along a line starting at the previous solution $\boldsymbol{w}_{t-1}^b$ and crossing the reduced problem's solution $\boldsymbol{w}_t$. This is implemented on lines 7 and 8.

3. The new cutting plane is computed to approximate the master objective

F at the point $\boldsymbol{w}_t^c \leftarrow (1 - \theta)\boldsymbol{w}_t^b + \theta\boldsymbol{w}_t$ (line 9), which lies on the line segment between the best-so-far solution $\boldsymbol{w}_t^b$ and the reduced problem's solution $\boldsymbol{w}_t$. $\theta \in (0, 1]$ is a prescribed parameter. Note that $\boldsymbol{w}_t^c$ must not be set directly to $\boldsymbol{w}_t^b$ in order to guarantee convergence (i.e., $\theta = 0$ is not allowed). It was found experimentally (Franc and Sonnenburg, 2010) that the value $\theta = 0.1$ works consistently well.

LS-BMRM converges to and ε-precise solution in $\mathcal{O}(\frac{1}{\varepsilon})$ iterations:

Theorem 7.2. *(Teo et al., 2010, theorem 7). Under the assumption of theorem 7.1 Algorithm 7.4 converges to the desired precision after*

$$T \leq \frac{8G^2 H^*}{\lambda \varepsilon}$$

iterations for any $\varepsilon < 4G^2 H^* \lambda^{-1}$.

At line 7 LS-BMRM requires solution of a line-search problem:

$$k^* = \operatorname*{argmin}_{k \geq 0} f(k) \quad \text{where} \quad f(k) = \lambda\Omega(\boldsymbol{w}' + k\boldsymbol{w}) + R(\boldsymbol{w}' + k\boldsymbol{w}). \quad (7.15)$$

Franc and Sonnenburg (2008, 2010) proposed a line-search algorithm which finds the exact solution of (7.15) if $\Omega(\boldsymbol{w}) = \frac{1}{2}\|\boldsymbol{w}\|_2^2$ and

$$R(\boldsymbol{w}) = \sum_{i=1}^m \max_{j=1,\ldots,p} \left(u_{ij} + \langle \boldsymbol{v}_{ij}, \boldsymbol{w} \rangle \right), \quad (7.16)$$

where $u_{ij} \in \mathbb{R}$ and $\boldsymbol{v}_{ij} \in \mathbb{R}^n$, $i = 1, \ldots, m$, $j = 1, \ldots, p$, are fixed scalars and vectors, respectively. In this case, the subdifferential of $\partial f(k)$ can be described by $\mathcal{O}(pm)$ line segments in 2D. Problem (7.15) can be replaced by solving $\partial f(k) \ni 0$ w.r.t. k, which is equivalent to finding among the line segments the one intersecting the x-axis. This line-search algorithm finds the exact solution of (7.15) in $\mathcal{O}(mp^2 + mp\log mp)$ time. The risk (7.16) emerges in most variants of the support vector machines learning algorithms, such as binary SVMs, multi-class SVMs, or SVM regression. Unfortunately, the algorithm is not applicable if p is huge, which excludes applications to structured-output SVM learning (Tsochantaridis et al., 2005).

7.2.3   Conclusions

A notable advantage of BMRM is its modularity and simplicity. One only needs to supply a procedure to compute the risk $R(\boldsymbol{w})$ and its subgradient $R'(\boldsymbol{w})$ at a point $\boldsymbol{w}$. The core part of BMRM, that is, solving the reduced problem, remains unchanged for a given regularizer Ω. Thus, many existing learning problems can be solved by a single optimization technique. Moreover, one can easily experiment with new learning formulations just by specifying the risk term R and its subgradient R', without spending time on development of a new solver for that particular problem.

The convergence speeds of BMRM and the accelerated LS-BMRM have been extensively studied on a variety of real-life problems in domains ranging from text classification, bioinformatics, and computer vision to computer security systems (Teo et al., 2007; Franc and Sonnenburg, 2008, 2010; Teo et al., 2010). Compared to the state-of-the-art dedicated solvers, BMRM is typically slightly slower, though, it is still competitive and practically useful. On the other hand, the LS-BMRM has proved to be among the fastest optimization algorithms for a variety of problems. Despite the similar theoretical convergence times, in practice the LS-BMRM is on average an order of magnitude faster than BMRM.

The most time-consuming part of BMRM, as well as of LS-BMRM, is the evaluation of the risk R and its subgradient R'. Fortunately, the risk, and thus also its subgradient, typically are additively decomposable, which allows for an efficient parallelization of their computation. The effect of the parallelization on the reduction of the computational time is empirically studied in Franc and Sonnenburg (2010) and Teo et al. (2010).

The relatively high memory requirements of BMRM/LS-BMRM may be the major deficiency if the method is applied to large-scale problems. The method stores in each iteration t a cutting plane of size $\mathcal{O}(n)$, where n is the dimension of the parameter vector $\boldsymbol{w} \in \mathbb{R}^n$, which leads to $\mathcal{O}(nt)$ memory complexity not counting the reduced problem, which is typically much less memory demanding. To alleviate the problem, Teo et al. (2010) propose a limited memory variant of BMRM maintaining up to K cutting planes aggregated from the original t cutting planes. Though that variant does not have an impact on the theoretical upper bound of the number of iterations, in practice it may significantly slow down the convergence.

The implementations of BMRM and LS-BMRM can be found in the SHOGUN machine learning toolbox (Sonnenburg et al., 2010) or in the open-source packages BMRM (`http://users.cecs.anu.edu.au/~chteo/BMRM.html`) and LIBOCAS (`http://cmp.felk.cvut.cz/~xfrancv/ocas/html/`).

7.3   Multiple Kernel Learning

Multiple kernel learning (MKL) (e.g., Bach et al., 2004) has recently become an active line of research. Given a mapping $\Phi : \mathcal{X} \mapsto \mathbb{R}^n$ that represents each object $x \in \mathcal{X}$ in n-dimensional feature space,[1] a kernel machine employs a kernel function

$$\mathbf{k}(x, x') = \langle \Phi(x), \Phi(x') \rangle$$

to compare two objects x and x' without *explicitly* computing $\Phi(x)$. Ultimately, a kernel machine learns an α-weighted linear combination of kernel functions with bias b

$$h(x) = \sum_{i=1}^{m} \alpha_i \mathbf{k}(x_i, x) + b, \tag{7.17}$$

where $x_1, \ldots, x_m$ is a set of training objects. For example, the support vector machine (SVM) classifier uses the sign of $h(x)$ to assign a class label $y \in \{-1, +1\}$ to the object x (e.g., Schölkopf and Smola, 2002).

Traditionally, just a single kernel function has been used. However, it has proved beneficial to consider not just a single kernel, but multiple kernels in various applications (see Section 7.3.4). Currently, the most popular way to combine kernels is via convex combinations, that is, introducing

$$B = \left\{ \boldsymbol{\beta} \in \mathbb{R}^K \,\middle|\, \|\boldsymbol{\beta}\|_1 = 1, \boldsymbol{\beta} \geq 0 \right\}, \tag{7.18}$$

the *composite kernel* is defined as

$$\mathbf{k}(x, x') = \sum_{k=1}^{K} \beta_k \mathbf{k}_k(x, x'), \qquad \boldsymbol{\beta} \in B, \tag{7.19}$$

where $\mathbf{k}_k \colon \mathcal{X} \times \mathcal{X} \to \mathbb{R}$, $k = 1, \ldots, K$ is a given set of positive-definite kernels (Schölkopf and Smola, 2002). Now, in contrast to single kernel algorithms, MKL learns, in addition to the coefficients $\boldsymbol{\alpha}$ and b, the weighting over kernels $\boldsymbol{\beta}$.

In Section 7.3.1, we review convex MKL for classification, and in Section 7.3.2, we show that this problem can be cast as minimization of a complicated convex function over a simple feasible set. In Section 7.3.3, we derive a CPA that transforms the MKL problem into a sequence of linear and quadratic programs, the latter of which can be efficiently solved by existing SVM solvers. Section 7.3.4 concludes this part.

---

1. For the sake of simplicity, we consider the n-dimensional Euclidean feature space. However, all the methods in this section can be applied even if the objects are mapped into arbitrary reproducing kernel Hilbert space (Schölkopf and Smola, 2002).

7.3.1 Convex Multiple Kernel Learning

Various MKL formulations have been proposed (Lanckriet et al., 2004a; Bach et al., 2004; Sonnenburg et al., 2006a; Varma and Babu, 2009; Kloft et al., 2009; Bach, 2008; Nath et al., 2009; Cortes et al., 2009). Here we focus solely on the convex optimization problem for classification as it was first stated by Zien and Ong (2007) and Rakotomamonjy et al. (2007). The same authors have shown that the mixed-norm approaches of Bach et al. (2004) and Sonnenburg et al. (2006a) are equivalent.

Let $\{(\boldsymbol{x}_1, y_1), \ldots, (\boldsymbol{x}_m, y_m)\} \in (\mathcal{X} \times \{-1, +1\})^m$ be a training set of examples of input $\boldsymbol{x}$ and output y assumed to be i.i.d. from an unknown distribution $p(\boldsymbol{x}, y)$. The input $\boldsymbol{x}$ is translated into a compositional feature vector $(\Phi_1(\boldsymbol{x}); \ldots; \Phi_K(\boldsymbol{x})) \in \mathbb{R}^{n_1 + \cdots + n_k}$ that is constructed by a set of K mappings $\Phi_k \colon \mathcal{X} \to \mathbb{R}^{n_k}$, $k = 1, \ldots, K$. The goal is to predict y from an unseen $\boldsymbol{x}$ by using a linear classifier,

$$y = \operatorname{sgn}\big(h(\boldsymbol{x})\big) \quad \text{where} \quad h(\boldsymbol{x}) = \sum_{k=1}^{K} \langle \boldsymbol{w}_k, \Phi_k(\boldsymbol{x}) \rangle + b, \tag{7.20}$$

whose parameters $\boldsymbol{w}_k \in \mathbb{R}^{n_k}$, $k = 1, \ldots, K$, $b \in \mathbb{R}$, are learned from the training examples. Using the definition $\frac{x}{0} = 0$ if $x = 0$ and ∞ otherwise, the parameters of the classifier (7.20) can be obtained by solving the following convex *primal MKL optimization problem* (Zien and Ong, 2007; Rakotomamonjy et al., 2007):

$$\min \quad \frac{1}{2} \sum_{k=1}^{K} \frac{1}{\beta_k} \|\boldsymbol{w}_k\|_2^2 + C \sum_{i=1}^{m} \xi_i \tag{7.21}$$

$$\text{w.r.t.} \quad \boldsymbol{\beta} \in B, \boldsymbol{w} = (\boldsymbol{w}_1, \ldots, \boldsymbol{w}_K) \in \mathbb{R}^{n_1 + \cdots + n_K}, \boldsymbol{\xi} \in \mathbb{R}^m, b \in \mathbb{R}$$

$$\text{s.t.} \quad \xi_i \geq 0 \text{ and } y_i \left(\sum_{k=1}^{K} \langle \boldsymbol{w}_k, \Phi_k(\boldsymbol{x}_i) \rangle + b \right) \geq 1 - \xi_i, \ i = 1, \ldots, m.$$

Analogously to the SVMs, the objective of (7.21) is composed of two terms. The first (regularization) term constrains the spaces of the parameters $\boldsymbol{w}_k$, $k = 1, \ldots, K$ in order to improve the generalization of the classifier (7.20). The second term, weighted by a prescribed constant $C > 0$, is an upper bound on the number of mistakes the classifier (7.20) makes on the training examples. In contrast to SVMs, positive weights $\boldsymbol{\beta}$ with ℓ_1-norm constraint (see (7.18)) are introduced to enforce block-wise sparsity, that is, rather few blocks of features Φ_k are selected (have non-zero weight $\boldsymbol{w}_k$). Since $\frac{1}{\beta_k} \gg 1$ for small β_k, non-zero components of $\boldsymbol{w}_k$ experience stronger penalization, and thus the smaller β_k is, the smoother $\boldsymbol{w}_k$ is. By definition, $\boldsymbol{w}_k = 0$ if $\beta_k = 0$. Note that for $K = 1$, the MKL problem (7.21) reduces to the standard two-class linear SVM classifier.

7.3.2   Min-Max Formulation of Multiple Kernel Learning

To apply kernels, the primal MKL problem (7.21) must be reformulated such that the feature vectors $\Phi_k(\boldsymbol{x}_i)$ appear in terms of dot products only. Following Rakotomamonjy et al. (2007), we can rewrite (7.21) as

$$\min\{F(\boldsymbol{\beta}) \mid \boldsymbol{\beta} \in B\}\,, \tag{7.22}$$

where $F(\boldsymbol{\beta})$ is a shortcut for solving the standard SVM primal on the $\boldsymbol{\beta}$-weighted concatenated feature space:

$$
\begin{aligned}
F(\boldsymbol{\beta}) = \min \quad & \frac{1}{2}\sum_{k=1}^{K}\frac{1}{\beta_k}\|\boldsymbol{w}_k\|_2^2 + C\sum_{i=1}^{m}\xi_i \\
\text{w.r.t.} \quad & \boldsymbol{w} = (\boldsymbol{w}_1,\ldots,\boldsymbol{w}_K) \in \mathbb{R}^{n_1+\cdots+n_K}, \boldsymbol{\xi} \in \mathbb{R}^m, b \in \mathbb{R} \\
\text{s.t.} \quad & \xi_i \geq 0 \text{ and } y_i\left(\sum_{k=1}^{K}\langle \boldsymbol{w}_k, \Phi_k(\boldsymbol{x}_i)\rangle + b\right) \geq 1 - \xi_i, i = 1,\ldots,m.
\end{aligned}
\tag{7.23}
$$

Note that in (7.23) the weights $\boldsymbol{\beta}$ are fixed and the minimization is over only $(\boldsymbol{w}, \boldsymbol{\xi}, b)$. The Lagrange dual of (7.23) reads (Rakotomamonjy et al., 2007)

$$D(\boldsymbol{\beta}) = \max\{S(\boldsymbol{\alpha}, \boldsymbol{\beta}) \mid \boldsymbol{\alpha} \in A\} \quad \text{where} \quad S(\boldsymbol{\alpha}, \boldsymbol{\beta}) = \sum_{k=1}^{K}\beta_k S_k(\boldsymbol{\alpha})\,, \tag{7.24}$$

and S_k and A are defined as follows:

$$
\begin{aligned}
S_k(\boldsymbol{\alpha}) &= \sum_{i=1}^{m}\alpha_i - \frac{1}{2}\sum_{i=1}^{m}\sum_{j=1}^{m}\alpha_i\alpha_j y_i y_j \langle \Phi_k(\boldsymbol{x}_i), \Phi_k(\boldsymbol{x}_j)\rangle \\
A &= \{\boldsymbol{\alpha} \in \mathbb{R}^m \mid 0 \leq \alpha_i \leq C, i = 1,\ldots,m, \sum_{i=1}^{m}\alpha_i y_i = 0\}\,.
\end{aligned}
\tag{7.25}
$$

Note that (7.24) is equivalent to solving the standard SVM dual with the composite kernel (7.19). Because (7.23) is convex and the Slater's qualification condition holds, the duality gap is zero, that is. $F(\boldsymbol{\beta}) = D(\boldsymbol{\beta})$. Substituting $D(\boldsymbol{\beta})$ for $F(\boldsymbol{\beta})$ in (7.22) leads to an equivalent *min-max MKL problem*:

$$\min\{D(\boldsymbol{\beta}) \mid \boldsymbol{\beta} \in B\}\,. \tag{7.26}$$

Let $\boldsymbol{\beta}^* \in \operatorname{argmax}_{\boldsymbol{\beta} \in B} D(\boldsymbol{\beta})$ and $\boldsymbol{\alpha}^* \in \operatorname{argmax}_{\boldsymbol{\alpha} \in A} S(\boldsymbol{\alpha}, \boldsymbol{\beta}^*)$. Then the solution of the primal MKL problem (7.21) can be computed analytically as

$$\boldsymbol{w}_k^* = \beta_k^* \sum_{i=1}^{m} \alpha_i^* y_i \Phi_k(\boldsymbol{x}_i) \quad \text{and} \quad b^* = y_i - \sum_{k=1}^{K} \langle \boldsymbol{w}_k^*, \Phi_k(\boldsymbol{x}_i) \rangle , \, i \in J , \quad (7.27)$$

where $J = \{j \in \{1, \ldots, m\} \mid 0 < \alpha_i^* < C\}$. The equalities (7.27) follow from the Karush-Kuhn-Tucker optimality conditions of problem (7.23) (e.g., Schölkopf and Smola, 2002). Note that in practice, b^* is computed as an average over all $|J|$ equalities, which is numerically more stable.

By substituting (7.27) and $\mathbf{k}_k(\boldsymbol{x}_i, \boldsymbol{x}) = \langle \Phi_k(\boldsymbol{x}_i), \Phi_k(\boldsymbol{x}) \rangle$ in the linear classification rule (7.20), we obtain the kernel classifier (7.17) with the composite kernel (7.19). In addition, after substituting $\mathbf{k}_k(\boldsymbol{x}_i, \boldsymbol{x}_j)$ for the dot products $\langle \Phi_k(\boldsymbol{x}_i), \Phi_k(\boldsymbol{x}_j) \rangle$ in (7.25) we can compute all the parameters of the kernel classifier without explicitly using the features $\Phi_k(\boldsymbol{x}_i)$.

7.3.3 Solving MKL via Cutting-planes

In this section, we will apply the cutting-plane Algorithm 7.2 to the min-max MKL problem (7.26).

It follows from (7.24) that the objective D is convex, since it is a pointwise maximum over an infinite number of functions $S(\boldsymbol{\alpha}, \boldsymbol{\beta})$, $\boldsymbol{\alpha} \in A$, which are linear in $\boldsymbol{\beta}$ (e.g., Boyd and Vandenberghe, 2004). By Danskin's theorem (Bertsekas, 1999, proposition B.25), the subgradient of D at point $\boldsymbol{\beta}$ equals the gradient $\nabla_{\boldsymbol{\beta}} S(\hat{\boldsymbol{\alpha}}, \boldsymbol{\beta})$ where $\hat{\boldsymbol{\alpha}} \in \operatorname{argmax}_{\boldsymbol{\alpha} \in A} S(\boldsymbol{\alpha}, \boldsymbol{\beta})$, that is, the subgradient reads

$$D'(\boldsymbol{\beta}) = [S_1(\hat{\boldsymbol{\alpha}}); \ldots; S_K(\hat{\boldsymbol{\alpha}})] \in \mathbb{R}^K . \quad (7.28)$$

Note that computing $D(\boldsymbol{\beta})$ and its subgradient $D'(\boldsymbol{\beta})$ requires solving the convex quadratic program (7.24) which is equivalent to the standard SVM dual computed on the composite kernel (7.19) with a fixed weighting $\boldsymbol{\beta}$ (Rakotomamonjy et al., 2007). Thus, existing SVM solvers are directly applicable.

Having the means to compute D and its subgradient D', we can approximate the objective D by its cutting-plane model

$$\begin{aligned} D_t(\boldsymbol{\beta}) &= \max_{i=0,\ldots,t-1} \left[D(\boldsymbol{\beta}_i) + \langle D'(\boldsymbol{\beta}_i), \boldsymbol{\beta} - \boldsymbol{\beta}_i \rangle \right] \\ &= \max_{i=0,\ldots,t-1} \langle \boldsymbol{\beta}, D'(\boldsymbol{\beta}_i) \rangle . \end{aligned} \quad (7.29)$$

The points $\{\boldsymbol{\beta}_0, \ldots, \boldsymbol{\beta}_{t-1}\}$ can be computed by Kelley's CPA (Algorithm 7.2)

Algorithm 7.5 Cutting-plane algorithm for solving the MKL problem. The algorithm requires solving a simple LP (line 7) and a convex QP (line 3) which is equivalent to the standard SVM dual.

1:   **Initialization:** $t \leftarrow 0$, $\boldsymbol{\beta}_0 \in B$ (e.g. $\boldsymbol{\beta}_0 = [\frac{1}{K}; \ldots; \frac{1}{K}])$, $\varepsilon > 0$
2:   **repeat**
3:       Let $\boldsymbol{\alpha}_t \in \mathrm{argmax}_{\boldsymbol{\alpha} \in A}\, S(\boldsymbol{\alpha}, \boldsymbol{\beta}_t)$
4:       Compute $D(\boldsymbol{\beta}_t) \leftarrow S(\boldsymbol{\alpha}_t, \boldsymbol{\beta}_t)$ and $D'(\boldsymbol{\beta}_t) = [S_1(\boldsymbol{\alpha}_t); \ldots; S_K(\boldsymbol{\alpha}_t)]$
5:       $t \leftarrow t + 1$
6:       Update the cutting plane model $D_t(\boldsymbol{\beta}) \leftarrow \max_{i=0,\ldots,t-1} \langle D'(\boldsymbol{\beta}_i), \boldsymbol{\beta} \rangle$
7:       Let $\boldsymbol{\beta}_t \in \mathrm{argmin}_{\boldsymbol{\beta} \in B}\, D_t(\boldsymbol{\beta})$
8:   **until** $D(\boldsymbol{\beta}_{t-1}) - D_t(\boldsymbol{\beta}_t) \leq \varepsilon$

---

as follows. Starting with $\boldsymbol{\beta}_0 \in B$, a new iterate is obtained by solving

$$\boldsymbol{\beta}_t \in \operatorname*{argmin}_{\boldsymbol{\beta} \in B} D_t(\boldsymbol{\beta}) \,, \tag{7.30}$$

which can be cast as a linear program. Note that since the feasible set B is bounded, so is the solution of (7.30). In each iteration t, the obtained point $\boldsymbol{\beta}_t$ is an estimate of the optimal $\boldsymbol{\beta}^*$, and it is also used to update the cutting-plane model (7.29). The process is repeated until the gap between $D(\boldsymbol{\beta}_{t-1})$ and $D_t(\boldsymbol{\beta}_t)$ falls below a prescribed ε, meaning that $D(\boldsymbol{\beta}_t) \leq D(\boldsymbol{\beta}^*) + \varepsilon$ holds. Algorithm 7.5 summarizes the method.

Originally, Sonnenburg et al. (2006a) converted problem (7.26) into a *semi-infinite linear problem* (SILP) that was solved by column generation. However, the SILP is equivalent to the epigraph form of (7.26) (see Section 7.1.1), and the column generation results in exactly the same Algorithm 7.5.

Since large-scale SVM training problems are usually solved by decomposition techniques such as chunking (e.g., used in Joachims, 1999), one may significantly speedup Algorithm 7.5 by alternately solving for $\boldsymbol{\alpha}$ and $\boldsymbol{\beta}$ within the SVM solver, avoiding solution of the full SVM model with high precision (Sonnenburg et al., 2006a). Furthermore, as noted in Section 7.2.1, potential oscillations occurring in cutting-plane methods can be reduced by the bundle methods, as has been done by Xu et al. (2009a).

7.3.4   Conclusions

Multiple kernel learning has been used in various applications across diverse fields such as bioinformatics, image analysis, signal processing, and biomedical applications like brain-computer interfaces. It is being applied to fusing heterogeneous data (Lanckriet et al., 2004b; Sonnenburg et al., 2006b; Zien and Ong, 2007; Rakotomamonjy et al., 2008; Varma and Babu, 2009), to understand the learned kernel classifier (Sonnenburg et al., 2005), to feature selection (Szafranski et al., 2008; Xu et al., 2009b; Subrahmanya and Shin, 2010), or to automated model selection (Sonnenburg et al., 2006a). In this section, we have illustrated that the min-max formulation of MKL problem (7.22) can be converted into a sequence of linear and quadratic programs, of which the LP is simple and the QP can be directly solved using any of the existing SVM solvers. There exist further extensions of this approach not discussed in this section, such as an infinite dimensional version of the min-max MKL which was proposed by Argyriou et al. (2006). We have provided efficient implementations of MKL in the SHOGUN machine learning toolbox (Sonnenburg et al., 2010).

7.4   MAP Inference in Graphical Models

MAP inference in graphical models (Wainwright and Jordan, 2008) leads to the following NP-hard combinatorial optimization problem: *given a set of variables and a set of functions of (small) subsets of the variables, maximize the sum of the functions over all the variables.* This is also known as the weighted constraint satisfaction problem (Rossi et al., 2006, chapter 9).

The problem has a natural LP relaxation, proposed independently by Shlezinger (1976), Koster et al. (1998), and Wainwright et al. (2005). It is crucial to optimize the LP in the dual because primal methods do not scale to large problems. The relaxation was extended by Wainwright et al. (2005), Wainwright and Jordan (2008), and Johnson et al. (2007) to a hierarchy of progressively tighter LP relaxations. Komodakis et al. (2007) pointed out that the LP approach can be seen as a dual decomposition of the problem to tractable subproblems.

Several authors have proposed to tighten the relaxation incrementally. First, primal methods were proposed by Koster et al. (1998), Sontag and Jaakkola (2007), and Sontag (2007). Then came dual methods (Werner, 2008a, 2010; Kumar and Torr, 2008; Sontag et al., 2008; Komodakis and Paragios, 2008). Let us remark that not all of the authors related these incremental schemes to cutting-plane methods.

We revisit here the approach of Werner (2008a, 2010), which, we believe, captures the very core of the dual cutting-plane approach to MAP inference in a clean and transparent way. It is a generalization of the dual LP relaxation approach of Shlezinger (1976) and the max-sum diffusion algorithm of Kovalevsky and Koval, which have recently been reviewed by Werner (2005, 2007).

The approach is surprisingly simple and general. Every subset of the variables is assigned a function ("interaction"), all of them except a small part (which defines the problem) being initially zero. Max-sum diffusion passes messages between pairs of the variable subsets, acting as reparameterizations of the problem which monotonically decrease its upper bound. While in the extreme case all pairs of variable subsets are coupled like this, coupling only some of them results in a relaxation of the problem. At any time during diffusion we can tighten the relaxation by coupling new pairs—this results in an incremental scheme, recognized as a dual cutting-plane method.

After introducing notation, we construct the integer hull of the problem and the hierarchy of its LP relaxations in Section 7.4.2. In Sections 7.4.3 and 7.4.4 we dualize the LP relaxation and describe the max-sum diffusion algorithm which optimizes the dual. In Section 7.4.5 we augment this to a dual cutting-plane algorithm and discuss the corresponding separation problem. Section 7.4.6 explains the geometry of this cutting-plane algorithm in the primal domain, relating it to the marginal polytope.

7.4.1   Notation and Problem Definition

Let V be an ordered set of *variables* (the order on V is used only for notation consistency). A variable $v \in V$ attains *states* $x_v \in X_v$, where X_v is the (finite) *domain* of the variable. The *joint domain* of a subset $A \subseteq V$ of the variables is the Cartesian product $X_A = \prod_{v \in A} X_v$, where the order of factors is given by the order on V. A tuple $x_A \in X_A$ is a *joint state* of variables A. An *interaction* with *scope* $A \subseteq V$ is a function $\theta_A \colon X_A \to \overline{\mathbb{R}} = \mathbb{R} \cup \{-\infty\}$.

Let $E \subseteq 2^V$ be a hypergraph on V (a set of subsets of V). Every variable subset $A \subseteq V$ is assigned an interaction, while θ_A is identically zero whenever $A \notin E$. Having to deal with so many interactions may seem scary—but it will always be evident that most of them do not contribute to sums and are never visited in algorithms. Our task is to compute

$$\max_{x_v \in X_V} \sum_{A \in E} \theta_A(x_A) = \max_{x_v \in X_V} \sum_{A \subseteq V} \theta_A(x_A) . \qquad (7.31)$$

For instance, if $V = (1,2,3,4)$ and $E = \{(1,3,4), (2,3), (2,4), (3)\}$, then (7.31) reads $\max\limits_{x_1,x_2,x_3,x_4} [\theta_{134}(x_1, x_3, x_4) + \theta_{23}(x_2, x_3) + \theta_{24}(x_2, x_4) + \theta_3(x_3)]$. Note, that since V is an ordered set, we use $(\cdots)$ rather than $\{\cdots\}$ to denote V and its subsets.

We will use $T = \{ (A, x_A) \mid A \subseteq V, \ x_A \in X_A \}$ to denote the set of all joint states of all variable subsets (T stands for "tuples"). All interactions θ_A, $A \subseteq V$, will be understood as a single vector $\boldsymbol{\theta} \in \overline{\mathbb{R}}^T$.

7.4.2 The Hierarchy of LP Relaxations

We define a mapping $\boldsymbol{\delta} \colon X_V \to \{0,1\}^T$ as follows: $\delta_A(y_A)(x_V)$ equals 1 if the joint state y_A is the restriction of joint state x_V on variables A, and 0 otherwise. Here, $\delta_A(y_A)(x_V)$ denotes the (A, y_A)-component of vector $\boldsymbol{\delta}(x_V) \in \{0,1\}^T$. This lets us write the objective function of (7.31) as a scalar product:

$$\sum_{A \subseteq V} \theta_A(x_A) = \sum_{A \subseteq V} \sum_{y_A} \theta_A(y_A)\, \delta_A(y_A)(x_V) = \langle \boldsymbol{\theta}, \boldsymbol{\delta}(x_V) \rangle \;.$$

Problem (7.31) can now be reformulated as

$$\max_{x_V \in X_V} \sum_{A \subseteq V} \theta_A(x_A) = \max_{x_V \in X_V} \langle \boldsymbol{\theta}, \boldsymbol{\delta}(x_V) \rangle = \max_{\boldsymbol{\mu} \in \boldsymbol{\delta}(X_V)} \langle \boldsymbol{\theta}, \boldsymbol{\mu} \rangle = \max_{\boldsymbol{\mu} \in \operatorname{conv} \boldsymbol{\delta}(X_V)} \langle \boldsymbol{\theta}, \boldsymbol{\mu} \rangle$$

where $\boldsymbol{\delta}(X_V) = \{\, \boldsymbol{\delta}(x_V) \mid x_V \in X_V \,\}$. This expresses problem (7.31) in the form (7.9), as a linear optimization over the integral hull $\operatorname{conv} \boldsymbol{\delta}(X_V) \subseteq [0,1]^T$.

Let $I = \{\, (A, B) \mid B \subseteq A \subseteq V \,\}$ denote the set of hyperedge pairs related by inclusion, that is, the inclusion relation on 2^V. For any $J \subseteq I$, we define a polytope $\mathcal{M}(J)$ to be the set of vectors $\boldsymbol{\mu} \in [0,1]^T$ satisfying

$$\sum_{x_{A \setminus B}} \mu_A(x_A) = \mu_B(x_B)\,, \qquad (A, B) \in J,\; x_B \in X_B\,, \tag{7.32a}$$

$$\sum_{x_A} \mu_A(x_A) = 1\,, \qquad\qquad A \subseteq V\,. \tag{7.32b}$$

What is this object? Any $\boldsymbol{\mu} \in \mathcal{M}(J)$ is a set of distributions $\mu_A \colon X_A \to [0,1]$ over every subset $A \subseteq V$ of the variables. Constraint (7.32b) normalizes the distributions. Constraint (7.32a) couples pairs of distributions, imposing μ_B as the marginal of μ_A whenever $(A, B) \in J$. For example, if $A = (1, 2, 3, 4)$ and $B = (2, 4)$, then (7.32a) reads $\sum_{x_1, x_3} \mu_{1234}(x_1, x_2, x_3, x_4) = \mu_{24}(x_2, x_4)$.

For brevity, we will use the shorthand $\mathcal{M}(I) = \mathcal{M}$. We claim that

$$\operatorname{conv} \boldsymbol{\delta}(X_V) = \mathcal{M}\,. \tag{7.33}$$

To see it, let us write a convex combination of the elements of $\boldsymbol{\delta}(X_V)$,

$$\boldsymbol{\mu} = \sum_{x_V} \mu_V(x_V)\, \boldsymbol{\delta}(x_V)\,, \tag{7.34}$$

where $\mu_V(x_V)$ denotes the coefficients of the convex combination. But μ_V is already part of $\boldsymbol{\mu}$. The (A, y_A)-component of vector (7.34) reads

$$\mu_A(y_A) = \sum_{x_V} \mu_V(x_V)\, \delta_A(y_A)(x_V) = \sum_{y_{V\backslash A}} \mu_V(y_V)\,.$$

But this is (7.32a) for $(A, B) = (V, A)$.

By imposing only a subset of all possible marginalization constraints (7.32a), an outer relaxation of the integral hull conv $\boldsymbol{\delta}(X_V) = \mathcal{M}$ is obtained. Namely, for any $J \subseteq I$ we have $\mathcal{M}(J) \supseteq \mathcal{M}$, and hence

$$\max\{\,\langle\boldsymbol{\theta}, \boldsymbol{\mu}\rangle \mid \boldsymbol{\mu} \in \mathcal{M}(J)\,\} \tag{7.35}$$

is a linear programming relaxation of problem (7.31), that is, its optimum is an upper bound on (7.31). All possible relaxations form a partially ordered hierarchy, indexed by $J \subseteq I$. Figure 7.2 shows examples.

The hierarchy could be made finer-grained by also selecting *individual* joint states, that is, by imposing marginalization equality (7.32a) for $(A, B, x_B) \in J$ where $J \subseteq I = \{\,(A, B, x_B) \mid B \subseteq A \subseteq V,\ x_B \in X_B\,\}$.

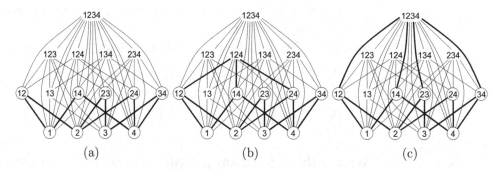

Figure 7.2: The Hasse diagram (precisely, its transitive closure) of the set 2^V of all subsets of $V = (1, 2, 3, 4)$. The nodes depict hyperedges $A \subseteq V$ (with hyperedge $\emptyset$ omitted) and the arcs depict hyperedge pairs $(A, B) \in I$. The hyperedges in circles form the problem hypergraph $E = \{(1), (2), (3), (4), (1,2), (1,4), (2,3), (2,4), (3,4)\}$, and the interactions over the non circled hyperedges are zero. Any subset $J \subseteq I$ of the arcs yields one possible relaxation (7.35) of problem (7.31). (a), (b), and (c) show three example relaxations, with J depicted as thick arcs.

7.4.3   The Dual of the LP Relaxation

Rather than solving the linear program (7.35) directly, it is much better to solve its dual. This dual is constructed as follows. Let matrices $\boldsymbol{A}$ and $\boldsymbol{B}$ be such that $\boldsymbol{A}\boldsymbol{\mu} = \boldsymbol{0}$ and $\boldsymbol{B}\boldsymbol{\mu} = \boldsymbol{1}$ are the sets of equalities (7.32a) and (7.32b), respectively. Then (7.35) can be written as the left linear program below:

$$\langle \boldsymbol{\theta}, \boldsymbol{\mu} \rangle \to \max \qquad\qquad \langle \boldsymbol{\psi}, \mathbf{1} \rangle \to \min \tag{7.36a}$$

$$\boldsymbol{A}\boldsymbol{\mu} = \mathbf{0} \qquad\qquad\qquad \boldsymbol{\varphi} \lessgtr \mathbf{0} \tag{7.36b}$$

$$\boldsymbol{B}\boldsymbol{\mu} = \mathbf{1} \qquad\qquad\qquad \boldsymbol{\psi} \lessgtr \mathbf{0} \tag{7.36c}$$

$$\boldsymbol{\mu} \ge \mathbf{0} \qquad\quad \boldsymbol{\varphi}\boldsymbol{A} + \boldsymbol{\psi}\boldsymbol{B} \ge \boldsymbol{\theta} \tag{7.36d}$$

On the right we wrote the LP dual, such that in (7.36b-d) a constraint and its Lagrange multiplier are always on the same line ($\lessgtr \mathbf{0}$ means that the variable vector is unconstrained). By eliminating the variables $\boldsymbol{\psi}$, the dual reads

$$\min_{\boldsymbol{\varphi}} \sum_{A \subseteq V} \max_{x_A} \theta_A^{\varphi}(x_A), \tag{7.37}$$

where we abbreviated $\boldsymbol{\theta}^{\varphi} = \boldsymbol{\theta} - \boldsymbol{\varphi}\boldsymbol{A}$. The components of vector $\boldsymbol{\theta}^{\varphi}$ read

$$\theta_A^{\varphi}(x_A) = \theta_A(x_A) - \sum_{B|(B,A)\in J} \varphi_{BA}(x_A) + \sum_{B|(A,B)\in J} \varphi_{AB}(x_B) \tag{7.38}$$

where $\boldsymbol{\varphi} = \{ \varphi_{AB}(x_B) \mid (A,B) \in J, \ x_B \in X_B \}$. Next we explain the meaning of (7.38) and (7.37).

A *reparameterization* is a transformation of $\boldsymbol{\theta}$ that preserves the objective function $\sum_{A \subseteq V} \theta_A$ of problem (7.31). The simplest reparameterization is done as follows: pick two interactions θ_A and θ_B with $B \subseteq A$, add an arbitrary function (a "message") $\varphi_{AB} \colon X_B \to \mathbb{R}$ to θ_A, and subtract the same function from θ_B:

$$\theta_A(x_A) \leftarrow \theta_A(x_A) + \varphi_{AB}(x_B), \qquad x_A \in X_A, \tag{7.39a}$$

$$\theta_B(x_B) \leftarrow \theta_B(x_B) - \varphi_{AB}(x_B), \qquad x_B \in X_B. \tag{7.39b}$$

For instance, if $A = (1,2,3,4)$ and $B = (2,4)$, then we add a function $\varphi_{1234,24}(x_2, x_4)$ to $\theta_{1234}(x_1, x_2, x_3, x_4)$ and subtract $\varphi_{1234,24}(x_2, x_4)$ from $\theta_{24}(x_2, x_4)$. This preserves $\theta_A + \theta_B$ (because φ_{AB} cancels out), and hence $\sum_{A \subseteq V} \theta_A$ as well. Applying reparameterization (7.39) to all pairs $(A, B) \in J$ yields (7.38).

Thus, (7.38) describes reparameterizations, that is, for every x_V and $\boldsymbol{\varphi}$ we have

$$\sum_{A \subseteq V} \theta_A(x_A) = \sum_{A \subseteq V} \theta_A^{\varphi}(x_A).$$

In addition (7.38) also preserves (for feasible $\boldsymbol{\mu}$) the objective of the primal program (7.36): $\boldsymbol{A}\boldsymbol{\mu} = \mathbf{0}$ implies $\langle \boldsymbol{\theta}^{\varphi}, \boldsymbol{\mu} \rangle = \langle \boldsymbol{\theta} - \boldsymbol{\varphi}\boldsymbol{A}, \boldsymbol{\mu} \rangle = \langle \boldsymbol{\theta}, \boldsymbol{\mu} \rangle$.

By the well-known max-sum dominance, for any $\boldsymbol{\theta}$ we have

$$\max_{x_V} \sum_{A \subseteq V} \theta_A(x_A) \leq \sum_{A \subseteq V} \max_{x_A} \theta_A(x_A) \,, \tag{7.40}$$

so the right-hand side of (7.40) is an upper bound on (7.31), which shows that the *dual (7.37) minimizes an upper bound on (7.31) by reparameterizations.*

Note that for each $(A, B) \in J$, marginalization constraint (7.32a) corresponds via duality to message φ_{AB}. The larger J is, the larger the set of reparameterizations (7.38) and hence the smaller the optimal value of (7.37).

When is inequality (7.40) (and hence the upper bound) tight? It happens if and only if the independent maximizers of the interactions agree on a common global assignment, that is, if there exists $y_V \in X_V$ such that

$$y_A \in \operatorname*{argmax}_{x_A} \theta_A(x_A) \,, \qquad A \subseteq V \,.$$

We will further refer to the set $\operatorname{argmax}_{x_A} \theta_A(x_A)$ as the *active joint states* of interaction θ_A. The test can be cast as the *constraint satisfaction problem* (CSP) (Mackworth, 1991; Rossi et al., 2006) formed by the active joint states of all the interactions (Shlezinger, 1976; Werner, 2007, 2010). Thus, if after solving (7.37) this CSP is satisfiable for $\boldsymbol{\theta}^\varphi$, the relaxation is tight and we have solved our instance of problem (7.31) exactly. Otherwise, we have only an upper bound on (7.31).

7.4.4   Max-sum diffusion

Max-sum diffusion is a simple convergent "message-passing" algorithm to tackle the dual LP. It seeks to reparameterize $\boldsymbol{\theta}$ such that

$$\max_{x_{A \setminus B}} \theta_A(x_A) = \theta_B(x_B) \,, \qquad (A, B) \in J, \ x_B \in X_B \,. \tag{7.41}$$

The algorithm repeats the following iteration:

Enforce (7.41) for a single pair $(A, B) \in J$ by reparameterization (7.39).

This is done by setting $\varphi_{AB}(x_B) = [\theta_B(x_B) - \max_{x_{A \setminus B}} \theta_A(x_A)]/2$ in (7.39). The algorithm converges to a fixed point when (7.41) holds for all $(A, B) \in J$.

Originally (Kovalevsky and Koval) max-sum diffusion was formulated for problems with only binary (no unary) interactions. The generalization (7.41) by Werner (2008a, 2010) is interesting because (7.41) has exactly the same form as (7.32a). This idea was pursued further in (Werner, 2008b).

Reparameterizing by messages rather than by modifying $\boldsymbol{\theta}$ yields Algorithm 7.6. To handle infinite weights correctly, the algorithm expects that $[\theta_B(x_B) > -\infty] \Leftrightarrow [\max_{x_{A \setminus B}} \theta_A(x_A) > -\infty]$ for every $(A, B) \in J$.

Algorithm 7.6 Max-sum diffusion

1: **repeat**
2:     **for** $(A, B) \in J$ and $x_B \in X_B$ such that $\theta_B(x_B) > -\infty$ **do**
3:         $\varphi_{AB}(x_B) \leftarrow \varphi_{AB}(x_B) + [\theta_B^{\varphi}(x_B) - \max_{x_{A \setminus B}} \theta_A^{\varphi}(x_A)]/2$
4:     **end for**
5: **until** convergence

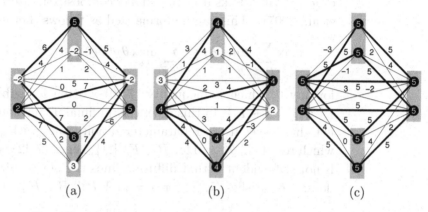

(a)   (b)   (c)

Figure 7.3: The visualization of a problem with $|X_v| = 2$ variable states and hypergraph E as in Figure 7.2. The variables are shown as boxes; their numbering is $\begin{smallmatrix}&1\\2&3&4\end{smallmatrix}$. Variable states are shown as circles, and joint states of variable pairs as edges. Weights $\theta_A(x_A)$, $A \in E$, are written in the circles and next to the edges. Active joint states are emphasized (black circles, thick edges). Example (a) is not a diffusion fixed point; (b) and (c) are diffusion fixed points for J from Figure 7.2a. Examples (a) and (b) are reparameterizations of each other (this is not obvious at first sight); (c) is not a reparameterization of (a) and (b). For (b), a global assignment x_V can be composed of the active joint states, and hence inequality (7.40) is tight. For (a) and (c), no global assignment x_V can be composed of the active joint states, hence inequality (7.40) is not tight.

The diffusion iteration decreases or preserves, but never increases, the upper bound. In general, the algorithm does not find the global minimum of (7.37) but only a certain local minimum (where "local" is meant w.r.t. block-coordinate moves), which is nevertheless very good in practice. These local minima are characterized by *local consistency* (Rossi et al., 2006, chapter 3) of the CSP formed by the active joint states.

Note that the only non trivial operation in Algorithm 7.6 is computing the max-marginals $\max_{x_{A \setminus B}} \theta_A^{\varphi}(x_A)$. By (7.38), this is an instance of problem (7.31). When $|A|$ is small (such as for a binary interaction), computing the max-marginals is trivial. But even when $|A|$ is large, depending on the function θ_A and on J, there may exist an algorithm polynomial in $|A|$ to compute $\max_{x_{A \setminus B}} \theta_A^{\varphi}(x_A)$. In that case, Algorithm 7.6 can still be used.

If $\theta_A = 0$, it depends only on J whether $\max_{x_{A\setminus B}} \theta_A^\varphi(x_A)$ can be computed in polynomial time. For instance, in Figure 7.2c we have $\theta_{1234} = 0$ and hence, by (7.38), $\theta_{1234}^\varphi(x_1, x_2, x_3, x_4) = \varphi_{1234,12}(x_1, x_2) + \varphi_{1234,23}(x_2, x_3) + \varphi_{1234,34}(x_3, x_4) + \varphi_{1234,14}(x_1, x_4)$. Thus we have a problem on a cycle, which can be solved more efficiently than by going through all states (x_1, x_2, x_3, x_4).

This suggests that in a sense diffusion solves certain small subproblems exactly (which links it to the dual decomposition interpretation (Komodakis et al., 2007)). This can be formalized as follows. Let $A \in F \subseteq 2^A$. Clearly,

$$\max_{x_A} \sum_{B \in F} \theta_B(x_B) \leq \sum_{B \in F} \max_{x_B} \theta_B(x_B) \qquad (7.42)$$

for any $\boldsymbol{\theta}$, which is inequality (7.40) written for subproblem F. Let $J = \{(A, B) \mid B \in F\}$. In this case, the minimal upper bound for subproblem F is tight. To see it, do reparameterization (7.39) with $\varphi_{AB} = \theta_B$ for $B \in F$, which results in $\theta_B = 0$ for $B \in F \setminus \{A\}$; hence (7.42) is trivially tight. What is not self-evident is that diffusion finds the *global* minimum in this case. It does: if $\boldsymbol{\theta}$ satisfies (7.41) for $J = \{(A, B) \mid B \in F\}$, then (7.42) is tight.

7.4.5   Dual Cutting-plane Algorithm

The relaxation can be tightened *incrementally* during dual optimization. At any time during algorithm 7.6, the current J can be extended by any $J' \subseteq I$, $J' \cap J = \emptyset$. The messages φ_{AB} for $(A, B) \in J'$ are initialized to zero. Clearly, this does not change the current upper bound. Future diffusion iterations can only preserve or improve the bound, so the scheme remains monotonic. This can be imagined as if the added variables φ_{AB} extended the space of possible reparameterizations, and diffusion is now trying to take advantage of it. If the bound does not improve, all we will have lost is the memory occupied by the added variables. Algorithm 7.7 describes this.

In the primal domain, this incremental scheme can be understood as a cutting-plane algorithm. We discuss this in Section 7.4.6.

Algorithm 7.7 Dual cutting-plane algorithm

1: **Initialization:** Choose $J \subseteq I$ and $\mathcal{J} \subseteq 2^I$
2: **repeat**
3:     Execute any number of iterations of algorithm 7.6
4:     *Separation oracle:* choose $J' \in \mathcal{J}$, $J \cap J' = \emptyset$
5:     $J \leftarrow J \cup J'$
6:     Allocate messages φ_{AB}, $(A, B) \in J'$, and set them to zero
7: **until** no suitable J' can be found

On line 4 of Algorithm 7.7 the separation oracle, which chooses a promising extension J' from some predefined set $\mathcal{J} \subseteq 2^I$ of candidate extensions, is called. We assume $|\mathcal{J}|$ is small so that it can be searched exhaustively. For that, we need a test to recognize whether a given J' would lead to a (good) bound improvement. We refer to this as the *separation test*.

Of course, a trivial necessary and sufficient separation test is to extend J by J' and run diffusion until convergence. One can easily invent a faster test:

> *Execute several diffusion iterations only on pairs J'. If this improves the bound, then running diffusion on $J \cup J'$ would inevitably improve the bound, too.*

This local test is sufficient but not necessary for improvement because even if running diffusion on J' does not improve the bound, it may change the problem such that future diffusion iterations on $J \cup J'$ improve it.

Even with a sufficient and necessary separation test, Algorithm 7.7 is "greedy" in the following sense. For $J'_1, J'_2 \subseteq I$, it can happen that extending J by J'_1 alone or by J'_2 alone does not lead to a bound improvement but extending J by $J'_1 \cup J'_2$ does. See (Werner, 2010) for an example.

The extension J' can be an arbitrary subset of I. One form of extension has a clear meaning: pick a hyperedge A not yet coupled to any other hyperedge, choose $F \subseteq 2^A$, and let $J' = \{ (A, B) \mid B \in F \}$. This can be seen as connecting a so far disconnected interaction θ_A to the problem.

An important special case is connecting a zero interaction, $\theta_A = 0$. Because, by (7.38), we have $\theta_A^\varphi(x_A) = \sum_{B \in F} \varphi_{AB}(x_B)$, we refer to this extension as *adding a zero subproblem F*. In this case, the separation test can be done more efficiently than by running diffusion on J'. This is based on the fact stated at the end of Section 7.4.4: if inequality (7.42) is not tight for current $\boldsymbol{\theta}^\varphi$, then running diffusion on J' will surely make it tight, that is, improve the bound. We do not need $A \in F$ here because $\theta_A = 0$. The gap in (7.42) is an estimate of the expected improvement.

This has a clear interpretation in CSP terms. Inequality (7.42) is tight if and only if the CSP formed by the active joint states of interactions F is satisfiable. If this CSP is unsatisfiable, then J' will improve the bound. Therefore, *the separation oracle needs to find a (small) unsatisfiable subproblem of the CSP formed by the active joint states*.

For instance, Figure 7.3c shows a problem after diffusion convergence, for J defined by Figure 7.2a. The CSP formed by the active joint states is not satisfiable because it contains an unsatisfiable subproblem, the cycle $F = \{(1, 2), (1, 4), (2, 4)\}$. Hence, adding zero subproblem F (which yields J from Figure 7.2b) and running diffusion would improve the bound. Adding the zero cycle $F = \{(1, 2), (1, 4), (2, 3), (3, 4)\}$ (yielding J from Figure 7.2c)

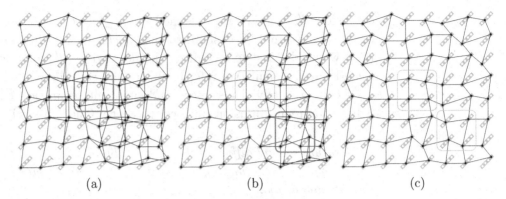

$$(a) \qquad\qquad (b) \qquad\qquad (c)$$

Figure 7.4: Two steps of the cutting-plane algorithm for a problem with an 8×8 grid graph E and $|X_v| = 4$ variable states. The set $\mathcal{J}$ of candidate extensions contains all cycles of length 4. Only the active joint states are shown. (a) shows the problem after diffusion has converged for $J = \{\,(A, B) \mid B \subseteq A; \ A, B \in E\,\}$. The upper bound is not tight because of the depicted unsatisfiable subproblem (an inconsistent cycle). Adding the cycle and letting diffusion reconverge results in (b) with a better bound. The original cycle is now satisfiable, but a new unsatisfiable cycle has occurred. Adding this cycle solves the problem (c).

or the whole zero problem $F = E$ would improve the bound too.

Figure 7.4 shows a more complex example.

Message-passing algorithms have a drawback: after extending J, they need a long time to reconverge. This can be partially alleviated by adding multiple subproblems at a time before full convergence. As some of the added subproblems might later turn out to be redundant, we found it helpful to remove redundant subproblems occasionally—which can be done without sacrificing monotonicity of bound improvement. This is a (dual) way of constraint management, often used in cutting-plane methods.

7.4.6   Zero Interactions as Projection, Marginal Polytope

In the beginning, formula (7.31), we added all possible zero interactions to our problem. This proved to be natural because the problem is, after all, defined only up to reparameterizations, and thus any zero interaction can become nonzero. Now, let us see what the LP relaxation would look like without adopting this abstraction. Let $T(E) = \{\,(A, x_A) \mid A \in E, \ x_A \in X_A\,\}$ denote the restriction of the set T to hypergraph E. Since zero interactions do not contribute to the objective function of (7.35), the latter can be written as

$$\max\{\,\langle \boldsymbol{\theta}, \boldsymbol{\mu} \rangle \mid \boldsymbol{\mu} \in \mathcal{M}(J)\,\} = \max\{\,\langle \pi_{T(E)}\boldsymbol{\theta}, \boldsymbol{\mu} \rangle \mid \boldsymbol{\mu} \in \pi_{T(E)}\mathcal{M}(J)\,\} \quad (7.43)$$

where $\pi_{D'}\boldsymbol{a} \in \overline{\mathbb{R}}^{D'}$ denotes the projection of a vector $\boldsymbol{a} \in \overline{\mathbb{R}}^{D}$ on dimensions $D' \subseteq D$; thus $\pi_{D'}$ deletes the components $D \setminus D'$ of $\boldsymbol{a}$. Applied to a set of vectors, $\pi_{D'}$ does this for every vector in the set. Informally, (7.43) shows that *zero interactions act as the projection of the feasible set onto the space of nonzero interactions*.

The set $\pi_{T(E)}\mathcal{M} \subseteq [0,1]^{T(E)}$ is recognized as the *marginal polytope* (Wainwright et al., 2005) of hypergraph E. Its elements $\boldsymbol{\mu}$ are the marginals over variable subsets E of some global distribution μ_V, which is not necessarily part of $\boldsymbol{\mu}$. The marginal polytope of the complete hypergraph $\pi_{T(2^V)}\mathcal{M} = \mathcal{M}$ is of fundamental importance because all other marginal polytopes are its projections. For $J \subseteq I$, the set $\pi_{T(E)}\mathcal{M}(J) \supseteq \pi_{T(E)}\mathcal{M}$ is a relaxation of the marginal polytope, which may contain elements $\boldsymbol{\mu}$ that no longer can be realized as the marginals of any global distribution μ_V.

Following Wainwright et al. (2005), we say a relaxation J is *local in E* if $A, B \in E$ for every $(A, B) \in J$. For instance, in Figure 7.2 only relaxation (a) is local. For local relaxations, the distributions μ_A, $A \notin E$, are not coupled to any other distributions and the action of $\pi_{T(E)}$ on $\mathcal{M}(J)$ is simple: it simply removes these superfluous coordinates. Thus, $\pi_{T(E)}\mathcal{M}(J)$ has an explicit description by a small (polynomial in $|E|$) number of linear constraints.

For nonlocal relaxations, the effect of the projection is in general complex and the number of facets of $\pi_{T(E)}\mathcal{M}(J)$ is exponential in $|E|$. It is well-known that computing the explicit description of a projection of a polyhedron can be extremely difficult—which suggests that directly looking for the facets of $\pi_{T(E)}\mathcal{M}$ might be a bad idea. Nonlocal relaxations can be seen as a *lift-and-project approach*: we lift from dimensions $T(E)$ to dimensions T, impose constraints in this lifted space, and project back onto dimensions $T(E)$.

Now the geometry of our cutting-plane algorithm in the primal space $[0,1]^{T(E)}$ is clear. Suppose max-sum diffusion has found a global optimum of the dual and let $\boldsymbol{\mu}^* \in [0,1]^{T(E)}$ be a corresponding primal optimum. A successful extension of J means that a set (perhaps exponentially large) of cutting planes is added to the primal that separates $\boldsymbol{\mu}^*$ from $\pi_{T(E)}\mathcal{M}$. However, $\boldsymbol{\mu}^*$ is not computed explicitly (and it is expensive to compute $\boldsymbol{\mu}^*$ from a dual optimum for large problems). In fact, $\boldsymbol{\mu}^*$ may not even exist because diffusion may find only a local optimum of the dual—we even need not run diffusion to full convergence.

7.4.7 Conclusions

We have presented the theory of the cutting-plane approach to the MAP inference problem, as well as a very general message-passing algorithm to implement this approach. In comparison with similar works, the theory, and Algorithm 7.6 in particular, is very simple. We have shown that for the case of adding subproblems, separation means finding a (small) unsatisfiable subproblem of the CSP formed by the active joint states.

We assumed, in Section 7.4.5, that the set $\mathcal{J}$ of candidate extensions is tractably small. Is there a polynomial algorithm to select an extension from an intractably large set $\mathcal{J}$? In particular, is there a polynomial algorithm to find a small, unsatisfiable subproblem (most interestingly, a cycle) in a given CSP? This is currently an open problem. An inspiration for finding such algorithms are local consistencies in CSP (Rossi et al., 2006, chapter 3).

Several polynomial algorithms are known to separate intractable families of cutting planes of the *max-cut polytope* (Deza and Laurent, 1997), which is closely related to the marginal polytope. Some of them have been applied to MAP inference by Sontag and Jaakkola (2007) and Sontag (2007). Since these algorithms work in the primal space, they cannot be used in our dual cutting-plane scheme—we need a *dual separation algorithm*.

Acknowledgments

Vojtěch Franc was supported by the Czech Ministry of Education project 1M0567 and by EC projects FP7-ICT-247525 HUMAVIPS and PERG04-GA-2008-239455 SEMISOL. Soeren Sonneburg was supported by the EU under the PASCAL2 Network of Excellence (ICT-216886) as well as DFG grants MU 987/6-1 and RA-1894/1-1. Tomáš Werner was supported by the EC grant 215078 (DIPLECS), Czech government grant MSM6840770038, and the Grant Agency of the Czech Republic grant P103/10/0783.

7.5 References

A. Argyriou, R. Hauser, C. A. Micchelli, and M. Pontil. A dc-programming algorithm for kernel selection. In *Proceedings of the International Conference on Machine Learning*, pages 41–48. ACM Press, 2006.

F. Bach. Exploring large feature spaces with hierarchical multiple kernel learning. In D. Koller, D. Schuurmans, Y. Bengio, and L. Bottou, editors, *Advances in Neural Information Processing Systems 21*, pages 105–112. MIT Press, 2008.

F. R. Bach, G. R. G. Lanckriet, and M. I. Jordan. Multiple kernel learning, conic duality, and the smo algorithm. In *Proceedings of the International Conference on Machine Learning*. ACM Press, 2004.

D. P. Bertsekas. *Nonlinear Programming*. Athena Scientific, Belmont, MA, second edition, 1999.

S. Boyd and L. Vandenberge.   Localization and cutting-plane methods. Unpublished lecture notes, Stanford University, California, USA, 2008. URL `http://see.stanford.edu/materials/lsocoee364b/05-localization_methods_notes.pdf`.

S. Boyd and L. Vandenberghe. *Convex Optimization*. Cambridge University Press, March 2004.

E. W. Cheney and A. A. Goldstein. Newton's method for convex programming and Tchebycheff approximation. *Numerische Mathematik*, 1:253–268, 1959.

C. Cortes and V. Vapnik. Support-vector networks. *Machine Learning*, 20(3): 273–297, 1995.

C. Cortes, M. Mohri, and A. Rostamizadeh. Learning non-linear combinations of kernels. In Y. Bengio, D. Schuurmans, J. Lafferty, C. Williams, and A. Culotta, editors, *Advances in Neural Information Processing Systems 22*, pages 396–404. MIT Press, 2009.

G. Dantzig, R. Fulkerson, and S. Johnson. Solution of a large-scale traveling-salesman problem. *Operations Research*, 2:393–410, 1954.

M. M. Deza and M. Laurent. *Geometry of Cuts and Metrics*. Springer, Berlin, 1997.

V. Franc and S. Sonnenburg. OCAS optimized cutting plane algorithm for support vector machines. In *Proceedings of the International Conference on Machine Learning*, pages 320–327. ACM Press, 2008.

V. Franc and S. Sonnenburg. Optimized cutting plane algorithm for large-scale risk minimization. *Journal of Machine Learning Research*, 10:2157–2192, 2010.

M. Grötschel, L. Lovász, and A. Schrijver. The ellipsoid method and its consequences in combinatorial optimization. *Combinatorica*, 1(2):169–197, 1981.

T. Joachims.   Making large–scale SVM learning practical.   In B. Schölkopf, C. Burges, and A. Smola, editors, *Advances in Kernel Methods: Support Vector Learning*, pages 169–184, Cambridge, MA, USA, 1999. MIT Press.

J. K. Johnson, D. M. Malioutov, and A. S. Willsky. Lagrangian relaxation for MAP estimation in graphical models. In *Allerton Conference on Communication, Control and Computing*, 2007.

J. E. Kelley. The cutting-plane method for solving convex programs. *Journal of the Society for Industrial and Applied Mathematics*, 8(4):703–712, 1960.

K. C. Kiwiel. An aggregate subgradient method for nonsmooth convex minimization. *Mathematical Programming*, 27(3):320–341, 1983.

M. Kloft, U. Brefeld, S. Sonnenburg, P. Laskov, K.-R. Müller, and A. Zien. Efficient and accurate Lp-norm multiple kernel learning. In Y. Bengio, D. Schuurmans, J. Lafferty, C. Williams, and A. Culotta, editors, *Advances in Neural Information Processing Systems 22*, pages 997–1005. MIT Press, 2009.

N. Komodakis and N. Paragios. Beyond loose LP-relaxations: Optimizing MRFs by repairing cycles. In *Proceedings of the European Conference on Computer Vision*, pages 806–820, 2008.

N. Komodakis, N. Paragios, and G. Tziritas. MRF optimization via dual decomposition: Message-passing revisited. In *Proceedings of the IEEE International Conference on Computer Vision*, 2007.

A. M. Koster, S. P. M. van Hoesel, and A. W. J. Kolen. The partial constraint satisfaction problem: Facets and lifting theorems. *Operations Research Letters*, 23(3–5):89–97, 1998.

V. A. Kovalevsky and V. K. Koval. A diffusion algorithm for decreasing the energy of the max-sum labeling problem. Unpublished, Glushkov Institute of Cybernetics, Kiev, USSR, circa 1975. Personally communicated to T. Werner by M. I. Schlesinger.

M. P. Kumar and P. H. S. Torr. Efficiently solving convex relaxations for MAP estimation. In *Proceedings of the International Conference on Machine Learning*, pages 680–687. ACM Press, 2008.

G. Lanckriet, N. Cristianini, L. E. Ghaoui, P. Bartlett, and M. I. Jordan. Learning the kernel matrix with semi-definite programming. *Journal of Machine Learning Research*, 5:27–72, 2004a.

G. Lanckriet, T. de Bie, N. Cristianini, M. Jordan, and W. Noble. A statistical framework for genomic data fusion. *Bioinformatics*, 20(16):2626–2635, 2004b.

C. Lemaréchal, A. Nemirovskii, and Y. Nesterov. New variants of bundle methods. *Mathematical Programming*, 69(1–3):111–147, 1995.

A. Mackworth. Constraint satisfaction. In *Encyclopaedia of Artificial Intelligence*, pages 285–292. John Wiley, 1991.

J. S. Nath, G. Dinesh, S. Raman, C. Bhattacharyya, A. Ben-Tal, and K. R. Ramakrishnan. On the algorithmics and applications of a mixed-norm based kernel learning formulation. In Y. Bengio, D. Schuurmans, J. Lafferty, C. Williams, and A. Culotta, editors, *Advances in Neural Information Processing Systems 22*, pages 844–852. MIT Press, 2009.

A. S. Nemirovskij and D. B. Yudin. *Problem Complexity and Method Efficiency in Optimization*. Wiley Interscience, New York, 1983.

A. Rakotomamonjy, F. Bach, S. Canu, and Y. Grandvalet. More efficiency in multiple kernel learning. In *Proceedings of the International Conference on Machine Learning*, pages 775–782, 2007.

A. Rakotomamonjy, F. R. Bach, S. Canu, and Y. Grandvalet. SimpleMKL. *Journal of Machine Learning Research*, 9:2491–2521, 2008.

F. Rossi, P. van Beek, and T. Walsh, editors. *Handbook of Constraint Programming*. Elsevier, 2006.

B. Schölkopf and A. Smola. *Learning with Kernels*. MIT Press, 2002.

A. Schrijver. *Combinatorial Optimization : Polyhedra and Efficiency*. Algorithms and Combinatorics. Springer, 2003.

M. I. Shlezinger. Syntactic analysis of two-dimensional visual signals in noisy conditions. *Cybernetics and Systems Analysis*, 12(4):612–628, 1976. Translated from the Russian.

S. Sonnenburg, G. Rätsch, and C. Schäfer. Learning interpretable SVMs for biological sequence classification. In S. Miyano, J. P. Mesirov, S. Kasif, S. Istrail, P. A. Pevzner, and M. Waterman, editors, *Research in Computational Molecular Biology, Proceedings of the 9th Annual International Conference (RECOMB)*, volume 3500 of *Lecture Notes in Computer Science*, pages 389–407. Springer-Verlag, 2005.

S. Sonnenburg, G. Rätsch, C. Schäfer, and B. Schölkopf. Large scale multiple kernel learning. *Journal of Machine Learning Research*, 7:1531–1565, 2006a.

S. Sonnenburg, A. Zien, and G. Rätsch. ARTS: Accurate recognition of transcription starts in human. *Bioinformatics*, 22(14):e472–e480, 2006b.

S. Sonnenburg, G. Rätsch, S. Henschel, C. Widmer, J. Behr, A. Zien, F. de Bona, A. Binder, C. Gehl, and V. Franc. The SHOGUN machine learning toolbox. *Journal of Machine Learning Research*, 11:1799–1802, June 2010. URL http://www.shogun-toolbox.org.

D. Sontag. Cutting plane algorithms for variational inference in graphical models. Master's thesis, Department of Electrical Engineering and Computer Science, MIT, 2007.

D. Sontag and T. Jaakkola. New outer bounds on the marginal polytope. In *Advances in Neural Information Processing Systems 20*. MIT Press, 2007.

D. Sontag, T. Meltzer, A. Globerson, T. Jaakkola, and Y. Weiss. Tightening LP relaxations for MAP using message passing. In *Proceedings of the 24th Conference on Uncertainty in Artificial Intelligence (UAI)*. AUAI Press, Corvallis, Oregon, 2008.

N. Subrahmanya and Y. C. Shin. Sparse multiple kernel learning for signal processing applications. *IEEE Transactions on Pattern Analysis and Machine Intelligence*, 32(5):788–798, 2010.

M. Szafranski, Y. Grandvalet, and A. Rakotomamonjy. Composite kernel learning. In *Proceedings of the International Conference on Machine Learning*, 2008.

C. Teo, Q. Le, A. Smola, and S. Vishwanathan. A scalable modular convex solver for regularized risk minimization. In *Proceedings of the 13th ACM SIGKDD International Conference on Knowledge Discovery and Data Mining*, pages 727–736, 2007.

C. Teo, S. Vishwanathan, A. Smola, and V. Quoc. Bundle methods for regularized risk minimization. *Journal of Machine Learning Research*, 11:311–365, 2010.

I. Tsochantaridis, T. Joachims, T. Hofmann, and Y. Altun. Large margin methods for structured and interdependent output variables. *Journal of Machine Learning Research*, 6:1453–1484, 2005.

M. Varma and B. R. Babu. More generality in efficient multiple kernel learning. In *Proceedings of the International Conference on Machine Learning*, pages 1065–1072, New York, NY, USA, 2009. ACM Press.

M. Wainwright, T. Jaakkola, and A. Willsky. MAP estimation via agreement on (hyper)trees: message passing and linear programming approaches. *IEEE Transactions on Information Theory*, 51(11):3697–3717, 2005.

M. J. Wainwright and M. I. Jordan. Graphical models, exponential families, and variational inference. *Foundations and Trends in Machine Learning*, 1(1-2):1–305, 2008.

T. Werner. A linear programming approach to max-sum problem: A review. Technical Report CTU–CMP–2005–25, Center for Machine Perception, Czech Technical University, 2005.

T. Werner. A linear programming approach to max-sum problem: A review. *IEEE Transactions on Pattern Analysis and Machine Intelligence*, 29(7):1165–1179, 2007.

T. Werner. High-arity interactions, polyhedral relaxations, and cutting plane algorithm for soft constraint optimisation (MAP-MRF). In *Proceedings of the IEEE Conference on Computer Vision and Pattern Recognition*, 2008a.

T. Werner. Marginal consistency: Unifying constraint propagation on commutative semirings. In *International Workshop on Preferences and Soft Constraints (held in conjunction with the 14th International Conference on Principles and Practice of Constraint Programming)*, pages 43–57, 2008b.

T. Werner. Revisiting the linear programming relaxation approach to Gibbs energy minimization and weighted constraint satisfaction. *IEEE Transactions on Pattern Analysis and Machine Intelligence*, 32(8):1474–1488, 2010.

Z. Xu, R. Jin, I. King, and M. Lyu. An extended level method for efficient multiple kernel learning. In Y. Bengio, D. Schuurmans, J. Lafferty, C. Williams, and A. Culotta, editors, *Advances in Neural Information Processing Systems 22*, pages 1825–1832, 2009a.

Z. Xu, R. Jin, J. Ye, M. R. Lyu, and I. King. Non-monotonic feature selection. In *Proceedings of the International Conference on Machine Learning*, pages 1145–1152, 2009b.

A. Zien and C. S. Ong. Multiclass multiple kernel learning. In *Proceedings of the International Conference on Machine Learning*, pages 1191–1198. ACM Press, 2007.

8   Introduction to Dual Decomposition for Inference

David Sontag   dsontag@csail.mit.edu
Microsoft Research New England
Cambridge, MA

Amir Globerson   gamir@cs.huji.ac.il
Hebrew University
Jerusalem, Israel

Tommi Jaakkola   tommi@csail.mit.edu
CSAIL, MIT
Cambridge, MA

Many inference problems with discrete variables result in a difficult combinatorial optimization problem. In recent years, the technique of dual decomposition, also called Lagrangian relaxation, has proved to be a powerful means of solving these inference problems by decomposing them into simpler components that are repeatedly solved independently and combined into a global solution. In this chapter, we introduce the general technique of dual decomposition through its application to the problem of finding the most likely (MAP) assignment in graphical models. We discuss both subgradient and block coordinate descent approaches to solving the dual problem. The resulting message-passing algorithms are similar to max-product, but can be shown to solve a linear programming relaxation of the MAP problem. We show how many of the MAP algorithms are related to each other, and also quantify when the MAP solution can and cannot be decoded directly from the dual solution.

8.1   Introduction

Many problems in engineering and the sciences require solutions to challenging combinatorial optimization problems. These include traditional problems such as scheduling, planning, fault diagnosis, or searching for molecular conformations. In addition, a wealth of combinatorial problems arise directly from probabilistic modeling (graphical models). Graphical models (Koller and Friedman, 2009) have been widely adopted in areas such as computational biology, machine vision, and natural language processing, and are increasingly being used as frameworks expressing combinatorial problems.

Consider, for example, a protein side-chain placement problem where the goal is to find the minimum energy conformation of amino acid side-chains along a fixed carbon backbone. The orientations of the side-chains are represented by discretized angles called rotamers. The combinatorial difficulty arises here from the fact that rotamer choices for nearby amino acids are energetically coupled. For globular proteins, for example, such couplings may be present for most pairs of side-chain orientations. This problem is couched in probabilistic modeling terms by associating molecular conformations with the setting of discrete random variables corresponding to the rotamer angles. The interactions between such random variables come from the energetic couplings between nearby amino acids. Finding the minimum energy conformation is then equivalently solved by finding the most probable assignment of states to the variables.

We will consider combinatorial problems that are expressed in terms of structured probability models (graphical models). A graphical model is defined over a set of discrete variables $x = \{x_j\}_{j \in V}$. Local interactions between the variables are modeled by functions $\theta_f(x_f)$ which depend only on a subset of variables $x_f = \{x_j\}_{j \in f}$. For example, if $x_f = (x_i, x_j)$, $\theta_f(x_f) = \theta_{ij}(x_i, x_j)$ may represent the coupling between rotamer angles x_i and x_j corresponding to nearby amino acids. The functions $\theta_f(x_f)$ are often represented as small tables of numbers with an adjustable value for each local assignment x_f. The joint probability distribution over all the variables x is then defined by combining all the local interactions,

$$\log P(x) = \sum_{i \in V} \theta_i(x_i) + \sum_{f \in F} \theta_f(x_f) + \text{const},$$

where we have included (singleton) functions biasing the states of individual variables. The local interactions provide a compact parametric description of the joint distribution, since we only need to specify the local functions (as small tables) rather than the full probability table involving one value for each complete assignment x.

Despite their compact description, graphical models are capable of describing complex dependencies among the variables. These dependences arise from the combined effect of all the local interactions. The problem of probabilistic inference is to reason about the underlying state of the variables in the model. Given the complex dependencies, it is not surprising that probabilistic inference is often computationally intractable. For example, finding the maximum probability assignment of a graphical model (also known as the MAP assignment) is NP-hard. Finding the MAP assignment remains hard even if the local functions depend on only two variables, as in the protein side-chain example. This has prompted extensive research into approximation algorithms, some of which often work well in practice.

One of the most successful approximation schemes has been to use relaxations of the original MAP problem. Relaxation methods take the original combinatorial problem and pose it as a constrained optimization problem. They then relax some of the constraints in an attempt to factor the problem into more independent subproblems, resulting in a tractable approximation of the original one. Two closely related relaxation schemes are dual decomposition (Johnson, 2008; Komodakis et al., 2011) and linear programming (LP) relaxations (Schlesinger, 1976; Wainwright et al., 2005). Although the approaches use different derivations of the approximation, they result in equivalent optimization problems.

Practical uses of MAP relaxations involve models with thousands of variables and constraints. While the relaxed optimization problems can generally be solved in polynomial time using a variety of methods and off-the-shelf solvers, most of these do not scale well to the problem sizes encountered in practice (e.g., see Yanover et al., 2006, for an evaluation of commercial LP solvers on inference problems). However, often the LP relaxations arising from graphical models have significant structure that can be exploited to design algorithms that scale well with the problem size.

This chapter introduces the dual decomposition approach, also known as Lagrangian relaxation, and focuses on efficient scalable algorithms for solving the relaxed problem. With dual decomposition, the original problem is broken up into smaller subproblems with the help of Lagrange multipliers. The smaller subproblems can be then be solved exactly by using combinatorial algorithms. The decomposition is subsequently optimized with respect to the Lagrange multipliers so as to encourage the subproblems to agree about the variables they share. For example, we could decompose the MAP problem into subproblems corresponding separately to each local function $\theta_f(\boldsymbol{x}_f)$. The Lagrange multipliers introduced in the decomposition would then modify the functions $\theta_f(\boldsymbol{x}_f)$ so that the local maximizing assignments agree across the subproblems. The decomposition may also involve larger components that nevertheless can be solved efficiently, such as a set of spanning trees that together cover all edges of a pairwise graphical model.

The chapter is organized as follows. We begin in section 8.2 with two example applications that illustrate the types of problems that can be solved using our techniques. We next define the MAP problem formally and introduce the dual decomposition approach in section 8.3. The next two sections, 8.4 and 8.5, describe algorithmic approaches to solving the dual decomposition optimization problem. In section 8.6 we discuss the relation between dual decomposition and LP relaxations of the MAP problem, showing that they are essentially equivalent. Finally, in section 8.7 we discuss how the MAP solution can be approximated from the dual decomposition solutions, and what formal guarantees we can make about it.

8.2   Motivating Applications

We will use two example applications to illustrate the role of local interactions, types of decompositions, and how they can be optimized. The first example, outlined earlier, is the protein side-chain placement problem (Yanover et al., 2008). The goal is to find the minimum energy conformation of amino acid side-chains in a fixed protein backbone structure. Each side-chain orientation is represented by a discrete variable x_i specifying the corresponding rotamer angles. Depending on the discretization and the side-chain (the number of dihedral angles), the variables may take on tens or hundreds of possible values (states). Energy functions for this problem typically separate into individual and pairwise terms; the pairwise terms take into consideration attractive and repulsive forces between side-chains that are near each other in the 3D structure.

The problem is equivalently represented as an inference problem in a pairwise Markov random field (MRF) model. Graphically, the MRF is an undirected graph with nodes $i \in V$ corresponding to variables (side-chain orientations x_i) and edges $ij \in E$ indicating interactions. Each pairwise energy term implies an edge $ij \in E$ in the model and defines a potential function $\theta_{ij}(x_i, x_j)$ over local assignments (x_i, x_j). Single node potential functions $\theta_i(x_i)$ can be included separately or absorbed by the edge potentials. The MAP inference problem is then to find the assignment $\boldsymbol{x} = \{x_i\}_{i \in V}$ that maximizes

$$\sum_{i \in V} \theta_i(x_i) + \sum_{ij \in E} \theta_{ij}(x_i, x_j).$$

Without additional restrictions on the choice of potential functions, or which

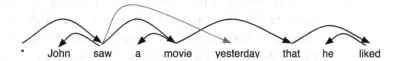

Figure 8.1: Example of dependency parsing for a sentence in English. Every word has one parent (i.e., a valid dependency parse is a directed tree). The red arc demonstrates a non-projective dependency.

edges to include, the problem is known to be NP-hard. Using the dual decomposition approach, we will break the problem into much simpler sub-problems involving maximizations of each single node potential $\theta_i(x_i)$ and each edge potential $\theta_{ij}(x_i, x_j)$ independently from the other terms. Although these local maximizing assignments are easy to obtain, they are unlikely to agree with each other without our modifying the potential functions. These modifications are provided by the Lagrange multipliers associated with agreement constraints.

Our second example is dependency parsing, a key problem in natural language processing (McDonald et al., 2005). Given a sentence, we wish to predict the dependency tree that relates the words in the sentence. A dependency tree is a directed tree over the words in the sentence where an arc is drawn from the head word of each phrase to words that modify it. For example, in the sentence shown in Fig. 8.1, the head word of the phrase "John saw a movie" is the verb "saw", and its modifiers are the subject "John" and the object "movie". Moreover, the second phrase "that he liked" modifies "movie". In many languages the dependency tree is *non-projective* in the sense that each word and its descendants in the tree do not necessarily form a contiguous subsequence.

Formally, given a sentence with m words, we have $m(m-1)$ binary arc selection variables $x_{ij} \in \{0, 1\}$. Since the selections must form a directed tree, the binary variables are governed by an overall function $\theta_T(\boldsymbol{x})$ with the idea that $\theta_T(\boldsymbol{x}) = -\infty$ is used to rule out any non-trees. The selections are further biased by weights on individual arcs, through $\theta_{ij}(x_{ij})$, which depend on the given sentence. In a simple arc-factored model, the predicted dependency structure is obtained by maximizing (McDonald et al., 2005)

$$\theta_T(\boldsymbol{x}) + \sum_{ij} \theta_{ij}(x_{ij}),$$

and can be found with directed maximum-weight spanning-tree algorithms.

More realistic dependency parsing models include additional higher-order interactions between the arc selections. For example, we may couple the modifier selections $\boldsymbol{x}_{|i} = \{x_{ij}\}_{j \neq i}$ (all outgoing edges) for a given word

i, expressed by a function $\theta_{i|}(\boldsymbol{x}_{|i})$. Finding the maximizing non-projective parse tree in a model that includes such higher-order couplings, without additional restrictions, is known to be NP-hard (McDonald and Satta, 2007). We consider here models where $\theta_{i|}(\boldsymbol{x}_{|i})$ can be individually maximized by dynamic programming algorithms (e.g., head-automata models), but become challenging as part of the overall dependency tree model:

$$\left(\theta_T(\boldsymbol{x})\right) + \left(\sum_{ij} \theta_{ij}(x_{ij}) + \sum_i \theta_{i|}(\boldsymbol{x}_{|i})\right) = \theta_1(\boldsymbol{x}) + \theta_2(\boldsymbol{x}).$$

The first component $\theta_1(\boldsymbol{x})$ ensures that we obtain a tree, and the second component $\theta_2(\boldsymbol{x})$ incorporates higher-order biases on the modifier selections. A natural dual decomposition in this case will be to break the problem into these two manageable components, which are then forced to agree on the arc selection variables (Koo et al., 2010).

8.3   Dual Decomposition and Lagrangian Relaxation

The previous section described several problems where we wish to maximize a sum over factors, each defined on some subset of the variables. Here we describe this problem in its general form, and introduce a relaxation approach for approximately maximizing it.

Consider a set of n discrete variables $x_1, \ldots, x_n$, and a set F of subsets on these variables (i.e., $f \in F$ is a subset of $V = \{1, \ldots, n\}$), where each subset corresponds to the domain of one of the factors. Also, assume we are given functions $\theta_f(\boldsymbol{x}_f)$ on these factors, as well as functions $\theta_i(x_i)$ on each of the individual variables.[1] The goal of the MAP problem is to find an assignment $\boldsymbol{x} = (x_1, \ldots, x_n)$ for all the variables which maximizes the sum of the factors:

$$\text{MAP}(\boldsymbol{\theta}) = \max_{\boldsymbol{x}} \sum_{i \in V} \theta_i(x_i) + \sum_{f \in F} \theta_f(\boldsymbol{x}_f). \tag{8.1}$$

The maximizing value is denoted by $\text{MAP}(\boldsymbol{\theta})$, and the maximizing assignment is called the MAP assignment. It can be seen that the examples in section 8.2 indeed correspond to this formulation.[2]

---

1. The singleton factors are not needed for generality, but we keep them for notational convenience and because one often has such factors.
2. Recall the protein side-chain placement problem, where we have an energy function defined on a pairwise MRF. Here, the set of factors F would simply be the edge potentials $\theta_{ij}(x_i, x_j)$ characterizing the couplings between orientations of nearby side-chains. In contrast, in non-projective dependency parsing we have just two factors, $\theta_1(\boldsymbol{x})$ and $\theta_2(\boldsymbol{x})$,

As mentioned earlier, the problem in Eq. 8.1 is generally intractable, and we thus need to resort to approximations. The key difficulty in maximizing Eq. 8.1 is that we need to find an x that maximizes the sum over factors rather than each factor individually. The key assumption we shall make in our approximations is that maximizing each of the factors $\theta_f(x_f)$ can be done efficiently, so that the only complication is their joint maximization.

Our approximation will proceed as follows: we will construct a *dual* function $L(\delta)$ with variables δ such that for all values of δ it holds that $L(\delta) \geq \mathrm{MAP}(\theta)$. In other words, $L(\delta)$ will be an upper bound on the value of the MAP assignment. We will then seek a δ that minimizes $L(\delta)$ so as to make this upper bound as tight as possible.

To describe the dual optimization problem, we first specify what the dual variables are. For every choice of $f \in F$, $i \in f$, and x_i, we will have a dual variable denoted by $\delta_{fi}(x_i)$.[3] This variable may be interpreted as the *message* that factor f sends to variable i about its state x_i. The dual function $L(\delta)$ and the corresponding optimization problem are then given by

$$\min_{\delta} L(\delta), \tag{8.2}$$

$$L(\delta) = \sum_{i \in V} \max_{x_i} \Big(\theta_i(x_i) + \sum_{f : i \in f} \delta_{fi}(x_i) \Big) + \sum_{f \in F} \max_{x_f} \Big(\theta_f(x_f) - \sum_{i \in f} \delta_{fi}(x_i) \Big).$$

The key property of the function $L(\delta)$ is that it involves maximization only over local assignments x_f, a task which we assume to be tractable. The dual thus decouples the original problem, resulting in a problem that can be optimized using local operations. Figure 8.2 illustrates this for a simple pairwise model.

We will introduce algorithms that minimize the approximate objective $L(\delta)$ using local updates. Each iteration of the algorithms repeatedly finds a maximizing assignment for the subproblems individually, using these to update the dual variables that glue the subproblems together. We describe two classes of algorithms, one based on a *subgradient method* (see section 8.4) and another based on *block coordinate descent* (see section 8.5). These dual algorithms are simple and widely applicable to combinatorial problems in machine learning such as finding MAP assignments of graphical models.

---

both defined on all of the variables, and the single node potentials $\theta_i(x_i)$ are identically zero (not used). The variables here are binary, one for each directed edge, denoting whether a particular dependency exists.

3. We use the notation $\delta_{fi}(x_i)$ to denote a variable indexed by i, f, and x_i. An alternative notation could have been $\delta(x_i, f, i)$, but $\delta_{fi}(x_i)$ is more compact.

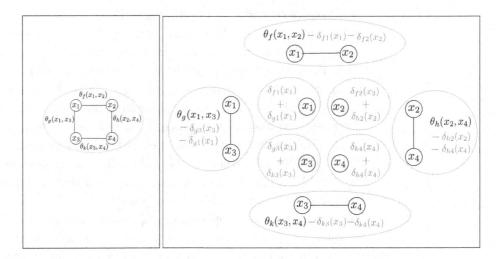

Figure 8.2: Illustration of the the dual decomposition objective. **Left:** The original pairwise model consisting of four factors. **Right:** The maximization problems corresponding to the objective $L(\boldsymbol{\delta})$. Each blue ellipse contains the factor to be maximized over. In all figures the singleton terms $\theta_i(x_i)$ are set to zero for simplicity.

8.3.1   Derivation of the Dual

In what follows we show how the dual optimization in Eq. 8.2 is derived from the original MAP problem in Eq. 8.1. We first slightly reformulate the problem by duplicating the x_i variables, once for each factor, and then enforcing that these are equal. Let x_i^f denote the copy of x_i used by factor f. Also, denote by $\boldsymbol{x}_f^f = \{x_i^f\}_{i \in f}$ the set of variables used by factor f, and, by $\boldsymbol{x}^F = \{\boldsymbol{x}_f^f\}_{f \in F}$, the set of all variable copies. This is illustrated graphically in Fig. 8.3. Then, our reformulated—but equivalent—optimization problem is

$$\begin{aligned} \max \quad & \textstyle\sum_{i \in V} \theta_i(x_i) + \sum_{f \in F} \theta_f(\boldsymbol{x}_f^f) \\ \text{s.t.} \quad & x_i^f = x_i, \quad \forall f, i \in f. \end{aligned} \tag{8.3}$$

If we did not have the constraints, this maximization would simply decompose into independent maximizations for each factor, each of which we assume can be done efficiently. To remove these complicating constraints, we use the technique of *Lagrangian relaxation* (Geoffrion, 1974; Schlesinger, 1976; Fisher, 1981; Lemaréchal, 2001; Guignard, 2003). First, introduce La-

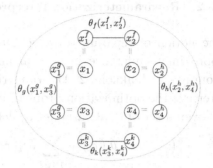

Figure 8.3: Illustration of the the derivation of the dual decomposition objective by creating copies of variables. The graph in Fig. 8.2 is shown here with the corresponding copies of the variables for each of the factors.

grange multipliers $\boldsymbol{\delta} = \{\delta_{fi}(x_i) : f \in F, i \in f, x_i\}$, and define the Lagrangian:

$$
\begin{aligned}
L(\boldsymbol{\delta}, \boldsymbol{x}, \boldsymbol{x}^F) &= \sum_{i \in V} \theta_i(x_i) + \sum_{f \in F} \theta_f(\boldsymbol{x}_f^f) \\
&+ \sum_{f \in F} \sum_{i \in f} \sum_{\hat{x}_i} \delta_{fi}(\hat{x}_i)\Big(1[x_i = \hat{x}_i] - 1[x_i^f = \hat{x}_i]\Big). \quad (8.4)
\end{aligned}
$$

The following problem is still equivalent to Eq. 8.3 for any value of $\boldsymbol{\delta}$,

$$
\begin{aligned}
\max_{\boldsymbol{x}, \boldsymbol{x}^F} \quad & L(\boldsymbol{\delta}, \boldsymbol{x}, \boldsymbol{x}^F) \\
\text{s.t.} \quad & x_i^f = x_i, \quad \forall f, i \in f.
\end{aligned} \quad (8.5)
$$

This follows because if the constraints in the above hold, then the last term in Eq. 8.4 is zero for any value of $\boldsymbol{\delta}$. In other words, the Lagrange multipliers are unnecessary if we already enforce the constraints.

Solving the maximization in Eq. 8.5 is as hard as the original MAP problem. To obtain a tractable optimization problem, we omit the constraint in Eq. 8.5 and define the function $L(\boldsymbol{\delta})$:

$$
\begin{aligned}
L(\boldsymbol{\delta}) &= \max_{\boldsymbol{x}, \boldsymbol{x}^F} L(\boldsymbol{\delta}, \boldsymbol{x}, \boldsymbol{x}^F) \\
&= \sum_{i \in V} \max_{x_i}\Big(\theta_i(x_i) + \sum_{f:i \in f} \delta_{fi}(x_i)\Big) + \sum_{f \in F} \max_{\boldsymbol{x}_f^f}\Big(\theta_f(\boldsymbol{x}_f^f) - \sum_{i \in f} \delta_{fi}(x_i^f)\Big).
\end{aligned}
$$

Since $L(\boldsymbol{\delta})$ maximizes over a larger space ($\boldsymbol{x}$ may not equal $\boldsymbol{x}^F$), we have that $\mathrm{MAP}(\boldsymbol{\theta}) \leq L(\boldsymbol{\delta})$. The *dual problem* is to find the tightest such upper bound by optimizing the Lagrange multipliers: solving $\min_{\boldsymbol{\delta}} L(\boldsymbol{\delta})$.

Note that the maximizations are now unambiguously independent. We obtain Eq. 8.2 by replacing $\boldsymbol{x}_f^f$ with $\boldsymbol{x}_f$.

8.3.2   Reparameterization Interpretation of the Dual

The notion of reparameterization has played a key role in understanding approximate inference in general and MAP approximations in particular (Sontag and Jaakkola, 2009; Wainwright et al., 2003). It can also be used to interpret the optimization problem in Eq. 8.2, as we briefly review here.

Given a set of dual variables $\boldsymbol{\delta}$, define new factors on x_i and $\boldsymbol{x}_f$ given by

$$
\begin{aligned}
\bar{\theta}_i^{\delta}(x_i) &= \theta_i(x_i) + \sum_{f:i\in f} \delta_{fi}(x_i) \\
\bar{\theta}_f^{\delta}(\boldsymbol{x}_f) &= \theta_f(\boldsymbol{x}_f) - \sum_{i\in f} \delta_{fi}(x_i) \; .
\end{aligned}
\tag{8.6}
$$

It is easy to see that these new factors define essentially the same function as the original factors $\boldsymbol{\theta}$. Namely, for all assignments $\boldsymbol{x}$,

$$
\sum_{i\in V} \theta_i(x_i) + \sum_{f\in F} \theta_f(\boldsymbol{x}_f) = \sum_{i\in V} \bar{\theta}_i^{\delta}(x_i) + \sum_{f\in F} \bar{\theta}_f^{\delta}(\boldsymbol{x}_f) \; .
\tag{8.7}
$$

We call $\bar{\boldsymbol{\theta}}$ a reparameterization of the original parameters $\boldsymbol{\theta}$. Next we observe that the function $L(\boldsymbol{\delta})$ can be written as

$$
L(\boldsymbol{\delta}) = \sum_{i\in V} \max_{x_i} \bar{\theta}_i^{\delta}(x_i) + \sum_{f\in F} \max_{\boldsymbol{x}_f} \bar{\theta}_f^{\delta}(\boldsymbol{x}_f).
\tag{8.8}
$$

To summarize the above, dual decomposition may be viewed as searching over a set of reparameterizations of the original factors $\boldsymbol{\theta}$, where each reparameterization provides an upper bound on the MAP and we are seeking to minimize this bound.

8.3.3   Formal Guarantees

In the previous section we showed that the dual always provides an upper bound on the optimum of our original optimization problem (Eq. 8.1),

$$
\max_{\boldsymbol{x}} \sum_{i\in V} \theta_i(x_i) + \sum_{f\in F} \theta_f(\boldsymbol{x}_f) \leq \min_{\boldsymbol{\delta}} L(\boldsymbol{\delta}).
\tag{8.9}
$$

We do not necessarily have strong duality, that is, equality in the above equation. However, for some functions $\theta(\boldsymbol{x})$ strong duality does hold, as stated in the following theorem[4]

Theorem 8.1. *Suppose that* $\exists \boldsymbol{\delta}^*, \boldsymbol{x}^*$ *where* $x_i^* \in \mathrm{argmax}_{x_i}\, \bar{\theta}_i^{\delta^*}(x_i)$ *and*

---

4. Versions of this theorem appear in multiple papers (e.g., see Geoffrion, 1974; Wainwright et al., 2005; Weiss et al., 2007).

$x_f^* \in \text{argmax}_{x_f} \ \bar{\theta}_f^{\delta^*}(x_f)$. *Then* x^* *is a solution to the maximization problem in Eq. 8.1 and hence* $L(\delta^*) = MAP(\theta)$.

Proof. For the given δ^* and x^* we have that

$$L(\delta^*) = \sum_i \bar{\theta}_i^{\delta^*}(x_i^*) + \sum_{f \in F} \bar{\theta}_f^{\delta^*}(x_f^*) = \sum_i \theta_i(x_i^*) + \sum_{f \in F} \theta_f(x_f^*), \qquad (8.10)$$

where the equalities follow from the maximization property of x^* and the reparameterization property of $\bar{\theta}$. On the other hand, from the definition of $MAP(\theta)$ we have that

$$\sum_i \theta_i(x_i^*) + \sum_{f \in F} \theta_f(x_f^*) \le MAP(\theta). \qquad (8.11)$$

Taking Eq. 8.10 and Eq. 8.11 together with $L(\delta^*) \ge MAP(\theta)$, we have equality in Eq. 8.11, so that x^* attains the MAP value and is therefore the MAP assignment, and $L(\delta^*) = MAP(\theta)$. $\qquad \square$

The conditions of theorem 8.1 correspond to the subproblems *agreeing* on a maximizing assignment. Since agreement implies optimality of the dual, it can occur only after our algorithms find the tightest upper bound. Although the agreement is not guaranteed, if we do reach such a state, then theorem 8.1 ensures that we have the *exact* solution to Eq. 8.1. The dual solution δ is said to provide a *certificate of optimality* in this case. In other words, if we find an assignment whose value matches the dual value, then the assignment has to be the MAP (strong duality).

For both the non-projective dependency parsing and protein side-chain placement problems, exact solutions (with certificates of optimality) are frequently found by using dual decomposition, in spite of the corresponding optimization problems being NP-complete (Koo et al., 2010; Sontag et al., 2008; Yanover et al., 2006).

We show in section 8.6 that equation (8.2) is the dual of an LP relaxation of the original problem. When the conditions of theorem 8.1 are satisfied, it means the LP relaxation is *tight* for this instance.

8.4 Subgradient Algorithms

For the remainder of this chapter, we show how to efficiently minimize the upper bound on the MAP assignment provided by the Lagrangian relaxation. Although $L(\delta)$ is convex and continuous, it is non-differentiable at all points δ where $\bar{\theta}_i^{\delta}(x_i)$ or $\bar{\theta}_f^{\delta}(x_f)$ have multiple optima (for some i or f). There are a large number of non-differentiable convex optimization techniques that could be applied in this setting (Fisher, 1981). In this section we describe the subgradient method, which has been widely applied to solving Lagrangian relaxation problems and is often surprisingly effective, in spite of being such a simple method. The subgradient method is similar to gradient descent, but is applicable to non-differentiable objectives.

A complete treatment of subgradient methods is beyond the scope of this chapter. Our focus will be to introduce the general idea so that we can compare and contrast it with the block coordinate descent algorithms described in the next section. We refer the reader to Komodakis et al. (2011) for a detailed treatment of subgradient methods as they relate to inference problems (see also Held et al. (1974) for an early application of the subgradient method for Lagrangian relaxation, and to Koo et al. (2010) for a recent application to the non-projective dependency parsing problem described earlier). Our dual decomposition differs slightly from these earlier works in that we explicitly included single node factors and enforced that all other factors agree with them. As a result, our optimization problem is unconstrained, regardless of the number of factors. In contrast, Komodakis et al. (2011) have constraints enforcing that some of the dual variables sum to zero, resulting in a projected subgradient method.

A subgradient of a convex function $L(\boldsymbol{\delta})$ at $\boldsymbol{\delta}$ is a vector $g_{\boldsymbol{\delta}}$ such that for all $\boldsymbol{\delta}'$, $L(\boldsymbol{\delta}') \geq L(\boldsymbol{\delta}) + g_{\boldsymbol{\delta}} \cdot (\boldsymbol{\delta}' - \boldsymbol{\delta})$. The subgradient method is very simple to implement, alternating between individually maximizing the subproblems (which provides the subgradient) and updating the dual parameters $\boldsymbol{\delta}$ using the subgradient. More specifically, the subgradient descent strategy is as follows. Assume that the dual variables at iteration t are given by $\boldsymbol{\delta}^t$. Then their value at iteration $t + 1$ is given by

$$\delta_{fi}^{t+1}(x_i) = \delta_{fi}^t(x_i) - \alpha_t\, g_{fi}^t(x_i)\,, \tag{8.12}$$

where $\boldsymbol{g}^t$ is a subgradient of $L(\boldsymbol{\delta})$ at $\boldsymbol{\delta}^t$ (i.e., $\boldsymbol{g} \in \partial L(\boldsymbol{\delta}^t)$) and α_t is a stepsize that may depend on t. We show one way to calculate this subgradient in section 8.4.1.

A well-known theoretical result is that the subgradient method is guaranteed to solve the dual to optimality whenever the stepsizes are chosen such that $\lim_{t \to \infty} \alpha_t = 0$ and $\sum_{t=0}^{\infty} \alpha_t = \infty$ (Anstreicher and Wolsey, 2009). One example of such a stepsize is $\alpha_t = \frac{1}{t}$. However, there are a large number of heuristic choices that can make the subgradient method faster. See Komodakis et al. (2011) for further possibilities of how to choose the step size, and for an empirical evaluation of these choices on inference problems arising from computer vision.

8.4.1   Calculating the Subgradient of $L(\delta)$

In this section we show how to calculate the subgradient of $L(\boldsymbol{\delta})$, completing the description of the subgradient algorithm. Given the current dual variables $\boldsymbol{\delta}^t$, we first choose a maximizing assignment for each subproblem. Let x_i^s be a maximizing assignment of $\bar{\theta}_i^{\delta^t}(x_i)$, and let $\boldsymbol{x}_f^f$ be a maximizing assignment of $\bar{\theta}_f^{\delta^t}(\boldsymbol{x}_f)$. The subgradient of $L(\boldsymbol{\delta})$ at $\boldsymbol{\delta}^t$ is then given by the following pseudocode:

$$g_{fi}^t(x_i) = 0, \qquad \forall f, i \in f, x_i$$

$\quad$**For** $f \in F$ and $i \in f$:

$\qquad$**If** $x_i^f \neq x_i^s$:

$$g_{fi}^t(x_i^s) \;=\; +1 \tag{8.13}$$
$$g_{fi}^t(x_i^f) \;=\; -1 \,. \tag{8.14}$$

Thus, each time that $x_i^f \neq x_i^s$, the subgradient update decreases the value of $\bar{\theta}_i^{\delta^t}(x_i^s)$ and increases the value of $\bar{\theta}_i^{\delta^t}(x_i^f)$. Similarly, for all $\boldsymbol{x}_{f\backslash i}$, the subgradient update decreases the value of $\bar{\theta}_f^{\delta^t}(x_i^f, \boldsymbol{x}_{f\backslash i})$ and increases the value of $\bar{\theta}_f^{\delta^t}(x_i^s, \boldsymbol{x}_{f\backslash i})$. Intuitively, as a result of the update, in the next iteration the factors are more likely to agree with one another on the value of x_i in their maximizing assignments.

Typically one runs the subgradient algorithm until either $L(\boldsymbol{\delta})$ stops decreasing significantly or we have reached some maximum number of iterations. If at any iteration t we find that $\boldsymbol{g}^t = 0$, then $x_i^f = x_i^s$ for all $f \in F, i \in f$. Therefore, by theorem 8.1, $\boldsymbol{x}^s$ must be an MAP assignment, and so we have solved the dual to optimality. However, the converse does not always hold: $\boldsymbol{g}^t$ may be non-zero even when $\boldsymbol{\delta}^t$ is dual optimal. We discuss these issues further in section 8.7.

In the *incremental* subgradient method, at each iteration one computes the subgradient using only some of the subproblems, $F' \subset F$, rather than using all factors in F (Bertsekas, 1995). This can significantly decrease the overall running time, and is also more similar to the block coordinate descent methods that we describe next, which make updates with respect to only one factor at a time.

8.4.2   Efficiently Maximizing over the Subproblems

To choose a maximizing assignment for each subproblem, needed for making the subgradient updates, we have to solve the following combinatorial optimization problem:

$$\max_{\boldsymbol{x}_f} \left[\theta_f(\boldsymbol{x}_f) - \sum_{i \in f} \delta_{fi}(x_i) \right] . \tag{8.15}$$

When the number of possible assignments $\boldsymbol{x}_f$ is small, this maximization can be done simply by enumeration. For example, in pairwise MRFs each factor $f \in F$ consists of just two variables, $|f| = 2$. Suppose each variable takes k states. Then this maximization takes k^2 time.

Often the number of assignments is large but the functions $\theta_f(\boldsymbol{x}_f)$ are *sparse*, allowing the maximization to be performed using dynamic programming or combinatorial algorithms. For example, in the non-projective dependency parsing problem, $\theta_1(\boldsymbol{x}) = -\infty$ if the set of edges specified by $\boldsymbol{x}$ include a cycle, and 0 otherwise. Maximization in Eq. 8.15 can be solved efficiently in this case as follows: We form a directed graph where the weight of the edge corresponding to x_i is $\delta_{1i}(1) - \delta_{1i}(0)$. Then, we solve Eq. 8.15 by finding a directed minimum-weight spanning tree on this graph.

The following are some of the sparse factors that are frequently found in inference problems, all of which can be efficiently maximized over:

- Tree structures (Wainwright et al., 2005; Komodakis et al., 2011)
- Matchings (Lacoste-Julien et al., 2006; Duchi et al., 2007; Yarkony et al., 2010)
- Supermodular functions (Komodakis et al., 2011)
- Cardinality and order constraints (Gupta et al., 2007; Tarlow et al., 2010)
- Functions with small support (Rother et al., 2009)

Consider, for example, a factor which enforces a cardinality constraint over binary variables: $\theta_f(\boldsymbol{x}_f) = 0$ if $\sum_{i \in f} x_i = L$, and $\theta_f(\boldsymbol{x}_f) = -\infty$ otherwise. Let $e_i = \delta_{fi}(1) - \delta_{fi}(0)$. To solve Eq. 8.15, we first sort e_i in ascending order for $i \in f$. Then, the maximizing assignment is obtained by setting the corresponding $x_i = 1$ for the first L values, and $x_i = 0$ for the remainder. Thus, computing the maximum assignment takes only $O(|f| \log |f|)$ time.

8.5 Block Coordinate Descent Algorithms

A different approach to solving the optimization problem in Eq. 8.2 is via coordinate descent. Coordinate-descent algorithms have a long history of being used to optimize Lagrangian relaxations (e.g., see Erlenkotter, 1978; Guignard and Rosenwein, 1989). Such algorithms work by fixing the values of all dual variables except for a set of variables, and then minimizing the objective as much as possible with respect to that set. The two key design choices to make are which variables to update and how to update them. In all the updates we consider below, the coordinates are updated to their optimal value (i.e., the one that minimizes $L(\boldsymbol{\delta})$), given the fixed coordinates.

There is a significant amount of flexibility with regard to choosing the coordinate blocks, that is, the variables to update. A first attempt at choosing such a block may be to focus only on coordinates for which the subgradient is non-zero. For example, if $x_i^s = \arg\max_{x_i} \bar{\theta}_i^{\delta^t}(x_i)$, we may choose to update only $\delta_{fi}(x_i^s)$, and not $\delta_{fi}(x_i)$, for $x_i \neq x_i^s$. However, this may result in too small a change in $L(\boldsymbol{\delta})$, and the coordinates we have not updated may require update at a later stage. The updates we consider below will update $\delta_{fi}(x_i)$ for all values of x_i regardless of the maximizing assignment. This is very different from the subgradient method, which updates only the dual variables corresponding to the maximizing assignments for the subproblems, $\delta_{fi}(x_i^s)$ and $\delta_{fi}(x_i^f)$.

We shall describe various block coordinate descent algorithms, each algorithm using an increasingly larger block size. The advantage of coordinate descent algorithms is that they are local, parameter free, simple to implement, and often provide faster convergence than subgradient methods. However, as we discuss later, there are cases where coordinate descent algorithms may not reach the dual optimum. In section 8.8 we discuss the issue of choosing between coordinate descent and subgradient based schemes.

For the practitioner interested in an algorithm to apply, in Fig. 8.4 we give pseudocode for one of the block coordinate descent algorithms.

8.5.1   The Max-Sum Diffusion algorithm

Suppose that we fix all of the dual variables $\boldsymbol{\delta}$ except $\delta_{fi}(x_i)$ for a specific f and i. We now wish to find the values of $\delta_{fi}(x_i)$ that minimize the objective $L(\boldsymbol{\delta})$, given the other fixed values. In general there is not a unique solution to this restricted optimization problem, and different update strategies will result in different overall running times.

The max-sum diffusion (MSD) algorithm (Kovalevsky and Koval; Werner, 2007, 2008) performs the following block coordinate descent update (for all x_i simultaneously):

$$\delta_{fi}(x_i) = -\tfrac{1}{2}\delta_i^{-f}(x_i) + \tfrac{1}{2}\max_{\boldsymbol{x}_{f\backslash i}} \left[\theta_f(\boldsymbol{x}_f) - \sum_{\hat{i}\in f\backslash i}\delta_{f\hat{i}}(x_{\hat{i}})\right], \tag{8.17}$$

where we define $\delta_i^{-f}(x_i) = \theta_i(x_i) + \sum_{\hat{f}\neq f}\delta_{\hat{f}i}(x_i)$. The algorithm iteratively

Inputs:

- A set of factors $\theta_i(x_i), \theta_f(\boldsymbol{x}_f)$

Output:

- An assignment $x_1, \ldots, x_n$ that approximates the MAP

Algorithm:

- Initialize $\delta_{fi}(x_i) = 0, \quad \forall f \in F, i \in f, x_i$
- Iterate until small enough change in $L(\boldsymbol{\delta})$ (see Eq. 8.2):
  For each $f \in F$, perform the updates

$$\delta_{fi}(x_i) = -\delta_i^{-f}(x_i) + \frac{1}{|f|} \max_{\boldsymbol{x}_{f \setminus i}} \left[\theta_f(\boldsymbol{x}_f) + \sum_{\hat{i} \in f} \delta_{\hat{i}}^{-f}(x_{\hat{i}}) \right] \tag{8.16}$$

  simultaneously for all $i \in f$ and x_i. We define $\delta_i^{-f}(x_i) = \theta_i(x_i) + \sum_{\hat{f} \neq f} \delta_{\hat{f}i}(x_i)$
- Return $x_i \in \arg\max_{\hat{x}_i} \bar{\theta}_i^{\boldsymbol{\delta}}(\hat{x}_i)$ (see Eq. 8.6).

Figure 8.4: Description of the MPLP block coordinate descent algorithm for minimizing the dual $L(\boldsymbol{\delta})$ (see section 8.5.2). Similar algorithms can be devised for different choices of coordinate blocks. See sections 8.5.1 and 8.5.3. The assignment returned in the final step follows the decoding scheme discussed in section 8.7.

chooses some f and sequentially performs these updates for each $i \in f$. In appendix 8.9.1 we show how to derive this algorithm as block coordinate descent on $L(\boldsymbol{\delta})$. The proof also illustrates the following *equalization* property: after the update, we have $\bar{\theta}_i^{\boldsymbol{\delta}}(x_i) = \max_{\boldsymbol{x}_{f \setminus i}} \bar{\theta}_f^{\boldsymbol{\delta}}(\boldsymbol{x}_f), \forall x_i$. In other words, the reparameterized factors for f and i agree on the utility of state x_i.

8.5.2   The MPLP algorithm

In this section, we show that it is possible to do coordinate descent in closed form over a significantly larger block of coordinates than that of section 8.5.1. The max-product linear programming (MPLP) algorithm was introduced as a coordinate-descent algorithm for LP relaxations of MAP problems (Globerson and Jaakkola, 2008). Here we show that it can also be interpreted as a block coordinate descent algorithm for Eq. 8.2.

Assume we fix all the variables $\boldsymbol{\delta}$ except $\delta_{fi}(x_i)$ for a specific f and *all i* (note that this differs from section 8.5.1, where only one i was not fixed). We now wish to find the values of $\delta_{fi}(x_i)$ that minimize the objective $L(\boldsymbol{\delta})$, given the other fixed values. We claim that the following update achieves this:

$$\delta_{fi}(x_i) = -\left(1 - \frac{1}{|f|}\right)\delta_i^{-f}(x_i) + \frac{1}{|f|}\max_{\boldsymbol{x}_{f\backslash i}}\left[\theta_f(\boldsymbol{x}_f) + \sum_{\hat{i}\in f\backslash i}\delta_{\hat{i}}^{-f}(x_{\hat{i}})\right]\!(8.18)$$

The update can be equivalently written as

$$\delta_{fi}(x_i) = -\delta_i^{-f}(x_i) + \frac{1}{|f|}\max_{\boldsymbol{x}_{f\backslash i}}\left[\theta_f(\boldsymbol{x}_f) + \sum_{\hat{i}\in f}\delta_{\hat{i}}^{-f}(x_{\hat{i}})\right]. \qquad (8.19)$$

In appendix 8.9.2 we show that this update indeed corresponds to the desired block coordinate descent on $L(\boldsymbol{\delta})$. Note that the update needs to be performed for all $i \in f$ and x_i simultaneously, unlike MSD, which updates for a particular i each time.

One iteration of MPLP (i.e., performing the updates once for every $f \in F$, $i \in f$, and x_i) has exactly the same running time as an iteration of MSD.[5] However, each MPLP update is much more effective, with bigger gains expected as $|f|$ grows. For example, consider a model with single node factors $\theta_i(x_i)$ and only one $\theta_f(\boldsymbol{x}_f)$, that is $|F| = 1$. For this model, MPLP exactly solves the dual in the first iteration, whereas MSD would take several iterations to solve the dual to optimality.

8.5.3 Larger Coordinate Blocks

Ideally we would like to optimize in closed form over as large a coordinate block as possible. It turns out that when the factors are pairwise, we can use considerably larger blocks than those that have been discussed above.

As an illustration of this approach, we now show a *star update* for pairwise models (Globerson and Jaakkola, 2008). Consider a model where factors correspond to pairs of variables, so that the elements of F are of the form $\{i, j\}$. Now, consider a block coordinate descent approach where the free coordinates are $\delta_{\{i,j\}j}(x_j), \delta_{\{i,j\}i}(x_i)$ for a given i and all its neighbors $j \in N(i)$. Although it is non-trivial to derive, we show in appendix 8.9.3 that a closed form update for these coordinates exists. The free coordinates

---

5. Assuming that we keep $\bar{\theta}_i^{\delta}(x_i)$ in memory and compute $\delta_i^{-f}(x_i) = \bar{\theta}_i^{\delta}(x_i) - \delta_{fi}(x_i)$.

that minimize the objective $L(\boldsymbol{\delta})$ are given by[6]

$$\delta_{\{i,j\}i}(x_i) = -\frac{1}{1+N_i}\gamma_i(x_i) + \gamma_{ji}(x_i) \qquad (8.20)$$

$$\delta_{\{i,j\}j}(x_j) = -\tfrac{1}{2}\delta_j^{-i}(x_j) + \tfrac{1}{2}\max_{x_i}\left[\theta_{ij}(x_i,x_j) + \frac{2}{1+N_i}\gamma_i(x_i) - \gamma_{ji}(x_i)\right]$$

where we define $\delta_j^{-i}(x_j) = \theta_j(x_j) + \sum_{k\in N(j)\backslash i}\delta_{\{k,j\}j}(x_j)$, and

$$\gamma_{ji}(x_i) = \max_{x_j}\left[\theta_{ij}(x_i,x_j) + \delta_j^{-i}(x_j)\right]$$

$$\gamma_i(x_i) = \theta_i(x_i) + \sum_{j\in N(i)}\gamma_{ji}(x_i)$$

where $N_i = |N(i)|$ is the number of neighbors of node i in the pairwise model. For a given node i, the update needs to be performed for all factors $\{i,j\}$, x_i, and x_j simultaneously, where $j \in N(i)$.

Often a closed-form update is not possible, but one can still efficiently compute the updates. For example, Sontag and Jaakkola (2009) give a linear time algorithm for coordinate descent in pairwise models using blocks corresponding to the edges of a tree-structured graph.

8.5.4   Efficiently Computing the Updates

Recall that the subgradient method needs to solve only one maximization per subproblem to perform the subgradient update (see Eq. 8.15). We showed in section 8.4.2 that these can be solved efficiently for a large number of *sparse* factors, even though the factors may involve a large number of variables.

Consider the MPLP update given in Eq. 8.16. To perform the updates, one must compute *max-marginals* for each subproblem, that is, the value of the optimal assignment after fixing each variable (individually) to one of its states. Specifically, for every f, i, and x_i, we must compute

$$\max_{\boldsymbol{x}_{f\backslash i}} h(\boldsymbol{x}_{f\backslash i}, x_i), \qquad h(\boldsymbol{x}_f) = \theta_f(\boldsymbol{x}_f) + \sum_{\hat{i}\in f}\delta_{\hat{i}}^{-f}(x_{\hat{i}}) \ .$$

Often this can be done in the same running time that it takes to find the maximizing assignment, that is, $\max_{\boldsymbol{x}_f} h(\boldsymbol{x}_f)$.[7] For example, if the subproblem is a tree structure, then with just two passes—propagating max-product messages from the leaves to the root, and then from the root back to the leaves—we can compute the max-marginals. Tarlow et al. (2010)

---

6. The star update that appeared in Globerson and Jaakkola (2008) under the name NMPLP, had an error. This is the correct version.
7. To perform the updates we need only the optimal *value*, not the maximizing assignment.

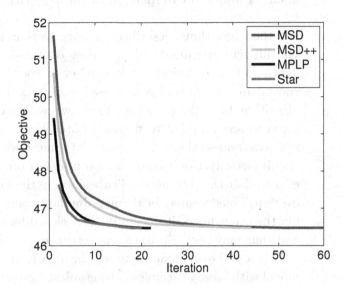

Figure 8.5: Comparison of three coordinate descent algorithms on a 10×10 two dimensional Ising grid. The dual objective $L(\boldsymbol{\delta})$ is shown as a function of iteration number. We multiplied the number of iterations for the star update by two, since each edge variable is updated twice.

show how max-marginals can be efficiently computed for factors enforcing cardinality and order constraints, and Duchi et al. (2007) show how to do this for matchings. In all of these cases, one iteration of MPLP has the same running time as one iteration of the subgradient method.

However, for some problems, computing the max-marginals can take longer. For example, it is not clear how to compute max-marginals efficiently for the factor $\theta_1(\boldsymbol{x})$ which enforced acyclicity in the non-projective dependency problem. The weight for the edge corresponding to x_i is $\delta_i^{-1}(0) - \delta_i^{-1}(1)$. The computational problem is to find, for each edge, the weight of the minimum directed spanning tree (MST) if we force it to include this edge ($x_i = 1$) and if we force it not to include this edge ($x_i = 0$).[8] We could compute the max-marginals by solving $2|V|$ separate directed MST problems, but this would increase the overall running time of each iteration. Thus, it is possible that the subgradient approach is more suitable here.

8.5.5   Empirical Comparison of the Algorithms

All the algorithms described in this section minimize $L(\boldsymbol{\delta})$, but do so via different coordinate blocks. How do these algorithms compare to one another? One criterion is the number of coordinates per block. The more coordinates we optimize in closed form, the faster we would expect our algorithm to make progress. Thus, we would expect the MPLP updates given in section 8.5.2 to minimize $L(\boldsymbol{\delta})$ in fewer iterations than the MSD updates from section 8.5.1, since the former optimizes over more coordinates simultaneously. Comparing the star update from section 8.5.3 to MPLP with edge updates is a bit more delicate since, in the former, the coordinate blocks overlap. Thus, we may be doing redundant computations in the star update. On the other hand, it does provide a closed-form update over larger blocks and thus may result in faster convergence.

To assess the difference between the algorithms, we test them on a pairwise model with binary variables. The graph structure is a two dimensional 10×10 grid and the interactions are Ising (see Globerson and Jaakkola, 2008, for a similar experimental setup). We compare three algorithms:

- MSD – At each iteration, for each edge, it updates the message from the edge to one of its endpoints (i.e., $\delta_{\{i,j\}i}(x_i)$ for all x_i), and then updates the message from the edge to its other endpoint.

- MPLP – At each iteration, for each edge, it updates the messages from the edge to both of its endpoints (i.e., $\delta_{\{i,j\}i}(x_i)$ *and* $\delta_{\{i,j\}j}(x_j)$, for all x_i, x_j).

- Star update – At each iteration, for each node i, it updates the messages from all edges incident on i to both of their endpoints (i.e., $\delta_{\{i,j\}i}(x_i)$ and $\delta_{\{i,j\}j}(x_j)$ for all $j \in N(i), x_i, x_j$).

- MSD++ – See section 8.5.6.

The running times per iteration of MSD and MPLP are identical. We let each iteration of the star update correspond to two iterations of the edge updates to make the running times comparable.

Results for a model with random parameters are shown in Fig. 8.5, and are representative of what we found across a range of model parameters. The results verify the arguments made above: MPLP is considerably faster than MSD, and the star update is somewhat faster than MPLP.

---

8. The corresponding max-marginal is the MST's weight plus the constant $-\sum_{i \in f} \delta_i^{-1}(0)$.

8.5.6   What Makes MPLP Faster Than MSD?

It is clear that MPLP uses a larger block size than MSD. However, the two algorithms also use different equalization strategies, which we will show is just as important. We illustrate this difference by giving a new algorithm, MSD++, which uses the same block size as MPLP, but an equalization method that is more similar to MSD's.

For a given factor f, assume all $\boldsymbol{\delta}$ variables are fixed except $\delta_{fi}(x_i)$ for all $i \in f$ and x_i (as in MPLP). We would like to minimize $L(\boldsymbol{\delta})$ with respect to the free variables. The part of the objective $L(\boldsymbol{\delta})$ that depends on the free variables is

$$\bar{L}(\boldsymbol{\delta}) = \sum_{i \in f} \max_{x_i} \bar{\theta}_i^{\boldsymbol{\delta}}(x_i) + \max_{\boldsymbol{x}_f} \bar{\theta}_f^{\boldsymbol{\delta}}(\boldsymbol{x}_f) \;,$$

where $\bar{\theta}$ are the reparameterizations defined in Eq. 8.6.

The MPLP update can be shown to give a $\boldsymbol{\delta}^{t+1}$ such that $\bar{L}(\boldsymbol{\delta}^{t+1}) = \max_{\boldsymbol{x}_f} h^{\boldsymbol{\delta}^t}(\boldsymbol{x}_f)$, where $h^{\boldsymbol{\delta}^t}(\boldsymbol{x}_f) = \sum_{i \in f} \bar{\theta}_i^{\boldsymbol{\delta}^t}(x_i) + \bar{\theta}_f^{\boldsymbol{\delta}^t}(\boldsymbol{x}_f)$. However, there are many possible choices for $\boldsymbol{\delta}^{t+1}$ that would achieve the same decrease in the dual objective. In particular, for any choice of non-negative weights α_f, α_i for $i \in f$ such that $\alpha_f + \sum_{i \in f} \alpha_i = 1$, we can choose $\boldsymbol{\delta}^{t+1}$ so that

$$\bar{\theta}_i^{\boldsymbol{\delta}^{t+1}}(x_i) \;=\; \alpha_i \cdot \max_{\boldsymbol{x}_{f \setminus i}} h^{\boldsymbol{\delta}^t}(\boldsymbol{x}_f) \qquad \forall i \in f,$$

$$\bar{\theta}_f^{\boldsymbol{\delta}^{t+1}}(\boldsymbol{x}_f) \;=\; h^{\boldsymbol{\delta}^t}(\boldsymbol{x}_f) - \sum_{i \in f} \alpha_i \cdot \max_{\hat{\boldsymbol{x}}_{f \setminus i}} h^{\boldsymbol{\delta}^t}(x_i, \hat{\boldsymbol{x}}_{f \setminus i}).$$

Thus, we have that $\max_{x_i} \bar{\theta}_i^{\boldsymbol{\delta}^{t+1}}(x_i) = \alpha_i \cdot \max_{\boldsymbol{x}_f} h^{\boldsymbol{\delta}^t}(\boldsymbol{x}_f)$ and $\max_{\boldsymbol{x}_f} \bar{\theta}_f^{\boldsymbol{\delta}^{t+1}}(\boldsymbol{x}_f) = \alpha_f \cdot \max_{\boldsymbol{x}_f} h^{\boldsymbol{\delta}^t}(\boldsymbol{x}_f)$. All of these statements can be proved using arguments analogous to those given in appendix 8.9.2.

The MPLP algorithm corresponds to the choice $\alpha_f = 0$, $\alpha_i = \frac{1}{|f|}$ for $i \in f$. The reparameterization results in as much of $h^{\boldsymbol{\delta}^t}(\boldsymbol{x}_f)$ being pushed to the single node terms $\bar{\theta}_i^{\boldsymbol{\delta}^{t+1}}(x_i)$ as possible. Thus, subsequent updates for other factors that share variables with f will be affected by this update as much as possible. Intuitively, this increases the amount of communication between the subproblems. MPLP's dual optimal fixed points will have $\max_{x_i} \bar{\theta}_i^{\boldsymbol{\delta}}(x_i) = \frac{L(\boldsymbol{\delta})}{|V|}$ for all i, and $\max_{\boldsymbol{x}_f} \bar{\theta}_f^{\boldsymbol{\delta}}(\boldsymbol{x}_f) = 0$ for all $f \in F$.

In contrast, each MSD update to $\delta_{fi}(x_i)$ divides the objective equally between the factor term and the node term (see appendix 8.9.1). As a result, MSD's dual optimal fixed points have $\max_{x_i} \bar{\theta}_i^{\boldsymbol{\delta}}(x_i) = \max_{\boldsymbol{x}_f} \bar{\theta}_f^{\boldsymbol{\delta}}(\boldsymbol{x}_f) = \frac{L(\boldsymbol{\delta})}{|V|+|F|}$ for all $i \in V$ and $f \in F$. Consider the alternative block coordinate descent update given by

$$\delta_{fi}(x_i) = \frac{-|f|}{|f|+1}\delta_i^{-f}(x_i) + \frac{1}{|f|+1}\max_{\boldsymbol{x}_{f\backslash i}}\left[\theta_f(\boldsymbol{x}_f) + \sum_{\hat{i}\in f\backslash i}\delta_{\hat{i}}^{-f}(x_{\hat{i}})\right] \quad (8.21)$$

for *all* $i \in f$ and x_i simultaneously. This update, which we call MSD++, corresponds to the choice $\alpha_i = \alpha_f = \frac{1}{|f|+1}$, and has similar fixed points.

We show in Fig. 8.5 that MSD++ is only slightly faster than MSD, despite using a block size that is the same as that of MPLP. This suggests that it is MPLP's choice of equalization method (i.e., the α_i's) that provides the substantial improvements over MSD, not simply the choice of block size. As $\alpha_f \to 1$, the number of iterations required to solve the dual to optimality increases even further. The extreme case, $\alpha_f = 1$, although still a valid block coordinate descent step, gets stuck after the first iteration.

8.5.7 Convergence of Dual Coordinate Descent

Although coordinate descent algorithms decrease the dual objective at every iteration, they are not generally guaranteed to converge to the dual optimum. The reason is that although the dual objective $L(\boldsymbol{\delta})$ is convex, it is not strictly convex. This implies that the minimizing coordinate value may not be unique, and thus convergence guarantees for coordinate descent algorithms do not hold (Bertsekas, 1995). Interestingly, for pairwise MRFs with binary variables, the fixed points of the coordinate descent algorithms do correspond to global optima (Kolmogorov and Wainwright, 2005; Globerson and Jaakkola, 2008).

One strategy to avoid the above problem is to replace the max function in the objective of Eq. 8.2 with a *soft-max* function (e.g., see Johnson, 2008; Hazan and Shashua, 2010) which is smooth and strictly convex. As a result, coordinate descent converges globally.[9] An alternative approach are the auction algorithms proposed by Bertsekas (1992). However, currently there does not seem to be a coordinate descent approach to Eq. 8.2 that is guaranteed to converge globally for general problems.

8.6 Relations to Linear Programming Relaxations

Linear programming relaxations are a popular approach to approximating combinatorial optimization problems. One of these relaxations is in fact well

---

9. To solve the original dual, the soft-max needs to be gradually changed to the true max.

known to be equivalent to the dual decomposition approach discussed in this chapter (Schlesinger, 1976; Guignard and Kim, 1987; Komodakis et al., 2011; Wainwright et al., 2005; Werner, 2007). In this section we describe the corresponding LP relaxation.

We obtain a relaxation of the discrete optimization problem given in Eq. 8.1 by replacing it with the following linear program:

$$\max_{\boldsymbol{\mu} \in M_L} \left\{ \sum_f \sum_{\boldsymbol{x}_f} \theta_f(\boldsymbol{x}_f) \mu_f(\boldsymbol{x}_f) + \sum_i \sum_{x_i} \theta_i(x_i) \mu_i(x_i) \right\} \tag{8.22}$$

where the *local marginal polytope* M_L enforces that $\{\mu_i(x_i), \forall x_i\}$ and $\{\mu_f(\boldsymbol{x}_f), \forall \boldsymbol{x}_f\}$ correspond to valid (local) probability distributions and that, for each factor f, $\mu_f(\boldsymbol{x}_f)$, is consistent with $\mu_i(x_i)$ for all $i \in f, x_i$:

$$M_L = \left\{ \boldsymbol{\mu} \geq 0 : \begin{array}{ll} \sum_{\boldsymbol{x}_{f \setminus i}} \mu_f(\boldsymbol{x}_f) = \mu_i(x_i) & \forall f, i \in f, x_i \\ \sum_{x_i} \mu_i(x_i) = 1 & \forall i \end{array} \right\}. \tag{8.23}$$

Standard duality transformations can be used to show that the convex dual of the LP relaxation (Eq. 8.22) is the Lagrangian relaxation (Eq. 8.2). The integral vertices of the local marginal polytope (i.e., $\boldsymbol{\mu}$ such that $\mu_i(x_i) \in \{0, 1\}$) can be shown to correspond 1-to-1 with assignments $\boldsymbol{x}$ to the variables of the graphical model. The local marginal polytope also has *fractional* vertices that do not correspond to any global assignment (Wainwright and Jordan, 2008). Since the relaxation optimizes over this larger space, the value of the LP solution always upper-bounds the value of the MAP assignment. We call the relaxation *tight* for an instance if the optimum of the LP has the same value as the MAP assignment.

The connection to LP relaxations is important for a number of reasons. First, it shows how the algorithms discussed in this chapter apply equally to dual decomposition and linear programming relaxations for the MAP problem. Second, it allows us to understand when two different dual decompositions are equivalent to each other in the sense that, for all instances, the Lagrangian relaxation provides the same upper bound (at optimality). Realizing that two formulations are equivalent allows us to attribute differences in empirical results to the algorithms used to *optimize* the duals, not the tightness of the LP relaxation.

The third reason why the connection to LP relaxations is important is that it allows us to tighten the relaxation by adding valid constraints that are guaranteed not to cut off any integer solutions. The dual decomposition that we introduced in section 8.3 used only single node *intersection sets*, enforcing that the subproblems are consistent with one another on the individual variables. To obtain tighter LP relaxations, we typically have to use larger intersection sets along with the new constraints.[10] Although further discussion of this is beyond the scope of this chapter, details can be found in Sontag (2010) and Werner (2008).

8.7   Decoding: Finding the MAP Assignment

Thus far, we have discussed how the subgradient method and block coordinate descent can be used to solve the Lagrangian relaxation. Although this provides an upper bound on the *value* of the MAP assignment, for most inference problems we actually want to find the assignment itself. In Fig. 8.4 we suggested one simple way to find an assignment from the dual beliefs, by locally decoding the single node reparameterizations:

$$x_i \;\leftarrow\; \operatorname*{argmax}_{\hat{x}_i} \bar{\theta}_i^{\boldsymbol{\delta}}(\hat{x}_i). \tag{8.24}$$

The node terms $\bar{\theta}_i^{\boldsymbol{\delta}}(x_i)$ may not have a unique maximum. However, when they do have a unique maximum, we say that $\boldsymbol{\delta}$ is *locally decodable* to $\boldsymbol{x}$. This section addresses the important question of when this and related approaches will succeed in finding the MAP assignment. Proofs are given in appendix 8.9.4.

Before the dual is solved (close) to optimality, it is typically not possible to give guarantees as to how good an assignment will be found by this local decoding scheme. However, once the algorithms have solved the dual to optimality, if the solution is locally decodable to $\boldsymbol{x}$, then $\boldsymbol{x}$ is both the MAP assignment and the only LP solution, as stated in the next theorem.

Theorem 8.2. *If a dual optimal $\boldsymbol{\delta}^*$ is locally decodable to $\boldsymbol{x}^*$, then the LP relaxation has a unique solution, and it is $\boldsymbol{x}^*$.*

Thus, this result gives a strong *necessary* condition for when locally decoding the dual solution can find the MAP assignments only when the LP relaxation is tight, the MAP assignment is unique, and there are no optimal fractional solutions.[11]

A natural question is whether this assumption that the LP relaxation has a unique solution that is integral, is *sufficient* for a dual solution to be locally decodable. As we illustrate later, in general the answer is no, as not all dual

---

10. Globerson and Jaakkola (2008) give the MPLP updates for larger intersection sets.
11. An alternative way to find the MAP assignment in this setting would be to directly solve the LP relaxation using a generic LP solver. However, recall that we use dual decomposition for reasons of computational efficiency.

solutions are locally decodable. We are, however, guaranteed that one exists.

Theorem 8.3. *If the LP relaxation has a unique solution and it is integral, there exists a dual optimal δ^* that is locally decodable.*

Ideally, our algorithms would be guaranteed to find such a locally decodable dual solution. We show in section 8.7.1 that the subgradient method is indeed guaranteed to find a locally decodable solution under this assumption. For coordinate descent methods there do exist cases where a locally decodable solution may not be found. However, in practice this is often not a problem, and there are classes of graphical models where locally decodable solutions are guaranteed to be found (see section 8.7.2).

Rather than locally decoding an assignment, we could try searching for a global assignment x^* which maximizes each of the local subproblems. We showed in theorem 8.1 that such an *agreeing* assignment would be an MAP assignment, and its existence would imply that the LP relaxation is tight. Thus, the LP relaxation being tight is a necessary requirement for this strategy to succeed. For pairwise models with binary variables, an agreeing assignment can be found in linear time when one exists (Johnson, 2008). However, this does not extend to larger factors or non-binary variables.

Theorem 8.4. *Finding an agreeing assignment is NP-complete even when the LP relaxation is tight and the MAP assignment is unique.*

The construction in the above theorem uses a model where the optimum of the LP relaxation is not unique; although there may be a unique MAP assignment, there are also fractional vertices that have the same optimal value. When the LP has a unique and integral optimum, the decoding problem is no longer hard (e.g., it can be solved by optimizing an LP). Thus, asking that the LP relaxation have a unique solution that is integral seems to be a very reasonable assumption for a decoding method to succeed. When the LP relaxation does not have a unique solution, a possible remedy is to perturb the objective function by adding a small (in magnitude) random vector. If the LP relaxation is not tight, we can attempt to tighten the relaxation, for example by using the approaches discussed in Sontag et al. (2008).

8.7.1 Subgradient Method

The problem of recovering a primal solution (i.e., a solution to the LP relaxation given in Eq. 8.22) when solving a Lagrangian relaxation using subgradient methods has been well studied (e.g., see Anstreicher and Wolsey, 2009; Nedić and Ozdaglar, 2009; Shor, 1985). We next describe one such approach (the simplest) and discuss its implications for finding the MAP assignment from the dual solution.

Given the current dual variables $\boldsymbol{\delta}^t$, define the following indicator functions for a maximizing assignment of the subproblems: $\mu_f^t(\boldsymbol{x}_f) = 1[\boldsymbol{x}_f = \operatorname{argmax}_{\hat{\boldsymbol{x}}_f} \bar{\theta}_f^{\boldsymbol{\delta}^t}(\hat{\boldsymbol{x}}_f)]$, and $\mu_i^t(x_i) = 1[x_i = \operatorname{argmax}_{\hat{x}_i} \bar{\theta}_i^{\boldsymbol{\delta}^t}(\hat{x}_i)]$. When the maximum is not unique, choose any one of the maximizing assignments.

Next, consider the average of these indicator functions across all subgradient iterations:

$$\bar{\mu}_f(\boldsymbol{x}_f) \;=\; \frac{1}{T} \sum_{t=1}^{T} \mu_f^t(\boldsymbol{x}_f)$$

$$\bar{\mu}_i(x_i) \;=\; \frac{1}{T} \sum_{t=1}^{T} \mu_i^t(x_i).$$

For many common choices of stepsizes, the estimate $\bar{\boldsymbol{\mu}}$ can be shown to converge to a solution of the LP relaxation given in Eq. 8.22, as $T \to \infty$.

When the LP relaxation has a unique solution and it is integral, then $\bar{\boldsymbol{\mu}}$ is guaranteed to converge to the unique MAP assignment. In particular, this implies that there must exist a subgradient iteration when $\boldsymbol{\mu}^t$ corresponds to the MAP assignment. At this iteration the subgradient is zero, and we obtain a certificate of optimality by theorem 8.1. When the LP relaxation is not tight, recovering a primal solution may be helpful for finding additional constraints to use in tightening the relaxation.

8.7.2   Coordinate Descent

As mentioned in section 8.5, the coordinate descent algorithms, while always providing a monotonic improvement in the dual objective, are not in general guaranteed to solve the Lagrangian relaxation to optimality. However, for some graphical models, such as pairwise models with binary variables, fixed points of the coordinate descent algorithms can be used to construct a solution to the LP relaxation (Kolmogorov and Wainwright, 2005; Globerson and Jaakkola, 2008). Thus, for these graphical models, all fixed points of the algorithms are dual optimal.

We additionally show that when the LP relaxation has a unique solution that is integral, then the fixed point must be locally decodable to the MAP assignment. On the other hand, for more general graphical models, there do exist degenerate cases when a fixed point of the coordinate descent algorithm is *not* locally decodable, even if it corresponds to a dual solution.

Theorem 8.5. *Suppose the LP relaxation has a unique solution and it is* *integral.*[12] *Then the following hold:*

1. *For binary pairwise MRFs, fixed points of the coordinate descent algorithms are locally decodable to* x^*.

2. *For pairwise tree-structured MRFs, fixed points of the coordinate descent algorithms are locally decodable to* x^*.

3. *There exist non-binary pairwise MRFs with cycles and dual optimal fixed points of the coordinate descent algorithms that are not locally decodable.*

The second result can be generalized to graphs with a single cycle. Despite the theoretical difficulty hinted at in the third result, empirically we have found that when the LP relaxation has a unique solution that is integral, local decoding nearly always finds the MAP assignment, when used with the coordinate descent algorithms (Sontag et al., 2008).

8.8 Discussion

The dual decomposition formulation that we introduced in this chapter is applicable whenever one can break an optimization problem into smaller subproblems that can be solved exactly by using combinatorial algorithms. The dual objective is particularly simple, consisting of a sum of maximizations over the individual subproblems, and provides an upper bound on the value of the MAP assignment. Minimizing the Lagrangian relaxation makes this bound tighter by pushing the subproblems to agree with one another on their maximizing assignment. If we succeed in finding an assignment that maximizes each of the individual subproblems, then this assignment is the MAP assignment and the dual provides a certificate of optimality. Even when not exact, the upper bound provided by the dual can be extremely valuable when used with a branch-and-bound procedure to find the MAP assignment (Geoffrion, 1974).

We described both subgradient methods and block coordinate algorithms for minimizing the Lagrangian relaxation. Both are notable for their simplicity and their effectiveness at solving real-world inference problems. A natural question is how to choose between the two for a specific problem. This question has been the subject of continual debate since the early work on Lagrangian relaxations (see Guignard, 2003, for a review).

Each approach has its advantages and disadvantages. The subgradient method is guaranteed to converge, but does not monotonically decrease

---

12. This implies that the MAP assignment x^* is unique.

the dual and requires a delicate choice of stepsize. The coordinate descent algorithms are typically much faster at minimizing the dual, especially in the initial iterations, and are also monotonic. However, to perform each update, one must compute max-marginals, and this may be impractical for some applications. We presented several coordinate update schemes. These vary both in the coordinates that they choose to minimize in each update, and in the way this minimization is performed. MPLP appears to be the best general choice.

It may be advantageous to use the two approaches together. For example, one could first use coordinate descent and then, when the algorithm is close to convergence, use the subgradient method to ensure global optimality. Another possibility is to alternate between the updates, as in the spacer step approach in optimization (see Bertsekas, 1995).

There are often many different ways to decompose a problem when applying dual decomposition. For example, in this chapter we suggested using a decomposition for pairwise models that uses one subproblem per edge. An alternative decomposition is to use a set of spanning trees that together cover all of the edge potentials (Wainwright et al., 2005). Still another approach is, before constructing the decomposition, to split some of the variables and introduce new potentials to enforce equality among all copies of a variable (Guignard, 2003). Using this technique, Yarkony et al. (2010) give a decomposition that uses a single tree spanning *all* of the original edge potentials. Although in all of these cases the decompositions correspond to the same LP relaxation, often a different decomposition can result in tighter or looser bounds being provided by the Lagrangian relaxation.

Dual decomposition methods have a wide range of potential applications in machine learning and, more broadly, engineering and the sciences. We have already observed some empirical successes for natural language processing, computational biology, and machine vision. We expect that as graphical models become more complex, techniques like the ones discussed in this chapter will become essential for performing fast and accurate inference.

Acknowledgments   We thank Michael Collins, Ce Liu, and the editors for their very useful feedback, and also M. Collins for Fig. 8.1. This work was partly supported by BSF grant 2008303.

Appendix: Technical Details

8.9.1   Derivation of the Max-Sum Diffusion Updates

For a given factor f and variable i, assume all δ variables are fixed except $\delta_{fi}(x_i)$. We would like to minimize $L(\delta)$ in Eq. 8.2 w.r.t. the free variables. Here we prove that the choice of $\delta_{fi}(x_i)$ in Eq. 8.17 corresponds to this minimization (other updates that achieve the same objective are possible since $L(\delta)$ is not strictly convex).

The part of the objective $L(\delta)$ that depends on the free parameters is

$$\max_{x_i}\Big(\theta_i(x_i) + \sum_{f:i\in f}\delta_{fi}(x_i)\Big) + \max_{\boldsymbol{x}_f}\Big(\theta_f(\boldsymbol{x}_f) - \sum_{\hat{i}\in f}\delta_{f\hat{i}}(x_{\hat{i}})\Big)$$

$$= \max_{x_i}\Big(\delta_{fi}(x_i) + \delta_i^{-f}(x_i)\Big) + \max_{x_i}\Big(-\delta_{fi}(x_i) + \max_{\boldsymbol{x}_{f\setminus i}}\Big[\theta_f(\boldsymbol{x}_f) - \sum_{\hat{i}\in f\setminus i}\delta_{f\hat{i}}(x_{\hat{i}})\Big]\Big)$$

$$\geq \max_{x_i}\Big(\delta_i^{-f}(x_i) + \max_{\boldsymbol{x}_{f\setminus i}}\Big[\theta_f(\boldsymbol{x}_f) - \sum_{\hat{i}\in f\setminus i}\delta_{f\hat{i}}(x_{\hat{i}})\Big]\Big) = \max_{x_i} g(x_i).$$

This lower bound can be achieved by choosing a $\delta_{fi}(x_i)$ such that

$$\delta_{fi}(x_i) + \delta_i^{-f}(x_i) = \tfrac{1}{2}g(x_i),$$

$$-\delta_{fi}(x_i) + \max_{\boldsymbol{x}_{f\setminus i}}\Big[\theta_f(\boldsymbol{x}_f) - \sum_{\hat{i}\in f\setminus i}\delta_{f\hat{i}}(x_{\hat{i}})\Big] = \tfrac{1}{2}g(x_i).$$

Indeed, the $\delta_{fi}(x_i)$ given by the update in Eq. 8.17 satisfies the above.

8.9.2   Derivation of the MPLP Updates

For a given factor f assume all δ variables are fixed except $\delta_{fi}(x_i)$ for all $i \in f$. We would like to minimize $L(\delta)$ in Eq. 8.2 w.r.t. the free variables. Here we prove that Eq. 8.18 does this optimally.

The part of the objective $L(\delta)$ that depends on the free parameters is

$$\bar{L}(\delta) = \sum_{i\in f}\max_{x_i}\Big(\theta_i(x_i) + \sum_{\hat{f}:i\in\hat{f}}\delta_{\hat{f}i}(x_i)\Big) + \max_{\boldsymbol{x}_f}\Big(\theta_f(\boldsymbol{x}_f) - \sum_{i\in f}\delta_{fi}(x_i)\Big). \quad (8.25)$$

Denote $A_i(\delta) = \max_{x_i}\Big(\theta_i(x_i) + \sum_{\hat{f}:i\in\hat{f}}\delta_{\hat{f}i}(x_i)\Big)$ and $A_f(\delta) = \max_{\boldsymbol{x}_f}\Big(\theta_f(\boldsymbol{x}_f) - \sum_{i\in f}\delta_{fi}(x_i)\Big)$. Then it follows that

$$\bar{L}(\delta) = \sum_{i\in f}A_i(\delta) + A_f(\delta) \geq \max_{\boldsymbol{x}_f}\Big(\theta_f(\boldsymbol{x}_f) + \sum_{i\in f}\delta_i^{-f}(x_i)\Big) = B.$$

This gives a lower bound B on the minimum of $\bar{L}(\boldsymbol{\delta})$. We next show that this lower bound is achieved by the MPLP update in Eq. 8.18. For each of the terms $A_i(\boldsymbol{\delta})$ we have (after the update)

$$
\begin{aligned}
A_i(\boldsymbol{\delta}) &= \max_{x_i} \frac{1}{|f|} \delta_i^{-f}(x_i) + \frac{1}{|f|} \max_{\boldsymbol{x}_{f\setminus i}} \left[\theta_f(\boldsymbol{x}_f) + \sum_{\hat{i} \in f\setminus i} \delta_{\hat{i}}^{-f}(x_{\hat{i}}) \right] \\
&= \frac{1}{|f|} \max_{\boldsymbol{x}_f} \left(\theta_f(\boldsymbol{x}_f) + \sum_{i \in f} \delta_i^{-f}(x_i) \right) = \frac{B}{|f|}.
\end{aligned}
\tag{8.26}
$$

The value of $A_f(\boldsymbol{\delta})$ after the MPLP update is

$$
\max_{\boldsymbol{x}_f} \left(\theta_f(\boldsymbol{x}_f) + \sum_{i \in f} \left\{ \frac{|f|-1}{|f|} \delta_i^{-f}(x_i) - \frac{1}{|f|} \max_{\hat{\boldsymbol{x}}_{f\setminus i}} \left[\theta_f(x_i, \hat{\boldsymbol{x}}_f) + \sum_{\hat{i} \in f\setminus i} \delta_{\hat{i}}^{-f}(\hat{x}_{\hat{i}}) \right] \right\} \right)
$$

$$
= \frac{1}{|f|} \max_{\boldsymbol{x}_f} \left(\sum_{i \in f} \left\{ \theta_f(\boldsymbol{x}_f) + \sum_{\hat{i} \in f\setminus i} \delta_{\hat{i}}^{-f}(x_{\hat{i}}) - \max_{\hat{\boldsymbol{x}}_{f\setminus i}} \left[\theta_f(x_i, \hat{\boldsymbol{x}}_f) + \sum_{\hat{i} \in f\setminus i} \delta_{\hat{i}}^{-f}(\hat{x}_{\hat{i}}) \right] \right\} \right)
$$

$$
\leq \frac{1}{|f|} \sum_{i \in f} \max_{\boldsymbol{x}_f} \left(\theta_f(\boldsymbol{x}_f) + \sum_{\hat{i} \in f\setminus i} \delta_{\hat{i}}^{-f}(x_{\hat{i}}) - \max_{\hat{\boldsymbol{x}}_{f\setminus i}} \left[\theta_f(x_i, \hat{\boldsymbol{x}}_f) + \sum_{\hat{i} \in f\setminus i} \delta_{\hat{i}}^{-f}(\hat{x}_{\hat{i}}) \right] \right)
$$

$$
= \frac{1}{|f|} \sum_{i \in f} \max_{x_i} \left(\max_{\boldsymbol{x}_{f\setminus i}} \left[\theta_f(\boldsymbol{x}_f) + \sum_{\hat{i} \in f\setminus i} \delta_{\hat{i}}^{-f}(x_{\hat{i}}) \right] - \max_{\boldsymbol{x}_{f\setminus i}} \left[\theta_f(\boldsymbol{x}_f) + \sum_{\hat{i} \in f\setminus i} \delta_{\hat{i}}^{-f}(x_{\hat{i}}) \right] \right)
$$

$$
= 0,
$$

where the first equality used $\sum_{i \in f} (|f| - 1) \delta_i^{-f}(x_i) = \sum_{i \in f} \sum_{\hat{i} \in f\setminus i} \delta_{\hat{i}}^{-f}(x_{\hat{i}})$. From the above and Eq. 8.26 it follows that $\bar{L}(\boldsymbol{\delta}) \leq B$. However, we know that B is a lower bound on $\bar{L}(\boldsymbol{\delta})$, so we must have $\bar{L}(\boldsymbol{\delta}) = B$, implying the optimality of the MPLP update.

8.9.3 Derivation of the Star Update

The MPLP update gives the closed form solution for block coordinate descent on coordinates $\delta_{\{i,j\}j}(x_j)$ and $\delta_{\{i,j\}i}(x_i)$ for a particular $\{i, j\} \in F$ (i.e., an edge update). We now wish to find the update where the free coordinates are $\delta_{\{i,j\}j}(x_j), \delta_{\{i,j\}i}(x_i)$ for a fixed i and all $j \in N(i)$. One way to find the optimal coordinates is by iterating the MPLP update for all the edges in the star until convergence. Notice that $\delta_j^{-i}(x_j)$, $\gamma_{ji}(x_i)$, and $\gamma_i(x_i)$ do not change after each edge update. We solve for $\delta_{\{i,j\}j}(x_j)$ and $\delta_{\{i,j\}i}(x_i)$ in terms of these quantities, using fixed point equations for MPLP.

The MPLP edge update for $f = \{i, j\}$ is given by

$$
\delta_{\{i,j\}i}(x_i) = -\tfrac{1}{2} \delta_i^{-j}(x_i) + \tfrac{1}{2} \gamma_{ji}(x_i).
\tag{8.27}
$$

Multiplying by 2, and then subtracting $\delta_{\{i,j\}i}(x_i)$ from both sides, we obtain

$$\delta_{\{i,j\}i}(x_i) = -\delta_i(x_i) + \gamma_{ji}(x_i), \tag{8.28}$$

where we define $\delta_i(x_i) = \theta_i(x_i) + \sum_{j \in N(i)} \delta_{\{i,j\}i}(x_i)$. Summing this over all neighbors of i, and adding $\theta_i(x_i)$ to both sides, we obtain

$$\begin{aligned} \delta_i(x_i) &= -N_i\delta_i(x_i) + \gamma_i(x_i) \\ \delta_i(x_i) &= \frac{1}{1+N_i}\gamma_i(x_i). \end{aligned} \tag{8.29}$$

Applying Eq. 8.28 and Eq. 8.29, we obtain the following for $\delta_i^{-j}(x_i)$:

$$\delta_i^{-j}(x_i) = \delta_i(x_i) - \delta_{\{i,j\}i}(x_i) = 2\delta_i(x_i) - \gamma_{ji}(x_i) = \frac{2\gamma_i(x_i)}{1+N_i} - \gamma_{ji}(x_i) .$$

Substituting this back into the MPLP update (Eq. 8.27) yields the update for $\delta_{\{i,j\}i}(x_i)$ given in Eq. 8.20. The update for $\delta_{\{i,j\}j}(x_j)$ is obtained by taking the MPLP update and substituting the above expression for $\delta_i^{-j}(x_i)$.

The dual variables given by the star update in Eq. 8.20 can be seen to be a fixed point of MPLP for all edges in the star. Since any fixed point of MPLP on a tree is dual optimal (see section 8.7.2), these updates provide the optimum for these coordinates.

8.9.4   Proofs of the Decoding Results

Duality in linear programming specifies *complementary slackness* conditions that every primal and dual solution must satisfy. In particular, it can be shown that for any optimal μ^* for the local LP relaxation given in Eq. 8.22 and any optimal δ^* for the Lagrangian relaxation $\min_\delta L(\delta)$:

$$\mu_i^*(x_i) > 0 \quad \Rightarrow \quad \bar{\theta}_i^{\delta^*}(x_i) = \max_{\hat{x}_i} \bar{\theta}_i^{\delta^*}(\hat{x}_i), \tag{8.30}$$

$$\mu_f^*(x_f) > 0 \quad \Rightarrow \quad \bar{\theta}_f^{\delta^*}(x_f) = \max_{\hat{x}_f} \bar{\theta}_f^{\delta^*}(\hat{x}_f). \tag{8.31}$$

Proof of theorem 8.2. Consider any $x_i \neq x_i^*$. Since $\bar{\theta}_i^{\delta^*}(x_i) < \bar{\theta}_i^{\delta^*}(x_i^*)$, complementary slackness (8.30) implies that $\mu_i^*(x_i) = 0$ for all optimal μ^*. Thus, the only solution to the LP relaxation corresponds to x^*.   □

Proof of theorem 8.3. This follows from strict complementary slackness (Vanderbei, 2007), which guarantees that a primal-dual pair (μ^*, δ^*) exists that satisfies the implication in (8.30) both ways. Since the LP relaxation has only one solution, and it corresponds to the MAP assignment, strict complementary slackness guarantees that such a δ^* is locally decodable.   □

Proof of theorem 8.4. We reduce from 3SAT. First, we encode 3SAT as an optimization problem of the form given in Eq. 8.1. The variables $x_i \in \{0, 1\}$ are the same as in the 3SAT formula. We have one factor $\theta_f(\boldsymbol{x}_f)$ for each clause f in the formula, defined on the corresponding three variables. $\theta_f(\boldsymbol{x}_f)$ is 0 if the clause is satisfied by $\boldsymbol{x}_f$, and $-\infty$ otherwise. $\theta_i(x_i) = 0$ for all i, x_i.

Suppose we could efficiently find an agreeing assignment when one exists. By theorem 8.1, this would be an MAP assignment (and the LP relaxation is tight), which in this case corresponds to a satisfying assignment of the 3SAT formula. Thus, if we could efficiently find an MAP assignment whenever the LP relaxation is tight, then we could efficiently solve 3SAT.

Finding the MAP assignment is hard even if it is unique, because the problem of finding a satisfying assignment when we are guaranteed that a formula has at most one satisfying assignment, called Unique-SAT, is also NP-hard, under randomized reductions (Valiant and Vazirani, 1985). $\qquad\square$

For 3SAT, the LP relaxation always has a fractional solution $\boldsymbol{\mu} \in M_L$ with objective value 0. Thus, by theorem 8.2 the dual solutions will never be locally decodable. Let $\mu_i(0) = \mu_i(1) = .5$ for all i. For each clause f, if it is satisfied by both $\boldsymbol{x}_f^1 = (0, 0, 0)$ and $\boldsymbol{x}_f^2 = (1, 1, 1)$, then let $\mu_f(\boldsymbol{x}_f^1) = \mu_f(\boldsymbol{x}_f^2) = .5$ and $\mu_f(\boldsymbol{x}_f) = 0$ for $\boldsymbol{x}_f \neq \boldsymbol{x}_f^1, \boldsymbol{x}_f^2$. Otherwise, f must be satisfied by $\boldsymbol{x}_f^1 = (0, 1, 1)$ and $\boldsymbol{x}_f^2 = (1, 0, 0)$, so we set $\mu_f(\boldsymbol{x}_f^1) = \mu_f(\boldsymbol{x}_f^2) = .5$.

Proof of theorem 8.5, part 1. The claim follows from (Globerson and Jaakkola, 2008, proposition 4). This result shows how to construct a fractional primal solution whenever at least one of the nodes i has $\bar{\theta}_i^{\boldsymbol{\delta}}(0) = \bar{\theta}_i^{\boldsymbol{\delta}}(1)$, that is, when $\boldsymbol{\delta}$ is not locally decodable. However, this would contradict our assumption that the LP relaxation has a unique solution and it is integral. Thus, $\boldsymbol{\delta}$ must be locally decodable. $\qquad\square$

All fixed points of the coordinate descent algorithms can be shown to satisfy *max-consistency*. Let A_i consist of all states $\hat{x}_i$ that maximize $\bar{\theta}_i^{\boldsymbol{\delta}}(x_i)$. By max-consistency, we mean that for all $f \in F, i \in f$, and $x_i \in A_i$, $\max_{\boldsymbol{x}_{f \backslash i}} \bar{\theta}_f^{\boldsymbol{\delta}}(\boldsymbol{x}_f) = \max_{\hat{\boldsymbol{x}}_f} \bar{\theta}_f^{\boldsymbol{\delta}}(\hat{\boldsymbol{x}}_f)$. This is trivial to see for MSD, since at a fixed point, $\bar{\theta}_i^{\boldsymbol{\delta}}(x_i) = \max_{\boldsymbol{x}_{f \backslash i}} \bar{\theta}_f^{\boldsymbol{\delta}}(\boldsymbol{x}_f) \ \forall f, i \in f$ and x_i. Thus, for $x_i \in A_i$, $\bar{\theta}_i^{\boldsymbol{\delta}}(x_i) = \max_{\hat{\boldsymbol{x}}_f} \bar{\theta}_f^{\boldsymbol{\delta}}(\hat{\boldsymbol{x}}_f)$. Putting these together shows max-consistency.

Proof of theorem 8.5, part 2. First, construct a reduced pairwise MRF with potentials $\boldsymbol{\theta}'$ where for each variable i we consider only those states x_i that maximize $\bar{\theta}_i^{\boldsymbol{\delta}}(x_i)$ (hereafter ignoring the other states). We let $\theta_{ij}'(\hat{x}_i, \hat{x}_j) = 0$ if $\hat{x}_i, \hat{x}_j$ maximize $\bar{\theta}_{ij}^{\boldsymbol{\delta}}(x_i, x_j)$, and $-\infty$ otherwise. By complementary slackness, all solutions of the LP relaxation for $\boldsymbol{\theta}'$ are also optimal for $\boldsymbol{\theta}$. By max-consistency, for every state x_i, $\exists x_j$ such that $\theta'(x_i, x_j) = 0$.

Suppose that there exist a vertex i and a state x_i' where $x_i' \neq x_i^*$ (that is, $\boldsymbol{\delta}$ is not locally decodable to $\boldsymbol{x}^*$). Then, by max-consistency we can construct

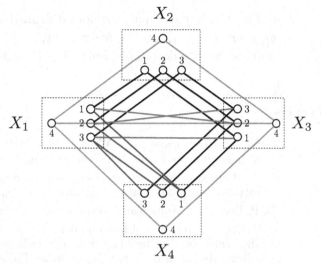

Figure 8.6: Illustration of the parameters of the pairwise MRF that we use in the proof of theorem 8.5, part 3. Each node represents a state $x_i \in \{1, 2, 3, 4\}$. An edge between x_i and x_j signifies that $\theta_{ij}(x_i, x_j) = 0$, whereas no edge between x_i and x_j signifies that $\theta_{ij}(x_i, x_j) = -1$.

an assignment (going one edge at a time in the tree rooted at i) $\boldsymbol{x}' \neq \boldsymbol{x}^*$ such that $\theta'(\boldsymbol{x}') = 0$. This shows that $\boldsymbol{x}'$ is an MAP assignment for $\boldsymbol{\theta}$. However, this contradicts the uniqueness of the MAP assignment. □

Proof of theorem 8.5, part 3. Consider a pairwise MRF on four variables with edges $E = \{(1, 2), (2, 3), (3, 4), (1, 4), (1, 3)\}$ and where each variable has four states. Let the parameters $\boldsymbol{\theta}$ be as specified in Fig. 8.6, and let the dual variables $\boldsymbol{\delta}$ be identically zero. Since the MAP assignment ($x_i = 4$ for all i) has value 0 and $L(\boldsymbol{\delta}) = 0$, $\boldsymbol{\delta}$ is dual optimal and the LP relaxation is tight. $\boldsymbol{\delta}$ is also a fixed point of the MPLP update given in Eq. 8.18.

What remains is to show that the LP relaxation has a *unique* solution (i.e., $x_i = 4$ for all i). First, note that for any primal optimal $\boldsymbol{\mu}^*$, $\mu_{ij}^*(x_i, x_j)$ must be 0 whenever $\theta_{ij}(x_i, x_j) = -1$. Subject to these constraints, we next show that $\mu_i^*(x_i)$ must be 0 for all i and $x_i \in \{1, 2, 3\}$. For $x \in \{1, 2, 3\}$, the local consistency constraints (see Eq. 8.23) along edges $(1, 2)$, $(2, 3)$ and $(3, 4)$ imply that $\mu_1^*(x) = \mu_{12}^*(x, x) = \mu_2^*(x) = \mu_{2,3}^*(x, x) = \mu_3^*(x) = \mu_{3,4}^*(x) = \mu_4^*(x)$. Similarly, the local consistency constraints along edge $(1, 3)$ imply that $\mu_1^*(1) = \mu_{1,3}^*(1, 2) = \mu_3^*(2)$ and $\mu_1^*(2) = \mu_{1,3}^*(2, 3) = \mu_3^*(3)$. Together, these imply that $\mu_i^*(x_i) = a$ for all i and $x_i \in \{1, 2, 3\}$. Now consider the local consistency constraints for edge $(1, 4)$. One of the constraints is that $\mu_1^*(3) = \mu_{1,4}^*(3, 3) + \mu_{1,4}^*(3, 2) = \mu_4^*(3) + \mu_4^*(2)$. Thus, we must have that $a = 2a$, which implies that $a = 0$. □

Note that there do exist locally decodable solutions for the example given in Fig. 8.6. In particular, consider δ defined as $\delta_{\{i,j\}i}(4) = \delta_{\{i,j\}j}(4) = 0$ and $\delta_{\{i,j\}i}(x) = \delta_{\{i,j\}j}(x) = \epsilon$ for $x \in \{1, 2, 3\}$. When $-.5 \leq \epsilon < 0$, δ is dual optimal, locally decodable, and a fixed point of MPLP.

8.10   References

K. M. Anstreicher and L. A. Wolsey. Two "well-known" properties of subgradient optimization. *Mathematical Programming*, 120(1):213–220, 2009.

D. P. Bertsekas. Auction algorithms for network flow problems: A tutorial introduction. *Computational Optimization and Applications*, 1:7–66, 1992.

D. P. Bertsekas. *Nonlinear Programming*. Athena Scientific, Belmont, MA, 1995.

J. Duchi, D. Tarlow, G. Elidan, and D. Koller. Using combinatorial optimization within max-product belief propagation. In B. Schölkopf, J. Platt, and T. Hoffman, editors, *Advances in Neural Information Processing Systems 19*, pages 369–376. MIT Press, 2007.

D. Erlenkotter. A dual-based procedure for uncapacitated facility location. *Operations Research*, 26(6):992–1009, 1978.

M. L. Fisher. The lagrangian relaxation method for solving integer programming problems. *Management Science*, 27(1):1–18, 1981.

A. M. Geoffrion. Lagrangean relaxation for integer programming. *Mathematical Programming Study*, 2:82–114, 1974.

A. Globerson and T. Jaakkola. Fixing max-product: Convergent message passing algorithms for MAP LP-relaxations. In J. Platt, D. Koller, Y. Singer, and S. Roweis, editors, *Advances in Neural Information Processing Systems 20*, pages 553–560. MIT Press, Cambridge, MA, 2008.

M. Guignard. Lagrangean relaxation. *TOP: An Official Journal of the Spanish Society of Statistics and Operations Research*, 11(2):151–200, 2003.

M. Guignard and S. Kim. Lagrangean decomposition: A model yielding stronger Lagrangean bounds. *Mathematical Programming*, 39(2):215–228, 1987.

M. Guignard and M. Rosenwein. An application-oriented guide for designing lagrangean dual ascent algorithms. *European Journal of Operational Research*, 43(2):197–205, 1989.

R. Gupta, A. Diwan, and S. Sarawagi. Efficient inference with cardinality-based clique potentials. In *Proceedings of the 24th International Conference on Machine Learning*, pages 329–336. ACM Press, New York, 2007.

T. Hazan and A. Shashua. Norm-product belief propagation: Primal-dual message-passing for approximate inference. *IEEE Transactions on Information Theory*, 56(12):6294–6316, 2010.

M. Held, P. Wolfe, and H. Crowder. Validation of subgradient optimization. *Mathematical Programming*, 6(1):62–88, 1974.

J. Johnson. *Convex Relaxation Methods for Graphical Models: Lagrangian and Maximum Entropy Approaches*. PhD thesis, Department of Electrical Engineering and Computer Science, MIT, 2008.

D. Koller and N. Friedman. *Probabilistic Graphical Models: Principles and Techniques*. MIT Press, 2009.

V. Kolmogorov and M. Wainwright. On the optimality of tree-reweighted max-product message-passing. In *Proceedings of the 21st Conference on Uncertainty in Artificial Intelligence*, pages 316–323. AUAI Press, Arlington, VA, 2005.

N. Komodakis, N. Paragios, and G. Tziritas. MRF energy minimization and beyond via dual decomposition. *IEEE Transactions on Pattern Analysis and Machine Intelligence*, 33(3):531–552, 2011.

T. Koo, A. M. Rush, M. Collins, T. Jaakkola, and D. Sontag. Dual decomposition for parsing with non-projective head automata. In *Proceedings of the 2010 Conference on Empirical Methods in Natural Language Processing*, pages 1288–1298, 2010.

V. A. Kovalevsky and V. K. Koval. A diffusion algorithm for decreasing the energy of the max-sum labeling problem. Unpublished, Glushkov Institute of Cybernetics, Kiev, USSR, circa 1975. Personally communicated to T. Werner by M. I. Schlesinger.

S. Lacoste-Julien, B. Taskar, D. Klein, and M. I. Jordan. Word alignment via quadratic assignment. In R. C. Moore, J. A. Bilmes, J. Chu-Carroll, and M. Sanderson, editors, *Proceedings of the Human Language Technology Conference of the North American Chapter of the Association of Computational Linguistics*, pages 112–119. The Association for Computational Linguistics, New York, 2006.

C. Lemaréchal. Lagrangian relaxation. In *Computational Combinatorial Optimization*, pages 112–156. Berlin, Springer, 2001.

R. McDonald and G. Satta. On the complexity of non-projective data-driven dependency parsing. In *Proceedings of the 10th International Conference on Parsing Technologies*, pages 121–132. Association for Computational Linguistics, Morristown, NJ, 2007.

R. McDonald, F. Pereira, K. Ribarov, and J. Hajic. Non-projective dependency parsing using spanning tree algorithms. In *Proceedings of the Conference on Human Language Technology and Empirical Methods in Natural Language Processing*, pages 523–530, 2005.

A. Nedić and A. Ozdaglar. Approximate primal solutions and rate analysis for dual subgradient methods. *SIAM Journal on Optimization*, 19(4):1757–1780, 2009.

C. Rother, P. Kohli, W. Feng, and J. Jia. Minimizing sparse higher order energy functions of discrete variables. *Proceedings of the IEEE Conference on Computer Vision and Pattern Recognition*, pages 1382–1389, 2009.

M. I. Schlesinger. Syntactic analysis of two-dimensional visual signals in noisy conditions. *Kibernetika*, 4:113–130, 1976. in Russian.

N. Z. Shor. *Minimization Methods for Non-Differentiable Functions*. Springer-Verlag, New York, NY, USA, 1985.

D. Sontag. *Approximate Inference in Graphical Models using LP Relaxations*. PhD thesis, Department of Electrical Engineering and Computer Science, MIT, 2010.

D. Sontag and T. Jaakkola. Tree block coordinate descent for MAP in graphical models. In *Proceedings of the 12th International Workshop on Artificial Intelligence and Statistics*, volume 9, pages 544–551. JMLR: W&CP, 2009.

D. Sontag, T. Meltzer, A. Globerson, T. Jaakkola, and Y. Weiss. Tightening LP relaxations for MAP using message passing. In *Proceedings of the 24th Conference on Uncertainty in Artificial Intelligence*, pages 503–510. AUAI Press, Arlington, VA, 2008.

D. Tarlow, I. Givoni, and R. Zemel. HOP-MAP: Efficient message passing with high order potentials. In *Proceedings of the 13th International Conference on Artificial Intelligence and Statistics*, volume 9, pages 812–819. JMLR: W&CP, 2010.

L. G. Valiant and V. V. Vazirani. NP is as easy as detecting unique solutions. In *Proceedings of the 17th Annual ACM Symposium on Theory of Computing*, pages 458–463, New York, 1985. ACM Press.

R. Vanderbei. *Linear Programming: Foundations and Extensions.* Springer, 3rd edition, 2007.

M. Wainwright and M. I. Jordan. Graphical models, exponential families, and variational inference. *Foundations and Trends in Machine Learning*, 1(1-2):1–305, 2008.

M. Wainwright, T. Jaakkola, and A. Willsky. Tree-based reparameterization framework for analysis of sum-product and related algorithms. *IEEE Transactions on Information Theory*, 49(5):1120–1146, 2003.

M. Wainwright, T. Jaakkola, and A. Willsky. MAP estimation via agreement on trees: message-passing and linear programming. *IEEE Transactions on Information Theory*, 51(11):3697–3717, 2005.

Y. Weiss, C. Yanover, and T. Meltzer. MAP estimation, linear programming and belief propagation with convex free energies. In *Proceedings of the 23rd Conference on Uncertainty in Artificial Intelligence*, pages 416–425. AUAI Press, Arlington, VA, 2007.

T. Werner. A linear programming approach to max-sum problem: A review. *IEEE Transactions on Pattern Analysis and Machine Intelligence*, 29(7):1165–1179, 2007.

T. Werner. High-arity interactions, polyhedral relaxations, and cutting plane algorithm for soft constraint optimisation (MAP-MRF). In *Proceedings of the IEEE Conference on Computer Vision and Pattern Recognition*, 2008.

C. Yanover, T. Meltzer, and Y. Weiss. Linear programming relaxations and belief propagation – an empirical study. *Journal of Machine Learning Research*, 7: 1887–1907, 2006.

C. Yanover, O. Schueler-Furman, and Y. Weiss. Minimizing and learning energy functions for side-chain prediction. *Journal of Computational Biology*, 15(7): 899–911, 2008.

J. Yarkony, C. Fowlkes, and A. Ihler. Covering trees and lower-bounds on quadratic assignment. In *Proceedings of the IEEE Conference on Computer Vision and Pattern Recognition*, 2010.

9   Augmented Lagrangian Methods for Learning, Selecting, and Combining Features

Ryota Tomioka   tomioka@mist.i.u-tokyo.ac.jp
The University of Tokyo
Tokyo, Japan

Taiji Suzuki   s-taiji@stat.t.u-tokyo.ac.jp
The University of Tokyo
Tokyo, Japan

Masashi Sugiyama   sugi@cs.titech.ac.jp
Tokyo Institute of Technology
Tokyo, Japan

We investigate the family of Augmented Lagrangian (AL) methods for minimizing the sum of two convex functions. In the context of machine learning, minimization of such a composite objective function is useful in enforcing various structures, such as sparsity, on the solution in a learning task. We introduce a particularly efficient instance of an augmented Lagrangian method called the Dual Augmented Lagrangian (DAL) algorithm, and discuss its connection to proximal minimization and operator splitting algorithms in the primal. Furthermore, we demonstrate that the DAL algorithm for the trace norm regularization can be used to learn features from multiple data sources and optimally combine them in a convex optimization problem.

9.1   Introduction

Sparse estimation has recently been attracting attention from both the theoretical side (Candès et al., 2006; Bach, 2008; Ng, 2004) and the practical side, for example, magnetic resonance imaging (Weaver et al., 1991; Lustig et al., 2007), natural language processing (Gao et al., 2007), and bioinformatics (Shevade and Keerthi, 2003).

Sparse estimation is commonly formulated in two ways: the regularized estimation (or MAP estimation) framework (Tibshirani, 1996), and the empirical Bayesian estimation (also known as the automatic relevance determination) (Neal, 1996; Tipping, 2001). Both approaches are based on optimizing some objective functions, though the former is usually formulated as a convex optimization and the later is usually nonconvex.

Recently, a connection between the two formulations has been discussed in Wipf and Nagarajan (2008) which showed that in some special cases the (nonconvex) empirical Bayesian estimation can be carried out by iteratively solving reweighted (convex) regularized estimation problems. Therefore, in this chapter we will focus on the convex approach.

A regularization-based sparse estimation problem can be formulated as

$$\min_{\boldsymbol{x} \in \mathbb{R}^n} \quad \underbrace{L(\boldsymbol{x}) + R(\boldsymbol{x})}_{=:f(\boldsymbol{x})}, \tag{9.1}$$

where $L : \mathbb{R}^n \to \mathbb{R}$ is called the loss term, which we assume to be convex and differentiable; $R : \mathbb{R}^n \to \mathbb{R}$ is called the regularizer, which is assumed to be convex but may be non-differentiable, and for convenience we denote the sum of the two by f. In addition, we assume that $f(\boldsymbol{x}) \to \infty$ as $\|\boldsymbol{x}\| \to \infty$.

Problem (9.1) is closely related to solving an operator equation

$$(A + B)(\boldsymbol{x}) \ni 0, \tag{9.2}$$

where A and B are nonlinear maximal monotone operators. In fact, if A and B are the subdifferentials of L and R, respectively, problems (9.1) and (9.2) are equivalent. Algorithms to solve the operator equation (9.2) are extensively studied and are called *operator splitting* methods (see Lions and Mercier (1979); Eckstein and Bertsekas (1992)). We will discuss their connections to minimization algorithms for (9.1) in sections 9.2.2 and 9.5.

We will distinguish between a *simple* sparse estimation problem, and a *structured* sparse estimation problem. A simple sparse estimation problem is written as

$$\min_{\boldsymbol{x} \in \mathbb{R}^n} \quad L(\boldsymbol{x}) + \phi_\lambda(\boldsymbol{x}), \tag{9.3}$$

where ϕ_λ is a closed proper convex function[1] and is "simple" in the sense of separability and sparsity, which we define in section 9.2.1. Examples of a simple sparse estimation problem include the Lasso (Tibshirani, 1996), also known as basis pursuit denoising (Chen et al., 1998); the group Lasso (Yuan and Lin, 2006); and the trace norm regularization (Fazel et al., 2001; Srebro et al., 2005; Tomioka and Aihara, 2007; Yuan et al., 2007).

A *structured* sparse estimation problem is written as

$$\min_{\boldsymbol{x} \in \mathbb{R}^n} \quad L(\boldsymbol{x}) + \phi_\lambda(\boldsymbol{B}\boldsymbol{x}), \tag{9.4}$$

where $\boldsymbol{B} \in \mathbb{R}^{l \times n}$ is a matrix and ϕ_λ is a simple sparse regularizer as in the simple sparse estimation problem (9.3). Examples of a structured sparse estimation problem include total variation denoising (Rudin et al., 1992), wavelet shrinkage (Weaver et al., 1991; Donoho, 1995), the fused Lasso (Tibshirani et al., 2005), and structured sparsity-inducing norms (Jenatton et al., 2009).

In this chapter, we present an augmented Lagrangian (AL) method (Hestenes, 1969; Powell, 1969) for the dual of the simple sparse estimation problem. We show that the proposed *dual* augmented Lagrangian (DAL) is equivalent to the proximal minimization algorithm in the primal, converges super-linearly[2], and each step is computationally efficient because DAL can exploit the *sparsity in the intermediate solution*. There has been a series of studies that derive AL approaches using Bregman divergence (see Yin et al. (2008); Cai et al. (2008); Setzer (2010)).

Although our focus will be mostly on the simple sparse estimation problem (9.3), the methods we discuss are also relevant for the structured sparse estimation problem (9.4). In fact, by taking the Fenchel dual (Rockafellar, 1970, theorem 31.2), we notice that solving the structured sparse estimation problem (9.4) is equivalent to solving the following minimization problem:

$$\min_{\boldsymbol{\beta} \in \mathbb{R}^l} \quad L^*(\boldsymbol{B}^T \boldsymbol{\beta}) + \phi_\lambda^*(-\boldsymbol{\beta}), \tag{9.5}$$

---

1. "Closed" means that the epigraph $\{(\boldsymbol{z}, y) \in \mathbb{R}^{m+1} : y \geq \phi_\lambda(\boldsymbol{z})\}$ is a closed set, and "proper" means that the function is not everywhere $+\infty$; see, e.g., Rockafellar (1970). In the sequel, we use the term "convex function" in the meaning of "closed proper convex function".

2. A sequence $\boldsymbol{x}^t$ $(t = 1, 2, \ldots)$ converges to $\boldsymbol{x}^*$ super-linearly, if $\|\boldsymbol{x}^{t+1} - \boldsymbol{x}^*\| \leq c_t \|\boldsymbol{x}^t - \boldsymbol{x}^*\|$, where $0 \leq c_t < 1$ and $c_t \to 0$ as $t \to \infty$.

where L^* and ϕ_λ^* are the convex conjugate functions of L and ϕ_λ, respectively. The above minimization problem resembles the simple sparse estimation problem (9.3) (the matrix $\boldsymbol{B}^T$ can be considered as part of the loss function). This fact was effectively used by Goldstein and Osher (2009) to develop the split Bregman iteration (SBI) algorithm (see also Setzer (2010)). See section 9.5 for more detailed discussion.

This chapter is organized as follows. In the next section, we introduce some simple sparse regularizers and review different types of sparsity they produce through the so-called proximity operator. A brief operator theoretic background for the proximity operator is also given. In section 9.3, we present the proximal minimization algorithm, which is the primal representation of the DAL algorithm. The proposed DAL algorithm is introduced in section 9.4 and we discuss both why the dual formulation is particularly suitable for the simple sparse estimation problem, and its rate of convergence. We discuss connections between approximate AL methods and two operator splitting algorithms, the forward-backward splitting and the Douglas-Rachford splitting, in section 9.5. In section 9.6, we apply the trace norm regularization to a real brain-computer interface data set for learning feature extractors and their optimal combination. The computational efficiency of the DAL algorithm is also demonstrated. Finally, we summarize the chapter in section 9.7. Some background material on convex analysis is given in the appendix.

9.2   Background

In this section, we define "simple" sparse regularizers through the associated proximity operators. In addition, section 9.2.2 provides some operator theoretic backgrounds, which we use in later sections, especially section 9.5.

9.2.1   Simple sparse regularizers

Here, we provide three examples of *simple* sparse regularizers: the ℓ_1-regularizer, the group Lasso regularizer, and the trace norm regularizer. Other regularizers obtained by applying these three regularizers in a blockwise manner will also be called simple; for example, the ℓ_1-regularizer for the first 10 variables and the group Lasso regularizer for the remaining variables. These regularizers share two important properties. First, they are *separable* (in some manner). Second, the so-called proximity operators they define return "sparse" vectors (with respect to their separability).

First, we need to define the proximity operator as below (see also Moreau (1965); Rockafellar (1970); Combettes and Wajs (2005)).

Definition 9.1. *The proximity operator corresponding to a convex function* $f : \mathbb{R}^n \to \mathbb{R}$ *over* $\mathbb{R}^n$ *is a mapping from* $\mathbb{R}^n$ *to itself and is defined as*

$$\text{prox}_f(\boldsymbol{z}) = \underset{\boldsymbol{x} \in \mathbb{R}^n}{\text{argmin}} \left(f(\boldsymbol{x}) + \frac{1}{2}\|\boldsymbol{x} - \boldsymbol{z}\|^2 \right), \tag{9.6}$$

where $\| \cdot \|$ *denotes the Euclidean norm.*

Note that the minimizer is unique because the objective is strongly convex. Although the above definition is given in terms of a function f over $\mathbb{R}^n$, the definition extends naturally to a function over a general Hilbert space (see Moreau (1965); Rockafellar (1970)).

The proximity operator (9.6) defines a unique decomposition of a vector $\boldsymbol{z}$ as

$$\boldsymbol{z} = \boldsymbol{x} + \boldsymbol{y},$$

where $\boldsymbol{x} = \text{prox}_f(\boldsymbol{z})$ and $\boldsymbol{y} = \text{prox}_{f^*}(\boldsymbol{z})$ (f^* is the convex conjugate of f). This is called Moreau's decomposition (see appendix 9.8.2). We denote Moreau's decomposition corresponding to the function f as follows:

$$(\boldsymbol{x}, \boldsymbol{y}) = \text{decomp}_f(\boldsymbol{z}). \tag{9.7}$$

Note that the above expression implies $\boldsymbol{y} \in \partial f(\boldsymbol{x})$ because $\boldsymbol{x}$ minimizes the objective (9.6) and $\partial f(\boldsymbol{x}) + \boldsymbol{x} - \boldsymbol{z} \ni 0$, where $\partial f(\boldsymbol{x})$ denotes the subdifferential of f at $\boldsymbol{x}$.

The first example of sparse regularizers is the ℓ_1-regularizer, or the Lasso regularizer (Tibshirani, 1996), which is defined as follows:

$$\phi_\lambda^{\ell_1}(\boldsymbol{x}) = \lambda\|\boldsymbol{x}\|_1 = \lambda \sum_{j=1}^{n} |x_j|, \tag{9.8}$$

where $|\cdot|$ denotes the absolute value. We can also allow each component to have a different regularization constant, which can be used to include an unregularized bias term.

The proximity operator corresponding to the ℓ_1-regularizer is known as the soft threshold operator (Donoho, 1995) and can be defined elementwise as follows:

$$\text{prox}_\lambda^{\ell_1}(\boldsymbol{z}) := \left(\max(|z_j| - \lambda, 0) \frac{z_j}{|z_j|} \right)_{j=1}^{n}, \tag{9.9}$$

where the ratio $z_j/|z_j|$ is defined to be zero if $z_j = 0$. The above expression can easily be derived because the objective (9.6) can be minimized for each component x_j independently for the ℓ_1-regularizer.

The second example of sparse regularizers is the group Lasso (Yuan and Lin, 2006) regularizer

$$\phi_\lambda^{\mathfrak{G}}(\boldsymbol{x}) = \lambda \sum_{\mathfrak{g} \in \mathfrak{G}} \|\boldsymbol{x}_{\mathfrak{g}}\|, \tag{9.10}$$

where $\mathfrak{G}$ is a nonoverlapping partition of $\{1, \ldots, n\}$, $\mathfrak{g} \in \mathfrak{G}$ is an index set $\mathfrak{g} \subseteq \{1, \ldots, n\}$, and $\boldsymbol{x}_{\mathfrak{g}}$ is a sub-vector of $\boldsymbol{x}$ specified by the indices in $\mathfrak{g}$. For example, the group Lasso regularizer arises when we are estimating a vector field on a grid over a two-dimensional vector space. Shrinking each component of the vectors individually through ℓ_1-regularization can produce vectors pointing along either the x-axis or the y-axis but not necessarily sparse as a vector field. We can group the x- and y-components of the vectors and apply the group Lasso regularizer (9.10) to shrink both components of the vectors simultaneously.

The proximity operator corresponding to the group Lasso regularizer can be written blockwise as follows:

$$\operatorname{prox}_\lambda^{\mathfrak{G}}(\boldsymbol{z}) := \left(\max(\|\boldsymbol{z}_{\mathfrak{g}}\| - \lambda, 0) \frac{\boldsymbol{z}_{\mathfrak{g}}}{\|\boldsymbol{z}_{\mathfrak{g}}\|} \right)_{\mathfrak{g} \in \mathfrak{G}}, \tag{9.11}$$

where the ratio $\boldsymbol{z}_{\mathfrak{g}}/\|\boldsymbol{z}_{\mathfrak{g}}\|_2$ is defined to be zero if $\|\boldsymbol{z}_{\mathfrak{g}}\|_2$ is zero. The above expression can be derived in a way analogous to the ℓ_1-case, because the objective (9.6) can be minimized for each block, and from the Cauchy-Schwarz inequality we have

$$\|\boldsymbol{x}_{\mathfrak{g}} - \boldsymbol{z}_{\mathfrak{g}}\|^2 + \lambda\|\boldsymbol{x}_{\mathfrak{g}}\| \geq (\|\boldsymbol{x}_{\mathfrak{g}}\| - \|\boldsymbol{z}_{\mathfrak{g}}\|)^2 + \lambda\|\boldsymbol{x}_{\mathfrak{g}}\|,$$

where the equality is obtained when $\boldsymbol{x}_{\mathfrak{g}} = c\boldsymbol{z}_{\mathfrak{g}}$; the coefficient c can be obtained by solving the one-dimensional minimization.

The last example of sparse regularizers is the trace-norm[3] regularizer, which is defined as

$$\phi_\lambda^{\mathrm{mat}}(\boldsymbol{x}) = \lambda\|\boldsymbol{X}\|_* = \lambda \sum_{j=1}^r \sigma_j(\boldsymbol{X}), \tag{9.12}$$

where $\boldsymbol{X}$ is a matrix obtained by rearranging the elements of $\boldsymbol{x}$ into a matrix of a prespecified size, $\sigma_j(\boldsymbol{X})$ is the jth largest singular value of $\boldsymbol{X}$, and r is the minimum of the number of rows and columns of $\boldsymbol{X}$. The proximity

---

3. The trace norm is also known as the nuclear norm (Boyd and Vandenberghe, 2004) and also as the Ky Fan r-norm (Yuan et al., 2007).

operator corresponding to the trace norm regularizer can be written as

$$\operatorname{prox}_\lambda^{\mathrm{mat}}(\boldsymbol{z}) := \operatorname{vec}\left(\boldsymbol{U} \max(\boldsymbol{S} - \lambda, 0)\boldsymbol{V}^T\right), \tag{9.13}$$

where $\boldsymbol{Z} = \boldsymbol{U}\boldsymbol{S}\boldsymbol{V}^T$ is the singular value decomposition of the matrix $\boldsymbol{Z}$ obtained by appropriately rearranging the elements of $\boldsymbol{z}$. The above expression can again be obtained by using the separability of ϕ_λ as follows:

$$\|\boldsymbol{X} - \boldsymbol{Z}\|_F^2 + \lambda \sum_{j=1}^r \sigma_j(\boldsymbol{X})$$

$$= \sum_{j=1}^r \sigma_j^2(\boldsymbol{X}) - 2\langle \boldsymbol{X}, \boldsymbol{Z} \rangle + \sum_{j=1}^r \sigma_j^2(\boldsymbol{Z}) + \lambda \sum_{j=1}^r \sigma_j(\boldsymbol{Z})$$

$$\geq \sum_{j=1}^r \sigma_j^2(\boldsymbol{X}) - 2 \sum_{j=1}^r \sigma_j(\boldsymbol{X})\sigma_j(\boldsymbol{Z}) + \sum_{j=1}^r \sigma_j^2(\boldsymbol{Z}) + \lambda \sum_{j=1}^r \sigma_j(\boldsymbol{Z})$$

$$= \sum_{j=1}^r \left((\sigma_j(\boldsymbol{X}) - \sigma_j(\boldsymbol{Z}))^2 + \lambda\sigma_j(\boldsymbol{Z})\right),$$

where $\|\cdot\|_F$ denotes the Frobenius norm and the inequality in the second line is due to von Neumann's trace theorem (Horn and Johnson, 1991), for which equality is obtained when the singular vectors of $\boldsymbol{X}$ and $\boldsymbol{Z}$ are the same. Singular values $\sigma_j(\boldsymbol{X})$ are obtained by the one-dimensional minimization in the last line.

Note again that the above three regularizers are *separable*. The ℓ_1-regularizer (9.8) decomposes into the sum of the absolute values of components of $\boldsymbol{x}$. The group Lasso regularizer (9.10) decomposes into the sum of the Euclidean norms of the groups of variables. Finally, the trace norm regularizer (9.12) decomposes into the sum of singular values. Moreover, the proximity operators they define sparsify vectors with respect to the separability of the regularizers, see equations (9.9), (9.11), and (9.13).

Note that the regularizer in the dual of the structured sparse estimation problem (9.5) is also separable, but the corresponding proximity operator does not sparsify a vector, see section 9.4.4 for more discussion.

The sparsity produced by the proximity operator (9.6) is a computational advantage of algorithms that iteratively compute the proximity operator, see Figueiredo and Nowak (2003); Daubechies et al. (2004); Combettes and Wajs (2005); Figueiredo et al. (2007); Wright et al. (2009); Beck and Teboulle (2009); Nesterov (2007). Other methods, such as interior point methods (Koh et al., 2007; Kim et al., 2007; Boyd and Vandenberghe, 2004), achieve sparsity only asymptotically.

9.2.2   Monotone Operator Theory Background

The proximity operator has been studied intensively in the context of monotone operator theory. This framework provides an alternative view on proximity operator-based algorithms and forms the foundation of operator splitting algorithms, which we discuss in section 9.5. In this section, we briefly provide background on monotone operator theory, see Rockafellar (1976a); Lions and Mercier (1979); Eckstein and Bertsekas (1992) for more details.

A nonlinear set-valued operator $T : \mathbb{R}^n \to 2^{\mathbb{R}^n}$ is called *monotone* if $\forall x, x' \in \mathbb{R}^n$,

$$\langle y' - y, x' - x \rangle \geq 0, \qquad \text{for all} \quad y \in T(x), y' \in T(x'),$$

where $\langle y, x \rangle$ denotes the inner product of two vectors $y, x \in \mathbb{R}^n$.

The graph of a set-valued operator T is the set $\{(x, y) : x \in \mathbb{R}^n, y \in T(x)\} \subseteq \mathbb{R}^n \times \mathbb{R}^n$. A monotone operator T is called *maximal* if the graph of T is not strictly contained in that of any other monotone operator on $\mathbb{R}^n$. The subdifferential of a convex function over $\mathbb{R}^n$ is an example of a maximal monotone operator. A set-valued operator T is called *single-valued* if the set $T(x)$ consists of a single vector for every $x \in \mathbb{R}^n$. With a slight abuse of notation we denote $y = T(x)$ in this case. The subdifferential of the function f defined over $\mathbb{R}^n$ is single-valued if and only if f is differentiable. The *sum* of two set-valued operators A and B is defined by the graph $\{(x, y + z) : y \in A(x), z \in B(x), x \in \mathbb{R}^n\}$. The *inverse* T^{-1} of a set-valued operator T is the operator defined by the graph $\{(x, y) : x \in T(y), y \in \mathbb{R}^n\}$.

Denoting the subdifferential of the function f by $T_f := \partial f$, we can rewrite the proximity operator (9.6) as

$$\text{prox}_f(z) = (I + T_f)^{-1}(z), \tag{9.14}$$

where I denotes the identity mapping. The above expression can be derived from the optimality condition $T_f(x) + x - z \ni 0$. Note that the above expression is single-valued, because the minimizer defining the proximity operator (9.6) is unique. Moreover, the monotonicity of the operator T_f guarantees that the proximity operator (9.14) is firmly nonexpansive[4]. Furthermore, $\text{prox}_f(z) = z$ if and only if $0 \in T_f(z)$, because if $z' = \text{prox}_f(z)$, then $z - z' \in T_f(z')$ and $0 \leq \langle z' - z, z - z' - y \rangle$ for all $y \in T_f(z)$.

---

4. An operator T is called *firmly nonexpansive* if $\|y' - y\|^2 \leq \langle x' - x, y' - y \rangle$ holds for all $y \in T(x)$, $y' \in T(x')$, $x, x' \in \mathbb{R}^n$. This is clearly stronger than the ordinary nonexpansiveness defined by $\|y' - y\| \leq \|x' - x\|$.

9.3 Proximal Minimization Algorithm

The proximal minimization algorithm (or the proximal point algorithm) iteratively applies the proximity operator (9.6) to obtain the minimizer of some convex function f. Although in practice it is probably never used in its original form, it functions as a foundation for the analysis of both AL algorithms and operator splitting algorithms.

Let $f : \mathbb{R}^n \to \mathbb{R} \cup \{+\infty\}$ be a convex function that we wish to minimize. Without loss of generality, we focus on unconstrained minimization of f; minimizing a function $f_0(\boldsymbol{x})$ in a convex set C is equivalent to minimizing $f(\boldsymbol{x}) := f_0(\boldsymbol{x}) + \delta_C(\boldsymbol{x})$ where $\delta_C(\boldsymbol{x})$ is the indicator function of C.

A *proximal minimization algorithm* for minimizing f starts from some initial solution $\boldsymbol{x}^0$ and iteratively solves the minimization problem

$$\boldsymbol{x}^{t+1} = \operatorname*{argmin}_{\boldsymbol{x} \in \mathbb{R}^n} \left(f(\boldsymbol{x}) + \frac{1}{2\eta_t} \|\boldsymbol{x} - \boldsymbol{x}^t\|^2 \right). \tag{9.15}$$

The second term in the iteration (9.15) keeps the next iterate $\boldsymbol{x}^{t+1}$ in the proximity of the current iterate $\boldsymbol{x}^t$; the parameter η_t controls the strength of the proximity term. From the above iteration, one can easily see that

$$f(\boldsymbol{x}^{t+1}) \leq f(\boldsymbol{x}^t) - \frac{1}{2\eta_t} \|\boldsymbol{x}^{t+1} - \boldsymbol{x}^t\|^2.$$

Thus, the objective value $f(\boldsymbol{x}^t)$ decreases monotonically as long as $\boldsymbol{x}^{t+1} \neq \boldsymbol{x}^t$.

The iteration (9.15) can also be expressed in terms of the proximity operator (9.6) as follows:

$$\boldsymbol{x}^{t+1} = \operatorname{prox}_{\eta_t f}(\boldsymbol{x}^t) = (I + \eta_t T_f)^{-1}(\boldsymbol{x}^t), \tag{9.16}$$

which is called the *proximal point algorithm* (Rockafellar, 1976a). Since each step is an application of the proximity operator (9.16), it is a firmly nonexpansive mapping for any choice of η_t. Actually, any iterative algorithm that uses a firmly nonexpansive mapping can be considered as a proximal point algorithm (Eckstein and Bertsekas, 1992). Moreover, $\boldsymbol{x}^{t+1} = \boldsymbol{x}^t$ if and only if $0 \in T_f(\boldsymbol{x}^t)$; that is, $\boldsymbol{x}^t$ is a minimizer of f. The connection between minimizing a convex function and finding a zero of a maximal monotone operator can be summarized as in table 9.1.

The iteration (9.15) can also be considered as an *implicit gradient step* because

$$\boldsymbol{x}^{t+1} - \boldsymbol{x}^t \in -\eta_t \partial f(\boldsymbol{x}^{t+1}). \tag{9.17}$$

Table 9.1: Comparison of the proximal minimization algorithm for convex optimization and the proximal point algorithm for solving operator equations

	Convex optimization	Operator equation
Objective	minimize $f(\boldsymbol{x})$	find $0 \in T_f(\boldsymbol{x})$
Algorithm	Proximal minimization algorithm $\boldsymbol{x}^{t+1} = \mathrm{prox}_{\eta_t f}(\boldsymbol{x}^t)$	Proximal point algorithm $\boldsymbol{x}^{t+1} = (I + \eta_t T_f)^{-1}(\boldsymbol{x}^t)$

Note that the subdifferential in the right-hand side is evaluated at the new point $\boldsymbol{x}^{t+1}$.

Rockafellar (1976a) has shown under mild assumptions, which also allow errors in the minimization (9.15), that the sequence $\boldsymbol{x}^0, \boldsymbol{x}^1, \boldsymbol{x}^2, \ldots$ converges[5] to a point $\boldsymbol{x}^\infty$ that satisfies $0 \in T_f(\boldsymbol{x}^\infty)$. Rockafellar (1976a) has also shown that the convergence of the proximal minimization algorithm is super-linear under the assumption that T_f^{-1} is locally Lipschitz around the origin.

The following theorem states the super-linear convergence of the proximal minimization algorithm in a non-asymptotic sense.

Theorem 9.1. *Let $\boldsymbol{x}^0, \boldsymbol{x}^1, \boldsymbol{x}^2 \ldots$ be the sequence generated by the exact proximal minimization algorithm (9.15) and let $\boldsymbol{x}^*$ be a minimizer of the objective function f. Assume that there is a positive constant σ and a scalar α $(1 \leq \alpha \leq 2)$ such that*

$$(\mathbf{A1}) \qquad f(\boldsymbol{x}^{t+1}) - f(\boldsymbol{x}^*) \geq \sigma \|\boldsymbol{x}^{t+1} - \boldsymbol{x}^*\|^\alpha \qquad (t = 0, 1, 2, \ldots).$$

Then the following inequality is true:

$$\|\boldsymbol{x}^{t+1} - \boldsymbol{x}^*\|^{\frac{1+(\alpha-1)\sigma\eta_t}{1+\sigma\eta_t}} \leq \frac{1}{1+\sigma\eta_t} \|\boldsymbol{x}^t - \boldsymbol{x}^*\|.$$

That is, $\boldsymbol{x}^t$ converges to $\boldsymbol{x}^*$ super-linearly if $\alpha < 2$ or $\alpha = 2$ and η_t is increasing, in a global and non-asymptotic sense.

Proof. See Tomioka et al. (2010a). □

Assumption **(A1)** is implied by assuming the strong convexity of f. However, it is weaker because we require **(A1)** only on the points generated by the algorithm. For example, the ℓ_1-regularizer (9.8) is not strongly convex, but it can be lower-bounded, as in assumption **(A1)**, inside any bounded set centered at the origin. In fact, the assumption on the Lipschitz continuity of ∂f^{-1} around the origin used in Rockafellar (1976b) implies assumption **(A1)**

---

5. The original statement was "converges in the weak topology", which is equivalent to strong convergence in a finite dimensional vector space.

due to the nonexpansiveness of the proximity operator (9.16), see Tomioka et al. (2010a) for a detailed discussion.

So far we have ignored the cost of the minimization (9.15). The convergence rate in the above theorem becomes faster as the proximity parameter η_t increases. However, typically the cost of the minimization (9.15) increases as η_t increases. In the next section, we focus on how we can carry out the update step (9.15) efficiently.

9.4   Dual Augmented Lagrangian (DAL) Algorithm

In this section, we introduce the Dual Augmented Lagrangian (DAL) (Tomioka and Sugiyama, 2009; Tomioka et al., 2010a) and show that it is equivalent to the proximal minimization algorithm discussed in the previous section. For the simple sparse estimation problem (9.3) each step in DAL is computationally efficient. Thus it is practical and can be analyzed through the proximal minimization framework.

9.4.1   DAL as Augmented Lagrangian Applied to the Dual Problem

DAL is an application of the augmented Lagrangian (AL) algorithm (Hestenes, 1969; Powell, 1969) to the dual of the simple sparse estimation problem

$$\text{(P)} \qquad \min_{\boldsymbol{x} \in \mathbb{R}^n} \quad f_\ell(\boldsymbol{A}\boldsymbol{x}) + \phi_\lambda(\boldsymbol{x}), \qquad (9.18)$$

where $f_\ell : \mathbb{R}^m \to \mathbb{R}$ is a loss function, which we assume to be a smooth convex function; $\boldsymbol{A} \in \mathbb{R}^{m \times n}$ is a design matrix. Note that we have further introduced a structure $L(\boldsymbol{x}) = f_\ell(\boldsymbol{A}\boldsymbol{x})$ from the simple sparse estimation problem (9.3). This is useful in decoupling the property of the loss function f_ℓ from that of the design matrix $\boldsymbol{A}$. In a machine learning problem, it is easy to discuss properties of the loss function (because we choose it), but we have to live with whatever property is possessed by the design matrix (the data matrix). For notational convenience we assume that for $\eta > 0$, $\eta \phi_\lambda(\boldsymbol{x}) = \phi_{\lambda\eta}(\boldsymbol{x})$; for example, see the ℓ_1-regularizer (9.8).

The dual problem of (P) can be written as the following minimization problem:

$$\text{(D)} \qquad \min_{\boldsymbol{\alpha} \in \mathbb{R}^m, \boldsymbol{v} \in \mathbb{R}^n} \quad f_\ell^*(-\boldsymbol{\alpha}) + \phi_\lambda^*(\boldsymbol{v}), \qquad (9.19)$$

$$\text{subject to} \qquad \boldsymbol{v} = \boldsymbol{A}^T \boldsymbol{\alpha}, \qquad (9.20)$$

where f_ℓ^* and ϕ_λ^* are the convex conjugate functions of f_ℓ and ϕ_λ, respectively.

Let η be a nonnegative real number. The *augmented Lagrangian* (AL) function $J_\eta(\boldsymbol{\alpha}, \boldsymbol{v}; \boldsymbol{x})$ is written as follows:

$$J_\eta(\boldsymbol{\alpha}, \boldsymbol{v}; \boldsymbol{x}) := f_\ell^*(-\boldsymbol{\alpha}) + \phi_\lambda^*(\boldsymbol{v}) + \langle \boldsymbol{x}, \boldsymbol{A}^T \boldsymbol{\alpha} - \boldsymbol{v} \rangle + \frac{\eta}{2} \|\boldsymbol{A}^T \boldsymbol{\alpha} - \boldsymbol{v}\|^2. \tag{9.21}$$

Note that the AL function is reduced to the ordinary Lagrangian if $\eta = 0$; the primal variable $\boldsymbol{x}$ appears in the AL function (9.21) as a Lagrangian multiplier vector; it is easy to verify that $\min_{\boldsymbol{\alpha}, \boldsymbol{v}} J_0(\boldsymbol{\alpha}, \boldsymbol{v}; \boldsymbol{x})$ gives the (sign inverted) primal objective function (9.18).

Similar to the proximal minimization approach discussed in the previous section, we choose a sequence of positive step size parameters $\eta_0, \eta_1, \dots$, and an initial Lagrangian multiplier $\boldsymbol{x}^0$. At every iteration, the DAL algorithm minimizes the AL function $J_{\eta_t}(\boldsymbol{\alpha}, \boldsymbol{v}; \boldsymbol{x}^t)$ (9.21) with respect to $(\boldsymbol{\alpha}, \boldsymbol{v})$, and the minimizer $(\boldsymbol{\alpha}^{t+1}, \boldsymbol{v}^{t+1})$ is used to update the Lagrangian multiplier $\boldsymbol{x}^t$ as follows:

$$(\boldsymbol{\alpha}^{t+1}, \boldsymbol{v}^{t+1}) := \underset{\boldsymbol{\alpha}, \boldsymbol{v}}{\operatorname{argmin}}\, J_{\eta_t}(\boldsymbol{\alpha}, \boldsymbol{v}; \boldsymbol{x}^t), \tag{9.22}$$

$$\boldsymbol{x}^{t+1} := \boldsymbol{x}^t + \eta_t(\boldsymbol{A}^T \boldsymbol{\alpha}^{t+1} - \boldsymbol{v}^{t+1}). \tag{9.23}$$

Intuitively speaking, we minimize an inner objective (9.22) and update (9.23) the Lagrangian multiplier $\boldsymbol{x}^t$ proportionally to the violation of the equality constraint (9.20). In fact, it can be shown that the direction $(\boldsymbol{A}^T \boldsymbol{\alpha}^{t+1} - \boldsymbol{v}^{t+1})$ is the negative gradient direction of the differentiable auxiliary function $f_{\eta_t}(\boldsymbol{x}) := -\min_{\boldsymbol{\alpha}, \boldsymbol{v}} J_{\eta_t}(\boldsymbol{\alpha}, \boldsymbol{v}; \boldsymbol{x})$, which coincides with $f(\boldsymbol{x})$ at the optimum (see Bertsekas (1982)).

Note that the terms in the AL function (9.21) that involve $\boldsymbol{v}$ are linear, quadratic, and the convex conjugate of the regularizer ϕ_λ. Accordingly, by defining *Moreau's envelope function* (see appendix 9.8.2 and also Moreau (1965); Rockafellar (1970)) Φ_λ^* as

$$\Phi_\lambda^*(\boldsymbol{y}) := \min_{\boldsymbol{y}' \in \mathbb{R}^n} \left(\phi_\lambda^*(\boldsymbol{y}') + \frac{1}{2} \|\boldsymbol{y} - \boldsymbol{y}'\|^2 \right), \tag{9.24}$$

we can rewrite the update equations (9.22) and (9.23) as follows:

$$\boldsymbol{\alpha}^{t+1} = \underset{\boldsymbol{\alpha} \in \mathbb{R}^m}{\operatorname{argmin}} \Big(\underbrace{f_\ell^*(-\boldsymbol{\alpha}) + \frac{1}{\eta_t} \Phi_{\lambda\eta_t}^*(\boldsymbol{x}^t + \eta_t \boldsymbol{A}^T \boldsymbol{\alpha})}_{=:\varphi_t(\boldsymbol{\alpha})} \Big), \tag{9.25}$$

$$\boldsymbol{x}^{t+1} := \operatorname{prox}_{\phi_{\lambda\eta_t}} \big(\boldsymbol{x}^t + \eta_t \boldsymbol{A}^T \boldsymbol{\alpha}^{t+1} \big), \tag{9.26}$$

where we use the identity $\operatorname{prox}_f(\boldsymbol{x}) + \operatorname{prox}_{f^*}(\boldsymbol{x}) = \boldsymbol{x}$ (see appendix 9.8.2). See Tomioka and Sugiyama (2009); Tomioka et al. (2010a) for the derivation.

9.4.2   DAL as a Primal Proximal Minimization

The following proposition states that the DAL algorithm is equivalent to the proximal minimization algorithm in the primal (and thus the algorithm is stable for any positive step size η_t); see also table 9.2.

Proposition 9.2. *The iteration* (9.25)-(9.26) *is equivalent to the proximal minimization algorithm* (9.15) *on the primal problem (P).*

Proof. The proximal minimization algorithm for the problem (9.18) is written as follows:

$$
\begin{aligned}
\boldsymbol{x}^{t+1} &:= \operatorname*{argmin}_{\boldsymbol{x}\in\mathbb{R}^n} \left(f_\ell(\boldsymbol{A}\boldsymbol{x}) + \phi_\lambda(\boldsymbol{x}) + \frac{1}{2\eta_t}\|\boldsymbol{x} - \boldsymbol{x}^t\|^2 \right)\\
&= \operatorname*{argmin}_{\boldsymbol{x}\in\mathbb{R}^n} \left(f_\ell(\boldsymbol{A}\boldsymbol{x}) + \frac{1}{\eta_t}\left(\phi_{\lambda\eta_t}(\boldsymbol{x}) + \frac{1}{2}\|\boldsymbol{x} - \boldsymbol{x}^t\|^2 \right) \right).
\end{aligned}
$$

Now we define

$$
\Phi_\lambda(\boldsymbol{x};\boldsymbol{x}_t) := \phi_\lambda(\boldsymbol{x}) + \frac{1}{2}\|\boldsymbol{x} - \boldsymbol{x}^t\|^2 \tag{9.27}
$$

and use the Fenchel duality to obtain

$$
\min_{\boldsymbol{x}\in\mathbb{R}^n}\left(f_\ell(\boldsymbol{A}\boldsymbol{x}) + \frac{1}{\eta_t}\Phi_{\lambda\eta_t}(\boldsymbol{x};\boldsymbol{x}^t) \right) = \max_{\boldsymbol{\alpha}\in\mathbb{R}^m}\left(-f_\ell^*(-\boldsymbol{\alpha}) - \frac{1}{\eta_t}\Phi_{\lambda\eta_t}^*(\eta_t\boldsymbol{A}^T\boldsymbol{\alpha};\boldsymbol{x}^t) \right), \tag{9.28}
$$

where f_ℓ^* and $\Phi_\lambda^*(\cdot;\boldsymbol{x}^t)$ are the convex conjugate functions of f_ℓ and $\Phi_\lambda(\cdot;\boldsymbol{x}^t)$, respectively. Here, since $\Phi_\lambda(\cdot;\boldsymbol{x}^t)$ is a sum of two convex functions, its convex conjugate is the infimal convolution (see appendix 9.8.1) of the convex conjugates, that is,

$$
\Phi_\lambda^*(\boldsymbol{y};\boldsymbol{x}^t) = \inf_{\tilde{\boldsymbol{v}}\in\mathbb{R}^n}\left(\phi_\lambda^*(\tilde{\boldsymbol{v}}) + \frac{1}{2}\|\boldsymbol{y} - \tilde{\boldsymbol{v}}\|^2 + \langle \boldsymbol{y} - \tilde{\boldsymbol{v}},\, \boldsymbol{x}^t \rangle \right). \tag{9.29}
$$

Since $\Phi_\lambda^*(\boldsymbol{y};\boldsymbol{x}^t) = \Phi_\lambda^*(\boldsymbol{x}^t + \boldsymbol{y};\boldsymbol{0}) = \Phi_\lambda^*(\boldsymbol{x}^t + \boldsymbol{y})$, ignoring a constant term that does not depend on $\boldsymbol{y}$, we have the inner minimization problem (9.25). In order to obtain the update equation (9.26), we turn back to the Fenchel duality theorem and notice that the minimizer $\boldsymbol{x}^{t+1}$ in the left-hand side of equation (9.28) satisfies

$$
\boldsymbol{x}^{t+1} \in \partial_{\boldsymbol{y}}\Phi_{\lambda\eta_t}^*(\boldsymbol{y};\boldsymbol{x}^t)\big|_{\boldsymbol{y}=\eta_t\boldsymbol{A}^T\boldsymbol{\alpha}^{t+1}}.
$$

Since $\Phi_\lambda^*(\boldsymbol{y};\boldsymbol{x}^t)$ is Moreau's envelope function of ϕ_λ^* (ignoring constants), it is differentiable and the derivative $\nabla_{\boldsymbol{y}}\Phi_{\lambda\eta_t}^*(\eta_t\boldsymbol{A}^T\boldsymbol{\alpha}^{t+1};\boldsymbol{x}^t)$ is given as follows

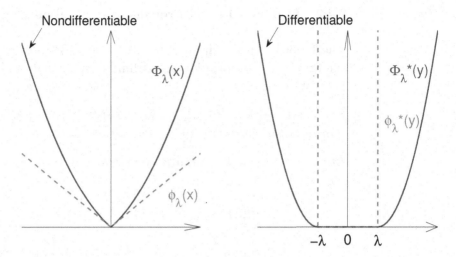

Figure 9.1: Comparison of $\Phi_\lambda(x; 0)$ (left) and $\Phi_\lambda^*(y; 0)$ (right) for the one-dimensional ℓ_1-regularizer $\phi_\lambda(x) = \lambda|x|$.

(see appendix 9.8.2):

$$
\begin{aligned}
\boldsymbol{x}^{t+1} &= \nabla_{\boldsymbol{y}} \Phi_{\lambda\eta_t}^* (\eta_t \boldsymbol{A}^T \boldsymbol{\alpha}^{t+1}; \boldsymbol{x}^t) \\
&= \nabla_{\boldsymbol{y}} \Phi_{\lambda\eta_t}^* (\boldsymbol{x}^t + \eta_t \boldsymbol{A}^T \boldsymbol{\alpha}^{t+1}; \boldsymbol{0}) = \text{prox}_{\phi_{\lambda\eta_t}} (\boldsymbol{x}^t + \eta_t \boldsymbol{A}^T \boldsymbol{\alpha}^{t+1}),
\end{aligned}
$$

from which we have the update equation (9.26).   □

The equivalence of proximal minimization and augmented Lagrangian we have shown above is not novel; it can be found, for example, in Rockafellar (1976b); Ibaraki et al. (1992). However, the above derivation can easily be generalized to the case when the loss function f_ℓ is not differentiable (Suzuki and Tomioka, 2010).

It is worth noting that $\Phi_\lambda(\cdot; \boldsymbol{x}^t)$ is not differentiable but $\Phi_\lambda^*(\cdot; \boldsymbol{x}^t)$ is. See figure 9.1 for a schematic illustration of the case of the one-dimensional ℓ_1-regularizer. Both the ℓ_1-regularizer ϕ_λ and its convex conjugate ϕ_λ^* are nondifferentiable at some points. The function $\Phi_\lambda(\boldsymbol{x}) := \Phi_\lambda(\boldsymbol{x}; \boldsymbol{0})$ is obtained by adding a quadratic proximity term to ϕ_λ (see equation (9.27)). Although Φ_λ is still nondifferentiable, its convex conjugate Φ_λ^* is differentiable due to the infimal convolution operator (see appendix 9.8.1) with the proximity term (see Equation (9.29)).

The differentiability of Moreau's envelope function Φ_λ^* makes the DAL approach (9.25)-(9.26) computationally efficient. At every step, we minimize a differentiable inner objective (9.25) and use the minimizer to compute the update step (9.26).

9.4.3   Exemplary Instance: ℓ_1-Regularizer

In order to understand the efficiency of minimizing the inner objective (9.25), let us consider the simplest sparse estimation problem: the ℓ_1-regularization.

For the ℓ_1-regularizer, $\phi_\lambda(\boldsymbol{x}) = \lambda\|\boldsymbol{x}\|_1$, the update equations (9.25) and (9.26) can be rewritten as follows:

$$\boldsymbol{\alpha}^{t+1} = \underset{\boldsymbol{\alpha}\in\mathbb{R}^m}{\operatorname{argmin}}\Big(\underbrace{f_\ell^*(-\boldsymbol{\alpha}) + \frac{1}{2\eta_t}\Big\|\operatorname{prox}_{\lambda\eta_t}^{\ell_1}(\boldsymbol{x}^t + \eta_t\boldsymbol{A}^T\boldsymbol{\alpha})\Big\|^2}_{=:\varphi_t(\boldsymbol{\alpha})}\Big), \qquad (9.30)$$

$$\boldsymbol{x}^{t+1} = \operatorname{prox}_{\lambda\eta_t}^{\ell_1}\big(\boldsymbol{x}^t + \eta_t\boldsymbol{A}^T\boldsymbol{\alpha}^{t+1}\big), \qquad (9.31)$$

where $\operatorname{prox}_\lambda^{\ell_1}$ is the soft threshold function (9.9); see Tomioka and Sugiyama (2009) and Tomioka et al. (2010a) for the derivation.

Note that the second term in the inner objective function $\varphi_t(\boldsymbol{\alpha})$ (9.30) is the squared sum of n one-dimensional soft thresholds. Thus we only need to compute the sum over the active components $\mathcal{J}^+ := \{j : |x_j^t(\boldsymbol{\alpha})| > \lambda\eta_t\}$ where $\boldsymbol{x}^t(\boldsymbol{\alpha}) := \boldsymbol{x}^t + \eta_t\boldsymbol{A}^T\boldsymbol{\alpha}$. In fact,

$$\Big\|\operatorname{prox}_{\lambda\eta_t}^{\ell_1}(\boldsymbol{x}^t(\boldsymbol{\alpha}))\Big\|^2 = \sum_{j=1}^n (\operatorname{prox}_{\lambda\eta_t}^{\ell_1}(x_j^t(\boldsymbol{\alpha})))^2 = \sum_{j\in\mathcal{J}^+}(\operatorname{prox}_{\lambda\eta_t}^{\ell_1}(x_j^t(\boldsymbol{\alpha})))^2.$$

Note that the flat area in the plot of $\Phi_\lambda^*(y)$ in figure 9.1 corresponds to an inactive component.

Moreover, the gradient and the Hessian of $\varphi_t(\boldsymbol{\alpha})$ can be computed as follows:

$$\nabla\varphi_t(\boldsymbol{\alpha}) = -\nabla f_\ell^*(-\boldsymbol{\alpha}) + \boldsymbol{A}\operatorname{prox}_{\lambda\eta_t}^{\ell_1}(\boldsymbol{x}^t + \eta_t\boldsymbol{A}^T\boldsymbol{\alpha}),$$

$$\nabla^2\varphi_t(\boldsymbol{\alpha}) = \nabla^2 f_\ell^*(-\boldsymbol{\alpha}) + \eta_t\boldsymbol{A}_+\boldsymbol{A}_+^T,$$

where $\boldsymbol{A}_+$ is the submatrix of $\boldsymbol{A}$ that consists of columns of $\boldsymbol{A}$ that correspond to the active components $\mathcal{J}^+$. Again, notice that only the active components enter the computation of the gradient and the Hessian.

Looking at figure 9.1 carefully, one might wonder what happens if the minimizer $\boldsymbol{\alpha}^{t+1}$ lands on a point where $\Phi_\lambda^*(\boldsymbol{y})$ starts to diverge from $\phi_\lambda^*(\boldsymbol{y})$ ($y = -\lambda, \lambda$ in figure 9.1). In fact, the second derivative of Φ_λ^* is discontinuous on such a point. Nevertheless, we can show that such an event is rare as in the following theorem.

Theorem 9.3. *Assume the regularizer $\phi_\lambda(\boldsymbol{x}) = \lambda\sum_{j=1}^n |x_j|$ (ℓ_1-regularizer). A minimizer $\boldsymbol{x}^*$ of the objective (9.18) has no component located exactly at the threshold λ for most λ in the sense that it can be avoided by an arbitrary small perturbation of λ.*

Proof. The optimality condition for the objective (9.18) with the ℓ_1-regularizer can be written as

$$\boldsymbol{x}^* = \text{prox}_\lambda^{\ell_1}(\boldsymbol{x}^* + \boldsymbol{v}^*), \qquad \boldsymbol{v}^* = -\boldsymbol{A}^T \nabla f_\ell(\boldsymbol{A}\boldsymbol{x}^*),$$

which implies $\|\boldsymbol{v}\|_\infty \leq \lambda$ and the complementary slackness conditions

$$x_j \geq 0 \quad \text{if} \quad v_j = \lambda, \tag{9.32a}$$

$$x_j = 0 \quad \text{if} \quad -\lambda < v_j < \lambda, \tag{9.32b}$$

$$x_j \leq 0 \quad \text{if} \quad v_j = -\lambda, \tag{9.32c}$$

for all $j = 1, \ldots, n$. Since the event $x_j = 0$ and $v_j = -\lambda$, or $x_j = 0$ and $v_j = \lambda$, can be avoided by an arbitrary small perturbation of λ for a generic design matrix $\boldsymbol{A}$ and a differentiable loss function f_ℓ, either $x_j^* + v_j^* > \lambda$ (9.32a), $-\lambda < x_j^* + v_j^* < \lambda$ (9.32b), or $x_j^* + v_j^* < -\lambda$ (9.32c) holds, which concludes the proof. $\qquad\qquad\qquad\qquad\qquad\qquad\qquad\square$

The above theorem guarantees that the inner objective (9.30) behaves like a twice differentiable function around the optimum for a generic choice of λ and $\boldsymbol{A}$. The theorem can immediately be generalized to the group Lasso regularizer (9.10) and the trace-norm regularizer (9.12) by appropriately defining the complementary slackness conditions (9.32a)–(9.32c).

9.4.4   Why Do We Apply the AL Method to the Dual?

One reason for applying the AL method to the dual problem (D) is that some loss functions are strongly convex only in the dual; for instance, the logistic loss, which is not strongly convex, becomes strongly convex by taking the convex conjugate. In general loss functions with Lipschitz continuous gradients become strongly convex in the dual (see also section 9.4.5).

Another reason is that the inner objective function does not have the sparsity discussed in section 9.4.3 when the AL method is applied to the primal. In fact, applying the AL method to the primal problem (P) is equivalent to applying the proximal minimization algorithm to the dual problem (D). Therefore, for the ℓ_1-case, the regularizer $\phi_\lambda(\boldsymbol{x})$ is defined as

$$\phi_\lambda(\boldsymbol{x}) := (\phi_\lambda^{\ell_1})^*(\boldsymbol{x}) = \begin{cases} 0 & (\text{if } \|\boldsymbol{x}\|_\infty \leq \lambda), \\ +\infty & (\text{otherwise}), \end{cases}$$

which is the convex conjugate of the ℓ_1-regularizer $\phi_\lambda^{\ell_1}$. Adding a quadratic proximity term, we obtain Φ_λ. By taking the convex conjugate of ϕ_λ and of Φ_λ, we obtain the ℓ_1-regularizer $\phi_\lambda^* := \phi_\lambda^{\ell_1}$ and Moreau's envelope function Φ_λ^* of the ℓ_1-regularizer (see figure 9.2).

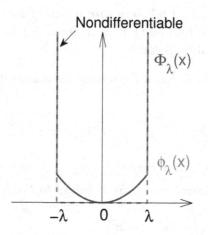
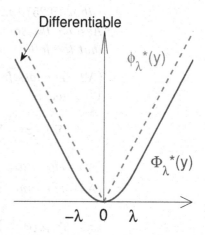

Figure 9.2: Comparison of $\Phi_\lambda(x)$ (left) and $\Phi_\lambda^*(y)$ (right) for the primal application of the AL method to the one-dimensional ℓ_1-problem.

Now, from figure 9.2, we can see that the envelope function $\Phi_\lambda^*(y)$ is quadratic for $|y| \leq \lambda$, which corresponds to inactive components and is linear for $|y| > \lambda$, which corresponds to active components. Thus, we need to compute the terms in the envelope function Φ_λ^* that correspond to both the active and the inactive components. Moreover, for the active components the envelope function behaves like a linear function around the minimum, which might be difficult to optimize, especially when combined with a loss function that is not strongly convex.

9.4.5 Super-linear Convergence of DAL

The asymptotic convergence rate of the DAL approach is guaranteed by classic results (see Rockafellar (1976a); Kort and Bertsekas (1976)) under mild conditions even when the inner minimization (9.25) is carried out only approximately. However, the condition to stop the inner minimization proposed in Rockafellar (1976a) is often difficult to check in practice. In addition, the analysis in Kort and Bertsekas (1976) assumes strong convexity of the objective. In our setting, the dual objective (9.19) is not necessarily strongly convex as a function of α and v; thus we cannot directly apply the result of Kort and Bertsekas (1976) to our problem, though the result is very similar to ours.

Here we provide a non-asymptotic convergence rate of DAL, which generalizes theorem 9.1 to allow for approximate inner minimization (9.25) with a practical stopping criterion.

Theorem 9.4. *Let $x^1, x^2, \ldots$ be the sequence generated by the DAL algo-*

rithm (9.25)-(9.26) and let $\boldsymbol{x}^*$ be a minimizer of the objective function f. Assume the same condition (A1) as in theorem 9.1 and in addition assume that the following conditions hold:

(A2) The loss function f_ℓ has a Lipschitz continuous gradient with modulus $1/\gamma$, that is,

$$\|\nabla f_\ell(\boldsymbol{z}) - \nabla f_\ell(\boldsymbol{z}')\| \le \frac{1}{\gamma}\|\boldsymbol{z} - \boldsymbol{z}'\| \qquad (\forall \boldsymbol{z}, \boldsymbol{z}' \in \mathbb{R}^m). \tag{9.33}$$

(A3) The proximity operator corresponding to ϕ_λ can be computed exactly.

(A4) The inner minimization (9.25) is solved to the following tolerance:

$$\|\nabla \varphi_t(\boldsymbol{\alpha}^{t+1})\| \le \sqrt{\frac{\gamma}{\eta_t}}\|\boldsymbol{x}^{t+1} - \boldsymbol{x}^t\|,$$

where γ is the constant in assumption (A2).

Under assumptions (A1)-(A4), the following inequality is true:

$$\|\boldsymbol{x}^{t+1} - \boldsymbol{x}^*\|^{\frac{1+\alpha\sigma\eta_t}{1+2\sigma\eta_t}} \le \frac{1}{\sqrt{1+2\sigma\eta_t}}\|\boldsymbol{x}^t - \boldsymbol{x}^*\|.$$

That is, $\boldsymbol{x}^t$ converges to $\boldsymbol{x}^*$ super-linearly if $\alpha < 2$ or $\alpha = 2$ and η_t is increasing.

Proof. See Tomioka et al. (2010a). $\qquad\qquad\qquad\qquad\qquad\qquad\qquad\square$

Note that the above stopping criterion (A4) is computable, since the Lipschitz constant γ depends only on the loss function used and not on the data matrix $\boldsymbol{A}$. Although the constant σ in assumption (A1) is difficult to quantify in practice, it is enough to know that it exists, because we do not need σ to compute the stopping criterion (A4). See Tomioka et al. (2010a) for more details.

9.5   Connections

The AL formulation in the dual is connected to various operator theoretic algorithms in the primal. We have already seen that the exact application of DAL corresponds to the proximal point algorithm in the primal (section 9.4.2). In this section, we show that two well-known *operator splitting* algorithms—forward-backward splitting and Douglas-Rachford splitting in the primal—can be regarded as approximate computations of the DAL approach. The results in this section are not novel and are based on Lions and Mercier (1979); Eckstein and Bertsekas (1992); Tseng (1991), see also recent

Table 9.2: Primal-dual correspondence of operator splitting algorithms and augmented Lagrangian algorithms

	Primal	Dual
Exact	Proximal minimization	Augmented Lagrangian algorithm (Rockafellar, 1976b)
Approximation	Forward-backward splitting	Alternating minimization algorithm (Tseng, 1991)
	Douglas-Rachford splitting	Alternating direction method of multipliers (Gabay and Mercier, 1976)

reviews in Yin et al. (2008); Setzer (2010); Combettes and Pesquet (2010). The methods we discuss in this section are summarized in table 9.2.

Note that these approximations are most valuable when the inner minimization problem (9.22) is not easy to minimize. In Goldstein and Osher (2009), an approximate AL method was applied to a *structured* sparse estimation problem, namely, the total variation denoising.

In this section we use the notation $L(\boldsymbol{x}) = f_\ell(\boldsymbol{A}\boldsymbol{x})$ for simplicity, since the discussions do not require the separation between the loss function and the design matrix as in section 9.4.

9.5.1   Forward-Backward Splitting

When the loss function L is differentiable, replacing the inner minimization (9.22) with the following sequential minimization steps

$$\boldsymbol{\alpha}^{t+1} = \underset{\boldsymbol{\alpha}\in\mathbb{R}^m}{\operatorname{argmin}} J_0(\boldsymbol{\alpha}, \boldsymbol{v}^t; \boldsymbol{x}^t), \tag{9.34}$$

$$\boldsymbol{v}^{t+1} = \underset{\boldsymbol{v}\in\mathbb{R}^n}{\operatorname{argmin}} J_{\eta_t}(\boldsymbol{\alpha}^{t+1}, \boldsymbol{v}; \boldsymbol{x}^t) \tag{9.35}$$

gives the forward-backward splitting (FBS) algorithm (Lions and Mercier, 1979; Combettes and Wajs, 2005; Combettes and Pesquet, 2010):

$$\boldsymbol{x}^{t+1} = \operatorname{prox}_{\phi_{\lambda\eta_t}}\left(\boldsymbol{x}^t - \eta_t\nabla L(\boldsymbol{x}^t)\right). \tag{9.36}$$

Note that in the first step (9.34), the ordinary Lagrangian ($\eta = 0$) is used and the augmented Lagrangian is used only in the second step (9.35). The above sequential procedure is proposed in Han and Lou (1988) and analyzed in Tseng (1991) under the name "alternating minimization algorithm".

The FBS algorithm was proposed in the context of finding a zero of the operator equation (9.2). When the operator A is single valued, the operator equation (9.2) implies

$$(I + \eta B)(\boldsymbol{x}) \ni (I - \eta A)(\boldsymbol{x}).$$

This motivates us to use the iteration

$$\boldsymbol{x}^{t+1} = (I + \eta B)^{-1}(I - \eta A)(\boldsymbol{x}^t).$$

The above iteration converges to the solution of the operator equation (9.2) if A is Lipschitz continuous and the step size η is small enough (see Lions and Mercier (1979); Combettes and Wajs (2005)). The iteration (9.36) is obtained by identifying $A = \nabla L$ and $B = \partial \phi_\lambda$. Intuitively, the FBS algorithm takes an *explicit* (forward) gradient step with respect to the differentiable term L and then takes an *implicit* (backward) gradient step (9.17) with respect to the nondifferentiable term ϕ_λ.

The FBS algorithm is also known as the iterative shrinkage/thresholding (IST) algorithm (see Figueiredo and Nowak (2003); Daubechies et al. (2004); Figueiredo et al. (2007); Wright et al. (2009); Beck and Teboulle (2009) and the references therein). The FBS algorithm converges as fast as the gradient descent on the loss term in problem (9.3). For example, when the loss term has a Lipschitz continuous gradient and is strongly convex, it converges linearly (Tseng, 1991). However, this is rarely the case in sparse estimation because typically the number of unknowns n is larger than the number of observations m. Beck and Teboulle (2009) proved that FBS converges as $O(1/k)$ without the strong convexity assumption. However, since the Lipschitz constant depends on the design matrix $\boldsymbol{A}$, it is difficult to quantify it for a machine learning problem. Nesterov (2007) and Beck and Teboulle (2009) proposed accelerated IST algorithms that converge as $O(1/k^2)$, which is also optimal under the first-order black-box model (Nesterov, 2007). The connection between the accelerated IST algorithm and the operator splitting framework is unknown.

9.5.2   Douglas-Rachford Splitting

Another commonly used approximation to minimize the inner objective function (9.22) is to perform minimization with respect to $\boldsymbol{\alpha}$ and $\boldsymbol{v}$ alternately, which is called the *alternating direction method of multipliers* (Gabay and Mercier, 1976). This approach is known to be equivalent to the Douglas-Rachford splitting (DRS) algorithm (Douglas Jr. and Rachford Jr., 1956; Lions and Mercier, 1979; Eckstein and Bertsekas, 1992; Combettes and Pesquet, 2010) when the proximity parameter η_t is chosen to be constant $\eta_t = \eta$.

Similar to the FBS algorithm, the DRS algorithm splits the operator equation (9.2) as follows:

$$(I + \eta B)(\boldsymbol{x}) \ni \boldsymbol{x} - \eta \boldsymbol{y}, \qquad (I + \eta A)(\boldsymbol{x}) \ni \boldsymbol{x} + \eta \boldsymbol{y}.$$

Accordingly, starting from some appropriate initial point $(\boldsymbol{x}^0, \boldsymbol{y}^0)$, the DRS algorithm performs the iteration

$$\left(\boldsymbol{x}^{t+1}, \eta \boldsymbol{y}^{t+1}\right) = \text{decomp}_{\eta A}\left((I + \eta B)^{-1}(\boldsymbol{x}^t - \eta \boldsymbol{y}^t) + \eta \boldsymbol{y}^t\right),$$

where with a slight abuse of notation, we denote by $(\boldsymbol{x}, \boldsymbol{y}) = \text{decomp}_A(\boldsymbol{z})$ the decomposition $\boldsymbol{x} + \boldsymbol{y} = \boldsymbol{z}$ with $\boldsymbol{x} = (I + A)^{-1}(\boldsymbol{z})$. Note that this implies $\boldsymbol{y} \in A(\boldsymbol{x})$; see the original definition (9.7).

Turning back to the DAL algorithm (9.22)-(9.23), due to the symmetry between $\boldsymbol{\alpha}$ and $\boldsymbol{v}$, there are two ways to convert the DAL algorithm to a DRS algorithm. First, by replacing the inner minimization (9.22) with the steps

$$\boldsymbol{v}^{t+1} = \underset{\boldsymbol{v} \in \mathbb{R}^n}{\text{argmin}}\, J_\eta(\boldsymbol{\alpha}^t, \boldsymbol{v}; \boldsymbol{x}^t), \qquad \boldsymbol{\alpha}^{t+1} = \underset{\boldsymbol{\alpha} \in \mathbb{R}^m}{\text{argmin}}\, J_\eta(\boldsymbol{\alpha}, \boldsymbol{v}^{t+1}; \boldsymbol{x}^t),$$

we obtain the (primal) DRS algorithm:

$$\left(\boldsymbol{x}^{t+1}, -\eta \boldsymbol{A}^T \boldsymbol{\alpha}^{t+1}\right) = \text{decomp}_{\eta L}\left(\text{prox}_{\phi_{\lambda\eta}}\left(\boldsymbol{x}^t + \eta \boldsymbol{A}^T \boldsymbol{\alpha}^t\right) - \eta \boldsymbol{A}^T \boldsymbol{\alpha}^t\right),$$
(9.37)

where $(\boldsymbol{x}, \boldsymbol{y}) = \text{decomp}_{\eta L}(\boldsymbol{z})$ denotes Moreau's decomposition (9.7). We can identify $A = \partial L$ and $B = \partial \phi_\lambda$ in update equation (9.37). This version of DRS (regularizer inside, loss outside) was considered in Combettes and Pesquet (2007) for image denoising with non-Gaussian likelihood models. When the loss function L is differentiable, the update equation (9.37) can be simplified as follows:

$$\boldsymbol{x}^{t+1} = \text{prox}_{\eta L}\left(\text{prox}_{\phi_{\lambda\eta}}(\boldsymbol{x}^t - \eta \nabla L(\boldsymbol{x}^t)) + \eta \nabla L(\boldsymbol{x}^t)\right),$$

which more closely resembles the FBS iteration (9.36).

On the other hand, by replacing the inner minimization (9.22) with the steps

$$\boldsymbol{\alpha}^{t+1} = \underset{\boldsymbol{\alpha} \in \mathbb{R}^m}{\text{argmin}}\, J_\eta(\boldsymbol{\alpha}, \boldsymbol{v}^t; \boldsymbol{x}^t),$$

$$\boldsymbol{v}^{t+1} = \underset{\boldsymbol{v} \in \mathbb{R}^n}{\text{argmin}}\, J_\eta(\boldsymbol{\alpha}^{t+1}, \boldsymbol{v}; \boldsymbol{x}^t),$$

we obtain another (primal) DRS algorithm:

$$\left(x^{t+1}, \eta v^{t+1}\right) = \mathrm{decomp}_{\phi_{\lambda\eta}}\left(\mathrm{prox}_{\eta L}(x^t - \eta v^t) + \eta v^t\right). \tag{9.38}$$

Here, we can identify $A = \partial \phi_\lambda$ and $B = \partial L$ in the update equation (9.38). This version of DRS (loss inside, regularizer outside) was proposed by Goldstein and Osher (2009) as an alternating direction method for the total variation denoising problem (9.5).

Each step of DRS is a firmly nonexpansive mapping, and thus DRS is unconditionally stable (Lions and Mercier, 1979), whereas the stability of FBS depends on the choice of the proximity parameter η. Moreover, DRS can be applied in both ways (see update equations (9.37) and (9.38)). In other words, both the loss function L and the regularizer ϕ_λ may be nondifferentiable, whereas FBS assumes that the loss function L is differentiable. However, this also means that both proximity operators need to be implemented for DRS, whereas FBS requires only one of them (Combettes and Pesquet, 2010).

9.6   Application

In this section, we demonstrate that the trace norm regularizer (9.12) can be used to learn features from multiple sources and combine them in an optimal way in a single optimization problem. We also demonstrate that DAL can efficiently optimize the associated minimization problem.

9.6.1   Problem setting

The problem we solve is a classification problem with multiple matrix-valued inputs (Tomioka et al., 2010b):

$$\min_{\substack{W^{(1)},\ldots,W^{(K)}, \\ b \in \mathbb{R}}} \sum_{i=1}^{m} \ell\left(\sum_{k=1}^{K} \langle X_i^{(k)}, W^{(k)} \rangle + b, y_i\right) + \lambda \sum_{k=1}^{K} \|W^{(k)}\|_*, \tag{9.39}$$

where the loss function ℓ is the logistic loss function

$$\ell(z, y) = \log(1 + \exp(-yz)), \tag{9.40}$$

and $\|\cdot\|_*$ denotes the trace norm (9.12).

By defining

$$\boldsymbol{x} = \left(\mathrm{vec}(\boldsymbol{W}^{(1)})^T, \ldots, \mathrm{vec}(\boldsymbol{W}^{(K)})^T, b\right)^T,$$

$$f_\ell(\boldsymbol{z}) = \sum_{i=1}^{m} \ell(z_i, y_i),$$

$\boldsymbol{A}$: an $m \times n$ matrix whose ith row is given as

$$\boldsymbol{A}_i = \left(\mathrm{vec}(\boldsymbol{X}_i^{(1)})^T, \ldots, \mathrm{vec}(\boldsymbol{X}_i^{(K)})^T, 1\right),$$

$$\phi_\lambda(\boldsymbol{x}) = \lambda \sum_{k=1}^{K} \|\boldsymbol{W}^{(k)}\|_*,$$

we can see that problem (9.39) is a special case of problem (9.18).

As a concrete example, we take a data set from a real brain-computer interface (BCI) experiment, where the task is to predict whether the upcoming voluntary finger movement is either right or left hand from the electroencephalography (EEG) measurements (Blankertz et al., 2002). The data set is publicly available through the BCI competition 2003 (data set IV) (Blankertz et al., 2004). More specifically, the data set consists of short segments of 28 channel multivariate signals of length 50 (500 ms long at 100 Hz sampling). The training set consists of $m = 316$ input segments (159 left and 157 right), and we tested the classifier on a separate test set consisting of 100 test segments.

Following the preprocessing used in Tomioka and Müller (2010), we compute three matrices from each segment. The first matrix is 28×50 and is obtained directly from the original signal by low-pass filtering at 20Hz. The second matrix is 28×28 and is derived by computing the covariance between the channels in the frequency band 7–15Hz (known as the α-band). Finally, the third matrix is 28×28 and is computed similarly to the second matrix in the frequency band 15–30Hz (known as the β-band). The total number of unknown variables is $n = 2969$.

We chose 20 log-linearly separated values of the regularization constant λ from 10 to 0.001. The proximity parameter is increased geometrically as $\eta_t = 1, 2, 4, 8, \ldots$; after 22 iterations it was as large as $2^{21} \simeq 2.1 \times 10^6$, which shows that DAL is stable across a wide range of η_t. The Lipschitz constant γ (see assumption (A2) in theorem 9.4) for the logistic loss (9.40) is $\gamma = 4$. We used the Newton method for the inner minimization problem (9.25). We implemented DAL in Matlab[6]. Each optimization was terminated when the

---

6. The code is available from http://www.ibis.t.u-tokyo.ac.jp/RyotaTomioka/Softwares.

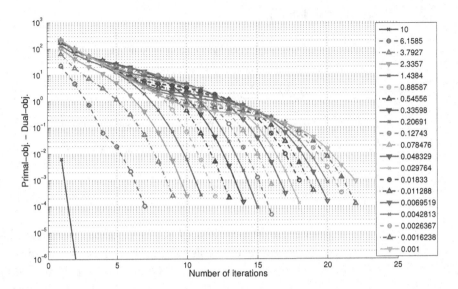

Figure 9.3: Convergence of DAL algorithm applied to a classification problem in BCI. The duality gap is plotted against the number of iterations. Each curve corresponds to a different regularization constant λ (shown on the right). Note that no warm start is used. Each iteration consumed roughly 1.2 seconds.

duality gap fell below 10^{-3}; see section 9.8.3.

9.6.2 Results

Figure 9.3 shows the sequence of the duality gap obtained by running the DAL algorithm on 20 different values of the regularization constant λ against the number of iterations. Note that the vertical axis is logarithmically scaled. We can see that the convergence of DAL becomes faster as the iteration proceeds; that is, it converges super-linearly. Each iteration consumed roughly 1.2 seconds on a Linux server with two 3.33 GHz Xeon processors, and the computation for 20 values of the regularization constant λ took about 350 seconds. Note that applying a simple warm start can significantly speedup the computation (about 70 percent reduction), but it is not used here because we are interested in the basic behavior of the DAL algorithm.

Figure 9.4 shows the singular value spectra of the coefficient matrices $W^{(1)}$, $W^{(2)}$, and $W^{(3)}$ obtained at the regularization constant $\lambda = 0.5456$, which achieved the highest test accuracy, 85 percent. The classifier has selected three components from the first data source (first-order component), four components from the second data source (second-order (α-band) component), and five components from the third data source (second-order (β-band) component). From the magnitude of the singular values, it seems

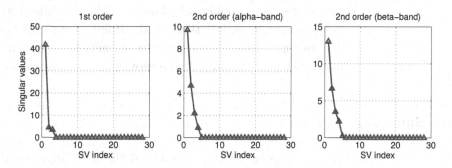

Figure 9.4: Singular value spectra of $\boldsymbol{W}^{(1)}$, $\boldsymbol{W}^{(2)}$, and $\boldsymbol{W}^{(3)}$, which correspond to the first-order component, the second-order (alpha) component, and the second-order (beta) component, respectively, obtained by solving optimization problem (9.39) at $\lambda = 0.5456$.

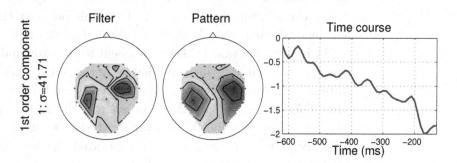

Figure 9.5: The visualization of the left singular vector (filter) and the right singular vector (time course) corresponding to the largest singular value of $\boldsymbol{W}^{(1)}$. Both filter and pattern are shown topographically on a head seen from above. The pattern shows the typical activity captured by the filter. See Tomioka and Müller (2010) for more details.

that the first-order component and the β-component are the most important for the classification, whereas the contribution of the α-component is less prominent (see Tomioka and Müller (2010)).

Within each data source, the trace norm regularization automatically learns feature extractors. Figure 9.5 visualizes the spatiotemporal profile of the learned feature extractor that corresponds to the leading singular value of $\boldsymbol{W}^{(1)}$ in figure 9.4. The filter (left) and the pattern (center) visualize the left singular-vector topographically according to the geometry of the EEG sensors. The time course (right) shows the right singular vector as a time series. Both the filter and the pattern show a clear lateralized bipolar structure. This bipolar structure, together with the downward trend in the time course is physiologically known as the lateralized readiness potential (or Bereitschaftspotential) (Cui et al., 1999). Note that the time course starts 630 ms and ends 130 ms *prior* to the actual movement because the task is to predict the laterality of the movement before it is executed.

9.7   Summary

In this chapter, we have presented the dual augmented Lagrangian (DAL) algorithm for sparse estimation problems, and discussed its connections to proximal minimization and other operator splitting algorithms.

The DAL algorithm is an augmented Lagrangian algorithm (Powell, 1969; Hestenes, 1969; Rockafellar, 1976b; Bertsekas, 1982) applied to the dual of the simple sparse estimation problem (9.3). For this problem, the sparsity of the intermediate solution can effectively be exploited to efficiently solve the inner minimization problem. This link between the sparsity and the efficiency distinguishes DAL from other AL algorithms.

We have shown that DAL is equivalent to the proximal minimization algorithm in the primal, which enabled us to rigorously analyze the convergence rate of DAL through the proximal minimization framework. We have shown that DAL converges superlinearly even in the case of inexact inner minimization. Importantly, the stopping criterion we used can be computed in practice; this is because we have separated the loss function f_ℓ from the design matrix $\boldsymbol{A}$ (see section 9.4.1).

The *structured* sparse estimation problem (9.4) can also be tackled through augmented Lagrangian algorithms in the primal (see Goldstein and Osher (2009); Lin et al. (2009)). However, as was discussed in section 9.4.4, for these algorithms the inner minimization is not easy to carry out exactly, because the convex conjugate regularizer ϕ_λ^* does not produce a sparse vector through the associated proximity operator.

Currently we are interested in how much the insights we gained about DAL transfer to *approximate* augmented Lagrangian algorithms, such as the alternating direction method, applied to the primal problem (structured sparse estimation) and the dual problem (simple sparse estimation), and the associated operator splitting methods in their respective dual problems. Application of augmented Lagrangian algorithms to kernel methods is another interesting direction (Suzuki and Tomioka, 2010).

Acknowledgment

We would like to thank Masakazu Kojima and Masao Fukushima for helpful discussions. This work was partially supported by MEXT KAKENHI 22700138 and 22700289, and the FIRST program.

Appendix: Mathematical Details

9.8.1 Infimal Convolution

Let $f : \mathbb{R}^n \to \mathbb{R}$ and $g : \mathbb{R}^n \to \mathbb{R}$ be two convex functions, and let f^* and g^* be their convex conjugate functions, respectively; That is,

$$f^*(\boldsymbol{y}) = \sup_{\boldsymbol{x}\in\mathbb{R}^n} \left(\langle \boldsymbol{y}, \boldsymbol{x}\rangle - f(\boldsymbol{x})\right), \quad g^*(\boldsymbol{y}) = \sup_{\boldsymbol{x}\in\mathbb{R}^n} \left(\langle \boldsymbol{y}, \boldsymbol{x}\rangle - g(\boldsymbol{x})\right).$$

Then,

$$(f+g)^*(\boldsymbol{y}) = \inf_{\boldsymbol{y}'\in\mathbb{R}^n} \left(f^*(\boldsymbol{y}') + g^*(\boldsymbol{y}-\boldsymbol{y}')\right) =: (f^*\Box g^*)(\boldsymbol{y}),$$

where $\Box$ denotes the *infimal convolution*.

See (Rockafellar, 1970, Theorem 16.4) for the proof.

9.8.2 Moreau's Theorem

Let $f : \mathbb{R}^n \to \mathbb{R}$ be convex and f^* its conjugate. Then, for $\boldsymbol{x} \in \mathbb{R}^n$

$$\text{prox}_f(\boldsymbol{x}) + \text{prox}_{f^*}(\boldsymbol{x}) = \boldsymbol{x}. \tag{9.41}$$

Moreover,

$$\widehat{f}(\boldsymbol{x}) + \widehat{f^*}(\boldsymbol{x}) = \frac{1}{2}\|\boldsymbol{x}\|^2, \tag{9.42}$$

where $\widehat{f}$ is Moreau's envelope function of f, namely,

$$\widehat{f}(\boldsymbol{x}) = \min_{\boldsymbol{x}'\in\mathbb{R}^n} \left(f(\boldsymbol{x}') + \frac{1}{2}\|\boldsymbol{x}' - \boldsymbol{x}\|^2\right).$$

Furthermore, the envelope $\widehat{f}$ is differentiable, and its gradient is:

$$\nabla \widehat{f}(\boldsymbol{x}) = \text{prox}_{f^*}(\boldsymbol{x}), \qquad \nabla \widehat{f^*}(\boldsymbol{x}) = \text{prox}_f(\boldsymbol{x}).$$

See Moreau (1965) and (Rockafellar, 1970, theorem 31.5) for the proof. Danskin's theorem (Bertsekas, 1999, proposition B.25) can also be used to show the result. Note that by differentiating both sides of equation (9.42), we obtain equation (9.41), which confirms the validity of the above statement.

9.8.3   Computation of the Duality Gap

We use the same strategy as in Koh et al. (2007) and Wright et al. (2009) to compute the duality gap as a stopping criterion for the DAL algorithm.

Let $\bar{\boldsymbol{\alpha}}^t := -\nabla f_\ell(\boldsymbol{A}\boldsymbol{x}^t)$. Note that the vector $\boldsymbol{A}^T \bar{\boldsymbol{\alpha}}^t$ does not necessarily lie in the domain of ϕ_λ^* in the dual problem (9.19). For trace norm regularization, the domain of ϕ_λ^* is matrices with maximum singular value equal to or smaller than λ. Thus we define $\tilde{\boldsymbol{\alpha}}^t = \bar{\boldsymbol{\alpha}}^t \min(1, \lambda/\|\boldsymbol{A}^T \bar{\boldsymbol{\alpha}}^t\|)$, where $\|\cdot\|$ is the spectral norm. Notice that $\|\boldsymbol{A}^T \tilde{\boldsymbol{\alpha}}^t\| \leq \lambda$ by construction. We compute the dual objective value as $d(\boldsymbol{x}^t) = -f_\ell^*(-\tilde{\boldsymbol{\alpha}}^t)$; and the duality gap is $\text{Gap}^t = f(\boldsymbol{x}^t) - d(\boldsymbol{x}^t)$, where f is the primal objective function (9.18).

9.9   References

F. R. Bach. Consistency of the group lasso and multiple kernel learning. *J. Mach. Learn. Res.*, 9:1179–1225, 2008.

A. Beck and M. Teboulle. A fast iterative shrinkage-thresholding algorithm for linear inverse problems. *SIAM J. Imaging Sciences*, 2(1):183–202, 2009.

D. P. Bertsekas. *Constrained Optimization and Lagrange Multiplier Methods*. Academic Press, 1982.

D. P. Bertsekas. *Nonlinear Programming*. Athena Scientific, second edition, 1999.

B. Blankertz, G. Curio, and K.-R. Müller. Classifying single trial EEG: Towards brain computer interfacing. In T. G. Diettrich, S. Becker, and Z. Ghahramani, editors, *Adv. in Neural Inf. Proc. Sys.*, volume 14, pages 157–164, 2002.

B. Blankertz, K.-R. Müller, G. Curio, T. M. Vaughan, G. Schalk, J. R. Wolpaw, A. Schlögl, C. Neuper, G. Pfurtscheller, T. Hinterberger, M. Schröder, and N. Birbaumer. The BCI competition 2003: Progress and perspectives in detection and discrimination of EEG single trials. *IEEE Trans. Biomed. Eng.*, 51(6):1044–1051, 2004.

S. Boyd and L. Vandenberghe. *Convex Optimization*. Cambridge University Press, 2004.

J.-F. Cai, E. J. Candès, and Z. Shen. A singular value thresholding algorithm for matrix completion. arXiv:0810.3286, 2008.

E. J. Candès, J. Romberg, and T. Tao. Robust uncertainty principles: Exact signal reconstruction from highly incomplete frequency information. *IEEE Trans. Inform. Theory*, 52(2):489–509, 2006.

S. Chen, D. Donoho, and M. Saunders. Atomic decomposition by basis pursuit. *SIAM J. Sci. Comput.*, 20(1):33–61, 1998.

P. L. Combettes and J.-C. Pesquet. A Douglas-Rachford splitting approach to nonsmooth convex variational signal recovery. *IEEE Journal on Selected Topics in Signal Processing*, 1(4):564–574, 2007.

P. L. Combettes and J.-C. Pesquet. Proximal splitting methods in signal processing. In H. H. Bauschke, R. Burachik, P. L. Combettes, V. Elser, D. R. Luke, and H. Wolkowicz, editors, *Fixed-Point Algorithms for Inverse Problems in Science and Engineering*. Springer, 2010.

P. L. Combettes and V. R. Wajs. Signal recovery by proximal forward-backward splitting. *Multiscale Modeling and Simulation*, 4(4):1168–1200, 2005.

R. Q. Cui, D. Huter, W. Lang, and L. Deecke. Neuroimage of voluntary movement: Topography of the bereitschaftspotential, a 64-channel DC current source density study. *Neuroimage*, 9(1):124–134, 1999.

I. Daubechies, M. Defrise, and C. De Mol. An iterative thresholding algorithm for linear inverse problems with a sparsity constraint. *Commun. Pur. Appl. Math.*, LVII:1413–1457, 2004.

D. L. Donoho. De-noising by soft-thresholding. *IEEE Trans. Inform. Theory*, 41 (3):613–627, 1995.

J. Douglas Jr. and H. H. Rachford Jr. On the numerical solution of heat conduction problems in two and three space variables. *Trans. Amer. Math. Soc.*, 82(2):421–439, 1956.

J. Eckstein and D. P. Bertsekas. On the Douglas-Rachford splitting method and the proximal point algorithm for maximal monotone operators. *Mathematical Programming*, 55(1):293–318, 1992.

M. Fazel, H. Hindi, and S. P. Boyd. A rank minimization heuristic with application to minimum order system approximation. In *Proc. of the American Control Conference*, volume 6, pages 4734–4738, 2001.

M. A. T. Figueiredo and R. Nowak. An EM algorithm for wavelet-based image restoration. *IEEE Trans. Image Process.*, 12(8):906–916, 2003.

M. A. T. Figueiredo, J. M. Bioucas-Dias, and R. D. Nowak. Majorization-minimization algorithm for wavelet-based image restoration. *IEEE Trans. Image Process.*, 16(12):2980–2991, 2007.

D. Gabay and B. Mercier. A dual algorithm for the solution of nonlinear variational problems via finite element approximation. *Computers and Mathematics with Applications*, 2(1):17–40, 1976.

J. Gao, G. Andrew, M. Johnson, and K. Toutanova. A comparative study of parameter estimation methods for statistical natural language processing. In *Proceedings of the 45th Annual Meeting of the Association for Computational Linguistics*, volume 45, pages 824–831, 2007.

T. Goldstein and S. Osher. The split Bregman method for L1 regularized problems. *SIAM Journal on Imaging Sciences*, 2(2):323–343, 2009.

S.-P. Han and G. Lou. A parallel algorithm for a class of convex programs. *SIAM J. Control Optimiz.*, 26(2):345–355, 1988.

M. R. Hestenes. Multiplier and gradient methods. *J. Optim. Theory Appl.*, 4:303–320, 1969.

R. A. Horn and C. R. Johnson. *Topics in matrix analysis*. Cambridge University Press, 1991.

S. Ibaraki, M. Fukushima, and T. Ibaraki. Primal-dual proximal point algorithm for linearly constrained convex programming problems. *Computational Optimization and Applications*, 1(2):207–226, 1992.

R. Jenatton, J.-Y. Audibert, and F. Bach. Structured variable selection with sparsity-inducing norms. Technical report, arXiv:0904.3523, 2009.

S.-J. Kim, K. Koh, M. Lustig, S. Boyd, and D. Gorinvesky. An interior-point method for large-scale l-regularized least squares. *IEEE Journal of Selected Topics in Signal Processing*, 1(4):606–617, 2007.

K. Koh, S.-J. Kim, and S. Boyd. An interior-point method for large-scale ℓ_1-regularized logistic regression. *Journal of Machine Learning Research*, 8:1519–1555, 2007.

B. W. Kort and D. P. Bertsekas. Combined primal–dual and penalty methods for convex programming. *SIAM Journal on Control and Optimization*, 14(2): 268–294, 1976.

Z. Lin, M. Chen, L. Wu, and Y. Ma. The augmented Lagrange multiplier method for exact recovery of corrupted low-rank matrices. *Mathematical Programming*, 2009. submitted.

P. L. Lions and B. Mercier. Splitting algorithms for the sum of two nonlinear operators. *SIAM Journal on Numerical Analysis*, 16(6):964–979, 1979.

M. Lustig, D. Donoho, and J. M. Pauly. Sparse MRI: The application of compressed sensing for rapid MR imaging, 2007. *Magn. Reson. Med.*, 58(6):1182–1195, 2007.

J. J. Moreau. Proximité et dualité dans un espace hilbertien. *Bulletin de la Soc. Math. France*, 93:273–299, 1965.

R. M. Neal. *Bayesian Learning for Neural Networks*. Springer, New York, 1996.

Y. Nesterov. Gradient methods for minimizing composite objective function. Technical Report 2007/76, Center for Operations Research and Econometrics (CORE), Catholic University of Louvain, 2007. Revised May 2010.

A. Y. Ng. Feature selection, L1 vs. L2 regularization, and rotational invariance. In *Proc. of the 21st International Conference on Machine Learning*, page 78, New York, NY, USA, 2004. ACM Press, New York.

M. J. D. Powell. A method for nonlinear constraints in minimization problems. In R. Fletcher, editor, *Optimization*, pages 283–298. Academic Press, London and New York, 1969.

R. T. Rockafellar. *Convex Analysis*. Princeton University Press, 1970.

R. T. Rockafellar. Monotone operators and the proximal point algorithm. *SIAM Journal on Control and Optimization*, 14:877–898, 1976a.

R. T. Rockafellar. Augmented Lagrangians and applications of the proximal point algorithm in convex programming. *Math. of Oper. Res.*, 1:97–116, 1976b.

L. I. Rudin, S. Osher, and E. Fatemi. Nonlinear total variation based noise removal algorithms. *Physica D*, 60:259–268, 1992.

S. Setzer. Operator splittings, Bregman methods and frame shrinkage in image processing. *International Journal of Computer Vision*, 92(3):265–280, 2010.

S. K. Shevade and S. S. Keerthi. A simple and efficient algorithm for gene selection using sparse logistic regression. *Bioinformatics*, 19(17):2246–2253, 2003.

N. Srebro, J. D. M. Rennie, and T. S. Jaakkola. Maximum-margin matrix factorization. In *Advances in Neural Information Processing Systems 17*, pages 1329–1336. MIT Press, Cambridge, MA, 2005.

T. Suzuki and R. Tomioka. SpicyMKL. *Machine Learning*, 2010. Submitted.

R. Tibshirani. Regression shrinkage and selection via the lasso. *J. Roy. Stat. Soc. series B*, 58(1):267–288, 1996.

R. Tibshirani, M. Saunders, S. Rosset, J. Zhu, and K. Knight. Sparsity and smoothness via the fused lasso. *J. Roy. Stat. Soc. series B*, 67(1):91–108, 2005.

M. E. Tipping. Sparse bayesian learning and the relevance vector machine. *J. Mach. Learn. Res.*, 1:211–244, 2001.

R. Tomioka and K. Aihara. Classifying matrices with a spectral regularization. In *Proc. of the 24th International Conference on Machine Learning*, pages 895–902. ACM Press, 2007.

R. Tomioka and K.-R. Müller. A regularized discriminative framework for EEG analysis with application to brain-computer interface. *Neuroimage*, 49(1):415–432, 2010.

R. Tomioka and M. Sugiyama. Dual augmented Lagrangian method for efficient sparse reconstruction. *IEEE Signal Processing Letters*, 16(12):1067–1070, 2009.

R. Tomioka, T. Suzuki, and M. Sugiyama. Super-linear convergence of dual augmented-Lagrangian algorithm for sparsity regularized estimation. Technical report, arXiv:0911.4046v2, 2010a.

R. Tomioka, T. Suzuki, M. Sugiyama, and H. Kashima. A fast augmented Lagrangian algorithm for learning low-rank matrices. In *Proc. of the 27th International Conference on Machine Learning*. Omnipress, 2010b.

P. Tseng. Applications of a splitting algorithm to decomposition in convex programming and variational inequalities. *SIAM J. Control Optimiz.*, 29(1):119–138, 1991.

J. B. Weaver, Y. Xu, D. M. Healy Jr, and L. D. Cromwell. Filtering noise from images with wavelet transforms. *Magnetic Resonance in Medicine*, 21(2):288–295, 1991.

D. Wipf and S. Nagarajan. A new view of automatic relevance determination. In *Advances in Neural Information Processing Systems 20*, pages 1625–1632. MIT Press, 2008.

S. J. Wright, R. D. Nowak, and M. A. T. Figueiredo. Sparse reconstruction by separable approximation. *IEEE Trans. Signal Process.*, 57(7):2479–2493, 2009.

W. Yin, S. Osher, D. Goldfarb, and J. Darbon. Bregman iterative algorithms for L1-minimization with applications to compressed sensing. *SIAM J. Imaging Sciences*, 1(1):143–168, 2008.

M. Yuan and Y. Lin. Model selection and estimation in regression with grouped variables. *J. Roy. Stat. Soc. series B*, 68(1):49–67, 2006.

M. Yuan, A. Ekici, Z. Lu, and R. Monteiro. Dimension reduction and coefficient estimation in multivariate linear regression. *J. Roy. Stat. Soc. series B*, 69(3):329–346, 2007.

10   The Convex Optimization Approach to Regret Minimization

Elad Hazan   ehazan@ie.technion.ac.il

Technion - Israel Institute of Technology
Haifa, Israel

A well-studied and general setting for prediction and decision making is regret minimization in games. Recently the design of algorithms in this setting has been influenced by tools from convex optimization. In this chapter we describe the recent framework of online convex optimization which naturally merges optimization and regret minimization. We describe the basic algorithms and tools at the heart of this framework, which have led to the resolution of fundamental questions of learning in games.

10.1   Introduction

In the online decision making scenario, a player has to choose from a pool of available decisions and then incurs a loss corresponding to the quality of the decision made. The regret minimization paradigm suggests the goal of incurring an average loss which approaches that of the best fixed decision in hindsight. Recently tools from convex optimization have given rise to algorithms which are more general, unifying previous results and many times giving new and improved regret bounds.

In this chapter we survey some of the recent developments in this exciting merger of optimization and learning. We start by describing two general templates for producing algorithms and proving regret bounds. The templates are very simple, and unify the analysis of many previous well-known and frequently used algorithms (i.e., multiplicative weights and gradient descent). For the setting of online linear optimization, we also prove that the two templates are equivalent.

After describing the framework and algorithmic templates, we describe some successful applications: characterization of regret bounds in terms of convexity of loss functions, bandit linear optimization, and variational regret bounds.

10.1.1   The Online Convex Optimization Model

In online convex optimization, an online player iteratively chooses a point from a set in Euclidean space denoted $\mathcal{K} \subseteq \mathbb{R}^n$. Following Zinkevich (2003), we assume that the set $\mathcal{K}$ is non-empty, bounded, and closed. For algorithmic efficiency reasons that will be apparent later, we also assume the set $\mathcal{K}$ to be convex.

We denote the number of iterations by T (which is unknown by the online player). At iteration t, the online player chooses $\mathbf{x}_t \in \mathcal{K}$. After committing to this choice, a convex cost function $\mathbf{f}_t : \mathcal{K} \mapsto \mathbb{R}$ is revealed. The cost incurred to the online player is the value of the cost function at the point she committed to $\mathbf{f}_t(\mathbf{x}_t)$. Henceforth we consider mostly *linear* cost functions, and abuse notation to write $\mathbf{f}_t(\mathbf{x}) = \mathbf{f}_t^\top \mathbf{x}$.

The feedback available to the player falls into two main categories. In the full information model, all information about the function $\mathbf{f}_t$ is observable by the player (after incurring the loss). In the "bandit" model, the player observes only the loss $\mathbf{f}_t(\mathbf{x}_t)$ itself.

The regret of the online player using algorithm $\mathcal{A}$ at time T is defined to be the total cost minus the cost of the best fixed single decision, where the best is chosen with the benefit of hindsight. We are usually interested in an upper bound on the worst-case guaranteed regret, denoted

$$\mathrm{Regret}_T(\mathcal{A}) = \sup_{\{\mathbf{f}_1,\ldots,\mathbf{f}_t\}} \left\{ \mathbf{E}\left[\sum_{t=1}^T \mathbf{f}_t(\mathbf{x}_t)\right] - \min_{\mathbf{x} \in \mathcal{K}} \sum_{t=1}^T \mathbf{f}_t(\mathbf{x}) \right\}.$$

Regret is the defacto standard in measuring the performance of learning algorithms.[1]

Intuitively, an algorithm performs well if its regret is sublinear in T, that is, $\mathrm{Regret}_T(\mathcal{A}) = o(T)$, since this implies that "on the average" the algorithm performs as well as the best fixed strategy in hindsight.

---

1. For some problems it is more natural to talk of the "payoff" given to the online player rather than the cost she incurs. If so, the payoff functions need to be concave and regret is defined analogously.

The running time of an algorithm for online game playing is defined to be the worst-case expected time to produce $\mathbf{x}_t$, for an iteration $t \in [T]$ [2] in a T iteration repeated game. Typically, the running time will depend on n, T and the parameters of the cost functions and underlying convex set.

10.1.2 Examples

10.1.2.1 Prediction from Experts Advice

Perhaps the best-known problem in prediction theory is the "experts problem". The decision maker has to choose from the advice of n given experts. After choosing one, a loss between zero and one is incurred. This scenario is repeated iteratively, and at each iteration the costs of the various experts are arbitrary. The goal is to do as well as the best expert in hindsight.

The online convex optimization problem captures this problem as a special case: the set of decisions is the set of all distributions over n elements (experts), that is the n-dimensional simplex $\mathcal{K} = \Delta_n = \{\mathbf{x} \in \mathbb{R}^n, \sum_i \mathbf{x}_i = 1, \mathbf{x}_i \geq 0\}$. Let the cost to the i'th expert at iteration t be denoted by $\mathbf{f}_t(i)$. Then the cost functions are given by $\mathbf{f}_t(x) = \mathbf{f}_t^\top \mathbf{x}$. This is the expected cost of choosing an expert according to distribution $\mathbf{x}$, and happens to be linear.

10.1.2.2 Online Shortest Paths

In the online shortest path problem, the decision maker is given a directed graph $G = (V, E)$ and a source-sink pair $s, t \in V$. At each iteration $t \in [T]$, the decision maker chooses a path $p_t \in \boldsymbol{P}_{s,t}$, where $\boldsymbol{P}_{s,t} \subseteq \{E\}^{|V|}$ is the set of all s, t-paths in the graph. The adversary independently chooses weights on the edges of the graph, given by a function from the edges to the reals $\mathbf{f}_t : E \mapsto \mathbb{R}$, which can be represented as a vector in m-dimensional space: $\mathbf{f}_t \in \mathbb{R}^m$. The decision maker suffers and observes loss, which is the weighted length of the chosen path $\sum_{e \in p_t} \mathbf{f}_t(e)$.

The discrete description of this problem as an experts problem, where we have an expert for every path, presents an efficiency challenge: there are potentially exponentially many paths in terms of the graph representation size. Much work has been devoted to resolving this efficiency issue, and efficient algorithms have been found in this discrete formulation, such as (Takimoto and Warmuth, 2003; Awerbuch and Kleinberg, 2008). However, the optimal regret bound for the bandit version of this problem eluded researchers for some time, and was finally resolved only within the online

---

2. Here and henceforth we denote the set of integers $\{1, ..., n\}$ by $[n]$.

convex optimization framework (Abernethy et al., 2008; Dani et al., 2008).

The online convex optimization framework suggests an inherently efficient model to capture this problem. Recall the standard description of the set of all distributions over paths (flows) in a graph as a convex set in $\mathbb{R}^m$, with $O(m + |V|)$ constraints. Denote this flow polytope by $\mathcal{K}$. The expected cost of a given flow $\mathbf{x} \in \mathcal{K}$ (distribution over paths) is then a linear function, given by $\mathbf{f}_t^\top \mathbf{x}$, where $\mathbf{f}_t(e)$ is the length of the edge $e \in E$.

10.1.2.3 Portfolio Selection

The universal portfolio selection problem which we briefly describe is due to Cover (1991). At each iteration $t = 1$ to T, the decision maker chooses a distribution of her wealth over n assets $\mathbf{x}_t \in \Delta_n$. The adversary independently chooses market returns for the assets, that is a vector $\mathbf{r}_t \in \mathbb{R}^n_+$ such that each coordinate $\mathbf{r}_t(i)$ is the price ratio for the i'th asset between the iterations t and $t + 1$. The ratio between the wealth of the investor at iterations $t+1$ and t is $\mathbf{r}_t^\top \mathbf{x}_t$, and hence the gain in this setting is defined to be the logarithm of this change ratio in wealth $\log(\mathbf{r}_t^\top \mathbf{x}_t)$. Notice that since $\mathbf{x}_t$ is the distribution of the investor's wealth, even if $\mathbf{x}_{t+1} = \mathbf{x}_t$, the investor may still need to trade in order to adjust for price changes.

The goal of regret minimization, which in this case corresponds to minimizing the difference $\max_{\mathbf{x} \in \Delta_n} \sum_{t=1}^T \log(\mathbf{r}_t^\top \mathbf{x}) - \sum_{t=1}^T \log(\mathbf{r}_t^\top \mathbf{x}_t)$, has an intuitive interpretation. The first term is the logarithm of the wealth accumulated by the distribution $\mathbf{x}^*$. Since this distribution is fixed, it corresponds to a strategy of rebalancing the position after every trading period, and hence is called a *constant rebalanced portfolio*. The second expression is the logarithm of the wealth accumulated by the online decision maker. Hence regret minimization corresponds to maximizing the ratio of investor wealth against wealth of the best benchmark from a pool of investing strategies.

A *universal* portfolio selection algorithm is defined to be one that attains regret converging to zero in this setting. Such an algorithm, albeit requiring exponential time, was first described in Cover (1991). The online convex optimization framework has given rise to much more efficient algorithms based on Newton's method (Hazan et al., 2007).

10.1.3 Algorithms for Online Convex Optimization

Algorithms for online convex optimization can be derived from rich algorithmic techniques developed for prediction in various statistical and machine learning settings. We describe two general algorithmic frameworks from which many previous algorithms can be derived as special cases.

Perhaps the most straightforward approach is for the online player to use whatever decision (point in the convex set) would have been optimal. Formally, let

$$\mathbf{x}_t = \arg\min_{\mathbf{x} \in \mathcal{K}} \sum_{i=1}^{t-1} \mathbf{f}_i(\mathbf{x}).$$

This type of strategy is known as "fictitious play" in economics, and was named "follow the leader" (FTL) by Kalai and Vempala (2005). As Kalai and Vempala point out, this strategy fails miserably in a worst-case sense. That is, its regret can be linear in the number of iterations, as the following example shows. Consider K to be the real line segment between -1 and +1, and $\mathbf{f}_0 = \frac{1}{2}\mathbf{x}$, and let $\mathbf{f}_i$ alternate between $-\mathbf{x}$ and $\mathbf{x}$. The FTL strategy will keep shifting between -1 and $+1$, always making the wrong choice.

Kalai and Vempala proceed to analyze a modification of FTL with added noise to "stabilize" the decision (this modification was originally due to Hannan (1957)). Similarly, much more general and varied twists on this basic FTL strategy can be conjured up, and, as we shall show, also analyzed successfully. This is the essence of the meta-algorithm defined in this section.

Another natural approach for online convex optimization is an iterative approach. Start with some decision $\mathbf{x} \in \mathcal{K}$, and iteratively modify it according to the cost functions that are encountered. Some natural update rules include the gradient update, updates based on a multiplicative rule, on Newton's method, and so forth. Indeed, all of these suggestions make for useful algorithms. But as we shall show, they can all be seen as special cases of the general methodology we analyze next.

10.2 The RFTL Algorithm and Its Analysis

Recall the caveat about straightforward use of follow-the-leader. As in the bad example we have considered, the prediction of FTL may vary wildly from one iteration to the next. This motivates the modification of the basic FTL strategy in order to stabilize the prediction. By adding a *regularization* term, we obtain the RFTL (regularized follow the leader) algorithm.

We proceed to formally describe the RFTL algorithmic template, and analyze it. While the analysis given is optimal asymptotically, we do not give the best constants possible, in order to simplify presentation.

In this section we consider only linear cost functions, $\mathbf{f}(\mathbf{x}) = \mathbf{f}^T\mathbf{x}$. The case of convex cost functions can be reduced to the linear case via the inequality $\mathbf{f}_t(\mathbf{x}_t) - \mathbf{f}_t(\mathbf{x}^*) \leq \nabla\mathbf{f}_t(\mathbf{x}_t)(\mathbf{x}_t - \mathbf{x}^*)$, and considering the function

$\hat{\mathbf{f}}_t(\mathbf{x}) = \nabla \mathbf{f}_t(\mathbf{x}_t)^\top \mathbf{x}$, which is now linear.

10.2.1   Algorithm Definition

The generic RFTL meta-algorithm is defined below. The regularization function $\mathcal{R}$ is assumed to be strongly convex and smooth such that it has a continuous second derivative.

Algorithm 10.1 RFTL

1: Input: $\eta > 0$, strongly convex regularizer function $\mathcal{R}$, and a convex compact set $\mathcal{K}$
2: Let $\mathbf{x}_1 = \arg\min_{\mathbf{x} \in \mathcal{K}} [\mathcal{R}(\mathbf{x})]$
3: **for** $t = 1$ to T **do**
4:     Predict $\mathbf{x}_t$
5:     Observe the payoff function $\mathbf{f}_t$
6:     Update

$$\mathbf{x}_{t+1} = \arg\min_{\mathbf{x} \in \mathcal{K}} \underbrace{\left[\eta \sum_{s=1}^{t} \mathbf{f}_s^\top \mathbf{x} + \mathcal{R}(\mathbf{x}) \right]}_{\Phi_t(\mathbf{x})} \tag{10.1}$$

7: **end for**

10.2.2   Special Cases: Multiplicative Updates and Gradient Descent

Two famous algorithms which are captured by algorithm 10.1 are called the multiplicative update algorithm and the gradient descent method. If $\mathcal{K} = \Delta_n = \{\mathbf{x} \geq 0 \,,\ \sum_i \mathbf{x}(i) = 1\}$, then taking $\mathcal{R}(\mathbf{x}) = \mathbf{x} \log \mathbf{x}$ gives a multiplicative update algorithm, in which

$$\mathbf{x}_{t+1}(i) = \frac{\mathbf{x}_t(i) \cdot e^{\eta \mathbf{f}_t(i)}}{\sum_{i=1}^{n} \mathbf{x}_t(i) \cdot e^{\eta \mathbf{f}_t(i)}}.$$

If $\mathcal{K}$ is the unit ball and $\mathcal{R}(\mathbf{x}) = \|\mathbf{x}\|_2^2$, we get the gradient descent algorithm, in which

$$\mathbf{x}_{t+1} = \frac{\mathbf{x}_t - \eta \mathbf{f}_t}{\|\mathbf{x}_t - \eta \mathbf{f}_t\|_2}.$$

It is possible to derive these special cases by the KKT optimality conditions of equation 10.1. However, we give an easier proof of these facts in the next section, in which we give an equivalent definition of RFTL for the case of linear cost functions.

10.2.3 The Regret Bound

Henceforth we make use of general matrix norms. A PSD matrix $A \succ 0$ gives rise to the norm $\|x\|_A = \sqrt{x^T A x}$. The *dual* norm of this matrix norm is $\|x\|_{A^{-1}} = \|x\|_A^*$. The generalized Cauchy-Schwartz theorem asserts that $x \cdot y \leq \|x\|_A \|y\|_A^*$. We usually take A to be the Hessian of the regularization function $\mathcal{R}(x)$, denoted $\nabla^2 \mathcal{R}(x)$. In this case, we shorthand the notation to be $\|x\|_{\nabla^2 \mathcal{R}(y)} = \|x\|_y$, and similarly $\|x\|_{\nabla^{-2}\mathcal{R}(y)} = \|x\|_y^*$. Denote

$$\lambda = \max_{t,\mathbf{x}\in\mathcal{K}} \mathbf{f}_t^\top [\nabla^2\mathcal{R}(\mathbf{x})]^{-1}\mathbf{f}_t \ , \quad D = \max_{\mathbf{u}\in\mathcal{K}} \mathcal{R}(\mathbf{u}) - \mathcal{R}(\mathbf{x}_1)$$

Notice that both λ and D depend on the regularization function, the convex decision set, and the magnitude of the cost functions.

Theorem 10.1. *Algorithm 10.1 achieves the following bound on the regret for every $\mathbf{u} \in \mathcal{K}$:*

$$Regret_T = \sum_{t=1}^T \mathbf{f}_t^\top (\mathbf{x}_t - \mathbf{u}) \leq 2\sqrt{2\lambda D T} \ .$$

Consider the expert problem, for example: the convex set is the simplex, $\mathcal{R}$ is taken to be the negative entropy function (which corresponds to the multiplicative update algorithm), and the costs are bounded by 1 in each coordinate. Then $\mathbf{f}^\top[\nabla^2\mathcal{R}(\mathbf{x})]^{-1}\mathbf{f} = \sum_i \mathbf{f}(i)^2 \mathbf{x}(i) \leq \sum_i \mathbf{x}(i) = 1$, which implies $\lambda \leq 1$. The parameter D in this case is bounded by $\max_{\mathbf{u}\in\Delta} \sum_i \mathbf{u}(i)\log\frac{1}{\mathbf{u}(i)} \leq \log n$. This gives the regret bound $O(\sqrt{T\log n})$, which is known to be tight.[3]

To prove theorem 10.1, we first relate the regret to the stability in prediction. This is formally captured by the FTL-BTL lemma, which holds in the general scenario.

Lemma 10.2 (FTL-BTL lemma). *For every $\mathbf{u} \in \mathcal{K}$, the algorithm defined by (10.1) enjoys the following regret guarantee*

$$\sum_{t=1}^T \mathbf{f}_t^\top(\mathbf{x}_t - \mathbf{u}) \ \leq \ \sum_{t=1}^T \mathbf{f}_t^\top(\mathbf{x}_t - \mathbf{x}_{t+1}) + \frac{1}{\eta}[\mathcal{R}(\mathbf{u}) - \mathcal{R}(\mathbf{x}_1)].$$

We defer the proof of this simple lemma to the appendix, and proceed with the (short) proof of the main theorem.

---

3. In the case of multiplicative updates, as well as in other regularization functions of interest, it is possible to obtain a tighter bound in theorem 10.1: the term λ can be redefined as $\lambda = \max_t \mathbf{f}_t^\top[\nabla^2\mathcal{R}(\mathbf{x}_t)]^{-1}\mathbf{f}_t$. The derivation is not in the scope of this survey; see Abernethy et al. (2008) for more details.

Main Theorem. Recall that $\mathcal{R}(x)$ is a convex function and $\mathcal{K}$ is convex. Then, by Taylor expansion (with its explicit remainder term via the mean value theorem) at $\mathbf{x}_{t+1}$, there exists a $\mathbf{z}_t \in [\mathbf{x}_{t+1}, \mathbf{x}_t]$ for which

$$\begin{aligned} \Phi_t(\mathbf{x}_t) &= \Phi_t(\mathbf{x}_{t+1}) + (\mathbf{x}_t - \mathbf{x}_{t+1})^\mathsf{T} \nabla \Phi_t(\mathbf{x}_{t+1}) + \frac{1}{2}\|\mathbf{x}_t - \mathbf{x}_{t+1}\|^2_{\mathbf{z}_t} \\ &\geq \Phi_t(\mathbf{x}_{t+1}) + \frac{1}{2}\|\mathbf{x}_t - \mathbf{x}_{t+1}\|^2_{\mathbf{z}_t} \end{aligned}$$

Recall our notation $\|\mathbf{y}\|^2_{\mathbf{z}} = \mathbf{y}^\mathsf{T} \nabla^2 \Phi_t(\mathbf{z})\mathbf{y}$, and it follows that $\|\mathbf{y}\|^2_{\mathbf{z}} = \mathbf{y}^\mathsf{T} \nabla^2 \mathcal{R}(\mathbf{z})\mathbf{y}$. The inequality above is true because $\mathbf{x}_{t+1}$ is a minimum of Φ_t over $\mathcal{K}$. Thus,

$$\begin{aligned} \|\mathbf{x}_t - \mathbf{x}_{t+1}\|^2_{\mathbf{z}_t} &\leq 2\,\Phi_t(\mathbf{x}_t) - 2\,\Phi_t(\mathbf{x}_{t+1}) \\ &= 2\,(\Phi_{t-1}(\mathbf{x}_t) - \Phi_{t-1}(\mathbf{x}_{t+1})) + 2\eta \mathbf{f}_t^\mathsf{T}(\mathbf{x}_t - \mathbf{x}_{t+1}) \\ &\leq 2\eta\,\mathbf{f}_t^\mathsf{T}(\mathbf{x}_t - \mathbf{x}_{t+1}). \end{aligned}$$

By the generalized Cauchy-Schwartz inequality,

$$\begin{aligned} \mathbf{f}_t^\mathsf{T}(\mathbf{x}_t - \mathbf{x}_{t+1}) &\leq \|\mathbf{f}_t\|^*_{\mathbf{z}_t} \cdot \|\mathbf{x}_t - \mathbf{x}_{t+1}\|_{\mathbf{z}_t} \qquad \text{general CS} \qquad (10.2) \\ &\leq \|\mathbf{f}_t\|^*_{\mathbf{z}_t} \cdot \sqrt{2\,\eta\,\mathbf{f}_t^\mathsf{T}(\mathbf{x}_t - \mathbf{x}_{t+1})}. \end{aligned}$$

Shifting sides and squaring, we get

$$\mathbf{f}_t^\mathsf{T}(\mathbf{x}_t - \mathbf{x}_{t+1}) \leq 2\,\eta\,\|\mathbf{f}_t\|^{*\,2}_{\mathbf{z}_t} \leq 2\,\eta\,\lambda.$$

Using this, together with the FTL-BTL lemma, and summing over T periods, we obtain the theorem. Choosing the optimal η, we obtain

$$R_T \leq \min_\eta \left\{ 2\,\eta\lambda\,T + \frac{1}{\eta}[\mathcal{R}(\mathbf{u}) - \mathcal{R}(\mathbf{x}_1)] \right\} \leq 2\sqrt{2\,D\,\lambda\,T}\;.$$

$\square$

10.3 The "Primal-Dual" Approach

The other approach for proving regret bounds, which we call primal-dual, originates from the link-function methodology, as introduced in Grove et al. (2001); Kivinen and Warmuth (2001), and is related to the mirrored descent paradigm in the optimization community. A central concept useful for this method are Bregman divergences, formally defined below.

Definition 10.1. *Denote by $B^\mathcal{R}(\mathbf{x}||\mathbf{y})$ the Bregman divergence with respect to the function $\mathcal{R}$, defined as*

$$B^\mathcal{R}(\mathbf{x}||\mathbf{y}) = \mathcal{R}(\mathbf{x}) - \mathcal{R}(\mathbf{y}) - (\mathbf{x} - \mathbf{y})^\mathsf{T}\nabla\mathcal{R}(\mathbf{y})\;.$$

The primal-dual algorithm is an iterative algorithm, which computes the next prediction using a simple update rule and the previous prediction. The generality of the method stems from the update being carried out in a dual space, where the duality notion is defined by the choice of regularization.

Algorithm 10.2 Primal-dual

1: Let $\mathcal{K}$ be a convex set
2: Input: parameter $\eta > 0$, regularizer function $\mathcal{R}(\mathbf{x})$
3: **for** $t = 1$ to T **do**
4:     If $t = 1$, choose $\mathbf{y}_1$ such that $\nabla \mathcal{R}(\mathbf{y}_1) = \mathbf{0}$
5:     If $t > 1$, choose $\mathbf{y}_t$ such that
        Lazy version: $\nabla \mathcal{R}(\mathbf{y}_t) = \nabla \mathcal{R}(\mathbf{y}_{t-1}) - \eta \, \mathbf{f}_{t-1}$.
        Active version: $\nabla \mathcal{R}(\mathbf{y}_t) = \nabla \mathcal{R}(\mathbf{x}_{t-1}) - \eta \, \mathbf{f}_{t-1}$.
6:     Project according to $B^{\mathcal{R}}$:

$$\mathbf{x}_t = \arg\min_{\mathbf{x} \in \mathcal{K}} B^{\mathcal{R}}(\mathbf{x} \| \mathbf{y}_t)$$

7: **end for**

10.3.1   Equivalence to RFTL in the Linear Setting

For the special case of linear cost functions, algorithm 10.2 (lazy version) and RFTL are identical, as we show now. The primal-dual algorithm, however, can be analyzed in a very different way, which is extremely useful in certain online scenarios.

Lemma 10.3. *For linear cost functions, the lazy primal-dual and RFTL algorithms produce identical predictions, that is,*

$$\arg\min_{x \in \mathcal{K}} \left(\mathbf{f}_t^{\mathsf{T}} \mathbf{x} + \frac{1}{\eta} \mathcal{R}(\mathbf{x}) \right) = \arg\min_{\mathbf{x} \in \mathcal{K}} B^{\mathcal{R}}(\mathbf{x} \| \mathbf{y}_t) \ .$$

Proof. First, observe that the unconstrained minimum

$$\mathbf{x}_t^* \equiv \arg\min_{\mathbf{x} \in \mathbb{R}^n} \left\{ \sum_{s=1}^{t-1} \mathbf{f}_s^{\mathsf{T}} \mathbf{x} + \frac{1}{\eta} \mathcal{R}(\mathbf{x}) \right\}$$

satisfies

$$\sum_{s=1}^{t-1} \mathbf{f}_s + \frac{1}{\eta} \nabla \mathcal{R}(\mathbf{x}_t^*) = \mathbf{0} \ .$$

Since $\mathcal{R}(\mathbf{x})$ is strictly convex, there is only one solution for the above equation and thus $\mathbf{y}_t = \mathbf{x}_t^*$. Hence,

$$
\begin{aligned}
B^{\mathcal{R}}(\mathbf{x}\|\mathbf{y}_t) &= \mathcal{R}(\mathbf{x}) - \mathcal{R}(\mathbf{y}_t) - (\nabla\mathcal{R}(\mathbf{y}_t))^\top(\mathbf{x}-\mathbf{y}_t) \\
&= \mathcal{R}(\mathbf{x}) - \mathcal{R}(\mathbf{y}_t) + \eta \sum_{s=1}^{t-1} \mathbf{f}_s^\top(\mathbf{x}-\mathbf{y}_t) \ .
\end{aligned}
$$

Since $\mathcal{R}(\mathbf{y}_t)$ and $\sum_{s=1}^{t-1} \mathbf{f}_s^\top \mathbf{y}_t$ are independent of $\mathbf{x}$, it follows that $B^{\mathcal{R}}(\mathbf{x}\|\mathbf{y}_t)$ is minimized at the point $\mathbf{x}$ that minimizes $\mathcal{R}(\mathbf{x}) + \eta \sum_{s=1}^{t-1} \mathbf{f}_s^\top \mathbf{x}$ over $\mathcal{K}$, which in turn implies that

$$
\arg\min_{\mathbf{x}\in\mathcal{K}} B^{\mathcal{R}}(\mathbf{x}\|\mathbf{y}_t) = \arg\min_{\mathbf{x}\in K} \left\{ \sum_{s=1}^{t-1} \mathbf{f}_s^\top \mathbf{x} + \frac{1}{\eta}\mathcal{R}(\mathbf{x}) \right\} .
$$

$\square$

10.3.2   Regret Bounds for the Primal-Dual Algorithm

Theorem 10.4. *Suppose that $\mathcal{R}$ is such that $B_{\mathcal{R}}(\mathbf{x},\mathbf{y}) \geq \|\mathbf{x}-\mathbf{y}\|^2$ for some norm $\|\cdot\|$. Let $\|\nabla\mathbf{f}_t(\mathbf{x}_t)\|^* \leq G_*$ for all t, and $\forall \mathbf{x} \in K \ B_{\mathcal{R}}(\mathbf{x},\mathbf{x}_1) \leq D^2$. Applying the primal-dual algorithm (active version) with $\eta = \frac{D}{2G_*\sqrt{T}}$, we have*

$$
Regret_T \leq DG_*\sqrt{T}
$$

Proof. Since the functions $\mathbf{f}_t$ are convex, for any $\mathbf{x}^* \in K$,

$$
\mathbf{f}_t(\mathbf{x}_t) - \mathbf{f}_t(\mathbf{x}^*) \leq \nabla\mathbf{f}_t(\mathbf{x}_t)^\top(\mathbf{x}_t - \mathbf{x}^*).
$$

The following property of Bregman divergences follows easily from the definition: for any vectors $\mathbf{x}, \mathbf{y}, \mathbf{z}$,

$$
(\mathbf{x}-\mathbf{y})^\top(\nabla\mathcal{R}(\mathbf{z}) - \nabla\mathcal{R}(\mathbf{y})) = B_{\mathcal{R}}(\mathbf{x},\mathbf{y}) - B_{\mathcal{R}}(\mathbf{x},\mathbf{z}) + B_{\mathcal{R}}(\mathbf{y},\mathbf{z}).
$$

Combining both observations,

$$
\begin{aligned}
2(\mathbf{f}_t(\mathbf{x}_t) - \mathbf{f}_t(\mathbf{x}^*)) &\leq 2\nabla\mathbf{f}_t(\mathbf{x}_t)^\top(\mathbf{x}_t - \mathbf{x}^*) \\
&= \frac{1}{\eta}(\nabla\mathcal{R}(\mathbf{y}_{t+1}) - \nabla\mathcal{R}(\mathbf{x}_t))^\top(\mathbf{x}^* - \mathbf{x}_t) \\
&= \frac{1}{\eta}[B_{\mathcal{R}}(\mathbf{x}^*,\mathbf{x}_t) - B_{\mathcal{R}}(\mathbf{x}^*,\mathbf{y}_{t+1}) + B_{\mathcal{R}}(\mathbf{x}_t,\mathbf{y}_{t+1})] \\
&\leq \frac{1}{\eta}[B_{\mathcal{R}}(\mathbf{x}^*,\mathbf{x}_t) - B_{\mathcal{R}}(\mathbf{x}^*,\mathbf{x}_{t+1}) + B_{\mathcal{R}}(\mathbf{x}_t,\mathbf{y}_{t+1})]
\end{aligned}
$$

where the last inequality follows from the generalized Pythagorean inequality (see Cesa-Bianchi and Lugosi (2006), lemma 11.3), as $\mathbf{x}_{t+1}$ is the projection w.r.t the Bregman divergence of $\mathbf{y}_{t+1}$ and $\mathbf{x}^* \in K$ is in the convex set. Summing over all iterations,

$$2\text{Regret} \quad \leq \quad \frac{1}{\eta}[B_{\mathcal{R}}(\mathbf{x}^*, \mathbf{x}_1) - B_{\mathcal{R}}(\mathbf{x}^*, \mathbf{x}_T)] + \sum_{t=1}^{T} \frac{1}{\eta} B_{\mathcal{R}}(\mathbf{x}_t, \mathbf{y}_{t+1})$$

$$\leq \quad \frac{1}{\eta}D^2 + \sum_{t=1}^{T} \frac{1}{\eta} B_{\mathcal{R}}(\mathbf{x}_t, \mathbf{y}_{t+1}). \tag{10.3}$$

We proceed to bound $B_{\mathcal{R}}(\mathbf{x}_t, \mathbf{y}_{t+1})$. By definition of the Bregman divergence, and the dual norm inequality stated before,

$$B_{\mathcal{R}}(\mathbf{x}_t, \mathbf{y}_{t+1}) + B_{\mathcal{R}}(\mathbf{y}_{t+1}, \mathbf{x}_t) = (\nabla \mathcal{R}(\mathbf{x}_t) - \nabla \mathcal{R}(\mathbf{y}_{t+1}))^\top (\mathbf{x}_t - \mathbf{y}_{t+1})$$

$$= 2\eta \nabla \mathbf{f}_t(\mathbf{x}_t)^\top (\mathbf{x}_t - \mathbf{y}_{t+1})$$

$$\leq \eta^2 G^{*\,2} + \|\mathbf{x}_t - \mathbf{y}_{t+1}\|^2.$$

Thus, by our assumption $B_{\mathcal{R}}(\mathbf{x}, \mathbf{y}) \geq \|\mathbf{x} - \mathbf{y}\|^2$, we have

$$B_{\mathcal{R}}(\mathbf{x}_t, \mathbf{y}_{t+1}) \leq \eta^2 G_*^2 + \|\mathbf{x}_t - \mathbf{y}_{t+1}\|^2 - B_{\mathcal{R}}(\mathbf{y}_{t+1}, \mathbf{x}_t) \leq \eta^2 G_*^2.$$

Plugging back into equation (10.3), and by non-negativity of the Bregman divergence, we get

$$\text{Regret} \leq \frac{1}{2}[\frac{1}{\eta}D^2 + \eta T G_{*\,2}] \leq D G_* \sqrt{T}$$

by taking $\eta = \frac{D}{2\sqrt{T} G_*}$.

$\square$

10.3.3 Deriving the Multiplicative Update and Gradient Descent Algorithms

We stated in section 10.3.2 that by taking $\mathcal{R}$ to be the negative entropy function over the simplex, the RFTL template specializes to become a multiplicative update algorithm. Since we have proved that RFTL is equivalent to the primal-dual algorithm, the same is true for the latter, and the same regret bound applies.

If $\mathcal{R}(\mathbf{x}) = \mathbf{x}\log\mathbf{x}$ is the negative entropy function, then $\nabla \mathcal{R}(\mathbf{x}) = \mathbf{1} + \log\mathbf{x}$, and hence the update rule for the primal-dual algorithm 10.2 (the lazy and adaptive versions are identical in this case) becomes

$$\log \mathbf{y}_t = \log \mathbf{x}_{t-1} - \eta \mathbf{f}_{t-1}$$

or $\mathbf{y}_t(i) = \mathbf{x}_{t-1}(i) \cdot e^{-\eta \mathbf{f}_{t-1}(i)}$. Since the entropy projection corresponds to scaling by the ℓ_1-norm, it follows that $\mathbf{x}_{t+1}(i) = \frac{\mathbf{x}_t(i) \cdot e^{\eta \mathbf{f}_t(i)}}{\sum_{i=1}^{n} \mathbf{x}_t(i) \cdot e^{\eta \mathbf{f}_t(i)}}$.

As for the regret bound, it is well-known that the entropy function satisfies that $B_R(\mathbf{x}, \mathbf{y}) \geq \frac{1}{4}\|\mathbf{x} - \mathbf{y}\|_1^2$ (which is essentially Pinsker's inequality (see (Cover and Thomas, 1991)). Thus, to apply theorem 10.4, we need to bound the the ℓ_∞-norm of the gradients, which corresponds to the maximal cost incurred by the experts. Assume this is bounded by 1, that is, $G_* \leq 1$. The Bregman divergence with respect to $\mathcal{R}$ is the relative entropy, and starting from $\mathbf{x}_1$ being the uniform distribution, it holds that $B_{\mathcal{R}}(\mathbf{x}, \mathbf{x}_1) \leq \log n$ for any $\mathbf{x}$ in the simplex. Thus, by theorem 10.4 the regret of the multiplicative weights algorithm for the experts problem is bounded by $O(\sqrt{T \log n})$.

To derive the online gradient descent algorithm, take $\mathcal{R} = \frac{1}{2}\|\mathbf{x}\|_2^2$. In this case, $\nabla\mathcal{R}(\mathbf{x}) = \mathbf{x}$, and hence the update rule for the primal-dual algorithm 10.2 becomes

$$\mathbf{y}_t = \mathbf{y}_{t-1} - \eta\mathbf{f}_{t-1},$$

and thus when $\mathcal{K}$ is the unit ball,

$$\mathbf{x}_{t+1} = \frac{\mathbf{x}_1 - \eta\sum_{\tau=2}^t \mathbf{f}_\tau}{\|\mathbf{x}_1 - \eta\sum_{\tau=2}^t \mathbf{f}_\tau\|_2} = \frac{\mathbf{x}_t - \eta\mathbf{f}_t}{\|\mathbf{x}_t - \eta\mathbf{f}_t\|_2}.$$

10.4   Convexity of Loss Functions

In this section we review one of the first consequences of the convex optimization approach to decision making: the characterization of attainable regret bounds in terms of convexity of loss functions. It has long been known that special kinds of loss functions permit tighter regret bounds than other loss functions. For example, in the portfolio selection problem, Cover's algorithm attained regret which depends on the number of iterations T as $O(\log T)$. This is in contrast to online linear optimization or the experts problem, in which $\Theta(\sqrt{T})$ is known to be tight.

In this section we give a simple gradient descent-based algorithm which attains logarithmic regret if the loss functions are *strongly convex*. Interestingly, the naive fictitious play (FTL) algorithm attains essentially the same regret bounds in this special case. Similar bounds are attainable under weaker conditions on the loss functions, which capture the portfolio selection problem, and have led to the efficient algorithm for Cover's problem (Hazan et al., 2007).

We say that a function is α-strongly convex if its second derivative is strictly bounded away from zero. In higher dimensions this corresponds to the matrix inequality $\nabla^2\mathbf{f}(\mathbf{x}) \succeq \alpha \cdot \mathbb{I}$, where $\nabla^2\mathbf{f}(\mathbf{x})$ is the Hessian of the function and $A \succeq B$ denotes that the matrix $A - B$ is positive semi-definite.

For example, the squared loss, that is, $\mathbf{f}(\mathbf{x}) = \|\mathbf{x} - \mathbf{a}\|_2^2$, is 1-strongly convex.

Algorithm 10.3 Online gradient descent

1:   Input: convex set $\mathcal{K}$, initial point $\mathbf{x}_0 \in \mathcal{K}$, learning rates $\eta_1, ..., \eta_t$
2:   **for** $t = 1$ to T **do**
3:       Let $\mathbf{y}_t = \mathbf{x}_{t-1} - \eta_{t-1} \nabla \mathbf{f}_{t-1}(\mathbf{x}_{t-1})$
4:       Project onto $\mathcal{K}$:

$$\mathbf{x}_t = \arg\min_{\mathbf{x} \in \mathcal{K}} \|\mathbf{x} - \mathbf{y}_t\|_2$$

5:   **end for**

The following theorem, proved in Hazan et al. (2007), establishes logarithmic bounds on the regret if the cost functions are strongly convex. Denote by G an upper bound on the Euclidean norm of the gradients.

Theorem 10.5. *The online gradient descent algorithm with stepsizes $\eta_t = \frac{1}{\alpha t}$ achieves the following guarantee for all $T \geq 1$:*

$$Regret_T(OGD) \leq \frac{G^2}{2\alpha}(1 + \log T).$$

Proof. Let $\mathbf{x}^* \in \arg\min_{\mathbf{x} \in P} \sum_{t=1}^T f_t(\mathbf{x})$. Recall the definition of regret:

$$\text{Regret}_T(OGD) = \sum_{t=1}^T \mathbf{f}_t(\mathbf{x}_t) - \sum_{t=1}^T \mathbf{f}_t(\mathbf{x}^*).$$

Denote $\nabla_t \triangleq \nabla \mathbf{f}_t(\mathbf{x}_t)$. By α-strong convexity, we have

$$f_t(\mathbf{x}^*) \geq f_t(\mathbf{x}_t) + \nabla_t^\top (\mathbf{x}^* - \mathbf{x}_t) + \frac{\alpha}{2}\|\mathbf{x}^* - \mathbf{x}_t\|^2$$

$$2(f_t(\mathbf{x}_t) - f_t(\mathbf{x}^*)) \leq 2\nabla_t^\top (\mathbf{x}_t - \mathbf{x}^*) - \alpha\|\mathbf{x}^* - \mathbf{x}_t\|^2. \tag{10.4}$$

Following Zinkevich's analysis, we upper-bound $\nabla_t^\top (\mathbf{x}_t - \mathbf{x}^*)$. Using the update rule for $\mathbf{x}_{t+1}$ and the generalized Pythagorian inequality (Cesa-Bianchi and Lugosi (2006), lemma 11.3), we get

$$\|\mathbf{x}_{t+1} - \mathbf{x}^*\|^2 = \|\Pi(\mathbf{x}_t - \eta_{t+1}\nabla_t) - \mathbf{x}^*\|^2 \leq \|\mathbf{x}_t - \eta_{t+1}\nabla_t - \mathbf{x}^*\|^2.$$

Hence,

$$\|\mathbf{x}_{t+1} - \mathbf{x}^*\|^2 \leq \|\mathbf{x}_t - \mathbf{x}^*\|^2 + \eta_{t+1}^2\|\nabla_t\|^2 - 2\eta_{t+1}\nabla_t^\top (\mathbf{x}_t - \mathbf{x}^*).$$

Then, shifting sides,

$$2\nabla_t^\top (\mathbf{x}_t - \mathbf{x}^*) \leq \frac{\|\mathbf{x}_t - \mathbf{x}^*\|^2 - \|\mathbf{x}_{t+1} - \mathbf{x}^*\|^2}{\eta_{t+1}} + \eta_{t+1}G^2. \tag{10.5}$$

Summing (10.5) from $t = 1$ to T. Set $\eta_{t+1} = 1/(\alpha t)$ and, using (10.4), we have

$$2 \sum_{t=1}^{T} \mathbf{f}_t(\mathbf{x}_t) - \mathbf{f}_t(\mathbf{x}^*) \leq \sum_{t=1}^{T} \|\mathbf{x}_t - \mathbf{x}^*\|^2 \left(\frac{1}{\eta_{t+1}} - \frac{1}{\eta_t} - \alpha \right) + G^2 \sum_{t=1}^{T} \eta_{t+1}$$

$$= G^2 \sum_{t=1}^{T} \frac{1}{\alpha t} \leq \frac{G^2}{\alpha}(1 + \log T). \quad \square$$

10.5   Recent Applications

In this section we describe two recent applications of the convex optimization view to regret minimization which have resolved open questions in the field.

10.5.1   Bandit Linear Optimization

The first application is to the bandit linear optimization problem. Online linear optimization is a special case of online convex optimization in which the loss functions are linear (such as analyzed for the RFTL algorithm). In the bandit version, called bandit linear optimization, the only feedback available to the decision maker is the loss (rather than the entire loss function). This general framework naturally captures important problems such as online routing and online ad-placement for search engine results.

This generalization was put forth by Awerbuch and Kleinberg (2008) in the context of the online shortest path problem. Awerbuch and Kleinberg (2008) gave an efficient algorithm for the problem with a suboptimal regret bound, and conjectured the existence of an efficient and optimal regret algorithm.

The problem attracted much attention in the machine learning community (Flaxman et al., 2005; Dani and Hayes, 2006; Dani et al., 2008; Bartlett et al., 2008). This question was finally resolved in (Abernethy et al., 2008) where an efficient and optimal expected regret algorithm was described. Later Abernethy and Rakhlin (2009) gave an efficient algorithm which also attains this optimal regret bound with high probability. The paper introduced the use of self-concordant barrier functions as a regularization in the RFTL framework. Self-concordant barriers are a powerful tool from optimization which has enabled researchers in operations research to develop efficient polynomial-time algorithms for (offline) convex optimization. The scope of this deep technical issue is beyond this survey, but the resolution of this open question is an excellent example of how the convex optimization approach to regret minimization led to the discovery of powerful tools which in turn resolved fundamental questions in machine learning.

10.5.2   Variational Regret Bounds

A cornerstone of modern machine learning are algorithms for prediction from expert advice, the first example of regret minimization we described. It is already well-established that there exist algorithms that, under fully adversarial cost sequences, attain average cost approaching that of the best expert in hindsight. More precisely, there exist efficient algorithms which attain regret of $O(\sqrt{T \log n})$ in the setting of prediction from expert advice with n experts.

However, a priori it is not clear why online learning algorithms should have high regret (growing with the number of iterations) in an unchanging environment. As an extreme example, consider a setting in which there are only two experts. Suppose that the first expert always incurs cost 1, whereas the second expert always incurs cost $\frac{1}{2}$. One would expect to figure out this pattern quickly, and focus on the second expert, thus incurring a total cost that is at most $\frac{T}{2}$ plus at most a constant extra cost (irrespective of the number of rounds T), thus having only constant regret. However, for a long time all analyses of expert learning algorithms gave only a regret bound of $\Theta(\sqrt{T})$ in this simple case (or very simple variations of it).

More generally, the natural bound on the regret of a "good" learning algorithm should depend on *variation* in the sequence of costs, rather than purely on the number of iterations. If the cost sequence has low variation, we expect our algorithm to be able to perform better.

This intuition has a direct analog in the stochastic setting: here, the sequence of experts' costs is independently sampled from a distribution. In this situation, a natural bound on the rate of convergence to the optimal expert is controlled by the variance of the distribution (low variance should imply faster convergence). This conjecture was formalized by Cesa-Bianchi, Mansour and Stoltz (henceforth the "CMS conjecture") in (Cesa-Bianchi et al., 2007), who assert that *"proving such a rate in the fully adversarial setting would be a fundamental result"*.

The CMS conjecture was proved in the more general case of online linear optimization in Hazan and Kale (2008). Again, the convex optimization view was instrumental in the solution, and taking the general linear optimization view, it was found that a simple geometric argument implies the result. Further work on variational bounds included an extension to the bandit linear optimization setting (Hazan and Kale, 2009a) and to exp-concave loss functions including the problem of portfolio selection (Hazan and Kale, 2009b).

10.6 References

J. Abernethy and A. Rakhlin. Beating the adaptive bandit with high probability. In *Proceedings of the 22nd Annual Conference on Learning Theory*, 2009.

J. Abernethy, E. Hazan, and A. Rakhlin. Competing in the dark: An efficient algorithm for bandit linear optimization. In *Proceedings of the 21st Annual Conference on Learning Theory*, pages 263–274, 2008.

B. Awerbuch and R. Kleinberg. Online linear optimization and adaptive routing. *J. Comput. Syst. Sci.*, 74(1):97–114, 2008.

P. L. Bartlett, V. Dani, T. P. Hayes, S. Kakade, A. Rakhlin, and A. Tewari. High-probability regret bounds for bandit online linear optimization. In *Proceedings of the 21st Annual Conference on Learning Theory*, pages 335–342, 2008.

N. Cesa-Bianchi and G. Lugosi. *Prediction, Learning, and Games.* Cambridge University Press, 2006.

N. Cesa-Bianchi, Y. Mansour, and G. Stoltz. Improved second-order bounds for prediction with expert advice. *Machine Learning*, 66(2–3):321–352, 2007.

T. Cover. Universal portfolios. *Math. Finance*, 1(1):1–19, 1991.

T. Cover and J. Thomas. *Elements of Information Theory.* John Wiley, 1991.

V. Dani and T. P. Hayes. Robbing the bandit: Less regret in online geometric optimization against an adaptive adversary. In *Proceedings of the 16th ACM-SIAM Symposium on Discrete Algorithms*, pages 937–943, 2006.

V. Dani, T. Hayes, and S. Kakade. The price of bandit information for online optimization. In J. Platt, D. Koller, Y. Singer, and S. Roweis, editors, *Advances in Neural Information Processing Systems 20*. MIT Press, Cambridge, MA, 2008.

A. Flaxman, A. T. Kalai, and H. B. McMahan. Online convex optimization in the bandit setting: Gradient descent without a gradient. In *Proceedings of the 16th Annual ACM-SIAM Symposium on Discrete Algorithms*, pages 385–394, 2005.

A. J. Grove, N. Littlestone, and D. Schuurmans. General convergence results for linear discriminant updates. *Machine Learning*, 43(3):173–210, 2001.

J. Hannan. Approximation to bayes risk in repeated play. *In M. Dresher, A. W. Tucker, and P. Wolfe, editors, Contributions to the Theory of Games, volume 3*, pages 97–139, 1957.

E. Hazan and S. Kale. Extracting certainty from uncertainty: Regret bounded by variation in costs. In *The 21st Annual Conference on Learning Theory (COLT)*, pages 57–68, 2008.

E. Hazan and S. Kale. Better algorithms for benign bandits. In C. Mathieu, editor, *ACM-SIAM Symposium on Discrete Algorithms (SODA).*, pages 38–47. SIAM, 2009a.

E. Hazan and S. Kale. On stochastic and worst-case models for investing. In *Advances in Neural Information Processing Systems 22*. MIT Press, 2009b.

E. Hazan, A. Agarwal, and S. Kale. Logarithmic regret algorithms for online convex optimization. *Machine Learning*, 69(2-3):169–192, 2007.

A. Kalai and S. Vempala. Efficient algorithms for online decision problems. *Journal of Computer and System Sciences*, 71(3):291–307, 2005.

J. Kivinen and M. K. Warmuth. Relative loss bounds for multidimensional regression problems. *Machine Learning*, 45(3):301–329, 2001.

E. Takimoto and M. K. Warmuth. Path kernels and multiplicative updates. *Journal of Machine Learning Research*, 4:773–818, 2003. special issue on learning theory.

M. Zinkevich. Online convex programming and generalized infinitesimal gradient ascent. In *Proceedings of the 20th International Conference on Machine Learning*, pages 928–936, 2003.

Appendix: The FTL-BTL Lemma

The following proof is essentially due to Kalai and Vempala (2005).

Proof of Lemma 10.2. For convenience, denote by $\mathbf{f}_0 = \frac{1}{\eta}\mathcal{R}$, and assume we start the algorithm from $t = 0$ with an arbitrary $\mathbf{x}_0$. The lemma is now proved by induction on T.

Induction base: Note that by definition, we have that $\mathbf{x}_1 = \arg\min_\mathbf{x}\{\mathcal{R}(\mathbf{x})\}$, and thus $\mathbf{f}_0(\mathbf{x}_1) \leq \mathbf{f}_0(\mathbf{u})$ for all $\mathbf{u}$, and $\mathbf{f}_0(\mathbf{x}_0) - \mathbf{f}_0(\mathbf{u}) \leq \mathbf{f}_0(\mathbf{x}_0) - \mathbf{f}_0(\mathbf{x}_1)$.

Induction step: Assume that that for T, we have

$$\sum_{t=0}^{T} \mathbf{f}_t(\mathbf{x}_t) - \mathbf{f}_t(\mathbf{u}) \leq \sum_{t=0}^{T} \mathbf{f}_t(\mathbf{x}_t) - \mathbf{f}_t(\mathbf{x}_{t+1}),$$

and let us prove for $T + 1$. Since $\mathbf{x}_{T+2} = \arg\min_\mathbf{x}\{\sum_{t=0}^{T+1} \mathbf{f}_t(\mathbf{x})\}$, we have

$$\sum_{t=0}^{T+1} \mathbf{f}_t(\mathbf{x}_t) - \sum_{t=0}^{T+1} \mathbf{f}_t(\mathbf{u})$$
$$\leq \sum_{t=0}^{T+1} \mathbf{f}_t(\mathbf{x}_t) - \sum_{t=0}^{T+1} \mathbf{f}_t(\mathbf{x}_{T+2})$$
$$= \sum_{t=0}^{T} (\mathbf{f}_t(\mathbf{x}_t) - \mathbf{f}_t(\mathbf{x}_{T+2})) + \mathbf{f}_{T+1}(\mathbf{x}_{T+1}) - \mathbf{f}_{T+1}(\mathbf{x}_{T+2})$$
$$\leq \sum_{t=0}^{T} (\mathbf{f}_t(\mathbf{x}_t) - \mathbf{f}_t(\mathbf{x}_{t+1})) + \mathbf{f}_{T+1}(\mathbf{x}_{T+1}) - \mathbf{f}_{T+1}(\mathbf{x}_{T+2})$$
$$= \sum_{t=0}^{T+1} \mathbf{f}_t(\mathbf{x}_t) - \mathbf{f}_t(\mathbf{x}_{t+1}),$$

where in the fourth line we used the induction hypothesis for $\mathbf{u} = \mathbf{x}_{T+2}$. We conclude that

$$\sum_{t=1}^{T} \mathbf{f}_t(\mathbf{x}_t) - \mathbf{f}_t(\mathbf{u})$$
$$\leq \sum_{t=1}^{T} \mathbf{f}_t(\mathbf{x}_t) - \mathbf{f}_t(\mathbf{x}_{t+1}) + [-\mathbf{f}_0(\mathbf{x}_0) + \mathbf{f}_0(\mathbf{u}) + \mathbf{f}_0(\mathbf{x}_0) - \mathbf{f}_0(\mathbf{x}_1)]$$
$$= \sum_{t=1}^{T} \mathbf{f}_t(\mathbf{x}_t) - \mathbf{f}_t(\mathbf{x}_{t+1}) + \frac{1}{\eta}[\mathcal{R}(\mathbf{u}) - \mathcal{R}(\mathbf{x}_1)].$$

$\square$

11   Projected Newton-type Methods in Machine Learning

Mark Schmidt   schmidtmarkw@gmail.com
University of British Columbia
Vancouver, BC, V6T 1Z4

Dongmin Kim   dmkim@cs.utexas.edu
University of Texas at Austin
Austin, Texas 78712

Suvrit Sra   suvrit@tuebingen.mpg.de
Max Planck Institute for Intelligent Systems
72076, Tübingen, Germany

We consider projected Newton-type methods for solving large-scale optimization problems arising in machine learning and related fields. We first introduce an algorithmic framework for projected Newton-type methods by reviewing a canonical projected (quasi-)Newton method. This method, while conceptually pleasing, has a high computation cost per iteration. Thus, we discuss two variants that are more scalable: two-metric projection and inexact projection methods. Finally, we show how to apply the Newton-type framework to handle nonsmooth objectives. Examples are provided throughout the chapter to illustrate machine learning applications of our framework.

11.1   Introduction

We study Newton-type methods for solving the optimization problem

$$\min_{\boldsymbol{x}} \quad f(\boldsymbol{x}) + r(\boldsymbol{x}), \quad \text{subject to} \quad \boldsymbol{x} \in \Omega, \tag{11.1}$$

where $f : \mathbb{R}^n \to \mathbb{R}$ is twice continuously differentiable and convex; $r : \mathbb{R}^n \to \mathbb{R}$ is continuous and convex, but not necessarily differentiable everywhere; and Ω is a simple convex constraint set. This formulation is general and captures numerous problems in machine learning, especially where f corresponds to a loss, and r to a regularizer. Let us, however, defer concrete examples of (11.1) until we have developed some theoretical background.

We propose to solve (11.1) via Newton-type methods, a certain class of second-order methods that are known to often work well for unconstrained problems. For constrained problems too, we may consider Newton-type methods that, akin to their unconstrained versions, iteratively minimize a quadratic approximation to the objective, this time subject to constraints. This idea dates back to Levitin and Polyak (1966, §7), and it is referred to as a *projected Newton* method.

Projected Newton methods for optimization over convex sets share many of the appealing properties of their unconstrained counterparts. For example, their iterations are guaranteed to improve the objective function for a small enough stepsize; global convergence can be shown under a variant of the Armijo condition; and rapid local convergence rates can be shown around local minima satisfying strong convexity (Bertsekas, 1999). In a similar vein, we may consider projected quasi-Newton methods, where we interpolate differences in parameter and gradient values to approximate the Hessian matrix. The resulting *Newton-type* methods are the subject of this chapter, and we will focus particularly on the limited memory Broyden-Fletcher-Goldfarb-Shanno (L-BFGS) quasi-Newton approximation. The main appeals of the L-BFGS approximation are its linear time iteration complexity and its strong empirical performance on a variety of problems.

The remainder of this chapter is organized as follows. We first restrict ourselves to smooth optimization, where $r(\boldsymbol{x}) = 0$. For this setting, we describe projected Newton-type methods (Section 11.2), covering basic implementation issues such as Hessian approximation and line-search. Then, we describe two-metric projection methods (Section 11.3), followed by inexact (or truncated) projected-Newton methods (Section 11.4). Finally, we discuss the nonsmooth setting, where $r(\boldsymbol{x}) \neq 0$, for which we describe two Newton-type methods (Section 11.5).

11.2   Projected Newton-type Methods

Projected Newton-type methods optimize their objective iteratively. At iteration k, they first approximate the objective function around the current iterate $\boldsymbol{x}^k$ by the quadratic model

$$Q^k(\boldsymbol{x}, \alpha) \triangleq f(\boldsymbol{x}^k) + (\boldsymbol{x} - \boldsymbol{x}^k)^T \nabla f(\boldsymbol{x}^k) + \frac{1}{2\alpha}(\boldsymbol{x} - \boldsymbol{x}^k)^T \boldsymbol{H}^k (\boldsymbol{x} - \boldsymbol{x}^k). \quad (11.2)$$

This model is parameterized by a positive stepsize α, and it uses a positive definite matrix $\boldsymbol{H}^k$ to approximate the Hessian $\nabla^2 f(\boldsymbol{x}^k)$. To generate the next iterate that decreases the objective while remaining feasible, the methods minimize the quadratic model (11.2) over the (*convex*) constraint set Ω. Thus, for a fixed $\alpha > 0$, they compute the unique element

$$\bar{\boldsymbol{x}}_\alpha^k = \underset{\boldsymbol{x} \in \Omega}{\operatorname{argmin}} \ Q^k(\boldsymbol{x}, \alpha), \quad (11.3)$$

which is then used to obtain the new iterate by simply setting

$$\boldsymbol{x}^{k+1} \leftarrow \boldsymbol{x}^k + \beta(\bar{\boldsymbol{x}}_\alpha^k - \boldsymbol{x}^k), \quad (11.4)$$

where $\beta \in (0, 1]$ is another stepsize. To ensure a sufficient decrease in the objective value, one typically begins by setting $\alpha = \beta = 1$, and then decreases one of them until $\boldsymbol{x}^{k+1}$ satisfies the following Armijo condition[1]

$$f(\boldsymbol{x}^{k+1}) \le f(\boldsymbol{x}^k) + \nu \langle \nabla f(\boldsymbol{x}^k), \boldsymbol{x}^{k+1} - \boldsymbol{x}^k \rangle, \qquad \nu \in (0, 1). \quad (11.5)$$

We collect the above described steps into Algorithm 11.1, which we present as the general framework for projected Newton-type methods.

Algorithm 11.1 A projected Newton-type method.

Given $\boldsymbol{x}^0 \in \Omega$, $\boldsymbol{H}^0 \succ \boldsymbol{0}$
for $k = 0, \dots,$ until some stopping criteria met **do**
    Step I: Build $Q^k(\boldsymbol{x}, \alpha)$ using (11.2)
    repeat
        Step IIa: Minimize $Q^k(\boldsymbol{x}, \alpha)$ over Ω
        Step IIb: Update $\boldsymbol{x}^{k+1} \leftarrow \boldsymbol{x}^k + \beta(\bar{\boldsymbol{x}}_\alpha^k - \boldsymbol{x}^k)$
        Step III: Update α and/or β
    until descent condition (11.5) is satisfied
end for

Convergence properties of various forms of this method are discussed, for example, in Bertsekas (1999, Section 2.3). In particular, convergence to a stationary point can be shown under the assumption that the eigenvalues of $\boldsymbol{H}^k$ are bounded between two positive constants. Also, if $\boldsymbol{x}^*$ is a minimizer of $f(\boldsymbol{x})$ over Ω satisfying certain conditions, and once $\boldsymbol{x}^k$ is sufficiently close

---
1. A typical value for the sufficient decrease parameter ν is 10^{-4}.

to $\boldsymbol{x}^*$, then $\alpha = 1$ and $\beta = 1$ are accepted as stepsizes and the sequence $\|\boldsymbol{x}^k - \boldsymbol{x}^*\|$ converges to zero at a superlinear rate.

Algorithm 11.1 is conceptually simple, and thus appealing. It can, however, have numerous variants, depending on how each step is implemented. For example, in Step I, which particular quadratic model is used; in Step II, how we minimize the model function; and in Step III, how we compute the stepsizes α and β. For each of these three steps there are multiple possible choices, and consequently, different combinations lead to methods of differing character. We describe some popular implementation choices below.

11.2.1 Building a Quadratic Model

When a positive definite Hessian is readily available, we can simply set $\boldsymbol{H}^k = \nabla^2 f(\boldsymbol{x}^k)$. By doing so, the quadratic model (11.2) becomes merely the approximation obtained via a second-order Taylor expansion of f. This model leads to computation of an *exact* Newton step at each iteration. At the other extreme, if we select $\boldsymbol{H}^k = \boldsymbol{I}$, the identity matrix of appropriate size, then the search direction of the resulting method reduces to the negative gradient, essentially yielding the projected gradient method. These two strategies often contrast with each other in terms of computing a search (descent) direction: the Newton step is considered one of the most sophisticated, while the gradient step is regarded as one of the simplest. In cases where we can efficiently compute the Euclidean projection operator, projected gradient steps have a low per-iteration computational cost. However, this benefit comes at the expense of linear convergence speed. The Newton step is usually more expensive; Step IIb will typically be costly to solve even if we can efficiently compute the Euclidean projection onto the constraint set. However, the more expensive Newton step generally enjoys a local superlinear convergence rate.

Despite its theoretical advantages, an exact Newton step often is resource intensive, especially when computing the exact Hessian is expensive. To circumvent some of the associated computational issues, one usually approximates the Hessian: this idea underlies the well-known quasi-Newton approximation. Let us therefore briefly revisit the BFGS update that approximates the exact Hessian.

11.2.1.1 BFGS *Update*

There exist several approximations to the Hessian, such as the Powell-Symmetric-Broyden (PSB), the Davidson-Fletcher-Powell (DFP), and the Broyden-Fletcher-Goldfarb-Shanno (BFGS). We focus on BFGS because it is believed to be the most effective in general (Gill et al., 1981; Bertsekas, 1999).

First, define the difference vectors g and s as follows:

$$g = \nabla f(x^{k+1}) - \nabla f(x^k), \quad \text{and} \quad s = x^{k+1} - x^k.$$

Now, assume we already have H^k, the current approximation to the Hessian. Then, the BFGS update adds a rank-two correction to H^k to obtain

$$H^{k+1} = H^k - \frac{H^k ss^T H^k}{s^T H^k s} + \frac{gg^T}{s^T g}. \tag{11.6}$$

We can plug H^{k+1} into (11.2) to obtain an updated model Q^{k+1}. But depending on the implementation of subsequent steps (11.3) and (11.4), it might be more convenient and computationally efficient to update an estimate to the inverse of H^k instead. For this case, we can apply the Sherman-Morrison-Woodbury formula to (11.6), thus obtaining the update

$$S^{k+1} = S^k + \left(1 + \frac{g^T S^k g}{s^T g}\right)\frac{ss^T}{s^T g} - \frac{(S^k gs^T + sg^T S^k)}{s^T g}, \tag{11.7}$$

where S^k is the inverse of H^k, also known as the *gradient scaling* matrix.

11.2.1.2 *Limited Memory* BFGS *Update.*

Though the BFGS update may greatly relieve the burden of Hessian computation, it still requires the same storage: $\mathcal{O}(n^2)$ for dense H^k or S^k, which is troublesome for large-scale problems. This difficulty is addressed by the limited memory BFGS (L-BFGS) update, where, instead of using full matrices H^k and S^k, a small number of vectors, say m, are used to approximate the Hessian or its inverse. The standard L-BFGS approach (Nocedal, 1980) can be implemented using the following formula (Nocedal and Wright, 2000)

$$\begin{aligned}
S^k = &\frac{s_{k-1}^T g_{k-1}}{g_{k-1}^T g_{k-1}} \bar{V}_{k-M}^T \bar{V}_{k-M} + \rho_{k-M} \bar{V}_{k-M+1}^T s_{k-M} s_{k-M}^T \bar{V}_{k-M+1} \\
&+ \rho_{k-M+1} \bar{V}_{k-M+2}^T s_{k-M+1} s_{k-M+1}^T \bar{V}_{k-M+2} \\
&+ \cdots \\
&+ \rho_{k-1} s_{k-1} s_{k-1}^T,
\end{aligned} \tag{11.8}$$

for $k \geq 1$; the scalars ρ_k and matrices $\bar{V}_{k-M}$ are defined by

$$\rho_k = 1/(s_k^T g_k), \quad \bar{V}_{k-M} = [V_{k-M} \cdots V_{k-1}], \quad \text{and} \quad V_k = I - \rho_k s_k g_k^T.$$

The L-BFGS approximation requires only $\mathcal{O}(mn)$ storage; moreover, multiplication of $\boldsymbol{S}^k$ by a vector can also be performed at this cost.

For both BFGS and L-BFGS, a choice that can significantly impact performance is the initial approximation $\boldsymbol{H}^0$. A typical strategy to select $\boldsymbol{H}^0$ is to set it to the negative gradient direction on the first iteration, and then to set $\boldsymbol{H}^0 = (\boldsymbol{g}^T\boldsymbol{g})/(\boldsymbol{g}^T\boldsymbol{s})\boldsymbol{I}$ on the next iteration. This choice was proposed by Shanno and Phua (1978) to optimize the condition number of the approximation. In the L-BFGS method we can reset $\boldsymbol{H}^0$ using this formula after *each* iteration (Nocedal and Wright, 2000, Section 7.2). With this strategy, the unit stepsizes of $\alpha = 1$ and $\beta = 1$ are typically accepted, which may remove the need for a line search on most iterations.

Provided that $\boldsymbol{H}^0$ is positive definite, the subsequent (implicit) Hessian approximations $\boldsymbol{H}^k$ generated by the L-BFGS update are guaranteed to be positive definite as long as $\boldsymbol{g}^T\boldsymbol{s}$ is positive (Nocedal and Wright, 2000, Section 6.1). This positivity is guaranteed if $f(\boldsymbol{x})$ is strongly convex, but when $f(\boldsymbol{x})$ is not strongly convex a more advanced strategy is required, see for instance Nocedal and Wright (2000, Section 18.3).

11.2.2 Solving the Subproblem

With our quadratic approximation $\mathcal{Q}^k(\boldsymbol{x}, \alpha)$ in hand, the next step is to solve the subproblem (11.3). For $\alpha \neq 0$, simple rearrangement shows that[2]

$$\bar{\boldsymbol{x}}_\alpha^k = \operatorname*{argmin}_{\boldsymbol{x} \in \Omega} \; \mathcal{Q}^k(\boldsymbol{x}, \alpha) = \operatorname*{argmin}_{\boldsymbol{x} \in \Omega} \; \frac{1}{2}\|\boldsymbol{x} - \boldsymbol{y}^k\|_{\boldsymbol{H}^k}^2, \qquad (11.9)$$

where $\|\boldsymbol{x}\|_{\boldsymbol{H}^k}$ is defined by the norm $\sqrt{\boldsymbol{x}^T\boldsymbol{H}^k\boldsymbol{x}}$, and $\boldsymbol{y}^k$ is the unconstrained Newton step: $\boldsymbol{y}^k = \boldsymbol{x}^k - \alpha[\boldsymbol{H}^k]^{-1}\nabla f(\boldsymbol{x}^k)$. In words, $\bar{\boldsymbol{x}}_\alpha^k$ is obtained by projecting the Newton step onto the constraint set Ω, where projection is with respect to the metric defined by the Hessian approximation $\boldsymbol{H}^k$.

One major drawback of (11.9) is that it can be computationally challenging, even when Ω has relatively simple structure. To ease the computational burden, instead of using the metric defined by $\boldsymbol{H}^k$, we could compute the projection under the standard Euclidean norm while slightly modifying the Newton step to ensure convergence. This is the subject of Section 11.3. Alternatively, in Section 11.4 we consider computing an *approximate* solution to (11.9) itself.

Note that if we replace $\boldsymbol{H}^k$ with $\boldsymbol{I}$, both as the projection metric and in

---

2. If we use $\boldsymbol{S}^k$, the inverse of the Hessian, then $\bar{\boldsymbol{x}}_\alpha^k$ may be equivalently obtained by solving $\operatorname{argmin}_{\boldsymbol{x} \in \Omega} \frac{1}{2}\|\boldsymbol{x} - (\boldsymbol{x}^k - \alpha\boldsymbol{S}^k\nabla f(\boldsymbol{x}^k))\|_{[\boldsymbol{S}^k]^{-1}}^2 = \operatorname{argmin}_{\boldsymbol{x} \in \Omega} \frac{1}{2}\|\boldsymbol{x} - \boldsymbol{y}^k\|_{[\boldsymbol{S}^k]^{-1}}^2$.

the Newton step, we recover gradient projection methods.

11.2.3 Computing the Stepsizes

Consider the stepsizes α and β in (11.3) and (11.4). Generally speaking, any positive α and β that generate x^{k+1} satisfying the descent condition (11.5) are acceptable. Practical choices are discussed below.

11.2.3.1 Backtracking

Suppose we fix $\alpha = 1$ for all k, and let $d^k = \bar{x}_1^k - x^k$. Then we obtain the following update:

$$x^{k+1} \leftarrow x^k + \beta d^k.$$

To select β, we can simply start with $\beta = 1$ and iteratively decrease β until the resulting x^{k+1} satisfies (11.5)[3]. More formally, we set $\beta = \tau \cdot \sigma^m$ for some $\tau > 0$ and $\sigma \in (0,1)$, where $m \geq 0$ is the first integer that satisfies

$$f(x^{k+1}) \leq f(x^k) + \tau \cdot \sigma^m \nabla f(x^k)(x^{k+1} - x^k).$$

Several strategies are available to reduce the number of backtracking iterations. For example, rather than simply dividing the stepsize by a constant, we can use information collected about the function during the line search to make a more intelligent choice. For example, if some trial value of β is not accepted, then we can set the next β to the minimum of the quadratic polynomial that has a value of $f(x^k)$ at zero, $f(x^k + \beta d^k)$ at β, and a slope of $\nabla f(x^k)^T d^k$ at zero (Nocedal and Wright, 2000, Section 3.5). This choice gives the optimal stepsize if $f(x)$ is a quadratic function, and often drastically reduces the number of backtracking iterations needed. For some functions, quadratic interpolation can also be used to give a more intelligent choice than $\beta = 1$ for the *first* trial value of β, while cubic interpolation can be used if we have tested more than one value of β or if we compute $\nabla f(x^k + \beta d^k)$ for the trial values of β (Nocedal and Wright, 2000, Section 3.5).

11.2.3.2 Backtracking (Armijo) along projection arc (Bertsekas, 1999)

Alternatively, we can set $\beta = 1$ for all k to obtain

$$x^{k+1} \leftarrow \bar{x}_\alpha^k,$$

---

3. Also known as Armijo backtracking along a feasible direction.

and then determine an α satisfying (11.5). Similar to simple backtracking, we compute $\alpha = s \cdot \sigma^m$ for some $s, \tau > 0$, and $\sigma \in (0,1)$, where $m \geq 0$ is the first integer that satisfies

$$f(\bar{x}^k_{s \cdot \sigma^m}) \leq f(x^k) + \tau \nabla f(x^k)(\bar{x}^k_{s \cdot \sigma^m} - x^k).$$

Unlike simple backtracking that searches along a line segment as β varies, this strategy searches along a potentially nonlinear path as α varies. Because of this, polynomial interpolation to select trial values of α is more tenuous than for simple backtracking, but on many problems polynomial interpolation still significantly decreases the number of trial values evaluated.

This stepsize computation might be more involved when computing projections onto Ω is expensive, since it requires solving an optimization problem to compute $\bar{x}^k_\alpha$ for each trial value of α. However, it can still be appealing because it is more likely to yield iterates that lie on the boundaries of the constraints. This property is especially useful when the boundaries of the constraints represent a solution of interest, such as *sparse* solutions with the constraint set $\Omega = \mathbb{R}^n_+$.

We next consider a few specific instantiations of the general framework introduced in this section. Specifically, we first consider two-metric projection methods for the specific case of bound-constrained problems (Section 11.3). Subsequently, we consider inexact projected Newton methods for optimization over more general simple convex sets (Section 11.4). Finally, we explore the versatility of the framework by extending it to problems with nonsmooth objective functions (Section 11.5).

11.3   Two-Metric Projection Methods

As mentioned earlier, computing the projection with respect to a quadratic norm defined by H^k can be computationally challenging. However, we often encounter problems with simple convex domains, onto which we can efficiently compute Euclidean projections. For optimization over such domains, we might therefore prefer projecting the Newton step under the Euclidean norm. Indeed, this choice is made by the well-known *two-metric projection* method, so named because it uses different matrices (metrics) for scaling the gradient and for computing the projection.

In two-metric projection algorithms, we can benefit from low iteration complexity if we use L-BFGS approximations. However, some problems still persist: the "obvious" procedure with an unmodified Newton step may not improve on the objective function, even for an arbitrarily small positive stepsize. Nevertheless, there are many cases where one can derive a two-metric projection method that can dodge this drawback without giving up the attractive properties of its unconstrained counterpart. A particular example is Bertsekas's projected Newton method (Bertsekas, 1982), and we discuss it below for the case where Ω consists of bound constraints.

The projected-Newton method may be viewed in light of Algorithm 11.1. Specifically, it takes the Hessian $\nabla^2 f(\boldsymbol{x}^k)$ and modifies its *inverse* so that the gradient scaling matrix $\boldsymbol{S}^k$ has a special structure. It subsequently invokes orthogonal projection in Step II, and then in Step III, it computes its stepsize using backtracking along the projection arc. The key variation from Algorithm 11.1 lies in how to modify the inverse Hessian to obtain a valid gradient scaling; the details follow below.

11.3.1 Bound Constrained Smooth Convex Problems

Consider the following special case of (11.1)

$$\min_{\boldsymbol{x}\in\mathbb{R}^n} \quad f(\boldsymbol{x}), \quad \text{subject to} \quad \boldsymbol{l} \le \boldsymbol{x} \le \boldsymbol{u}, \tag{11.10}$$

where $\boldsymbol{l}$ and $\boldsymbol{u}$ are fixed vectors, and inequalities are taken componentwise (which can be set to ∞ or $-\infty$ if the variables are unbounded). The function f is assumed to be convex and twice continuously differentiable. Such bound-constrained problems arise as Lagrangian duals of problems with convex inequality constraints, or when we have natural restrictions (e.g., nonnegativity) on the variables. For bound-constrained problems the projection under the Euclidean norm is the standard orthogonal projection obtained by taking componentwise medians among l_i, u_i, and x_i:

$$[\mathcal{P}(\boldsymbol{x})]_i \triangleq \operatorname{mid}\{l_i, x_i, u_i\}.$$

At each iteration, we partition the variables into two groups: *free* and *restricted*. Restricted variables are defined as a particular subset of the variables close to their bounds, based on the sign of the corresponding components in the gradient. Formally, the set of *restricted* variables is

$$\mathcal{I}^k \triangleq \left\{ i \mid x_i^k \le l_i + \varepsilon \wedge \partial_i f(\boldsymbol{x}^k) > 0, \quad \text{or} \quad x_i^k \ge u_i - \varepsilon \wedge \partial_i f(\boldsymbol{x}^k) < 0 \right\},$$

for some small positive ε. The set $\mathcal{I}^k$ collects variables that are near their bounds, *and* for which the objective $f(\boldsymbol{x})$ can be decreased by moving the variables toward (or past) their bounds. The set of *free* variables, denoted $\mathcal{F}^k$, is simply defined as the complement of $\mathcal{I}^k$ in the set $\{1, 2, \ldots, n\}$.

Without loss of generality, let us assume that $\mathcal{F}^k = \{1, 2, \cdots, N\}$ and $\mathcal{I}^k = \{N+1, \cdots, n\}$. Now define a diagonal matrix $\boldsymbol{D}^k \in \mathbb{R}^{n-N \times n-N}$ that scales the *restricted* variables, a typical choice being the identity matrix. We denote the scaling with respect to the free variables as $\bar{\boldsymbol{S}}^k \in \mathbb{R}^{N \times N}$, which, for the projected-Newton method, is given by the principal submatrix of the inverse of the Hessian $\nabla^2 f(\boldsymbol{x}^k)$, as induced by the free variables. In symbols, this is

$$\bar{\boldsymbol{S}}^k \leftarrow [\nabla^2 f(\boldsymbol{x}^k)]^{-1}_{\bar{\mathfrak{F}}^k}. \tag{11.11}$$

With these definitions, we are now ready to present the main step of the two-metric projection algorithm. This step can be written as the Euclidean projection of a Newton step that uses a gradient scaling $\boldsymbol{S}^k$ of the form

$$\boldsymbol{S}^k \triangleq \begin{bmatrix} \bar{\boldsymbol{S}}^k & \boldsymbol{0} \\ \boldsymbol{0} & \boldsymbol{D}^k \end{bmatrix}. \tag{11.12}$$

The associated stepsize α can be selected by backtracking along the projection arc until the Armijo condition is satisfied. Note that this choice of the stepsize computation does not increase the computational complexity of the method, since computing the orthogonal projection after each backtracking step is trivial. Combining this gradient scaling with orthogonal projection, we obtain the projected Newton update:

$$
\begin{aligned}
\boldsymbol{x}^{k+1} \leftarrow \bar{\boldsymbol{x}}^k_\alpha &= \underset{\boldsymbol{l} \leq \boldsymbol{x} \leq \boldsymbol{u}}{\operatorname{argmin}} \frac{1}{2} \|\boldsymbol{x} - (\boldsymbol{x}^k - \alpha \boldsymbol{S}^k \nabla f(\boldsymbol{x}^k))\|^2_{[\boldsymbol{S}^k]^{-1}} \\
&\approx \underset{\boldsymbol{l} \leq \boldsymbol{x} \leq \boldsymbol{u}}{\operatorname{argmin}} \frac{1}{2} \|\boldsymbol{x} - (\boldsymbol{x}^k - \alpha \boldsymbol{S}^k \nabla f(\boldsymbol{x}^k))\|^2_{\boldsymbol{I}} \\
&= \mathcal{P}[\boldsymbol{x}^k - \alpha^k \boldsymbol{S}^k \nabla f(\boldsymbol{x}^k)],
\end{aligned}
\tag{11.13}
$$

where α^k is computed by backtracking along the projection arc.

This algorithm has been shown to be globally convergent (Bertsekas, 1982; Gafni and Bertsekas, 1984), and under certain conditions achieves local superlinear convergence.

Theorem 11.1 (Convergence). *Assume that ∇f is Lipschitz continuous on Ω, and $\nabla^2 f$ has bounded eigenvalues. Then every limit point of $\{\boldsymbol{x}^k\}$ generated by iteration (11.13) is a stationary point of (11.10).*

Theorem 11.2 (Convergence rate). *Let f be strictly convex and twice continuously differentiable. Let $\boldsymbol{x}^*$ be the non degenerate optimum of Problem (11.13) and assume that for some $\delta > 0$, $\nabla^2 f(\boldsymbol{x})$ has bounded eigenvalues for all $\boldsymbol{x}$ that satisfy $\|\boldsymbol{x} - \boldsymbol{x}^*\| < \delta$. Then the sequence $\{\boldsymbol{x}^k\}$ generated by iteration (11.13) converges to $\boldsymbol{x}^*$, and the rate of convergence in $\{\|\boldsymbol{x}^k - \boldsymbol{x}^*\|\}$ is superlinear.*

Although the convergence rate of the two-metric projection method has been shown for $\boldsymbol{S}^k$ derived from the Hessian, the convergence itself merely requires a positive definite gradient scaling $\boldsymbol{S}^k$ with bounded eigenvalues for all k (Bertsekas, 1982). Thus, the quasi-Newton approximations introduced in Section 11.2 are viable choices to derive convergent methods,[4] and we present such variations of the two-metric method in the following example.

Example 11.1 (Nonnegative least-squares). *A problem of considerable importance in the applied sciences is the nonnegative least-squares (NNLS):*

$$\min_{\boldsymbol{x}} \quad \tfrac{1}{2}\|\boldsymbol{A}\boldsymbol{x} - \boldsymbol{b}\|_2^2, \quad \textit{subject to } \boldsymbol{x} \geq 0, \tag{11.14}$$

where $\boldsymbol{A} \in \mathbb{R}^{m \times n}$. This problem is essentially an instance of (11.10).

Given Algorithm 11.1, one can simply implement the update (11.13) and then use BFGS or L-BFGS to obtain $\boldsymbol{S}^k$. However, we can further exploit the simple constraint $\boldsymbol{x} \geq \boldsymbol{0}$ and improve the computational (empirical) efficiency of the algorithm. To see how, consider the restricted variables in the update (11.13). When variable $i \in \mathfrak{I}^k$ and ε becomes sufficiently small, we obtain

$$\mathcal{P}[\boldsymbol{x}^k - \alpha^k \boldsymbol{S}^k \nabla f(\boldsymbol{x}^k)]_i = \mathcal{P}[x_i^k - \alpha^k [\boldsymbol{D}^k]_{ii} \cdot \partial_i f(\boldsymbol{x}^k)] = 0.$$

In other words, if $i \in \mathfrak{I}^k$, then $x_i^{k+1} = 0$, whereby we can safely ignore these variables throughout the update. In an implementation, this means that we can confine computations to free *variables, which can save a large number of floating point operations, especially when $|\mathfrak{F}^k| \ll |\mathfrak{I}^k|$.*

Example 11.2 (Linear SVM). *Consider the standard binary classification task with inputs $(\boldsymbol{x}_i, y_i)_{i=1}^m$, where $\boldsymbol{x}_i \in \mathbb{R}^n$ and $y_i \in \pm 1$. Assume for simplicity that we wish to learn a bias-free decision function $f(\boldsymbol{x}) = sgn(\boldsymbol{w}^T \boldsymbol{x})$ by solving either the SVM* primal

$$\begin{aligned} \underset{\boldsymbol{w}}{\textit{minimize}} \quad & \tfrac{1}{2}\boldsymbol{w}^T\boldsymbol{w} + C\sum_{i=1}^m \xi_i \\ \textit{subject to} \quad & y_i(\boldsymbol{w}^T\boldsymbol{x}_i) \geq 1 - \xi_i, \quad \xi_i \geq 0, \quad 1 \leq i \leq m, \end{aligned} \tag{11.15}$$

or its (more familiar) dual

$$\begin{aligned} \underset{\boldsymbol{\alpha}}{\textit{minimize}} \quad & \tfrac{1}{2}\boldsymbol{\alpha}^T\boldsymbol{Y}\boldsymbol{X}^T\boldsymbol{X}\boldsymbol{Y}\boldsymbol{\alpha} - \boldsymbol{\alpha}^T\mathbf{1} \\ \textit{subject to} \quad & 0 \leq \alpha_i \leq C, \end{aligned} \tag{11.16}$$

---

4. In a simpler but still globally convergent variation of the two-metric projection method, we could simply set $\boldsymbol{S}^k$ to be a diagonal matrix with positive diagonal elements.

where $\boldsymbol{Y} = \mathrm{Diag}(y_1, \ldots, y_m)$ *and* $\boldsymbol{X} = [\boldsymbol{x}_1, \ldots, \boldsymbol{x}_m] \in \mathbb{R}^{n \times m}$. *The dual (11.16) is a special case of (11.10), and can be solved by adapting the two-metric projection method in a manner similar to that for NNLS.*

Example 11.3 (Sparse Gaussian graphical models). *The dual to a standard formulation for learning sparse Gaussian graphical models takes the form (Banerjee et al., 2006)*

$$\min_{\tilde{\Sigma} + \boldsymbol{X} \succ 0} \quad -\log\det(\tilde{\Sigma} + \boldsymbol{X}), \quad \textit{subject to } |\boldsymbol{X}_{ij}| \leq \lambda_{ij}, \quad \forall_{ij}. \qquad (11.17)$$

There has been substantial recent interest in solving this problem (Banerjee et al., 2006). Here $\tilde{\Sigma}$ represents the empirical covariance of a data set, and the bound constraints on the elements of the matrix encourage the associated graphical model to be sparse for sufficiently large values of the λ_{ij} variables.

Notice that the constraint $|\boldsymbol{X}_{ij}| \leq \lambda_{ij}$ is equivalent to the bound constraints $-\lambda_{ij} \leq \boldsymbol{X}_{ij} \leq \lambda_{ij}$. Thus, provided $\tilde{\Sigma} + \boldsymbol{X}$ is positive definite for the initial $\boldsymbol{X}$, we can apply a simplified two-metric projection algorithm to this problem in which we use projection to address the bound constraints and backtracking to modify the iterates when they leave the positive definite cone.

11.4   Inexact Projection Methods

The previous section focused on examples with bound constraints. For optimizing over more general but still simple convex sets, an attractive choice is *inexact* projected Newton methods. These methods represent a natural generalization of methods for unconstrained optimization that are alternatively referred to as *Hessian-free, truncated,* or *inexact* Newton methods. In inexact projected Newton methods, rather than finding the exact minimizer in Step IIa of Algorithm 11.1, we find an approximate minimizer using an iterative solver. That is, we use a single-metric projection, but solve the projection inexactly. Note that the iterative solver can be a first-order optimization strategy, and thus can take advantage of an efficient Euclidean projection operator. Under only mild conditions on the iterative solver, this approximate projection algorithm still leads to an improvement in the objective function. There are many ways to implement an inexact projected Newton strategy, but in this section we focus on the one described in Schmidt et al. (2009). In this method, we use the L-BFGS Hessian approximation, which we combine with simple Armijo backtracking and a variant of the projected gradient algorithm for iteratively solving subproblems. In Section 11.4.1 we review an effective iterative solver, and Section 11.4.2 we discuss using it within an inexact projected-Newton method.

11.4.1   Spectral Projected Gradient

The traditional motivation for examining projected Newton methods is that the basic gradient projection method may take a very large number of iterations to reach an acceptably accurate solution. However, there has been substantial recent interest in variants of gradient projection that exhibit much better empirical convergence properties. For example, Birgin et al. (2000) presented several *spectral* projected gradient (SPG) methods. In SPG methods, either α or β is set to 1, and the other stepsize is set to one of the stepsizes proposed by Barzilai and Borwein (1988). For example, we might set $\beta = 1$ and α to

$$\alpha^{bb} \triangleq \frac{\boldsymbol{g}^T \boldsymbol{s}}{\boldsymbol{g}^T \boldsymbol{g}}, \quad \text{where} \quad \boldsymbol{g} = \nabla f(\boldsymbol{x}^{k+1}) - \nabla f(\boldsymbol{x}^k), \text{ and } \boldsymbol{s} = \boldsymbol{x}^{k+1} - \boldsymbol{x}^k. \quad (11.18)$$

Subsequently, backtracking along one of the two stepsizes is used to satisfy the *non monotonic* Armijo condition (Grippo et al., 1986):

$$f(\boldsymbol{x}^{k+1}) \leq \max_{i=k-m:k} \{f(\boldsymbol{x}^i)\} + \tau \nabla f(\boldsymbol{x}^k)(\boldsymbol{x}^{k+1} - \boldsymbol{x}^k), \quad \tau \in (0,1).$$

Unlike the ordinary Armijo condition (11.5), allows some temporary increase in the objective. This non monotonic Armijo condition typically accepts the initial step length, even if it increases the objective function, while still ensuring global convergence of the method.[5] Experimentally, these two simple modifications lead to large improvements in the convergence speed of the method. Indeed, due to its strong empirical performance, SPG has recently been explored in several other applications (Dai and Fletcher, 2005; Figueiredo et al., 2007; van den Berg and Friedlander, 2008).

An alternative to SPG for accelerating the basic projected gradient method is the method of Nesterov (2004, Section 2.2.4). In this strategy, an extra extrapolation step is added to the iteration, thereby allowing the method to achieve the optimal worst-case convergence rate among a certain class of algorithms. Besides SPG and this optimal gradient algorithm, there can be numerous alternative iterative solvers. But we restrict our discussion to an SPG-based method and consider some implementation and theoretical details for it.

---

5. A typical value for the number m of previous function values to consider is 10.

11.4.2 SPG-based Inexact Projected Newton

Recall Step IIa in the general framework of Algorithm 11.1:[6]

$$\bar{\boldsymbol{x}}_1^k = \operatorname*{argmin}_{\boldsymbol{x}\in\Omega} \mathcal{Q}^k(\boldsymbol{x},1), \tag{11.19}$$

where the quadratic model is

$$\mathcal{Q}^k(\boldsymbol{x},1) = f(\boldsymbol{x}^k) + (\boldsymbol{x}-\boldsymbol{x}^k)^T \nabla f(\boldsymbol{x}^k) + \frac{1}{2}(\boldsymbol{x}-\boldsymbol{x}^k)^T \boldsymbol{H}^k(\boldsymbol{x}-\boldsymbol{x}^k).$$

For the remainder of this section, we denote $\mathcal{Q}^k(\boldsymbol{x},1)$ by $\mathcal{Q}^k$ when there is no confusion. Inexact projected Newton methods solve the subproblem (11.19) only approximately; we denote this approximate solution by $\boldsymbol{z}^k$ below.

At each iteration of an SPG-based inexact projected Newton method, we first compute the gradient $\nabla f(\boldsymbol{x}^k)$ and (implicitly) compute the quadratic term $\boldsymbol{H}^k$ in $\mathcal{Q}^k$. Subsequently, we try to minimize this $\mathcal{Q}^k$ over the feasible set, using iterations of an SPG algorithm. Even if f or ∇f is difficult to compute, this SPG subroutine can be efficient if $\mathcal{Q}^k$ and $\nabla \mathcal{Q}^k$ can be evaluated rapidly. Given $f(\boldsymbol{x}^k)$ and $\nabla f(\boldsymbol{x}^k)$, the dominant cost in evaluating $\mathcal{Q}^k$ and $\nabla \mathcal{Q}^k$ is pre-multiplication by $\boldsymbol{H}^k$. By taking the compact representation of Byrd et al. (1994),

$$\boldsymbol{H}^k = \sigma^k \mathrm{I} - \boldsymbol{N}\boldsymbol{M}^{-1}\boldsymbol{N}^T, \quad \text{where} \quad \boldsymbol{N} \in \mathbb{R}^{n\times 2m},\ \boldsymbol{M} \in \mathbb{R}^{2m\times 2m}, \tag{11.20}$$

we can compute $\mathcal{Q}^k$ and $\nabla \mathcal{Q}^k$ in $\mathcal{O}(mn)$ under the L-BFGS Hessian approximation.

In addition to $\mathcal{Q}^k$ and $\nabla \mathcal{Q}^k$, the SPG subroutine also requires computing the Euclidean projection $\mathcal{P}_\Omega$ onto the feasible set Ω. However, note that the SPG subroutine does not evaluate f or ∇f. Hence, the SPG-based inexact projected Newton method is most effective on problems where computing the projection is much less expensive than evaluating the objective function.[7]

Although in principle we could use SPG to solve the problem (11.19) exactly, in practice this is expensive and ultimately unnecessary. Thus, we terminate the SPG subroutine before the exact solution is found. One might be concerned about terminating the SPG subroutine early, especially because an approximate solution to (11.3) will in general not be a descent

---

6. We assume that $\alpha = 1$ and backtrack along β so that the iterative solver is invoked only once for each iteration; however, the inexact Newton method does not rule out the possibility of fixing β (and invoking the iterative solver for each backtracking step).
7. This is different from many classical optimization problems such as quadratic programming, where evaluating the objective function may be relatively inexpensive but computing the projection may be as difficult as solving the original problem.

direction. Fortunately, we can guarantee that the SPG subroutine yields a descent direction even under early termination if we initialize it with $\boldsymbol{x}^k$ and we perform at least one SPG iteration. To see this, note that positive definiteness of $\boldsymbol{H}^k$ implies that a sufficient condition for $\boldsymbol{z}^k - \boldsymbol{x}^k$ to be a descent direction for some vector $\boldsymbol{z}^k$ is that $\mathcal{Q}^k(\boldsymbol{z}^k, \alpha^k) < f(\boldsymbol{x}^k)$, since this implies the inequality

$$(\boldsymbol{z}^k - \boldsymbol{x}^k)^T \nabla f(\boldsymbol{x}^k) < 0.$$

By substituting $\mathcal{Q}^k(\boldsymbol{x}^k, \alpha^k) = f(\boldsymbol{x}^k)$, we see that

$$\mathcal{Q}^k(\boldsymbol{z}^k, \alpha^k) < \mathcal{Q}^k(\boldsymbol{x}^k, \alpha^k) = f(\boldsymbol{x}^k),$$

where $\boldsymbol{z}^k$ is the first point satisfying the Armijo condition when we initialize SPG with $\boldsymbol{x}^k$.[8] In other words, if we initialize the SPG subroutine with $\boldsymbol{x}^k$, then the SPG iterate gives a descent direction after the first iteration, and every subsequent iteration. Thus, it can be safely terminated early. In an implementation we can parameterize the maximum number of the SPG iterations by c, which results in an $\mathcal{O}(mnc)$ iteration cost for the inexact Newton method, assuming that projection requires $\mathcal{O}(n)$ time.

Example 11.4 (Blockwise-sparse Gaussian graphical models). *Consider a generalization of Example 11.3 where instead of constraining the absolute values of matrix elements, we constrain the norms of a disjoint set of groups (indexed by g) of elements:*

$$\min_{\tilde{\Sigma} + \boldsymbol{X} \succ 0} \quad \tfrac{1}{2} \log \det(\tilde{\Sigma} + \boldsymbol{X}), \quad \textit{subject to } \|\boldsymbol{X}_g\|_2 \le \lambda_g, \forall_g. \qquad (11.21)$$

This generalization is similar to those examined in Duchi et al. (2008) and Schmidt et al. (2009), and it encourages the Gaussian graphical model to be sparse across groups of variables (i.e., all edges in a group g will be either included in or excluded from the graph). Thus, formulation (11.21) encourages the precision matrix to have a blockwise sparsity pattern. Unfortunately, this generalization can no longer be written as a problem with bound constraints, nor can we characterize the feasible set with a finite number of linear constraints (though it is possible to write the feasible set using quadratic constraints). Nevertheless, it is easy to compute the projection onto the norm constraints; to project a matrix $\boldsymbol{X}$ onto the feasible set with respect to the norm constraints we simply set $\boldsymbol{X}_g = \lambda_g / \|\boldsymbol{X}_g\|_2$ for each group g. Considering the potentially high cost of evaluating the log-determinant

---

8. We will be able to satisfy the Armijo condition, provided that $\boldsymbol{x}^k$ is not already a minimizer.

function (and its derivative), this simple projection suggests that inexact projected Newton methods are well suited for solving (11.21).

11.5   Toward Nonsmooth Objectives

In this section we reconsider Problem (11.1), but unlike previous sections, we now allow $r(\boldsymbol{x}) \neq 0$. The resulting *composite optimization* problem occurs frequently in machine learning and statistics, especially with $r(\boldsymbol{x})$ being a *sparsity-promoting* regularizer (see e.g., Chapter 2).

How should we deal with the nondifferentiability of $r(\boldsymbol{x})$ in the context of Newton-like methods? While there are many possible answers to this question, we outline two simple but effective solutions that align well with the framework laid out so far.

11.5.1   Two-Metric Subgradient Projection Methods

We first consider the following special case of (11.1):

$$\min_{\boldsymbol{x} \in \mathbb{R}^n} \quad \mathcal{F}(\boldsymbol{x}) = f(\boldsymbol{x}) + \sum_i r_i(x_i), \tag{11.22}$$

where $r(\boldsymbol{x})$ has the separable form $r(\boldsymbol{x}) = \sum_i r_i(x_i)$ and each $r_i : \mathbb{R} \to \mathbb{R}$ is continuous and convex but not necessarily differentiable. A widely used instance of this problem is when we have $r_i(x_i) = \lambda_i |x_i|$ for fixed $\lambda_i > 0$, corresponding to ℓ_1-regularization. Note that this problem has a structure similar to the bound-constrained optimization problem (11.10); the latter has *separable constraints*, while problem (11.22) has a *separable nonsmooth* term. We can use separability of the nonsmooth term to derive a *two-metric subgradient projection* method for (11.22), that is analogous to the two-metric gradient projection method discussed in Section 11.3. The main idea is to choose an appropriately defined *steepest descent* direction and then to take a step resembling a two-metric projection iteration in this direction.

To define an appropriate steepest descent direction, we note that even though the objective in (11.22) is not differentiable, its directional derivatives always exist. Thus, analogous to the differentiable case, we can define the steepest descent direction as the direction that minimizes the directional derivative; among all vectors with unit norm, the steepest descent direction locally decreases the objective most quickly. This direction is closely related to the element of the subdifferential of a function $\mathcal{F}(\boldsymbol{x})$ with minimum norm.

Definition 11.1 (Mininum-norm subgradient). *Let*

$$z^k = \operatorname*{argmin}_{z \in \partial \mathcal{F}(x)} \|z\|_2. \tag{11.23}$$

Following an argument outlined in (Bertsekas et al., 2003, Section 8.4),[9] the steepest descent direction for a convex function $\mathcal{F}(x)$ at a point x^k is $-z^k$, where the subdifferential of (11.22) is given by

$$\partial \mathcal{F}(x) = \partial\{f(x^k) + r(x^k)\} = \nabla f(x^k) + \partial r(x^k).$$

Using the separability of $r(x)$, we see that the minimum-norm subgradient (11.23) with respect to a variable x_i, is given by

$$z_i^k = \begin{cases} 0, & \text{if } -\nabla_i f(x^k) \in \left\{\partial^- r_i(x_i^k),\ \partial^+ r_i(x_i^k)\right\}, \\ \min\left\{\left|\nabla_i f(x^k) + \partial^- r_i(x_i^k)\right|,\ \left|\nabla_i f(x^k) + \partial^+ r_i(x_i^k)\right|\right\}, & \text{otherwise,} \end{cases}$$

where the directional derivative $\partial^+ r_i(x_i^k)$ is given by

$$\partial^+ r_i(x_i^k) = \lim_{\delta \to 0^+} \frac{r_i(x_i^k + \delta) - r_i(x_i^k)}{\delta}.$$

The directional derivative $\partial^- r_i(x_i^k)$ is defined similarly, with δ going to zero from below. Thus, it is easy to compute z^k given $\nabla f(x^k)$, as well as the left and right partial derivatives ($\partial^- r_i(x_i^k)$ and $\partial^+ r_i(x_i^k)$) for each $r_i(x_i^k)$. Observe that when $r_i(x_i^k)$ is differentiable, $\partial^- r_i(x_i^k) = \partial^+ r_i(x_i^k)$, whereby the minimum norm subgradient is simply $\nabla_i f(x^k) + \nabla_i r(x^k)$. Further, note that $z^k = 0$ at a global optimum; otherwise $-z^k$ yields a descent direction and we can use it in place of the negative gradient within a line search method.

Similar to steepest descent for smooth functions, a generalized steepest descent for nonsmooth functions may converge slowly, and thus we seek a Newton-like variant. A natural question is whether we can merely use a scaling matrix S^k to scale the steepest descent direction. Similar to the two-metric projection algorithm, the answer is "no" for essentially the same reason: in general a scaled version of the steepest descent direction may turn out to be an *ascent* direction.

However, we can still use a similar solution to the problem. If we make the positive definite scaling matrix S^k *diagonal* with respect to the variables x_i that are close to locations where $r_i(x_i)$ is nondifferentiable, then we can still ensure that the method generates descent directions. Thus, we obtain

---

9. Replacing maximization with minimization and concavity with convexity.

a simple Newton-like method for nonsmooth optimization that uses iterates of the form

$$\boldsymbol{x}^{k+1} \leftarrow \boldsymbol{x}^k - \alpha \boldsymbol{S}^k \boldsymbol{z}^k. \tag{11.24}$$

Here, matrix $\boldsymbol{S}^k$ has the same structure as (11.12), but now the variables that receive a diagonal scaling are variables close to nondifferentiable values. Formally, the set of *restricted* variables is:

$$\mathcal{I}^k \triangleq \big\{ i \ \big| \min_{d_i \in \mathcal{D}_i} |d_i - x_i| \le \varepsilon \big\}, \tag{11.25}$$

where $\mathcal{D}_i$ is the (countable) set containing all locations where $r_i(x_i)$ is nondifferentiable.

In many applications where we seek to solve a problem of the form (11.22), we expect the function to be nondifferentiable with respect to several of the variables at a solution. Further, it may be desirable that intermediate iterations of the algorithm lie at nondifferentiable points. For example, these might represent sparse solutions if a nondifferentiability occurs at zero. In these cases, we can add a projection step to the iteration that encourages intermediate iterates to lie at points of nondifferentiability. Specifically, if a variable x_i crosses a point of nondifferentiability, we project onto the point of nondifferentiability. Since we use a diagonal scaling with respect to the variables that are close to points of nondifferentiability, this projection reduces to computing the Euclidean projection onto bound constraints, where the upper and lower bounds are given by the nearest upper and lower points of nondifferentiability. Thus, each iteration is effectively a two-metric subgradient projection iteration. To make our description concrete, let us look at a specific example below.

Example 11.5 (ℓ_1-Regularization). *A prototypical composite minimization problem in machine learning is the ℓ_1-regularized task*

$$\min_{\boldsymbol{x} \in \mathbb{R}^n} \quad f(\boldsymbol{x}) + \sum_{i=1}^{n} \lambda_i |x_i|. \tag{11.26}$$

The scalars $\lambda_i \ge 0$ control the degree of regularization, and for sufficiently large λ_i, the parameter x_i is encouraged to be exactly zero.

To apply our framework, we need to efficiently compute the minimum norm subgradient $\boldsymbol{z}^k$ for (11.26); this gradient may be computed as

$$z_i^k \triangleq \begin{cases} \nabla_i f(\boldsymbol{x}) + \lambda_i \, \mathrm{sgn}(x_i), & |x_i| > 0 \\ \nabla_i f(\boldsymbol{x}) + \lambda_i, & x_i = 0, \nabla_i f(\boldsymbol{x}) < -\lambda_i \\ \nabla_i f(\boldsymbol{x}) - \lambda_i, & x_i = 0, \nabla_i f(\boldsymbol{x}) > \lambda_i \\ 0, & x_i = 0, -\lambda_i \le \nabla_i f(\boldsymbol{x}) \le \lambda_i. \end{cases} \tag{11.27}$$

For this problem, the restrictied variable set (11.25) corresponds to those variables sufficiently close to zero, $\{i|\,|x_i| \leq \varepsilon\}$. Making S^k partially diagonal with respect to the restricted variables as before, we define the two-metric projection step for ℓ_1-regularized optimization as

$$x^{k+1} = \mathcal{P}_\mathcal{O}[x^k - \alpha S^k z^k, x^k]. \tag{11.28}$$

Here, the orthant projection (that sets variables to exactly zero) is defined as

$$\mathcal{P}_\mathcal{O}(y, x)_i \triangleq \begin{cases} 0, & \text{if } x_i y_i < 0, \\ y_i, & \text{otherwise.} \end{cases}$$

Applying this projection is effective at sparsifying the parameter vector since it sets variables that change sign to exactly zero, and it also ensures that the line search does not cross points of nondifferentiability. Provided that x^k is not stationary, the steps in (11.28) are guaranteed to improve the objective for sufficiently small α. The stepsize α is selected by a backtracking line search along the projection arc to satisfy a variant of the Armijo condition where the gradient is replaced by the minimum norm subgradient. If at some iteration the algorithm identifies the correct set of nonzero variables and then maintains the orthant of the optimal solution, then the algorithm essentially reduces to an unconstrained Newton-like method applied to the nonzero variables.

In the two-metric projection algorithm for bound-constrained optimization the choice of the diagonal scaling matrix D^k simply controls the rate at which very small variables move toward zero, and does not have a significant impact on the performance of the algorithm. However, the choice of D^k in the algorithm for ℓ_1-regularization can have a significant effect on the performance of the method, since if D^k is too large, we may need to perform several backtracking steps before the step length is accepted, while too small a value will require many iterations to set very small variables to exactly zero. One possibility is to compute the Barzilai-Borwein scaling α^{bb} of the variables given by (11.18), and set D^k to $\alpha^{bb} I$.

11.5.2   Proximal Newton-like Methods

The method of Section 11.5.1 crucially relies on separability of the nonsmooth function $r(x)$. For more general nonsmooth $r(x)$, an attractive choice is to tackle the nondifferentiability of $r(x)$ via proximity operators (Moreau, 1962; Combettes and Wajs, 2005; Combettes and Pesquet, 2009). These operators are central to forward-backward splitting methods (Combettes and Pesquet, 2009),[10] as well as to methods based on surrogate optimization (Figueiredo and Nowak, 2003; Daubechies et al., 2004), separable approximation (Wright et al., 2009), gradient-mapping (Nesterov, 2007), or to a proximal trust-region framework (Kim et al., 2010).

The idea of a proximity operator is simple. Let $r : X \subseteq \mathbb{R}^d \to (-\infty, \infty]$ be a lower semicontinuous, proper convex function. For a point $\boldsymbol{y} \in X$, the *proximity operator* for r applied to $\boldsymbol{y}$ is defined as

$$\mathrm{prox}_r(\boldsymbol{y}) = \underset{\boldsymbol{x} \in X}{\mathrm{argmin}} \quad \tfrac{1}{2}\|\boldsymbol{x} - \boldsymbol{y}\|_2^2 + r(\boldsymbol{x}). \qquad (11.29)$$

This operator generalizes the projection operator, since when $r(\boldsymbol{x})$ is the indicator function for a convex set C, (11.29) reduces to projection onto C. This observation suggests that we might be able to replace projection operators with proximity operators. Indeed, this replacement is done in *forward-backward splitting* methods, where one iterates

$$\boldsymbol{x}^{k+1} = \mathrm{prox}_{\alpha^k r}(\boldsymbol{x}^k - \alpha^k \nabla f(\boldsymbol{x}^k));$$

the iteration "splits" the update into differentiable (forward) and nondifferentiable (proximal or backward) steps. This method generalizes first-order projected gradient methods, and under appropriate assumptions it can be shown that the sequence $\{f(\boldsymbol{x}^k) + r(\boldsymbol{x}^k)\}$ converges to $f(\boldsymbol{x}^*) + r(\boldsymbol{x}^*)$, where x^* is a stationary point.

At this point, the reader may already suspect how we might use proximity operators in our Newton-like methods. The key idea is simple: build a quadratic model, but *only* for the differentiable part, and tackle the nondifferentiable part via a suitable proximity operator. This simple idea was also previously exploited by Wright et al. (2009) and Kim et al. (2010). Formally, we consider the regularized quadratic model

$$\mathcal{Q}^k(\boldsymbol{x}, \alpha) \triangleq f(\boldsymbol{x}^k) + (\boldsymbol{x} - \boldsymbol{x}^k)^T \nabla f(\boldsymbol{x}^k) + \frac{1}{2\alpha}(\boldsymbol{x} - \boldsymbol{x}^k)^T \boldsymbol{H}^k (\boldsymbol{x} - \boldsymbol{x}^k) + r(\boldsymbol{x}),$$
$$(11.30)$$

whose minimizer can be recast as the *generalized proximity operator*:

$$\mathrm{prox}_{\alpha \cdot r}^{\boldsymbol{H}^k}(\boldsymbol{y}^k) = \underset{\boldsymbol{x} \in \mathbb{R}^n}{\mathrm{argmin}} \quad \tfrac{1}{2}\|\boldsymbol{x} - \boldsymbol{y}^k\|_{\boldsymbol{H}^k}^2 + \alpha r(\boldsymbol{x}), \qquad (11.31)$$

where $\boldsymbol{y}^k = \boldsymbol{x}^k - \alpha[\boldsymbol{H}^k]^{-1}\nabla f(\boldsymbol{x}^k)$; observe that under the transformation

---

10. Including *iterative soft-thresholding* as a special case.

$x \to [\boldsymbol{H}^k]^{1/2}\boldsymbol{x}$, (11.31) may be viewed as a standard proximity operator.

Using this generalized proximity operator, our Newton-like algorithm becomes

$$\boldsymbol{x}^{k+1} = \text{prox}_{\alpha \cdot r}^{\boldsymbol{H}^k}(\boldsymbol{x}^k - \alpha[\boldsymbol{H}^k]^{-1}\nabla f(\boldsymbol{x}^k)). \tag{11.32}$$

If instead of the true inverse Hessian, we use $\boldsymbol{H}^k = \boldsymbol{I}$, iteration (11.32) degenerates to the traditional forward-backward splitting algorithm. Furthermore, it is equally straightforward to implement a quasi-Newton variant of (11.32), where, for example, $\boldsymbol{H}^k$ is obtained by an L-BFGS approximation to $\nabla^2 f(\boldsymbol{x}^k)$. Another practical choice might be an inexact quasi-Newton variant, where we use iterations of the SPG-like method of Wright et al. (2009) to approximately minimize $\mathcal{Q}^k(\boldsymbol{x}, \alpha)$ under an L-BFGS approximation of $f(\boldsymbol{x})$; in other words, the generalized proximity operator (11.31) is computed inexactly.

Similar to inexact projected Newton methods, under only mild assumptions we can guarantee that an inexact solution to the generalized proximity operator yields an improvement on the original objective for a sufficiently small stepsize α. For example, assume that $\nabla f(\boldsymbol{x})$ is Lipschitz continuous and that we find a value $\boldsymbol{y}$ such that $\mathcal{Q}^k(\boldsymbol{y}, \alpha) < \mathcal{Q}^k(\boldsymbol{x}^k, \alpha)$ in (11.30). Then we have

$$
\begin{aligned}
f(\boldsymbol{x}^k) + r(\boldsymbol{x}^k) &= \mathcal{Q}^k(\boldsymbol{x}^k, \alpha) \\
&> \mathcal{Q}^k(\boldsymbol{y}, \alpha) \\
&= f(\boldsymbol{x}^k) + (\boldsymbol{y} - \boldsymbol{x}^k)^T \nabla f(\boldsymbol{x}^k) + \frac{1}{2\alpha}(\boldsymbol{y} - \boldsymbol{x}^k)^T \boldsymbol{H}^k (\boldsymbol{y} - \boldsymbol{x}^k) + r(\boldsymbol{y}) \\
&\geq f(\boldsymbol{x}^k) + (\boldsymbol{y} - \boldsymbol{x}^k)^T \nabla f(\boldsymbol{x}^k) + \frac{m}{2\alpha}\|\boldsymbol{y} - \boldsymbol{x}^k\|_2^2 + r(\boldsymbol{y}) \\
&\geq f(\boldsymbol{y}) + r(\boldsymbol{y}) \quad (\text{for } 0 < \alpha \leq m/\mathcal{L}),
\end{aligned}
$$

where m is the smallest eigenvalue of $\boldsymbol{H}^k$ and the last inequality follows from Lipschitz continuity of the gradient (Bertsekas, 1999, Proposition A.24), where $\mathcal{L}$ is the Lipschitz constant of the gradient of $f(\boldsymbol{x})$. A similar property holds if $\nabla f(\boldsymbol{x})$ is only locally Lipschitz continuous.

Example 11.6 (Group ℓ_1-regularization). *Consider a generalization of Example 11.5 where instead of penalizing the absolute values of each element of $\boldsymbol{x}$, we penalize the ℓ_2 norms of a set of disjoint groups indexed by g:*

$$\min_{\boldsymbol{x} \in \mathbb{R}^n} \quad f(\boldsymbol{x}) + \sum_g \lambda_g \|\boldsymbol{x}_g\|_2. \tag{11.33}$$

The regularizer in (11.33) is referred to as a group regularizer ($\ell_{1,2}$-regularizer), since it encourages sparsity in terms of groups of variables, *and dates back to Bakin (1999). The regularization term is nonsmooth when an entire group of variables is set to $\mathbf{0}$. However, the proximal operator for this regularizer is easily computed: given a vector $\boldsymbol{y}$, with groups $\boldsymbol{y}_g$, we simply set $\boldsymbol{x}_g = (\boldsymbol{y}_g/\|\boldsymbol{y}_g\|_2)\max\{0,\ \|\boldsymbol{y}_g\|_2 - \alpha\lambda_g\}$. Thus, inexact proximal Newton methods are well suited to solving (11.33).*

Example 11.7 (Group nuclear norm regularization). *A related problem is optimizing a smooth function of several matrix inputs with regularization of the nuclear norms of the matrices:*

$$\min_{\boldsymbol{X}_1,\boldsymbol{X}_2,\ldots,\boldsymbol{X}_n} \quad f(\boldsymbol{X}_1,\boldsymbol{X}_2,\ldots,\boldsymbol{X}_n) + \sum_{i=1}^{n}\lambda_i\|\boldsymbol{X}_i\|_*. \tag{11.34}$$

Here we use $\|\boldsymbol{X}\|_*$ to the denote the nuclear norm (or trace norm), the sum of the singular values of $\boldsymbol{X}$. This regularization not only encourages sparsity across individual matrices, but also encourages each matrix to be low rank. The proximal operator for the nuclear norm can be computed by soft-thresholding the singular values of each $\boldsymbol{X}_i$ (Cai et al., 2010). That is, to compute the proximal operator, we replace each singular value σ_j of each $\boldsymbol{X}_i$ with $\sigma_j = \max\{0,\ \sigma_j - \alpha\lambda_i\}$, where α is the parameter of the quadratic approximation (11.30). Thus, inexact proximal Newton methods are well suited to solving (11.34) too, especially when it is more expensive to evaluate f and ∇f than it is to compute the singular value decomposition of each $\boldsymbol{X}_i$.

11.6   Summary and Discussion

In this chapter, we have concentrated on minimizing twice-differentiable convex functions, both when their exact Hessian is feasible to use and when quasi-Newton choices are more practical. Note that the quasi-Newton approach can also be applied when the objective function is only once differentiable. Furthermore, we may relax the assumption of convexity if we concede that the stationary point found by the method may not be a local or global minimum.

As for implementational strategies, while we have focused on L-BFGS methods, an alternative restricted memory strategy is to use implicit Hessian-vector products. For example, in the two-metric projection strategy we can use Hessian-vector products within a linear conjugate gradient iteration to solve the scaling with respect to the free variables, as in Nocedal and Wright (2000, Section 7.1), while we can use Hessian-vector products within the SPG subroutine for inexact projected Newton methods. An alternative means to optimize nonsmooth objectives with an L-BFGS approximation is given by Yu et al. (2010). Variants of the L-BFGS approximation that apply in stochastic scenarios are examined in Sunehag et al. (2009).

We close by noting some open issues regarding convergence of the methods discussed in this section. First, global convergence of methods based on the minimum norm subgradient without a diminishing stepsize can be tenuous because of the lack of continuity in the derivatives of sequences. For example, see the counterexample in Bertsekas (1999, Exercise 6.3.8). Andrew and Gao (2007) give a proof of global convergence of a method related to the two-metric subgradient projection method we discuss in Section 11.5.1, but as pointed out by Yu et al. (2010), their proof does not account for this lack of continuity. Thus, while the algorithms of Andrew and Gao (2007) and Section 11.5.1 appear to be very effective in practice, it remains to be shown whether they are globally convergent in general without additional assumptions.

A related issue is showing whether the two-metric subgradient projection method identifies the correct set of nonzero variables after a finite number of iterations, and then has a superlinear convergence rate when exact second-order information is available. Also, in Examples 1.5 and 1.6, we use a projection with respect to a subset of the constraints and do not project with respect to the positive definite constraint that is known not to be active at the solution. Although this strategy has been used by several authors, and seems not to significantly affect empirical convergence of the method when given a suitable starting point, formally examining convergence under this heuristic deserves some theoretical attention.

Finally, there are not yet formal proofs of global and local convergence for inexact projected Newton methods, but this appears to be a simpler task than showing convergence of the methods discussed in the previous paragraph. For example, it is likely that global convergence can be proved by showing that a suitable gradient-related condition (Bertsekas, 1999, Section 1.2) applies to the first iteration in the SPG subroutine that satisfies the Armijo condition, while a local convergence rate can likely be shown by using a forcing sequence (Nocedal and Wright, 2000, Section 7.1) on the solution accuracy of the SPG subroutine.

11.7   References

G. Andrew and J. Gao. Scalable training of L_1-regularized log-linear models. In *Proceedings of the 24th International Conference on Machine Learning*, pages 33–40, 2007.

S. Bakin. *Adaptive regression and model selection in data mining problems*. PhD thesis, Australian National University, Canberra, 1999.

O. Banerjee, L. El Ghaoui, A. d'Aspremont, and G. Natsoulis. Convex optimization techniques for fitting sparse Gaussian graphical models. In *Proceedings of the 23rd International Conference on Machine Learning*, pages 89–96, 2006.

J. Barzilai and J. Borwein. Two-point step size gradient methods. *IMA Journal of Numerical Analysis*, 8(1):141–148, 1988.

D. P. Bertsekas. Projected Newton methods for optimization problems with simple constraints. *SIAM Jounal on Control and Optimization*, 20(2):221–246, 1982.

D. P. Bertsekas. *Nonlinear Programming*. Athena Scientific, second edition, 1999.

D. P. Bertsekas, A. Nedic, and A. E. Ozdaglar. *Convex Analysis and Optimization*. Athena Scientific, 2003.

E. G. Birgin, J. M. Martínez, and M. Raydan. Nonmonotone spectral projected gradient methods on convex sets. *SIAM Journal on Optimization*, 10(4):1196–1211, 2000.

R. H. Byrd, J. Nocedal, and R. B. Schnabel. Representations of quasi-Newton matrices and their use in limited memory methods. *Mathematical Programming*, 63(1):129–156, 1994.

J. F. Cai, E. J. Candès, and Z. Shen. A singular value thresholding algorithm for matrix completion. *SIAM Journal on Optimization*, 20(4):1956–1982, 2010.

P. L. Combettes and J. Pesquet. Proximal Splitting Methods in Signal Processing. *arXiv:0912.3522v2*, December 2009.

P. L. Combettes and V. R. Wajs. Signal recovery by proximal forward-backward splitting. *Multiscale Modeling and Simulation*, 4(4):1168–1200, 2005.

Y. H. Dai and R. Fletcher. Projected Barzilai-Borwein methods for large-scale box-constrained quadratic programming. *Numerische Mathematik*, 100(1):21–47, 2005.

I. Daubechies, M. Defrise, and C. De Mol. An iterative thresholding algorithm for linear inverse problems with a sparsity constraint. *Communications on Pure and Applied Mathematics*, 57(11):1413–1457, 2004.

J. Duchi, S. Gould, and D. Koller. Projected subgradient methods for learning sparse gaussians. In *Proceedings of the 24th Conference on Uncertainty in Artificial Intelligence*, pages 145–152, 2008.

M. Figueiredo, R. Nowak, and S. Wright. Gradient projection for sparse reconstruction: Application to compressed sensing and other inverse problems. *IEEE Journal of Selected Topics in Signal Processing*, 1(4):586–597, 2007.

M. A. T. Figueiredo and R. D. Nowak. An EM algorithm for wavelet-based image restoration. *IEEE Transactions on Image Processing*, 12(8):906–916, 2003.

E. M. Gafni and D. P. Bertsekas. Two-metric projection methods for constrained optimization. *SIAM Journal on Control and Optimization*, 22(6):936–964, 1984.

P. E. Gill, W. Murray, and M. H. Wright. *Practical Optimization*. Academic Press, 1981.

L. Grippo, F. Lampariello, and S. Lucidi. A nonmonotone line search technique for Newton's method. *SIAM Journal on Numerical Analysis*, 23(4):707–716, 1986.

D. Kim, S. Sra, and I. S. Dhillon. A scalable trust-region algorithm with application to mixed-norm regression. In *Proceedings of the 27th International Conference on Machine Learning*, pages 519–526, 2010.

E. S. Levitin and B. T. Polyak. Constrained minimization methods. *USSR Computational Mathematics and Mathematical Physics*, 6:1–50, 1966. English translation of a paper in Zh. Vȳchisl. Mat. i Mat. Fiz. 6, 5, 787-823, 1966.

J.-J. Moreau. Fonctions convexes duales et points proximaux dans un espace hilbertien. *C. R. Acad. Sci. Paris*, 255:2897–2899, 1962.

Y. Nesterov. *Introductory Lectures on Convex Optimization: A Basic Course.* Springer, 2004.

Y. Nesterov. Gradient methods for minimizing composite objective function. Technical report, Université Catholique de Louvain, 2007.

J. Nocedal. Updating quasi-Newton matrices with limited storage. *Mathematics of Computation*, 35(151):773–782, 1980.

J. Nocedal and S. J. Wright. *Numerical Optimization.* Springer, second edition, 2000.

M. Schmidt, E. van den Berg, M. Friedlander, and K. Murphy. Optimizing costly functions with simple constraints: A limited-memory projected quasi-Newton algorithm. In *Proceedings of the 12th International Conference on Artificial Intelligence and Statistics*, volume 5, pages 456–463, 2009.

D. F. Shanno and K. H. Phua. Matrix conditioning and nonlinear optimization. *Mathematical Programming*, 14(1):149–160, 1978.

P. Sunehag, J. Trumpf, S. V. N. Vishwanathan, and N. N. Schraudolph. Variable metric stochastic approximation theory. In *Proceedings of the 12th International Conference on Artificial Intelligence and Statistics*, volume 5, pages 560–566, 2009.

E. van den Berg and M. P. Friedlander. Probing the Pareto frontier for basis pursuit solutions. *SIAM Journal on Scientific Computing*, 31(2):890–912, 2008.

S. J. Wright, R. D. Nowak, and M. A. T. Figueiredo. Sparse reconstruction by separable approximation. *IEEE Transactions on Signal Processing*, 57(7):2479–2493, 2009.

J. Yu, S. V. N. Vishwanathan, S. Günter, and N. N. Schraudolph. A quasi-newton approach to nonsmooth convex optimization. *Journal of Machine Learning Research*, 11:1145–1200, 2010.

12 Interior-Point Methods in Machine Learning

Jacek Gondzio   J.Gondzio@ed.ac.uk

School of Mathematics and Maxwell Institute for Mathematical Sciences
The University of Edinburgh, Mayfield Road, Edinburgh EH9 3JZ, United Kingdom

Interior-point methods for linear and (convex) quadratic programming display several features which make them particularly attractive for very large-scale optimization. They have an impressive low-degree polynomial worst-case complexity. In practice, they display an unrivalled ability to deliver optimal solutions in an almost constant number of iterations which depends very little, if at all, on the problem's dimension. Since many problems in machine learning can be recast as linear or quadratic optimization problems and it is common for them to have large or huge sizes, interior-point methods are natural candidates to be applied in this context.

In this chapter we will discuss several major issues related to interior point methods, including the worst-case complexity result, the features responsible for their ability to solve very large problems, and their existing and potential applications in machine learning.

12.1 Introduction

Soon after Karmarkar (1984) had published his seminal paper, interior-point methods (IPMs) were claimed to have unequalled efficiency when applied to large-scale problems. Karmarkar's first worst-case complexity proof was based on the use of projective geometry and cleverly chosen potential function, but was rather complicated. It generated huge interest in the optimization community and soon led to improvements and clarifications of the theory. A major step in this direction was made by Gill et al. (1986), who drew the community's attention to a close relation between Karmarkar's projective method and the projected Newton barrier method. The impressive effort of Lustig, Marsten, Shanno, and their collaborators in the late 1980s provided a better understanding of the computational aspects of IPMs for linear programming, including the central role played by the logarithmic barrier functions in the theory (Marsten et al., 1990). In the early 1990s sufficient evidence was already gathered to justify claims of the spectacular efficiency of IPMs for very large-scale linear programming (Lustig et al., 1994) and their ability to compete with a much older rival, the simplex method (Dantzig, 1963).

The simplex method has also gone through major developments over the last 25 years (Forrest and Goldfarb, 1992; Maros, 2003; Hall and McKinnon, 2005). It remains competitive for solving linear optimization problems and certainly provides a healthy pressure for further development of IPMs. It is widely accepted nowadays that there exist classes of problems for which one method may significantly outperform the other. The large size of the problem generally seems to favor interior-point methods. However, the structure of the problem, and in particular the sparsity pattern of the constraint matrix which determines the cost of linear algebra operations, may occasionally render one of the approaches impractical. The simplex method exploits well the hypersparsity of the problem (Hall and McKinnon, 2005). On the other hand, interior-point methods have a well-understood ability to take advantage of any block-matrix structure in the linear algebra operations, and therefore are significantly easier to parallelize (Gondzio and Grothey, 2006).

Many machine learning (ML) applications are formulated as optimization problems. Although, the vast majority of them lead to (easy) unconstrained optimization problems, certain classes of ML applications require dealing with linear constraints or variable nonnegativity constraints. interior-point methods are well suited to solve such problems because of their ability to handle inequality constraints very efficiently by using the logarithmic barrier functions.

The support vector machine training problems form an important class of ML applications which lead to constrained optimization formulations, and therefore can take a full advantage of IPMs. The early attempts to apply IPMs in the support vector machine training context (Ferris and Munson, 2003; Fine and Scheinberg, 2002; Goldfarb and Scheinberg, 2004) were very successful and generated further interest among the optimization community, stimulating several new developments (Gertz and Griffin, 2009; Jung et al., 2008; Woodsend and Gondzio, 2009, 2010). They relied on the ability of IPMs at taking advantage of the problem's special structure to reduce the cost of linear algebra operations. In this chapter we will concentrate on the support vector machine training problems and will use them to demonstrate the main computational features of interior-point methods.

The chapter is organized as follows. In Section 12.2 we will introduce the quadratic optimization problem and define the notation used. In Section 12.3 we will comment on the worst-case complexity result of a particular interior-point algorithm for convex quadratic programming, the feasible algorithm operating in a small neighborhood of the central path induced by the 2-norm. In Section 12.4 we will discuss several applications of interior-point methods which have been developed since about 2000 for solving different constrained optimization problems arising in support vector machine training. In Section 12.5 we will discuss existing and potential techniques which may accelerate the performance of interior-point methods in this context. Finally, in Section 12.6 we will give our conclusions and comment on possible further developments of interior-point methods.

12.2 Interior-Point Methods: Background

Consider the primal-dual pair of convex quadratic programming (QP) problems

Primal	Dual
$\min \quad c^T x + \frac{1}{2} x^T Q x$	$\max \quad b^T y - \frac{1}{2} x^T Q x$
s.t. $\quad Ax = b,$	s.t. $\quad A^T y + s - Qx = c,$
$\quad x \geq 0;$	$\quad y$ free, $s \geq 0,$

$$(12.1)$$

where $A \in \mathcal{R}^{m \times n}$ has full row rank $m \leq n$, $Q \in \mathcal{R}^{n \times n}$ is a positive semidefinite matrix, $x, s, c \in \mathcal{R}^n$, and $y, b \in \mathcal{R}^m$. Using Lagrangian duality theory (see Bertsekas, 1995), the first-order optimality conditions for these problems can be written as

$$
\begin{aligned}
Ax &= b \\
A^T y + s - Qx &= c \\
XSe &= 0 \\
(x, s) &\geq 0,
\end{aligned}
$$

$$(12.2)$$

where X and S are diagonal matrices in $\mathcal{R}^{n \times n}$ with elements of vectors x and s spread across the diagonal, respectively, and $e \in \mathcal{R}^n$ is the vector of ones. The third equation, $XSe = 0$, called the complementarity condition, can be rewritten as $x_j s_j = 0, \forall j = \{1, 2, \ldots, n\}$ and implies that at least one of the two variables x_j and s_j has to be zero at the optimum. The complementarity condition is often a source of difficulty when solving optimization problems, and the optimization approaches differ essentially in the way they deal with this condition.

Active set methods and their prominent example, the simplex method for linear programming, make an intelligent guess that either $x_j = 0$ or $s_j = 0$. They choose a subset of indices $j \in \mathcal{B} \subset \{1, 2, \ldots, n\}$ such that x_j is allowed to be nonzero and force the corresponding $s_j = 0$, while for the remaining indices $j \in \mathcal{N} = \{1, 2, \ldots, n\} \setminus \mathcal{B}$ they force $x_j = 0$ and allow s_j to take nonzero values. Such a choice simplifies the linear algebra operations which can be reduced (in the LP case) to consider only a submatrix of A induced by columns from set $\mathcal{B}$. The simplex method allows only one index to be swapped between $\mathcal{B}$ and $\mathcal{N}$ at each iteration. (In the more general context of active set methods, only one index of variable and/or active constraint can be exchanged at each iteration.) Hence an inexpensive update is performed to refresh active/inactive matrices, and this is reflected in a very low (almost negligible) cost of a single iteration. However, active set methods may require a huge number of iterations to be performed. This is a consequence of the difficulty in guessing the correct partition of indices into basic-nonbasic (active-inactive) subsets. The simplex method for linear programming is not a polynomial algorithm. Klee and Minty (1972) constructed a problem of dimension n, the solution of which requires 2^n iterations of the simplex method. However, in practice it is very rare for the simplex method to perform more than $m + n$ iterations on its way to an optimal solution (Forrest and Goldfarb, 1992; Maros, 2003; Hall and McKinnon, 2005).

Interior-point methods perturb the complementarity condition and replace $x_j s_j = 0$ with $x_j s_j = \mu$, where the parameter μ is driven to zero. This removes the need to "guess" the partitioning into active and inactive inequality constraints: the algorithm gradually reduces μ, and the partition of vectors x and s into zero and nonzero elements is gradually revealed as the algorithm progresses. Removing the need to "guess" the optimal partition is at the origin of the proof of the polynomial worst-case complexity of the interior-point method. Indeed, the best IPM algorithm known to date finds the ε-accurate solution of an LP or convex QP problem in $\mathcal{O}(\sqrt{n}\log(1/\varepsilon))$ iterations (Renegar, 1988). Again, in practice IPMs perform much better than that and converge in a number of iterations which is almost a constant, independent of the problem dimension (Colombo and Gondzio, 2008). However, one iteration of an IPM may be costly. Unlike the simplex method, which works with a submatrix of A, IPM involves the complete matrix A to compute the Newton direction for the perturbed first-order optimality conditions, and for nontrivial sparsity patterns in A this operation may be expensive and occasionally prohibitive.

The derivation of an interior-point method for optimization relies on three basic ideas:

1. Logarithmic barrier functions are used to "replace" the inequality constraints

2. Duality theory is applied to barrier subproblems to derive the first-order optimality conditions which take the form of a system of nonlinear equations, and

3. Newton's method is employed to solve this system of nonlinear equations.

To avoid the need to guess the activity of inequality constraints $x \geq 0$, interior-point methods employ the logarithmic barrier function of the form $-\mu \sum_{j=1}^{n} \log x_j$ added to the objective of the primal problem in (12.1). The barrier parameter μ weighs the barrier in relation to the QP objective. A large value of μ means that the original objective is less important, and the optimization focuses on minimizing the barrier term. The repelling force of the barrier prevents any of the components x_j from approaching their boundary value of zero. In other words, the presence of the barrier keeps the solution x in the interior of the positive orthant. Reducing the barrier term changes the balance between the original QP objective and the penalty for approaching the boundary. The smaller μ is the stronger the role of the original QP objective is. Much of the theory and practice of IPMs concentrates on clever ways of reducing the barrier term from a large initial value, used to promote centrality at the beginning of the optimization, to small values needed to weaken the barrier and to allow the algorithm to approach an optimal solution. In the linear programming case, the optimal solution lies on the boundary of the feasible region and many components of vector x are zero.

Applying Lagrangian duality theory to the barrier QP subproblem

$$\min \quad c^T x + \frac{1}{2} x^T Q x - \mu \sum_{j=1}^{n} \log x_j \quad \text{s.t.} \quad Ax = b \tag{12.3}$$

gives the following first-order optimality conditions:

$$\begin{aligned} Ax &= b \\ A^T y + s - Qx &= c \\ XSe &= \mu e \\ (x, s) &\geq 0. \end{aligned} \tag{12.4}$$

Comparison of (12.2) and (12.4) reveals that the only difference is a perturbation of the complementarity constraint which for the barrier subproblem requires all complementarity products $x_j s_j$ to take the same value μ. Observe that the perturbed complementarity condition is the only nonlinear equation in (12.4). For any $\mu > 0$, system (12.4) has a unique solution, $(x(\mu), y(\mu), s(\mu))$, $x(\mu) > 0$, $s(\mu) > 0$, which is called a μ-center. A family of these solutions for all positive values of μ determines a (continuous) path $\{(x(\mu), y(\mu), s(\mu)) : \mu > 0\}$ which is called the primal-dual *central path* or *central trajectory*.

interior-point algorithms apply Newton's method to solve the system of nonlinear equations (12.4). There is no need to solve this system to a high degree of accuracy. Recall that (12.4) is only an approximation of (12.2) corresponding to a specific choice of the barrier parameter μ. There is no need to solve it exactly because the barrier μ will have to be reduced anyway. IPMs apply only *one* iteration of the Newton method to this system of nonlinear equations and immediately reduce the barrier. Driving μ to zero is a tool which enforces convergence of (12.4) to (12.2), and takes iterates of IPM toward an optimal solution of (12.1). To perform a step of Newton's method for (12.4), the Newton direction $(\Delta x, \Delta y, \Delta s)$ is computed by solving the following system of linear equations,

$$\begin{bmatrix} A & 0 & 0 \\ -Q & A^T & I_n \\ S & 0 & X \end{bmatrix} \cdot \begin{bmatrix} \Delta x \\ \Delta y \\ \Delta s \end{bmatrix} = \begin{bmatrix} \xi_p \\ \xi_d \\ \xi_\mu \end{bmatrix} = \begin{bmatrix} b - Ax \\ c + Qx - A^T y - s \\ \mu e - XSe \end{bmatrix}, \quad (12.5)$$

where I_n denotes the identity matrix of dimension n.

The theory of interior-point methods requires careful control of the error in the perturbed complementarity condition $XSe \approx \mu e$. Take an arbitrary $\theta \in (0,1)$, compute $\mu = x^T s/n$, and define

$$N_2(\theta) = \{(x, y, s) \in \mathcal{F}^0 \mid \|XSe - \mu e\| \leq \theta\mu\}, \quad (12.6)$$

where $\mathcal{F}^0 = \{(x, y, s) \mid Ax = b, A^T y + s - Qx = c, (x, s) > 0\}$ denotes the primal-dual strictly feasible set. (Unless explicitly stated otherwise, the vector norm $\|\cdot\|$ will always denote the Euclidean norm.) Observe that all points in $N_2(\theta)$ exactly satisfy the first two (linear) equations in (12.4) and approximately satisfy the third (nonlinear) equation. In fact, $N_2(\theta)$ defines a neighborhood of the central path. Interestingly, the size of this neighborhood reduces with the barrier parameter μ. The theory of IPMs requires all the iterates to stay in this neighborhood. This explains why an alternative name to IPMs is path-following methods: indeed, these algorithms follow the central path on their way to optimality.

In the next section we will comment on an impressive feature of the interior-point method: it is possible to prove that an algorithm operating in the $N_2(\theta)$ neighborhood that is applied to a convex QP converges to an ε-accurate solution in $\mathcal{O}(\sqrt{n}\log(1/\varepsilon))$ iterations.

12.3   Polynomial Complexity Result

A detailed proof of the complexity result is beyond the scope of this chapter. The reader interested in the proof may consult an excellent textbook on IPMs by Wright (1997) in which a proof for the linear programming case is given. An extension to an IPM for quadratic programming requires some extra effort, and care has to be taken of terms which result from the quadratic objective.

The proof heavily uses the fact that all iterates belong to an $N_2(\theta)$ neighborhood (12.6) of the central path. Consequently, all iterates are strictly primal-dual feasible which simplifies the right-hand-side vector in the linear system defining the Newton direction (12.5):

$$\begin{bmatrix} A & 0 & 0 \\ -Q & A^T & I_n \\ S & 0 & X \end{bmatrix} \cdot \begin{bmatrix} \Delta x \\ \Delta y \\ \Delta s \end{bmatrix} = \begin{bmatrix} 0 \\ 0 \\ \sigma \mu e - XSe \end{bmatrix}. \tag{12.7}$$

A systematic (though very slow) reduction of the complementarity gap is imposed by forcing a decrease of the barrier term in each iteration l. The required reduction of μ may seem very small: $\mu^{l+1} = \sigma\mu^l$, where $\sigma = 1 - \beta/\sqrt{n}$ for some $\beta \in (0,1)$. However, after a sufficiently large number of iterations, proportional to $\sqrt{n}$, the achieved reduction is already noticeable because

$$\frac{\mu^l}{\mu^0} = (1 - \beta/\sqrt{n})^{\sqrt{n}} \approx e^{-\beta}.$$

After $C \cdot \sqrt{n}$ iterations, the reduction achieves $e^{-C\beta}$. For a sufficiently large constant C the reduction can thus be arbitrarily large (i.e., the complementarity gap can become arbitrarily small). In other words, after a number of iterations proportional to $\sqrt{n}$, the algorithm gets arbitrarily close to a solution. In the parlance of complexity theory, the algorithm converges in $\mathcal{O}(\sqrt{n})$ iterations. We state the complexity result but omit the proof.

Theorem 12.3.1. *Given $\varepsilon > 0$, suppose that a feasible starting point $(x^0, y^0, s^0) \in N_2(0.1)$ satisfies $(x^0)^T s^0 = n\mu^0$, where $\mu^0 \leq 1/\varepsilon^\kappa$, for some positive constant κ. Then there exists an index L with $L = \mathcal{O}(\sqrt{n}\ln(1/\varepsilon))$, such that $\mu^l \leq \varepsilon, \quad \forall l \geq L$.*

The very good worst-case complexity result of IPM for quadratic program-

ming is beyond any competition in the field of optimization. Two features
in particular are unprecedented. First, the number of iterations is bounded
by the square root of the problem dimension. The computational experience
of Colombo and Gondzio (2008) shows a much better practical iteration
complexity which displays a logarithmic dependence on the problem dimen-
sion. Second, the complexity result reveals the dependence $\mathcal{O}(\ln(1/\varepsilon))$ on the
required precision ε. Unlike IPMs, gradient methods (Nesterov, 2005) can
provide only complexity results of $\mathcal{O}(1/\varepsilon)$ or $\mathcal{O}(1/\varepsilon^2)$. If one solves problems
to merely 1- or 2-digit exact solution ($\varepsilon = 10^{-1}$ or $\varepsilon = 10^{-2}$), the terms
$1/\varepsilon$ or $1/\varepsilon^2$ in the complexity result may seem acceptable. However, for a
higher accuracy, say, $\varepsilon = 10^{-3}$ or smaller, the superiority of IPMs becomes
obvious. (In the author's opinion, this outstanding feature of IPMs is not
appreciated enough by the machine learning community.)

The practical implementation of IPMs differs in several points from the
algorithm which possesses the best theoretical worst-case complexity. First,
the most efficient primal-dual method is the *infeasible* algorithm. Indeed,
there is no reason to force the algorithm to stay within the primal-dual
strictly feasible set $\mathcal{F}^0$ and unnecessarily limit its space to maneuver. IPMs
deal easily with any infeasibility in the primal and dual equality constraints
by taking them into account in the Newton system (12.5). Second, there is
no reason to restrict the iterates to the (very small) $N_2(\theta)$ neighborhood of
the central path. Practical algorithms (Colombo and Gondzio, 2008) use a
symmetric neighborhood $N_S(\gamma) = \{(x, y, s) \in \mathcal{F}^0 \,|\, \gamma\mu \leq x_j s_j \leq 1/\gamma\mu, \forall j\}$,
where $\gamma \in (0, 1)$ or a so-called infinity neighborhood $N_\infty(\gamma) = \{(x, y, s) \in$
$\mathcal{F}^0 \,|\, \gamma\mu \leq x_j s_j, \forall j\}$, in which only too-small complementarity products are
forbidden. Third, there is no reason to be overcautious in reducing the
complementarity gap by a term $\sigma = 1 - \beta/\sqrt{n}$ which is so close to 1. Practical
algorithms allow σ to be any number from the interval $(0, 1]$ and, indeed, the
author's experience (Colombo and Gondzio, 2008) shows that the average
reduction of the complementarity gap achieved in each IPM iteration $\sigma_{average}$
is usually in the interval $(0.1, 0.5)$. Deviation from the (close to 1) value of
σ allowed by the theory requires the extra safeguards to make sure x and s
remain nonnegative. This means that Newton steps have to be damped and
stepsize α takes values smaller than 1.

12.4   Interior-Point Methods for Machine Learning

The main difficulty and the main computational effort in IPM algorithms is
the solution of the Newton equation system: either (12.7) if we use a feasible
algorithm of theoretical interest, or (12.5) if we use a practical infeasible

algorithm. A common approach is to eliminate $\Delta s = X^{-1}(\xi_\mu - S\Delta x)$ and get the following symmetric but indefinite augmented system,

$$\begin{bmatrix} -Q - \Theta^{-1} & A^T \\ A & 0 \end{bmatrix} \begin{bmatrix} \Delta x \\ \Delta y \end{bmatrix} = \begin{bmatrix} f \\ h \end{bmatrix} = \begin{bmatrix} \xi_d - X^{-1}\xi_\mu \\ \xi_p \end{bmatrix}, \qquad (12.8)$$

where $\Theta = XS^{-1}$, or make one more elimination step $\Delta x = (Q + \Theta^{-1})^{-1}(A^T\Delta y - f)$ and get the symmetric and positive definite normal equations system

$$(A(Q + \Theta^{-1})^{-1}A^T)\Delta y = g = A(Q + \Theta^{-1})^{-1}f + h. \qquad (12.9)$$

For linear optimization problems (when $Q = 0$) the normal equations system (12.9) is usually the preferred (and default) option. For quadratic optimization problems with nontrivial matrix Q, an augmented system (12.8) is the best option. Indeed, the inversion of $(Q + \Theta^{-1})$ might completely destroy the sparsity in (12.9) and make the solution of this system very inefficient. There exists an important class of *separable* quadratic optimization problems in which Q is a diagonal matrix, and therefore the operation $(Q + \Theta^{-1})^{-1}$ produces a diagonal matrix and allows for the reduction to normal equations.

Several well-known reformulation tricks allow the extension of the class of separable problems and the conversion of certain nonseparable problems into separable ones (see Vanderbei, 1997). This is possible, for example, when matrix Q can be represented as $Q = Q_0 + VDV^T$, where Q_0 is easily invertible (say, diagonal) and $V \in \mathcal{R}^{n \times k}$, $D \in \mathcal{R}^{k \times k}$ with $k \ll n$ defining a low-rank correction. By introducing an extra variable $u = V^T x$, the quadratic objective term in problem (12.1) can be rewritten as $x^T Q x = x^T Q_0 x + u^T D u$ and the following quadratic optimization problem equivalent to (12.1) is obtained:

$$\begin{aligned} \min \quad & c^T x + \tfrac{1}{2}x^T Q_0 x + u^T D u \\ \text{s.t.} \quad & Ax = b, \\ & V^T x - u = 0, \\ & x \geq 0,\ u \text{ free.} \end{aligned} \qquad (12.10)$$

Although this new problem has more constraints ($m + k$ as opposed to m in (12.1)) and has $n + k$ variables, while (12.1) had only n, it is significantly easier to solve because its quadratic form $\begin{bmatrix} Q_0 & 0 \\ 0 & D \end{bmatrix}$ is easily invertible (diagonal) and allows for the use of the normal equations formulation in the computation of Newton direction.

Numerous classification problems in support vector machine training applications benefit from the above transformation. They include, for example, 1- or 2-norm classification, universum classification, and ordinal and ε-insensitive regressions. To demonstrate how the technique works, we will consider a 2-norm classification with support vector machines using the simplest linear kernel. Let a training set of n points $p_j \in \mathcal{R}^k$, $j = 1, 2, ..., n$ with binary labels $r_j \in \{-1, 1\}$, $j = 1, 2, ..., n$ be given. We look for a hyperplane $w^T p + w_0 = 0$ which best separates the points with different labels, namely, it maximizes the separation margin and minimizes the overall 1-norm error of misclassifications. The corresponding quadratic optimization problem and its dual have the following forms:

Primal   Dual

$$
\begin{array}{ll}
\min & \tfrac{1}{2} w^T w + \tau e^T \xi \\
\text{s.t.} & R(P^T w + w_0 e) \geq e - \xi \\
& \xi \geq 0;
\end{array}
\qquad
\begin{array}{ll}
\max & e^T z - \tfrac{1}{2} z^T (R P^T P R) z \\
\text{s.t.} & r^T z = 0 \\
& 0 \leq z \leq \tau e,
\end{array}
\qquad (12.11)
$$

where $P \in \mathcal{R}^{k \times n}$ is a matrix the columns of which are formed by the points $p_j \in \mathcal{R}^k$, $R \in \mathcal{R}^{n \times n}$ is a diagonal matrix with labels r_j on the diagonal, $\xi \in \mathcal{R}^n$ are errors of misclassification, and τ is a positive parameter measuring the penalty of misclassifications.

Direct application of IPM to any of these problems would be challenging because of the expected very large size of the data set n. The primal problem has an easy, separable quadratic objective but a large number of linear constraints. The dual problem, on the other hand, has only a single equality constraint but its Hessian matrix $R P^T P R \in \mathcal{R}^{n \times n}$ is completely dense. The dual form is preferred by the ML community because it can easily accommodate any general kernel K. (The dual problem in (12.11) corresponds to a linear kernel $K = P^T P$.)

To provide a better understanding of where the difficulty is hidden, we give forms of augmented equation systems which would be obtained if an IPM was applied directly to the primal or to the dual in (12.11):

$$
\begin{bmatrix}
-I_k & 0 & PR \\
0 & -\Theta_\xi^{-1} & I_n \\
R P^T & I_n & 0
\end{bmatrix}
\begin{bmatrix}
\Delta w \\
\Delta \xi \\
\Delta y
\end{bmatrix}
=
\begin{bmatrix}
f_w \\
f_\xi \\
h
\end{bmatrix}
\qquad (12.12)
$$

and

$$
\begin{bmatrix}
-(R P^T P R + \Theta_z^{-1}) & r \\
r^T & 0
\end{bmatrix}
\begin{bmatrix}
\Delta z \\
\Delta y
\end{bmatrix}
=
\begin{bmatrix}
f_z \\
h
\end{bmatrix}.
\qquad (12.13)
$$

To simplify the discussion, we keep using the notation of (12.8) and always denote Lagrange multipliers associated with the linear constraints as y and the right-hand-side vectors in these equations as (f, h). The dimensions of these vectors have to be derived from the formulations of the primal and dual problems in (12.11). For example, for the primal problem and equation

(12.12), $\Delta y \in \mathcal{R}^n, f_w \in \mathcal{R}^k, f_\xi \in \mathcal{R}^n$, and $h \in \mathcal{R}^n$; for the dual problem and equation (12.13), $\Delta y \in \mathcal{R}, f_z \in \mathcal{R}^n$, and $h \in \mathcal{R}$. It is easy to verify that the elimination of diagonal block $\mathrm{diag}\{I_k, \Theta_\xi^{-1}\}$ in (12.12) (which corresponds to the elimination of Δw and $\Delta \xi$) would create a dense normal equations matrix of form $RP^T PR + \Theta_\xi$, producing a dense linear equation system with difficulty comparable to that of (12.13).

Although the matrix $RP^T PR + \Theta_\xi \in \mathcal{R}^{n \times n}$ (or $RP^T PR + \Theta_z^{-1} \in \mathcal{R}^{n \times n}$ in (12.13)) is completely dense and is expected to be large, its inversion can be computed efficiently using the Sherman-Morrison-Woodbury (SMW) formula, which exploits the low-rank representation of this matrix. Indeed, since $PR \in \mathcal{R}^{k \times n}$ and $\Theta_\xi \in \mathcal{R}^{n \times n}$ is invertible, we can write

$$(RP^T PR + \Theta_\xi)^{-1} = \Theta_\xi^{-1} - \Theta_\xi^{-1} RP^T (I_k + PR\Theta_\xi^{-1} RP^T) PR\Theta_\xi^{-1} \quad (12.14)$$

and then replace equation $(RP^T PR + \Theta_\xi)\Delta y = g$ with a sequence of operations:

Step 1: calculate $\quad t_1 = PR\Theta_\xi^{-1} g$,

Step 2: solve $\quad (I_k + PR\Theta_\xi^{-1} RP^T) t_2 = t_1$,

Step 3: calculate $\quad \Delta y = \Theta_\xi^{-1}(g - RP^T t_2)$.

Since we expect $k \ll n$, the application of the SMW formula offers a major improvement over a direct inversion of the large and dense matrix $RP^T PR + \Theta_\xi$. Indeed, SMW requires several matrix-vector multiplications with $PR \in \mathcal{R}^{k \times n}$ which involve only kn flops, and building and inversion of the Schur complement matrix

$$S = I_k + PR\Theta_\xi^{-1} RP^T, \quad\quad\quad\quad\quad\quad (12.15)$$

which needs $\mathcal{O}(k^2 n + k^3)$ flops. In contrast, building and inversion of $RP^T PR + \Theta_\xi$ would require $\mathcal{O}(kn^2 + n^3)$ flops. An additional and very important advantage of the SMW algorithm is its storage efficiency: the matrix $RP^T PR + \Theta_\xi$ does not have to be formulated and stored; we only need to store original data $PR \in \mathcal{R}^{k \times n}$ and the $k \times k$ Schur complement matrix (12.15).

Ferris and Munson (2003) considered a variety of formulations of linear support vector machines and applied interior-point methods to solve them. They used the OOQP solver of Gertz and Wright (2003) as a basic tool for their developments. The Newton equation systems were solved using the SMW formula. The results of their efforts very clearly demonstrated the IPM's ability to deal with problems in which the number of data points n was large, reaching millions. Their test examples had a moderate number of features $k = 34$.

The efficiency of an SMW-based IPM implementation is determined by the linear algebra operation of solving (12.12) (or (12.13)). This approach is very easy to parallelize (Ferris and Munson, 2003) because the bulk of the work lies in the matrix-vector multiplications operating on PR and its transpose. Indeed, significant speedups may be achieved simply by splitting the storage of this matrix between different processors and reducing the number n_i of points stored on a given processor $i = 1, 2, ..., p$, $(\sum_{i=1}^{p} n_i = n)$ to improve data locality.

The Schur complement approach has an inherent weakness that is difficult to overcome. Its numerical accuracy critically depends on the stability of the easily invertible matrix (Θ_ξ in (12.14)) and the scaling of columns in the low-rank corrector (PR in (12.14)). It is actually a general weakness of SMW that is unrelated to IPM applications. In our case, when SMW is applied in the interior-point method for support vector machines, only one of these two potential weaknesses can be remedied. It is possible to scale the original problem data P and improve the properties of the low-rank corrector PR. However, to the best of the author's knowledge, there is no easy way to control the behavior of matrix Θ_ξ. The entries of this matrix display a disastrous difference in magnitude: as IPM approaches optimality, elements in one subset go to infinity while elements in the other subset go to zero. Consequently, the inversion of Θ_ξ is very unstable and always adversely affects the accuracy of the solution which can be obtained using the SMW formula (12.14).

Goldfarb and Scheinberg (2008) constructed a small artificial dataset on which a Schur complement-based IPM implementation ran into numerical difficulties and could not attain the required accuracy of solution. The product-form Cholesky factorization (PFCF) approach of Goldfarb and Scheinberg (2004) can handle such cases in a stable way. Instead of computing an explicit Cholesky decomposition, their approach builds the Cholesky matrix through a sequence of updates of an initial factor. The approach is well suited to dealing with matrices of the form $Q = Q_0 + VV^T$, such as the matrix $\Theta_\xi + RP^T PR$ in (12.14). It starts from a decomposition of Q_0 and updates it after adding every rank-1 corrector $V_i V_i^T$ from the matrix VV^T. The approach has been implemented in two solvers, SVM-QP and SVM-QP-presolve (Goldfarb and Scheinberg, 2008), and when applied to medium-scale problems it has demonstrated numerical stability in practice. It is not clear whether the PFCF can be implemented in parallel and this seems to question its applicability to large-scale machine learning problems (see Woodsend and Gondzio, 2009).

Bearing in mind the need to develop parallel implementation to tackle very large problems, Woodsend and Gondzio (2010) have exploited the separable

QP formulations of several support vector machine problems and solved them directly with an interior-point method. Their approach avoids the use of the SMW formula, which could introduce instability but still relies on parallelism-friendly block-matrix operations. We will illustrate the key idea by considering the dual in (12.11).

As we have already observed, the matrix of the quadratic form in this problem, $RP^T PR$, is dense. However, it is a low-rank matrix and we will exploit its known decomposition. Namely, we define $u = PRz$ and observe that $z^T RP^T PRz = u^T u$, so the problem can be reformulated as

$$
\begin{aligned}
\min \quad & -e^T z + \tfrac{1}{2} u^T u \\
\text{s.t.} \quad & r^T z = 0 \\
& PRz - u = 0 \\
& 0 \le z \le \tau e, \ u \text{ free.}
\end{aligned}
\tag{12.16}
$$

Unlike the dual in (12.11), which had n variables and only one constraint, the new problem has $n+k$ variables and $k+1$ constraints. It is slightly larger than (12.11) but is separable, and the linear equation system to compute the Newton direction

$$
\begin{bmatrix}
-\Theta_z^{-1} & 0 & r & RP^T \\
0 & -I_k & 0 & -I_k \\
r^T & 0 & 0 & 0 \\
PR & -I_k & 0 & 0
\end{bmatrix}
\begin{bmatrix}
\Delta z \\
\Delta u \\
\Delta y_1 \\
\Delta y_2
\end{bmatrix}
=
\begin{bmatrix}
f_z \\
f_u \\
h_1 \\
h_2
\end{bmatrix},
\tag{12.17}
$$

has an easy-to-invert $(n+k) \times (n+k)$ diagonal block at position $(1,1)$. After the elimination of this leading block (which corresponds to the elimination of Δz and Δu), we obtain the normal equations

$$
\left(
\begin{bmatrix}
r^T & 0 \\
PR & -I_k
\end{bmatrix}
\begin{bmatrix}
\Theta_z & 0 \\
0 & I_k
\end{bmatrix}
\begin{bmatrix}
r & RP^T \\
0 & -I_k
\end{bmatrix}
\right)
\begin{bmatrix}
\Delta y_1 \\
\Delta y_2
\end{bmatrix}
=
\begin{bmatrix}
g_1 \\
g_2
\end{bmatrix},
\tag{12.18}
$$

which form a system of only $k+1$ linear equations with $k+1$ unknowns. Forming the matrix involved in this system can easily be parallelized. It suffices to split the matrix $P \in \mathcal{R}^{k \times n}$ into blocks $P_i \in \mathcal{R}^{k \times n_i}$, $i = 1, 2, ..., p$ with $\sum_{i=1}^{p} n_i = n$ and gather the partial summation results in the operation

$$
PR\Theta_z RP^T = \sum_{i=1}^{p} P_i R_i \Theta_{zi} R_i P_i^T,
\tag{12.19}
$$

executed on p independent blocks. The separability-exploiting IPM approach of Woodsend and Gondzio (2009) described above has been implemented using OOPS (Gondzio and Grothey, 2006) and tested on very large-scale problems from the PASCAL Challenge,

http://largescale.first.fraunhofer.de/.
The implementation is available for research use from
http://www.maths.ed.ac.uk/ERGO/software.html.

It is worth mentioning important advantages of the separable QP formulation which distinguish it from two approaches discussed earlier: the one based on the SMW formula (Ferris and Munson, 2003) and the one employing the product form Cholesky factorization (Goldfarb and Scheinberg, 2004, 2008). Unlike the SMW approach which can easily lose accuracy due to multiple inversions of Θ_ξ in (12.14), the separable formulation (12.16) avoids such operations and does not suffer from any instability. In contrast to the PFCF approach, which is inherently sequential, the separable formulation (12.16) allows for an easy parallelization of its major computational tasks.

In summary, interior-point methods provide an attractive alternative to a plethora of other approaches in machine learning. In the context of support vector machines, extensive tests on large instances from the PASCAL Challenge demonstrated (Woodsend and Gondzio, 2009) that IPMs compete very well with the other approaches in terms of CPU time efficiency, and outperform the other approaches in terms of reliability. This is consistent with a general reputation of IPMs as very stable and reliable optimization algorithms.

12.5   Accelerating Interior-Point Methods

Stability and reliability of IPMs have their origin in the use of the Newton method for barrier subproblems and a very "mild" nonlinearity introduced by the logarithmic barrier function. In some applications these features come at too high a price. Numerous optimization problems, including those arising in machine learning, do not have to be solved to a high degree of accuracy. Therefore, fast algorithms are sought which could provide a very rough approximation to the solution of the optimization problem in no time at all. This is one of the reasons why the ML community is so keen on very simple first-order (gradient)-based optimization techniques.

There have been several attempts to improve interior-point methods by reducing the cost of a single iteration. Two of them have been specifically developed for support vector machine applications. They share a common feature and try to guess the activity of inequality constraints, then use only a subset of these constraints when computing Newton directions. Consider again the primal problem in (12.11) and the corresponding Newton equation system in (12.12). The elimination of a large but easily invertible block

$$
\begin{bmatrix} -\Theta_\xi^{-1} & I_n \\ I_n & 0 \end{bmatrix},
$$

which corresponds to the elimination of $\Delta\xi$ and Δy from the equation system (12.12), produces the small $k \times k$ system

$$(I_k + PR\Theta_\xi RP^T)\Delta w = g_w. \tag{12.20}$$

We have already mentioned that the magnitude of elements of matrix Θ_ξ may vary significantly. Indeed, $\Theta_{\xi j} = \xi_j/\eta_j$, where η_j is the Lagrange multiplier associated with the simple inequality constraint $\xi_j \geq 0$. The complementarity condition requires that $\xi_j\eta_j = 0$ at optimality. IPM uses a perturbed complementarity condition $\xi_j\eta_j = \mu$ and forces ξ_j and η_j to be strictly positive. However, when IPM approaches optimality, one of these variables necessarily has to become very small. Consequently, the ratio ξ_j/η_j goes either to infinity or to zero. Although this might be the source of numerical difficulties when solving systems (12.12) and (12.20), it may also be exploited as a feature to simplify these equations. The matrix in (12.20) can be written in the outer product form

$$M = I_k + \sum_{j=1}^n r_j^2 \Theta_{\xi j} p_j p_j^T, \tag{12.21}$$

where $r_j^2 = 1$ (because $r_j \in \{-1,+1\}$) and p_j denotes column j of P, that is, point j in the training set. Since many elements of Θ_ξ are very small, their corresponding outer product contributions to M may be neglected. An approximation of M may be formed as follows:

$$\tilde{M} = I_k + \sum_{\{j:\Theta_{\xi j} \geq \delta\}} \Theta_{\xi j} p_j p_j^T, \tag{12.22}$$

where δ is a prescribed tolerance.

Jung et al. (2008) use information on complementarity products of ξ_j and η_j to determine small elements of $\Theta_{\xi j}$ which may be dropped in the summation. The constraints $r_j(p_j^T w + w_0) \geq 1 - \xi_j$ in the primal problem (12.11), which correspond to indices j associated with small terms $\Theta_{\xi j}$, are likely to be active at optimality. Jung et al. (2008) use the approximation $\tilde{M}$ of M to compute an approximate Newton step. Gertz and Griffin (2009) use the same approximation for a different purpose. They employ a conjugate gradient algorithm to solve (12.20) and use $\tilde{M}$ as a preconditioner of M. In summary, both approaches try to simplify the computations and replace M with its approximation $\tilde{M}$, exploiting obvious savings resulting from replacing the summation over $j \in \{1, 2, ..., n\}$ with the summation over a subset of indices $\{j : \Theta_{\xi j} \geq \delta\}$. However, both approaches have to deal with certain computational overheads: Jung et al. (2008) have to accept a significant increase of the number of iterations resulting from the use of inexact directions, while Gertz and Griffin (2009) need to bear an extra effort of matrix-vector multiplications in the conjugate gradient algorithm.

To conclude the discussion of different acceleration techniques applicable in the IPM context, we need to draw the reader's attention to a recent development of a *matrix-free* variant of the interior-point method (Gondzio, 2010). This approach has been developed with the purpose of solving very large and huge optimization problems for which storage of the problem data alone may already be problematic, and constructing and factoring any of the matrices in the Newton equations (augmented system or normal equations) is expected to be prohibitive. The approach works in a matrix-free regime: Newton equations are never formulated explicitly. Instead, an inexact Newton method (Dembo et al., 1982; Bellavia, 1998) is used, that is, the Newton direction is computed using an iterative approach from the Krylov subspace family. The key feature of the new method which distinguishes it from other matrix-free approaches is that the preconditioner for the iterative method is constructed using the matrix-free regime as well. The method has been described in Gondzio (2010) as a general-purpose one. However, it should be straightforward to specialize it to machine learning problems. We discuss it briefly below.

Consider a problem such as the separable reformulation (12.16) of the dual problem (12.11) and assume that the number of rows, $k + 1$, and the number of columns, $n + k$, are large. One might think of the number of features k being of the order 10^4 or larger, and the number of training points n going into millions or larger. The otherwise very efficient separable formulation (12.16) would demonstrate its limitations for such dimensions because the $(k + 1) \times (k + 1)$ normal equation matrix (12.18) would be excessively expensive to form and factor. Following Woodsend and Gondzio (2010), building the matrix would need $\mathcal{O}(nk^2)$ flops, and factoring it would require an additional $\mathcal{O}(k^3)$ flops. The matrix-free approach (Gondzio, 2010) solves (12.18) *without* forming and factoring the normal equation matrix. It uses the conjugate gradient method, which does not require the normal equation matrix

$$H = \bar{A}\bar{D}\bar{A}^T = \begin{bmatrix} r^T & 0 \\ PR & -I_k \end{bmatrix} \begin{bmatrix} \Theta_z & 0 \\ 0 & I_k \end{bmatrix} \begin{bmatrix} r & RP^T \\ 0 & -I_k \end{bmatrix} \qquad (12.23)$$

to be explicitly formulated but needs only to perform matrix-vector multiplications with it. These operations can be executed as a sequence of matrix-vector multiplications with the constraint matrix $\bar{A}$, its transpose, and the diagonal scaling matrix $\bar{D}$. Matrix Θ_z in the diagonal part is always very ill-conditioned, and consequently so is H. The conjugate gradient algorithm will never converge unless an appropriate preconditioner is used. The preconditioner proposed by Gondzio (2010) is a low-rank partial Cholesky factorization of H which is also constructed in the matrix-free regime.

12.6   Conclusions

In this chapter we have discussed the main features of interior-point methods which make them attractive for very large-scale optimization and for application in the machine learning context. IPMs offer an unequalled worst-case complexity: they converge to an ε-accurate solution in $\mathcal{O}(\sqrt{n}\log(1/\varepsilon))$ iterations. In practice they perform much better than the worst-case analysis predicts, and solve linear or convex quadratic problems in a number of iterations which very slowly (logarithmically) grows with the problem dimension. Since machine learning applications are usually very large, IPMs offer an attractive solution methodology for them. We have illustrated the use of IPMs in a particular class of ML problems: support vector machine training. IPMs display excellent stability and robustness, which makes them very competitive in this context. A novel matrix-free variant of the interior-point method is a promising approach for solving very large and huge optimization problems arising in machine learning applications.

Acknowledgment

The author is grateful to Marco Colombo, Pedro Munari and Kristian Woodsend for reading a draft of this chapter and offering useful suggestions which led to its improvement.

12.7   References

S. Bellavia. An inexact interior-point method. *Journal of Optimization Theory and Applications*, 96(1):109–121, 1998.

D. P. Bertsekas. *Nonlinear Programming*. Athena Scientific, 1995.

M. Colombo and J. Gondzio. Further development of multiple centrality correctors for interior point methods. *Computational Optimization and Applications*, 41(3): 277–305, 2008.

G. B. Dantzig. *Linear Programming and Extensions*. Princeton University Press, Princeton, N.J., 1963.

R. S. Dembo, S. C. Eisenstat, and T. Steihaug. Inexact Newton methods. *SIAM Journal on Numerical Analysis*, 19:400–408, 1982.

M. C. Ferris and T. S. Munson. Interior point methods for massive support vector machines. *SIAM Journal on Optimization*, 13(3):783–804, 2003.

S. Fine and K. Scheinberg. Efficient SVM training using low-rank kernel representations. *Journal of Machine Learning Research*, 2:243–264, 2002.

J. J. H. Forrest and D. Goldfarb. Steepest-edge simplex algorithms for linear programming. *Mathematical Programming*, 57:341–374, 1992.

E. M. Gertz and J. Griffin. Using an iterative linear solver in an interior-point method for generating support vector machines. *Computational Optimization and Applications*, 47(3):431–453, 2009.

E. M. Gertz and S. J. Wright. Object-oriented software for quadratic programming. *ACM Transactions on Mathematical Software*, 29(1):58–81, 2003.

P. E. Gill, W. Murray, M. A. Saunders, J. A. Tomlin, and M. H. Wright. On the projected Newton barrier methods for linear programming and an equivalence to Karmarkar's projective method. *Mathematical Programming*, 36:183–209, 1986.

D. Goldfarb and K. Scheinberg. A product-form Cholesky factorization method for handling dense columns in interior point methods for linear programming. *Mathematical Programming*, 99(1):1–34, 2004.

D. Goldfarb and K. Scheinberg. Numerically stable LDL^T factorizations in interior point methods for convex quadratic programming. *IMA Journal of Numerical Analysis*, 28(4):806–826, 2008.

J. Gondzio. Matrix-free interior point method. Technical Report ERGO-2009-012, School of Mathematics, University of Edinburgh, Edinburgh EH9 3JZ, Scotland, UK, April 2010.

J. Gondzio and A. Grothey. Direct solution of linear systems of size 10^9 arising in optimization with interior point methods. In R. Wyrzykowski, J. Dongarra, N. Meyer, and J. Wasniewski, editors, *Parallel Processing and Applied Mathematics*, volume 3911 of *Lecture Notes in Computer Science*, pages 513–525. Springer-Verlag, Berlin, 2006.

J. A. J. Hall and K. I. M. McKinnon. Hyper-sparsity in the revised simplex method and how to exploit it. *Computational Optimization and Applications*, 32(3):259–283, 2005.

J. H. Jung, D. O'Leary, and A. Tits. Adaptive constraint reduction for training support vector machines. *Elecronic Transactions on Numerical Analysis*, 31:156–177, 2008.

N. K. Karmarkar. A new polynomial–time algorithm for linear programming. *Combinatorica*, 4(4):373–395, 1984.

V. Klee and G. Minty. How good is the simplex algorithm? In O. Shisha, editor, *Inequalities-III*, pages 159–175. Academic Press, 1972.

I. J. Lustig, R. E. Marsten, and D. F. Shanno. Interior point methods for linear programming: Computational state of the art. *ORSA Journal on Computing*, 6(1):1–14, 1994.

I. Maros. *Computational Techniques of the Simplex Method*. Kluwer Academic, Boston, 2003.

R. E. Marsten, R. Subramanian, M. J. Saltzman, I. J. Lustig, and D. F. Shanno. Interior point methods for linear programming: Just call Newton, Lagrange, and

Fiacco and McCormick! *Interfaces*, 20(4):105–116, 1990.

Y. Nesterov. Smooth minimization of non-smooth functions. *Mathematical Programming, Series A*, 103:127–152, 2005.

J. Renegar. A polynomial-time algorithm, based on Newton's method, for linear programming. *Mathematical Programming*, 40:59–93, 1988.

R. J. Vanderbei. *Linear Programming: Foundations and Extensions*. Kluwer Academic, Boston, 1st edition, 1997.

K. Woodsend and J. Gondzio. Hybrid MPI/OpenMP parallel linear support vector machine training. *Journal of Machine Learning Research*, 10:1937–1953, 2009.

K. Woodsend and J. Gondzio. Exploiting separability in large-scale linear support vector machine training. *Computational Optimization and Applications*, 49: 241–269, 2011.

S. J. Wright. *Primal-Dual Interior-Point Methods*. SIAM, Philadelphia, 1997.

13 The Tradeoffs of Large-Scale Learning

Léon Bottou   leon@bottou.org
NEC Laboratories of America
Princeton, NJ, USA

Olivier Bousquet   olivier.bousquet@m4x.org
Google
Zurich, Switzerland

This chapter develops a theoretical framework that takes into account the effect of approximate optimization on learning algorithms. The analysis shows distinct tradeoffs for the case of small-scale and large-scale learning problems. Small-scale learning problems are subject to the usual approximation–estimation tradeoff. Large-scale learning problems are subject to a qualitatively different tradeoff involving the computational complexity of the underlying optimization algorithm in non-trivial ways. For instance, a mediocre optimization algorithm, stochastic gradient descent, is shown to perform very well on large-scale learning problems.

13.1   Introduction

The computational complexity of learning algorithms has seldom been taken into account by the learning theory. Valiant (1984) states that a problem is "learnable" when there exists a "probably approximately correct" learning algorithm *with polynomial complexity*. Whereas much progress has been made on the statistical aspect (e.g., Vapnik, 1982; Boucheron et al., 2005; Bartlett and Mendelson, 2006), very little has been said about the complexity side of this proposal (e.g., Judd, 1988).

Computational complexity becomes the limiting factor when one envisions large amounts of training data. Two important examples come to mind:

- Data mining exists because competitive advantages can be achieved by analyzing the masses of data that describe the life of our computerized society. Since virtually every computer generates data, the data volume is proportional to the available computing power. Therefore, one needs learning algorithms that scale roughly linearly with the total volume of data.

- Artificial intelligence attempts to emulate the cognitive capabilities of human beings. Our biological brains can learn quite efficiently from the continuous streams of perceptual data generated by our senses, using limited amounts of sugar as a source of power. This observation suggests that there are learning algorithms whose computing time requirements scale roughly linearly with the total volume of data.

This chapter develops the ideas initially proposed by Bottou and Bousquet (2008). Section 13.2 proposes a decomposition of the test error where an additional term represents the impact of approximate optimization. In the case of small-scale learning problems, this decomposition reduces to the well-known tradeoff between approximation error and estimation error. In the case of large-scale learning problems, the tradeoff is more complex because it involves the computational complexity of the learning algorithm. Section 13.3 explores the asymptotic properties of the large-scale learning tradeoff for various prototypical learning algorithms under various assumptions regarding the statistical estimation rates associated with the chosen objective functions. This part clearly shows that the best optimization algorithms are not necessarily the best learning algorithms. Maybe more surprisingly, certain algorithms perform well regardless of the assumed rate of the statistical estimation error. Section 13.4 reports experimental results supporting this analysis.

13.2   Approximate Optimization

13.2.1   Setup

Following Duda and Hart (1973) and Vapnik (1982), we consider a space of input-output pairs $(x, y) \in \mathcal{X} \times \mathcal{Y}$ endowed with a probability distribution $P(x, y)$. The conditional distribution $P(y|x)$ represents the unknown relationship between inputs and outputs. The discrepancy between the predicted output $\hat{y}$ and the real output y is measured with a loss function $\ell(\hat{y}, y)$. Our benchmark is the function f^* that minimizes the expected risk

$$E(f) = \int \ell(f(x), y) \, dP(x, y) = \mathbb{E}\left[\ell(f(x), y)\right],$$

that is,

$$f^*(x) = \arg\min_{\hat{y}} \mathbb{E}\left[\ell(\hat{y}, y) \mid x\right].$$

Although the distribution $P(x, y)$ is unknown, we are given a sample $\mathcal{S}$ of n independently drawn training examples (x_i, y_i), $i = 1 \ldots n$. We define the empirical risk

$$E_n(f) = \frac{1}{n} \sum_{i=1}^{n} \ell(f(x_i), y_i) = \mathbb{E}_n[\ell(f(x), y)].$$

Our first learning principle is choosing a family $\mathcal{F}$ of candidate prediction functions and finding the function $f_n = \arg\min_{f \in \mathcal{F}} E_n(f)$ that minimizes the empirical risk. Well-known combinatorial results (e.g., Vapnik, 1982) support this approach, provided that the chosen family $\mathcal{F}$ is sufficiently restrictive. Since the optimal function f^* is unlikely to belong to the family $\mathcal{F}$, we also define $f_{\mathcal{F}}^* = \arg\min_{f \in \mathcal{F}} E(f)$. For simplicity, we assume that f^*, $f_{\mathcal{F}}^*$, and f_n are well defined and unique.

We can then decompose the excess error as

$$
\begin{aligned}
\mathcal{E} \;&=\; \mathbb{E}\left[E(f_{\mathcal{F}}^*) - E(f^*)\right] \;+\; \mathbb{E}\left[E(f_n) - E(f_{\mathcal{F}}^*)\right] \\
&=\; \mathcal{E}_{\text{app}} + \mathcal{E}_{\text{est}},
\end{aligned}
\tag{13.1}
$$

where the expectation is taken with respect to the random choice of training set. The *approximation error* $\mathcal{E}_{\text{app}}$ measures how closely functions in $\mathcal{F}$ can approximate the optimal solution f^*. The *estimation error* $\mathcal{E}_{\text{est}}$ measures the effect of minimizing the empirical risk $E_n(f)$ instead of the expected risk $E(f)$. The estimation error is determined by the number of training examples and by the capacity of the family of functions (Vapnik, 1982). Large families[1] of functions have *smaller approximation errors* but lead to *higher estimation errors*. This tradeoff has been extensively discussed in the literature (Vapnik, 1982; Boucheron et al., 2005) and has led to excess errors that scale between the inverse and the inverse square root of the number of

---

1. We often consider nested families of functions of the form $F_c = \{f \in \mathcal{H}, \; \Omega(f) \leq c\}$. Then, for each value of c, function f_n is obtained by minimizing the regularized empirical risk $E_n(f) + \lambda\Omega(f)$ for a suitable choice of the Lagrange coefficient λ. We can then control the estimation-approximation tradeoff by choosing λ instead of c.

examples (Zhang, 2004; Steinwart and Scovel, 2005).

13.2.2 Optimization Error

Finding f_n by minimizing the empirical risk $E_n(f)$ is often a computationally expensive operation. Since the empirical risk $E_n(f)$ is already an approximation of the expected risk $E(f)$, it should not be necessary to carry out this minimization with great accuracy. For instance, we could stop an iterative optimization algorithm long before its convergence.

Let us assume that our minimization algorithm returns an approximate solution $\tilde{f}_n$ such that $E_n(\tilde{f}_n) < E_n(f_n) + \rho$ where $\rho \geq 0$ is a predefined tolerance. An additional term $\mathcal{E}_{opt} = \mathbb{E}\big[E(\tilde{f}_n) - E(f_n)\big]$ then appears in the decomposition of the excess error $\mathcal{E} = \mathbb{E}\big[E(\tilde{f}_n) - E(f^*)\big]$:

$$\begin{aligned} \mathcal{E} &= \mathbb{E}\left[E(f_{\mathcal{F}}^*) - E(f^*)\right] + \mathbb{E}\left[E(f_n) - E(f_{\mathcal{F}}^*)\right] + \mathbb{E}\left[E(\tilde{f}_n) - E(f_n)\right] \\ &= \mathcal{E}_{app} + \mathcal{E}_{est} + \mathcal{E}_{opt}. \end{aligned} \tag{13.2}$$

We call this additional term the *optimization error*. It reflects the impact of the approximate optimization on the generalization performance. Its magnitude is comparable to ρ (see section 13.3.1).

13.2.3 The Approximation–Estimation–Optimization Tradeoff

This decomposition leads to a more complicated compromise. It involves three variables and two constraints. The constraints are the maximal number of available training examples and the maximal computation time. The variables are the size of the family of functions $\mathcal{F}$, the optimization accuracy ρ, and the number of examples n. This is formalized by the following optimization problem:

$$\min_{\mathcal{F},\rho,n} \mathcal{E} = \mathcal{E}_{app} + \mathcal{E}_{est} + \mathcal{E}_{opt} \quad \text{subject to} \left\{ \begin{array}{rcl} n &\leq& n_{max} \\ T(\mathcal{F},\rho,n) &\leq& T_{max} \end{array} \right. \tag{13.3}$$

The number n of training examples is a variable because we could choose to use only a subset of the available training examples in order to complete the optimization within the alloted time. This happens often in practice. Table 13.1 summarizes the typical evolution of the quantities of interest as the three variables $\mathcal{F}$, n, and ρ increase.

The solution of the optimization program (13.3) depends critically on which budget constraint is active: constraint $n < n_{max}$ on the number of examples, or constraint $T < T_{max}$ on the training time.

Table 13.1: Typical variations when $\mathcal{F}$, n, and ρ increase

		$\mathcal{F}$	n	ρ
$\mathcal{E}_{app}$	(approximation error)	↘		
$\mathcal{E}_{est}$	(estimation error)	↗	↘	
$\mathcal{E}_{opt}$	(optimization error)	$\cdots$	$\cdots$	↗
T	(computation time)	↗	↗	↘

- We speak of a *small-scale learning problem* when (13.3) is constrained by the maximal number of examples $n_{\max}$. Since the computing time is not limited, we can reduce the optimization error $\mathcal{E}_{opt}$ to insignificant levels by choosing a ρ that is arbitrarily small. The excess error is then dominated by the approximation and estimation errors, $\mathcal{E}_{app}$ and $\mathcal{E}_{est}$. Taking $n = n_{\max}$, we recover the approximation-estimation tradeoff that is the object of abundant literature.

- We speak of a *large-scale learning problem* when (13.3) is constrained by the maximal computing time $T_{\max}$. Approximate optimization, that is, choosing $\rho > 0$, possibly can achieve better generalization because more training examples can be processed during the allowed time. The specifics depend on the computational properties of the chosen optimization algorithm through the expression of the computing time $T(\mathcal{F}, \rho, n)$.

13.3 Asymptotic Analysis

In section 13.2.2, we extended the classical approximation–estimation trade-off by taking the optimization error into account. We gave an objective criterion to distiguish small-scale and large-scale learning problems. In the small-scale case, we recovered the classical tradeoff between approximation and estimation. The large-scale case is substantially different because it involves the computational complexity of the learning algorithm. In order to clarify the large-scale learning tradeoff with sufficient generality, this section makes several simplifications:

- We are studying upper bounds of the approximation, estimation, and optimization errors (13.2). It is often accepted that these upper bounds give a realistic idea of the actual convergence rates (Vapnik et al., 1994; Bousquet, 2002; Tsybakov, 2004; Bartlett et al., 2006). Another way to find comfort in this approach is to say that we study guaranteed convergence rates instead of the possibly pathological special cases.

- We are studying the asymptotic properties of the tradeoff when the problem size increases. Instead of carefully balancing the three terms, we write $\mathcal{E} = \mathcal{O}(\mathcal{E}_{app}) + \mathcal{O}(\mathcal{E}_{est}) + \mathcal{O}(\mathcal{E}_{opt})$ and only need to ensure that the three terms decrease with the same asymptotic rate.

- We are considering a fixed family of functions $\mathcal{F}$, and therefore avoid taking into account the approximation error $\mathcal{E}_{app}$. This part of the tradeoff covers a wide spectrum of practical realities, such as choosing models and features. In the context of this work, we do not believe we can meaningfully address this without discussing, for instance, the thorny issue of feature selection. Instead, we focus on the choice of optimization algorithm.

- Finally, in order to keep this chapter short, we consider that the family of functions $\mathcal{F}$ is linearly parameterized by a vector $w \in \mathbb{R}^d$. We also assume that x, y, and w are bounded, ensuring that there is a constant B such that $0 \leq \ell(f_w(x), y) \leq B$ and $\ell(\cdot, y)$ is Lipschitz.

We first explain how the uniform convergence bounds provide convergence rates that take the optimization error into account. Then we discuss and compare the asymptotic learning properties of several optimization algorithms.

13.3.1   Convergence of the Estimation and Optimization Errors

The optimization error $\mathcal{E}_{opt}$ depends on the optimization accuracy ρ. However, the accuracy ρ involves the empirical quantity $E_n(\tilde{f}_n) - E_n(f_n)$, whereas the optimization error $\mathcal{E}_{opt}$ involves its expected counterpart $E(\tilde{f}_n) - E(f_n)$. This section discusses the impact of the optimization error $\mathcal{E}_{opt}$ and of the accuracy ρ on generalization bounds that leverage the uniform convergence concepts pioneered by Vapnik and Chervonenkis (e.g., Vapnik, 1982).

Following Massart (2000), in the following discussion we use the letter c to refer to any positive constant. Successive occurrences of the letter c do not necessarily imply that the constants have identical values.

13.3.1.1   *Simple Uniform Convergence Bounds*

Recall that we assume that $\mathcal{F}$ is linearly parameterized by $w \in \mathbb{R}^d$. Elementary uniform convergence results then state that

$$\mathbb{E}\left[\sup_{f \in \mathcal{F}} |E(f) - E_n(f)|\right] \leq c\sqrt{\frac{d}{n}},$$

where the expectation is taken with respect to the random choice of the training set.[2] This result immediately provides a bound on the estimation error:

$$
\begin{aligned}
\mathcal{E}_{\text{est}} &= \mathbb{E}\left[\left(E(f_n) - E_n(f_n)\right) + \left(E_n(f_n) - E_n(f_{\mathcal{F}}^*)\right) + \left(E_n(f_{\mathcal{F}}^*) - E(f_{\mathcal{F}}^*)\right)\right] \\
&\leq 2\,\mathbb{E}\left[\sup_{f \in \mathcal{F}} |E(f) - E_n(f)|\right] \leq c\sqrt{\frac{d}{n}}.
\end{aligned}
$$

This same result also provides a combined bound for the estimation and optimization errors:

$$
\begin{aligned}
\mathcal{E}_{\text{est}} + \mathcal{E}_{\text{opt}} &= \mathbb{E}\left[E(\tilde{f}_n) - E_n(\tilde{f}_n)\right] + \mathbb{E}\left[E_n(\tilde{f}_n) - E_n(f_n)\right] \\
&\quad + \mathbb{E}\left[E_n(f_n) - E_n(f_{\mathcal{F}}^*)\right] + \mathbb{E}\left[E_n(f_{\mathcal{F}}^*) - E(f_{\mathcal{F}}^*)\right] \\
&\leq c\sqrt{\frac{d}{n}} + \rho + 0 + c\sqrt{\frac{d}{n}} = \mathcal{O}\left(\rho + \sqrt{\frac{d}{n}}\right).
\end{aligned}
$$

Unfortunately, this convergence rate is known to be pessimistic in many important cases. More sophisticated bounds are required.

13.3.1.2 *Faster Rates in the Realizable Case*

When the loss function $\ell(\hat{y}, y)$ is positive, with probability $1 - e^{-\tau}$ for any $\tau > 0$, relative uniform convergence bounds (e.g., Vapnik, 1982) state that

$$
\sup_{f \in \mathcal{F}} \frac{E(f) - E_n(f)}{\sqrt{E(f)}} \leq c\sqrt{\frac{d}{n}\log\frac{n}{d} + \frac{\tau}{n}}.
$$

This result is very useful because it provides faster convergence rates $\mathcal{O}(\log n/n)$ in the *realizable case*, that is, when $\ell(f_n(x_i), y_i) = 0$ for all training examples (x_i, y_i). We then have $E_n(f_n) = 0$, and $E_n(\tilde{f}_n) \leq \rho$, and we can write

$$
E(\tilde{f}_n) - \rho \leq c\sqrt{E(\tilde{f}_n)}\sqrt{\frac{d}{n}\log\frac{n}{d} + \frac{\tau}{n}}.
$$

Viewing this as a second-degree polynomial inequality in variable $\sqrt{E(\tilde{f}_n)}$, we obtain

---

2. Although the original Vapnik-Chervonenkis bounds have the form $c\sqrt{\frac{d}{n}\log\frac{n}{d}}$, the logarithmic term can be eliminated using the "chaining" technique (e.g., Bousquet, 2002).

$$E(\tilde{f}_n) \le c \left(\rho + \frac{d}{n} \log \frac{n}{d} + \frac{\tau}{n} \right).$$

Integrating this inequality using a standard technique (see, e.g., Massart, 2000), we obtain a better convergence rate of the combined estimation and optimization error:

$$\mathcal{E}_{\text{est}} + \mathcal{E}_{\text{opt}} = \mathbb{E}\left[E(\tilde{f}_n) - E(f_{\mathcal{F}}^*) \right] \le \mathbb{E}\left[E(\tilde{f}_n) \right] = c \left(\rho + \frac{d}{n} \log \frac{n}{d} \right).$$

13.3.1.3 Fast Rate Bounds

Many authors (e.g., Bousquet, 2002; Bartlett and Mendelson, 2006; Bartlett et al., 2006) obtain fast statistical estimation rates in more general conditions. These bounds have the general form

$$\mathcal{E}_{\text{app}} + \mathcal{E}_{\text{est}} \le c \left(\mathcal{E}_{\text{app}} + \left(\frac{d}{n} \log \frac{n}{d} \right)^{\alpha} \right) \quad \text{for } \frac{1}{2} \le \alpha \le 1. \tag{13.4}$$

This result holds when one can establish the following variance condition:

$$\forall f \in \mathcal{F} \quad \mathbb{E}\left[\left(\ell(f(X), Y) - \ell(f_{\mathcal{F}}^*(X), Y) \right)^2 \right] \le c \left(E(f) - E(f_{\mathcal{F}}^*) \right)^{2 - \frac{1}{\alpha}}. \tag{13.5}$$

The convergence rate of (13.4) is described by the exponent α, which is determined by the quality of the variance bound (13.5). Works on fast statistical estimation identify two main ways to establish such a variance condition.

■ Exploiting the strict convexity of certain loss functions (Bartlett et al., 2006, theorem 12). For instance, Lee et al. (1998) establish a $\mathcal{O}(\log n/n)$ rate using the squared loss $\ell(\hat{y}, y) = (\hat{y} - y)^2$.

■ Making assumptions on the data distribution. In the case of pattern recognition problems, for instance, the Tsybakov condition indicates how cleanly the posterior distributions $P(y|x)$ cross near the optimal decision boundary (Tsybakov, 2004; Bartlett et al., 2006). The realizable case discussed in section 13.3.1.2 can be viewed as an extreme example of this.

Despite their much greater complexity, fast rate estimation results can accommodate the optimization accuracy ρ, using essentially the methods illustrated in sections 13.3.1.1 and 13.3.1.2. We then obtain a bound of the form

$$\mathcal{E} = \mathcal{E}_{\text{app}} + \mathcal{E}_{\text{est}} + \mathcal{E}_{\text{opt}} = \mathbb{E}\left[E(\tilde{f}_n) - E(f^*)\right] \leq c\left(\mathcal{E}_{\text{app}} + \left(\frac{d}{n}\log\frac{n}{d}\right)^{\alpha} + \rho\right).$$

$$(13.6)$$

For instance, a general result with $\alpha = 1$ is provided by Massart (2000, theorem 4.2). Combining this result with standard bounds on the complexity of classes of linear functions (e.g., Bousquet, 2002) yields the following result:

$$\mathcal{E} = \mathcal{E}_{\text{app}} + \mathcal{E}_{\text{est}} + \mathcal{E}_{\text{opt}} = \mathbb{E}\left[E(\tilde{f}_n) - E(f^*)\right] \leq c\left(\mathcal{E}_{\text{app}} + \frac{d}{n}\log\frac{n}{d} + \rho\right). \quad (13.7)$$

See also Mendelson (2003), and Bartlett and Mendelson (2006) for more bounds taking the optimization accuracy into account.

13.3.2   Gradient Optimization Algorithms

We now discuss and compare the asymptotic learning properties of four gradient optimization algorithms. Recall that the family of function $\mathcal{F}$ is linearly parameterized by $w \in \mathbb{R}^d$. Let $w_{\mathcal{F}}^*$ and w_n correspond to the functions $f_{\mathcal{F}}^*$ and f_n defined in section 13.2.1. In this section, we assume that the functions $w \mapsto \ell(f_w(x), y)$ are convex and twice differentiable with continuous second derivatives. For simplicity we also assume that the empirical const function $C(w) = E_n(f_w)$ has a single minimum, w_n.

Two matrices play an important role in the analysis: the Hessian matrix H and the gradient covariance matrix G, both measured at the empirical optimum w_n:

$$H = \frac{\partial^2 C}{\partial w^2}(w_n) = \mathbb{E}_n\left[\frac{\partial^2 \ell(f_{w_n}(x), y)}{\partial w^2}\right], \quad (13.8)$$

$$G = \mathbb{E}_n\left[\left(\frac{\partial \ell(f_{w_n}(x), y)}{\partial w}\right)\left(\frac{\partial \ell(f_{w_n}(x), y)}{\partial w}\right)'\right]. \quad (13.9)$$

The relation between these two matrices depends on the chosen loss function. In order to summarize them, we assume that there are constants $\lambda_{\max} \geq \lambda_{\min} > 0$ and $\nu > 0$ such that, for any $\eta > 0$, we can choose the number of examples n large enough to ensure that the following assertion is true with probability greater than $1 - \eta$:

$$\text{tr}(G H^{-1}) \leq \nu \quad \text{and} \quad \text{EigenSpectrum}(H) \subset [\lambda_{\min}, \lambda_{\max}]. \quad (13.10)$$

The condition number $\kappa = \lambda_{\max}/\lambda_{\min}$ provides a convenient measure of the difficulty of the optimization problem (Dennis Jr. and Schnabel, 1983).

The assumption $\lambda_{\min} > 0$ avoids complications with stochastic gradient algorithms. This assumption is weaker than strict convexity because it applies only in the vicinity of the optimum. For instance, consider a loss function obtained by smoothing the well-known hinge loss $\ell(z, y) = \max\{0, 1 - yz\}$ in a small neighborhood of its non-differentiable points. Function $C(w)$ is then piecewise linear with smoothed edges and vertices. It is not strictly convex. However, its minimum is likely to be on a smoothed vertex with a non singular Hessian. When we have strict convexity, the argument of Bartlett et al. (2006, theorem 12) yields fast estimation rates $\alpha \approx 1$ in (13.4) and (13.6). That is not necessarily the case here.

The four algorithms considered in this chapter use information about the gradient of the cost function to iteratively update their current estimate $w(t)$ of the parameter vector.

- *Gradient descent (GD)* iterates

$$w(t+1) \;=\; w(t) - \eta \frac{\partial C}{\partial w}(w(t)) \;\;=\; w(t) - \eta \frac{1}{n} \sum_{i=1}^{n} \frac{\partial}{\partial w} \ell\big(f_{w(t)}(x_i), y_i\big)$$

where $\eta > 0$ is a small enough gain. GD is an algorithm with *linear convergence* (Dennis Jr. and Schnabel, 1983): when $\eta = 1/\lambda_{\max}$, this algorithm requires $\mathcal{O}(\kappa \log(1/\rho))$ iterations to reach accuracy ρ. The exact number of iterations depends on the choice of the initial parameter vector.

- *Second-order gradient descent (2GD)* iterates

$$w(t+1) \;=\; w(t) - H^{-1} \frac{\partial C}{\partial w}(w(t)) \;=\; w(t) - \frac{1}{n} H^{-1} \sum_{i=1}^{n} \frac{\partial}{\partial w} \ell\big(f_{w(t)}(x_i), y_i\big)$$

where matrix H^{-1} is the inverse of the Hessian matrix (13.8). This is more favorable than Newton's algorithm because we do not evaluate the local Hessian at each iteration, but optimistically assume that an oracle has revealed in advance the value of the Hessian at the optimum. 2GD is a superlinear optimization algorithm with *quadratic convergence* (Dennis Jr. and Schnabel, 1983). When the cost is quadratic, a single iteration is sufficient. In the general case, $\mathcal{O}(\log\log(1/\rho))$ iterations are required to reach accuracy ρ.

- *Stochastic gradient descent (SGD)* picks a random training example (x_t, y_t) at each iteration and updates the parameter w on the basis of this example only:

$$w(t+1) \;=\; w(t) - \frac{\eta}{t}\,\frac{\partial}{\partial w}\ell\big(f_{w(t)}(x_t), y_t\big).$$

Murata (1998, section 2.2) characterizes the mean $\mathbb{E}_s[w(t)]$ and variance $\mathrm{Var}_s[w(t)]$ with respect to the distribution implied by the random examples drawn from a given training set $\mathcal{S}$ at each iteration. Applying this result to the discrete training set distribution for $\eta = 1/\lambda_{\min}$, we have $\delta w(t)^2 = \mathcal{O}(1/t)$ where $\delta w(t)$ is a shorthand notation for $w(t) - w_n$. We can then write

$$
\begin{aligned}
\mathbb{E}_s\big[\,C(w(t)) - \inf C\,\big] \;&=\; \mathbb{E}_s\big[\mathrm{tr}\big(H\,\delta w(t)\,\delta w(t)'\big)\big] + \mathrm{o}\big(\tfrac{1}{t}\big) \\
&=\; \mathrm{tr}\big(\,H\,\mathbb{E}_s[\delta w(t)]\,\mathbb{E}_s[\delta w(t)]' + H\,\mathrm{Var}_s[w(t)]\,\big) + \mathrm{o}\big(\tfrac{1}{t}\big) \\
&\leq\; \tfrac{\mathrm{tr}(GH)}{t} + \mathrm{o}\big(\tfrac{1}{t}\big) \;\leq\; \tfrac{\nu\kappa^2}{t} + \mathrm{o}\big(\tfrac{1}{t}\big).
\end{aligned}
\tag{13.11}
$$

Therefore, the SGD algorithm reaches accuracy ρ after less than $\nu\kappa^2/\rho + \mathrm{o}(1/\rho)$ iterations on average. The SGD convergence is essentially limited by the stochastic noise induced by the random choice of one example at each iteration. Neither the initial value of the parameter vector w nor the total number of examples n appears in the dominant term of this bound! When the training set is large, one could reach the desired accuracy ρ measured on the whole training set without even visiting all the training examples. This is in fact a kind of generalization bound.

• *Second-order stochastic gradient descent (2SGD)* replaces the gain η with the inverse of the Hessian matrix H:

$$w(t+1) \;=\; w(t) - \frac{1}{t}\,H^{-1}\,\frac{\partial}{\partial w}\ell\big(f_{w(t)}(x_t), y_t\big).$$

Unlike standard gradient algorithms, using the second-order information does not change the influence of ρ on the convergence rate but improves the constants. Again using (Murata, 1998, theorem 4), accuracy ρ is reached after $\nu/\rho + \mathrm{o}(1/\rho)$ iterations.

For each of the four gradient algorithms, the first three columns of table 13.2 report the time for a single iteration, the number of iterations needed to reach a predefined accuracy ρ, and their product, the time needed to reach accuracy ρ. These asymptotic results are valid with probability 1, since the probability of their complement is smaller than η for any $\eta > 0$.

The fourth column bounds the time necessary to reduce the excess error $\mathcal{E}$ below $c\,(\mathcal{E}_{\mathrm{app}} + \varepsilon)$ where c is the constant from (13.6). This is computed by

Algorithm	Cost of one iteration	Iterations to reach ρ	Time to reach accuracy ρ	Time to reach $\mathcal{E} \leq c\,(\mathcal{E}_{\mathrm{app}} + \varepsilon)$
GD	$\mathcal{O}(nd)$	$\mathcal{O}\left(\kappa \log \frac{1}{\rho}\right)$	$\mathcal{O}\left(nd\kappa \log \frac{1}{\rho}\right)$	$\mathcal{O}\left(\frac{d^2\kappa}{\varepsilon^{1/\alpha}} \log^2 \frac{1}{\varepsilon}\right)$
2GD	$\mathcal{O}(d^2 + nd)$	$\mathcal{O}\left(\log \log \frac{1}{\rho}\right)$	$\mathcal{O}\left((d^2 + nd) \log \log \frac{1}{\rho}\right)$	$\mathcal{O}\left(\frac{d^2}{\varepsilon^{1/\alpha}} \log \frac{1}{\varepsilon} \log \log \frac{1}{\varepsilon}\right)$
SGD	$\mathcal{O}(d)$	$\frac{\nu\kappa^2}{\rho} + o\left(\frac{1}{\rho}\right)$	$\mathcal{O}\left(\frac{d\nu\kappa^2}{\rho}\right)$	$\mathcal{O}\left(\frac{d\,\nu\,\kappa^2}{\varepsilon}\right)$
2SGD	$\mathcal{O}(d^2)$	$\frac{\nu}{\rho} + o\left(\frac{1}{\rho}\right)$	$\mathcal{O}\left(\frac{d^2\nu}{\rho}\right)$	$\mathcal{O}\left(\frac{d^2\,\nu}{\varepsilon}\right)$

Table 13.2: Asymptotic results for gradient algorithms (with probability 1). Compare the second-to-last column (time to optimize) with the last column (time to reach the excess test error ε). n–number of examples; d–parameter dimension; for κ, ν see equation (13.10).

observing that choosing $\rho \sim \left(\frac{d}{n} \log \frac{n}{d}\right)^\alpha$ in (13.6) achieves the fastest rate for ε, with minimal computation time. We can then use the asymptotic equivalences $\rho \sim \varepsilon$ and $n \sim \frac{d}{\varepsilon^{1/\alpha}} \log \frac{1}{\varepsilon}$. Setting the fourth column expressions to $T_{\max}$ and solving for ε yields the *best excess error achieved by each algorithm* within the limited time $T_{\max}$. This provides the asymptotic solution of the estimation–optimization tradeoff (13.3) for large-scale problems satisfying our assumptions.

These results clearly show that the generalization performance of *large-scale learning systems* depends on both the statistical properties of the objective function and the computational properties of the chosen optimization algorithm. Their combination leads to surprising consequences:

- *The SGD and 2SGD results do not depend on the estimation rate α.* When the estimation rate is poor, there is less need to optimize accurately. That leaves time to process more examples. A potentially more useful interpretation leverages the fact that (13.11) is already a kind of generalization bound: its fast rate trumps the slower rate assumed for the estimation error.

- *Second-order algorithms bring few asymptotical improvements in ε.* Although the superlinear 2GD algorithm improves the logarithmic term, all four algorithms are dominated by the polynomial term in $(1/\varepsilon)$. However, there are important variations in the influence of the constants d, κ, and ν. These constants are very important in practice.

- *Stochastic algorithms (SGD, 2SGD) yield the best generalization performance despite showing the worst optimization performance* on the empirical cost. This phenomenon has already been described and observed in experiments (e.g., Bottou and Le Cun, 2004).

In contrast, since the optimization error $\mathcal{E}_{\mathrm{opt}}$ of *small-scale learning systems* can be reduced to insignificant levels, their generalization performance is

Model	Algorithm	Training Time	Objective	Test Error
Hinge loss $\lambda = 10^{-4}$	SVMLight	23,642 secs	0.2275	6.02%
	SVMPerf	66 secs	0.2278	6.03%
	SGD	**1.4 secs**	0.2275	6.02%
Logistic loss $\lambda = 10^{-5}$	TRON ($\rho = 10^{-2}$)	30 secs	0.18907	5.68%
	TRON ($\rho = 10^{-3}$)	44 secs	0.18890	5.70%
	SGD	**2.3 secs**	0.18893	5.66%

Table 13.3: Results with linear Support Vector Machines on the RCV1 dataset.

determined solely by the statistical properties of the objective function.

13.4 Experiments

This section empirically compares SGD with other optimization algorithms on two well known machine learning tasks. The SGD C++ source code is available from `http://leon.bottou.org/projects/sgd`.

13.4.1 SGD for Support Vector Machines

We first consider a well-known text categorization task, the classification of documents belonging to the CCAT category in the RCV1-v2 dataset (Lewis et al., 2004). In order to collect a large training set, we swap the RCV1-v2 official training and testing sets. The resulting training sets and test sets contain 781,265 and 23,149 examples, respectively. The 47,152 TF/IDF features are recomputed on the basis of this new split. We use a simple linear model with the usual hinge loss Support Vector Machine objective function

$$\min_{w} \quad C(w, b) = \frac{\lambda}{2} + \frac{1}{n} \sum_{i=1}^{n} \ell(y_t(wx_t + b)) \quad \text{with} \quad \ell(z) = \max\{0, 1 - z\}.$$

The first two rows of table 13.3 replicate the results reported by Joachims (2006) for the same data and the same value of the hyperparameter λ.

The third row of table 13.3 reports results obtained with the SGD algorithm:

$$w_{t+1} = w_t - \eta_t \left(\lambda w + \frac{\partial \ell(y_t(wx_t + b))}{\partial w} \right) \quad \text{with} \quad \eta_t = \frac{1}{\lambda(t + t_0)}.$$

The bias b is updated similarly. Since λ is a lower bound of the smallest eigenvalue of the Hessian, our choice of gains η_t approximates the optimal schedule (see section 13.3.2). The offset t_0 was chosen to ensure that the

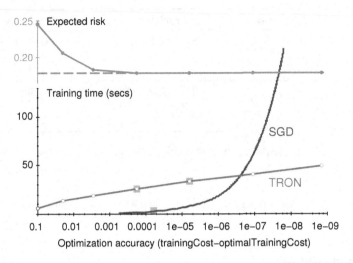

Figure 13.1: Training time and testing loss as a function of the optimization accuracy ρ for SGD and TRON (Lin et al., 2007)

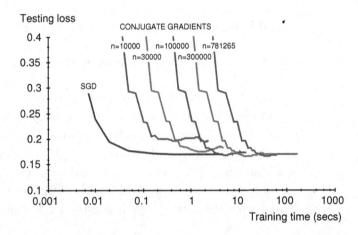

Figure 13.2: Testing loss versus training time for SGD, and for conjugate gradients running on subsets of the training set

initial gain is comparable with the expected size of the parameter w. The results clearly indicate that SGD offers a good alternative to the usual Support Vector Machine solvers.

Comparable results were obtained by Shalev-Shwartz et al. (2007), using an algorithm that essentially amounts to a stochastic gradient corrected by a projection step. Our results indicate that the projection step is not an essential component of this performance.

Table 13.3 also reports results obtained with the logistic loss $\ell(z) = \log(1 + e^{-z})$ in order to avoid the issues related to the nondifferentiability of the hinge loss. Note that this experiment uses a much better value for λ. Our comparison points were obtained with a state-of-the-art superlinear optimizer (Lin et al., 2007), using the stopping criteria $\rho = 10^{-2}$ and $\rho = 10^{-3}$. The very simple SGD algorithm clearly learns faster.

Figure 13.1 shows how much time each algorithm takes to reach a given optimization accuracy. The superlinear algorithm TRON reaches the optimum with 10 digits of accuracy in less than one minute. The stochastic gradient starts more quickly but is unable to deliver such a high accuracy. The upper part of the figure clearly shows that the testing set loss stops decreasing long before the superlinear algorithm overcomes the SGD algorithm.

Figure 13.2 shows how the testing loss evolves with the training time. The stochastic gradient descent curve can be compared with the curves obtained using conjugate gradients[3] on subsets of the training examples with increasing sizes. Assume, for instance, that our computing time budget is one second. Running the conjugate gradient algorithm on a random subset of 30,000 training examples achieves a much better performance than running it on the whole training set. How to guess the right subset size a priori remains unclear. Meanwhile, running the SGD algorithm on the full training set reaches the same testing set performance much faster.

13.4.2 SGD for Conditional Random Fields

The CoNLL 2000 chunking task (Tjong Kim Sang and Buchholz, 2000) consists of dividing a sentence into syntactically correlated segments such as noun phrase, verb phrase, and so on. The training set contains 8936 sentences divided into 106,978 segments. Error measurements are performed using a separate set of 2012 sentences divided into 23,852 segments. Results are

---

3. This experimental setup was suggested by Olivier Chapelle (personal communication). His variant of the conjugate gradient algorithm performs inexact line searches using a single inexpensive Newton step. This is effective because exact line searches usually demand many function evaluations which are expensive when the training set is large.

Algorithm	Training Time	Training Cost	Test F1 Score
CRF++/L-BFGS	4335 secs	**9042**	93.74%
CRF SGD	**568 secs**	9098	93.75%

Table 13.4: Results for Conditional Random Fields on the CoNLL 2000 chunking task

traditionally reported using an F1 measure that takes into account both the segment boundaries and the segment classes.

The chunking task has been successfully approached using Conditional Random Fields (Lafferty et al., 2001; Sha and Pereira, 2003) to tag the words with labels indicating the class and the boundaries of each segment. Our baseline is the Conditional Random Field model provided with the CRF++ software (Kudo, 2007). Our CRF SGD implementation replicates the features of the CRF++ software but uses SGD to optimize the Conditional Random Field objective function. The model contains 1,679,700 parameters in both cases.

Table 13.4 compares the training time, the final training cost, and the test performance of the model when trained using the standard CRF++ L-BFGS optimizer and the SGD implementation. The SGD version runs considerably faster.

Comparable speeds were obtained by Vishwanathan et al. (2006), using a stochastic gradient with a novel adaptive gain scheduling method. Our results indicate that this adaptive gain is not the essential component of this performance. The main cause lies with the fundamental tradeoffs outlined in this chapter.

13.5   Conclusion

Taking into account budget constraints on both the number of examples and the computation time, we find *qualitative differences* between the generalization performance of small-scale learning systems and large-scale learning systems. The generalization properties of large-scale learning systems depend on both the statistical properties of the objective function and the computational properties of the optimization algorithm. We illustrate this fact with some asymptotic results on gradient algorithms.

This framework leaves room for considerable refinements. Shalev-Shwartz and Srebro (2008) rigorously extend the analysis to regularized risk formulations with linear parameterization and find again that, for learning purposes, SGD algorithms are often more attractive than standard primal or dual algorithms with good optimization complexity (Joachims, 2006; Hush et al., 2006). It could also be interesting to investigate how the choice of a surrogate loss function (Zhang, 2004; Bartlett et al., 2006) impacts the large-scale case.

13.6 References

P. L. Bartlett and S. Mendelson. Empirical minimization. *Probability Theory and Related Fields*, 135(3):311–334, 2006.

P. L. Bartlett, M. I. Jordan, and J. D. McAuliffe. Convexity, classification and risk bounds. *Journal of the American Statistical Association*, 101(473):138–156, 2006.

L. Bottou and O. Bousquet. The tradeoffs of large scale learning. In J. Platt, D. Koller, Y. Singer, and S. Roweis, editors, *Advances in Neural Information Processing Systems 20*, pages 161–168. MIT Press, 2008.

L. Bottou and Y. Le Cun. Large scale online learning. In S. Thrun, L. K. Saul, and B. Schölkopf, editors, *Advances in Neural Information Processing Systems 16*. MIT Press, Cambridge, MA, 2004.

S. Boucheron, O. Bousquet, and G. Lugosi. Theory of classification: a survey of recent advances. *ESAIM: Probability and Statistics*, 9:323–375, 2005.

O. Bousquet. *Concentration Inequalities and Empirical Processes Theory Applied to the Analysis of Learning Algorithms*. PhD thesis, Ecole Polytechnique, 2002.

J. E. Dennis Jr. and R. B. Schnabel. *Numerical Methods for Unconstrained Optimization and Nonlinear Equations*. Prentice-Hall, Englewood Cliffs, NJ, 1983.

R. O. Duda and P. E. Hart. *Pattern Classification and Scene Analysis*. John Wiley, 1973.

D. Hush, P. Kelly, C. Scovel, and I. Steinwart. QP algorithms with guaranteed accuracy and run time for support vector machines. *Journal of Machine Learning Research*, 7:733–769, 2006.

T. Joachims. Training linear SVMs in linear time. In *Proceedings of the 12th ACM SIGKDD International Conference on Knowledge Discovery and Data Mining*, pages 217–226, Philadelphia, PA, August 2006. ACM Press.

J. S. Judd. On the complexity of loading shallow neural networks. *Journal of Complexity*, 4(3):177–192, 1988.

T. Kudo. CRF++: Yet another CRF toolkit, 2007. http://crfpp.sourceforge.net.

J. D. Lafferty, A. McCallum, and F. C. N. Pereira. Conditional random fields: Probabilistic models for segmenting and labeling sequence data. In C. E. Brodley and A. P. Danyluk, editors, *Proceedings of the Eighteenth International Conference on Machine Learning*, pages 282–289. Morgan Kaufmann, 2001.

W. S. Lee, P. L. Bartlett, and R. C. Williamson. The importance of convexity in learning with squared loss. *IEEE Transactions on Information Theory*, 44(5):1974–1980, 1998.

D. D. Lewis, Y. Yang, T. G. Rose, and F. Li. RCV1: A new benchmark collection for text categorization research. *Journal of Machine Learning Research*, 5:361–397, 2004.

C.-J. Lin, R. C. Weng, and S. S. Keerthi. Trust region Newton methods for large-scale logistic regression. In Z. Ghahramani, editor, *Proceedings of the 24th International Conference on Machine Learning*, pages 561–568, Corvallis, OR, June 2007. ACM Press.

P. Massart. Some applications of concentration inequalities to statistics. *Annales de la Faculté des Sciences de Toulouse*, series 6, 9(2):245–303, 2000.

S. Mendelson. A few notes on statistical learning theory. In S. Mendelson and A. J. Smola, editors, *Advanced Lectures in Machine Learning*, volume 2600 of *Lecture Notes in Computer Science*, pages 1–40. Springer-Verlag, New York, 2003.

N. Murata. A statistical study of on-line learning. In D. Saad, editor, *Online Learning and Neural Networks*. Cambridge University Press, 1998.

F. Sha and F. Pereira. Shallow parsing with conditional random fields. In *Proceedings of the 2003 Human Language Technology Conference and 4th Meeting of the North American Chapter of the Association for Computational Linguistics*, 2003.

S. Shalev-Shwartz and N. Srebro. SVM optimization: inverse dependence on training set size. In *Proceedings of the 25th International Conference on Machine Learning*, pages 928–935. ACM Press, 2008.

S. Shalev-Shwartz, Y. Singer, and N. Srebro. Pegasos: Primal estimated subgradient solver for SVM. In Z. Ghahramani, editor, *Proceedings of the 24th International Conference on Machine Learning*, pages 807–814. ACM Press, June 2007.

I. Steinwart and C. Scovel. Fast rates for support vector machines. In P. Auer and R. Meir, editors, *Proceedings of the 18th Conference on Learning Theory (COLT 2005)*, volume 3559 of *Lecture Notes in Computer Science*, pages 279–294, Bertinoro, Italy, June 2005. Springer-Verlag.

E. F. Tjong Kim Sang and S. Buchholz. Introduction to the CoNLL-2000 Shared Task: Chunking. In C. Cardie, W. Daelemans, C. Nedellec, and E. Tjong Kim Sang, editors, *Proceedings of CoNLL-2000 and LLL-2000*, pages 127–132, 2000.

A. B. Tsybakov. Optimal aggregation of classifiers in statistical learning. *Annals of Statistics*, 32(1):135–166, 2004.

L. G. Valiant. A theory of the learnable. *Proceedings of the 16th Annual ACM Symposium on the Theory of Computing*, pages 436–445, 1984.

V. N. Vapnik. *Estimation of Dependences Based on Empirical Data*. Springer-Verlag, Berlin, 1982.

V. N. Vapnik, E. Levin, and Y. LeCun. Measuring the VC-dimension of a learning machine. *Neural Computation*, 6(5):851–876, 1994.

S. V. N. Vishwanathan, N. N. Schraudolph, M. W. Schmidt, and K. P. Murphy. Accelerated training of conditional random fields with stochastic gradient methods. In W. W. Cohen and A. Moore, editors, *Proceedings of the 23rd International Conference on Machine Learning*, pages 969–976. ACM Press, 2006.

T. Zhang. Statistical behavior and consistency of classification methods based on convex risk minimization. *The Annals of Statistics*, 32:56–85, 2004.

14 Robust Optimization in Machine Learning

Constantine Caramanis   caramanis@mail.utexas.edu
The University of Texas at Austin
Austin, Texas

Shie Mannor   shie@ee.technion.ac.il
Technion, the Israel Institute of Technology
Haifa, Israel

Huan Xu   huan.xu@mail.utexas.edu
The University of Texas at Austin
Austin, Texas

Robust optimization is a paradigm that uses ideas from convexity and duality to immunize solutions of convex problems against bounded uncertainty in the parameters of the problem. Machine learning is fundamentally about making decisions under uncertainty, and optimization has long been a central tool; thus, at a high level there is no surprise that robust optimization should have a role to play. Indeed, the first part of the story told in this chapter is about specializing robust optimization to specific optimization problems in machine learning. Yet, beyond this, there have been several surprising and deep developments in the use of robust optimization and machine learning, connecting consistency, generalization ability, and other properties (such as sparsity and stability) to robust optimization.

In addition to surveying the direct applications of robust optimization to machine learning, important in their own right, this chapter explores some of these deeper connections, and points the way toward opportunities for applications and challenges for further research.

14.1   Introduction

Learning, optimization, and decision making from data must cope with uncertainty introduced both implicitly and explicitly. Uncertainty can be explicitly introduced when the data collection process is noisy, or when some data are corrupted. It may be introduced when the model specification is wrong, assumptions are missing, or factors are overlooked. Uncertainty is also implicitly present in pristine data, insofar as a finite sample empirical distribution, or function thereof, cannot exactly describe the true distribution in most cases. In the optimization community, it has long been known that the effect of even small uncertainty can be devastating in terms of the quality or feasibility of a solution. In machine learning, overfitting has long been recognized as a central challenge, and a plethora of techniques, many of them regularization-based, have been developed to combat this problem. The theoretical justification for many of these techniques lies in controlling notions of complexity, such as metric entropy or VC-dimension.

This chapter considers both uncertainty in optimization, and overfitting, from a unified perspective: robust optimization. In addition to introducing a novel technique for designing algorithms that are immune to noise and do not overfit data, robust optimization also provides a theoretical justification for the success of these algorithms: algorithms have certain properties, such as consistency, good generalization, or sparsity, *because they are robust.*

Robust optimization (e.g., Soyster, 1973; El Ghaoui and Lebret, 1997; Ben-Tal and Nemirovski, 2000; Bertsimas and Sim, 2004; Bertsimas et al., 2010; Ben-Tal et al., 2009, and many others) is designed to deal with parameter uncertainty in convex optimization problems. For example, one can imagine a linear program, $\min : \{\mathbf{c}^\top \mathbf{x} \,|\, A\mathbf{x} \le \mathbf{b}\}$, where there is uncertainty in the constraint matrix A, the objective function, $\mathbf{c}$, or the right-hand-side vector, $\mathbf{b}$. Robust optimization develops immunity to a deterministic or set-based notion of uncertainty. Thus, in the face of uncertainty in A, instead of solving $\min : \{\mathbf{c}^\top \mathbf{x} \,|\, A\mathbf{x} \le \mathbf{b}\}$, one solves $\min : \{\mathbf{c}^\top \mathbf{x} \,|\, A\mathbf{x} \le \mathbf{b}, \ \forall A \in \mathcal{U}\}$, for some suitably defined *uncertainty set* $\mathcal{U}$. We give a brief introduction to robust optimization in Section 14.2.

The remainder of this chapter is organized as follows. In Section 14.2 we provide a brief review of robust optimization. In Section 14.3 we discuss direct applications of robust optimization to constructing algorithms that are resistant to data corruption. This is a direct application not only of the methodology of robust optimization, but also of the motivation behind the development of robust optimization. The focus is on developing computationally efficient algorithms, resistant to bounded but otherwise arbitrary (even adversarial) noise. In Sections 14.4–14.6, we show that robust optimization's impact on machine learning extends far outside the originally envisioned scope as developed in the optimization literature. In Section 14.4, we show that many existing machine learning algorithms that are based on regularization, including support vector machines (SVMs), ridge regression, and Lasso, are special cases of robust optimization. Using this reinterpretation, their success can be understood from a unified perspective. We also show how the flexibility of robust optimization paves the way for the design of new regularization-like algorithms. Moreover, we show that robustness can be used directly to prove properties such as regularity and sparsity. In Section 14.5, we show that robustness can be used to prove statistical consistency. Then, in Section 14.6, we extend the results of Section 14.5, showing that an algorithm's generalization ability and its robustness are related in a fundamental way.

In summary, we show that robust optimization has deep connections to machine learning. In particular it yields a unified paradigm that (a) explains the success of many existing algorithms; (b) provides a prescriptive algorithmic approach to creating new algorithms with desired properties; and (c) allows us to prove general properties of an algorithm.

14.2 Background on Robust Optimization

In this section we provide a brief background on robust optimization, and refer the reader to the survey by Bertsimas et al. (2010), the textbook of Ben-Tal et al. (2009), and references to the original papers therein for more details.

Optimization affected by parameter uncertainty has long been a focus of the mathematical programming community. As has been demonstrated in compelling fashion (Ben-Tal and Nemirovski, 2000), solutions to optimization problems can exhibit remarkable sensitivity to perturbations in the problem parameters, thus often rendering a computed solution highly infeasible, suboptimal, or both. This parallels developments in related fields, particularly robust control (refer to Zhou et al., 1996; Dullerud and Paganini, 2000, and the references therein).

Stochastic programming (e.g., Prékopa, 1995; Kall and Wallace, 1994) assumes that the uncertainty has a probabilistic description. In contrast, robust optimization is built on the premise that the parameters vary arbitrarily in some a priori known bounded set, called the *uncertainty set*. Suppose we are optimizing a function $f_0(\mathbf{x})$, subject to the m constraints $f_i(\mathbf{x}, \mathbf{u}_i) \leq 0$, $i = 1, \ldots, m$, where $\mathbf{u}_i$ denotes the parameters of function i. Then, whereas the nominal optimization problem solves $\min\{f_0(\mathbf{x}) : f_i(\mathbf{x}, \mathbf{u}_i) \leq 0, i = 1, \ldots, m\}$, assuming that the $\mathbf{u}_i$ are known, robust optimization solves

$$\min_{\mathbf{x}} : \quad f_0(\mathbf{x}) \tag{14.1}$$
$$\text{s.t.} : \quad f_i(\mathbf{x}, \mathbf{u}_i) \leq 0, \ \forall \mathbf{u}_i \in \mathcal{U}_i, \ i = 1, \ldots, m.$$

14.2.1 Computational Tractability

The tractability of robust optimization, subject to standard and mild Slater-like regularity conditions, amounts to separation for the convex set: $\mathcal{X}(\mathcal{U}) \triangleq \{\mathbf{x} : f_i(\mathbf{x}, \mathbf{u}_i) \leq 0, \ \forall \mathbf{u}_i \in \mathcal{U}_i, \ i = 1, \ldots, m\}$. If there is an efficient algorithm that asserts $\mathbf{x} \in \mathcal{X}(\mathcal{U})$ or otherwise provides a separating hyperplane, then (14.2) can be solved in polynomial time. While the set $\mathcal{X}(\mathcal{U})$ is a convex set as long as each function f_i is convex in $\mathbf{x}$, it is not in general true that there is an efficient separation algorithm for the set $\mathcal{X}(\mathcal{U})$. However, in many cases of broad interest and application, solving the robust problem can be done efficiently—the robustified problem may be of complexity comparable to that of the nominal one. We outline some of the main complexity results below.

14.2.1.1 An Example: Linear Programs with Polyhedral Uncertainty

When the uncertainty set, $\mathcal{U}$, is polyhedral, the separation problem is not only efficiently solvable, it is also in fact linear; thus the robust counterpart is equivalent to a linear optimization problem. To illustrate this, consider the problem with uncertainty in the constraint matrix:

$$\min_{\mathbf{x}} : \quad \mathbf{c}^\top \mathbf{x}$$
$$\text{s.t.} : \quad \max_{\{\mathbf{a}_i : \mathbf{D}_i \mathbf{a}_i \leq \mathbf{d}_i\}} [\mathbf{a}_i^\top \mathbf{x}] \leq b_i, \quad i = 1, \ldots, m.$$

The dual of the subproblem (recall that $\mathbf{x}$ is not a variable of optimization in the inner max) again becomes a linear program

$$\begin{bmatrix} \max_{\mathbf{a}_i} : & \mathbf{a}_i^\top \mathbf{x} \\ \text{s.t.} : & \mathbf{D}_i \mathbf{a}_i \leq \mathbf{d}_i \end{bmatrix} \longleftrightarrow \begin{bmatrix} \min_{\mathbf{p}_i} : & \mathbf{p}_i^\top \mathbf{d}_i \\ \text{s.t.} : & \mathbf{p}_i^\top \mathbf{D}_i = \mathbf{x} \\ & \mathbf{p}_i \geq 0 \end{bmatrix},$$

and therefore the robust linear optimization now becomes:

$$\min_{\mathbf{x}, \mathbf{p}_1, \ldots, \mathbf{p}_m} : \quad \mathbf{c}^\top \mathbf{x}$$
$$\text{s.t.} : \quad \mathbf{p}_i^\top \mathbf{d}_i \leq b_i, \quad i = 1, \ldots, m$$
$$\mathbf{p}_i^\top \mathbf{D}_i = \mathbf{x}, \quad i = 1, \ldots, m$$
$$\mathbf{p}_i \geq 0, \quad i = 1, \ldots, m.$$

Thus the size of such problems grows polynomially in the size of the nominal problem and the dimensions of the uncertainty set.

14.2.1.2 *Some General Complexity Results*

We now list a few of the complexity results that are relevant to this chapter. The reader may refer to Bertsimas et al. (2010); Ben-Tal et al. (2009), and references therein for further details. The robust counterpart for a linear program (LP) with polyhedral uncertainty is again an LP. For an LP with ellipsoidal uncertainty, the counterpart is a second order cone program (SOCP). A convex quadratic program with ellipsoidal uncertainty has a robust counterpart that is a semidefinite program (SDP). An SDP with ellipsoidal uncertainty has an NP-hard robust counterpart.

14.2.2 Probabilistic Interpretations and Results

The computational advantage of robust optimization is largely due to the fact that the formulation is deterministic, and one deals with uncertainty sets rather than probability distributions. While the paradigm makes sense when the disturbances are not stochastic, or the distribution is not known, tractability advantages have made robust optimization an appealing computational framework even when the uncertainty is stochastic and the distribution is fully or partially known. A major success of robust optimization has been the ability to derive a priori probability guarantees—for instance, probability of feasibility—that the solution to a robust optimization will satisfy, under a variety of probabilistic assumptions. Thus robust optimization is a tractable framework one can use to build solutions with probabilistic guarantees such as minimum probability of feasibility, or maximum probability of hinge loss beyond some threshold level, and so on. This probabilistic interpretation of robust optimization is used throughout this chapter.

14.3 Robust Optimization and Adversary Resistant Learning

In this section we overview some of the direct applications of robust optimization to coping with uncertainty (adversarial or stochastic) in machine learning problems. The main themes are (a) the formulations one obtains when using different uncertainty sets and (b) the probabilistic interpretation and results one can derive by using robust optimization. Using ellipsoidal uncertainty, we show that the resulting robust problem is tractable. Moreover, we show that this robust formulation has interesting probabilistic interpretations. Then, using a polyhedral uncertainty set, we show that sometimes it is possible to tractably model combinatorial uncertainty, such as missing data.

Robust optimization-based learning algorithms have been proposed for various learning tasks, such as learning and planning (Nilim and El Ghaoui, 2005), Fisher linear discriminant analysis (Kim et al., 2005), PCA (d'Aspremont et al., 2007), and many others. Instead of providing a comprehensive survey, we use support vector machines (SVMs; e.g., Vapnik and Lerner, 1963; Boser et al., 1992; Cortes and Vapnik, 1995) to illustrate the methodology of robust optimization.

Standard SVMs consider the standard binary classification problem, where we are given a finite number of training samples $\{\mathbf{x}_i, y_i\}_{i=1}^{m} \subseteq \mathbb{R}^n \times \{-1, +1\}$, and must find a linear classifier, specified by the function $h^{\mathbf{w},b}(\mathbf{x}) = \operatorname{sgn}(\langle \mathbf{w}, \mathbf{x} \rangle + b)$, where $\langle \cdot, \cdot \rangle$ denotes the standard inner product. The parameters $(\mathbf{w}, b)$ are obtained by solving the following convex optimization problem:

$$
\begin{aligned}
\min_{\mathbf{w}, b, \boldsymbol{\xi}} : \quad & r(\mathbf{w}, b) + C \sum_{i=1}^{m} \xi_i \\
\text{s.t.} : \quad & \xi_i \geq \left[1 - y_i(\langle \mathbf{w}, \mathbf{x}_i \rangle + b) \right], \quad i = 1, \dots, m; \\
& \xi_i \geq 0, \quad i = 1, \dots, m;
\end{aligned}
\tag{14.2}
$$

where $r(\mathbf{w}, b)$ is a regularization term, e.g., $r(\mathbf{w}, b) = \frac{1}{2}\|\mathbf{w}\|_2^2$. There are a number of related formulations, some focusing on controlling VC-dimension, promoting sparsity, or some other property (see Schölkopf and Smola (2001); Steinwart and Christmann (2008), and references therein).

There are three natural ways that uncertainty affects the input data: corruption in the location, $\mathbf{x}_i$; corruption in the label, y_i; and corruption via altogether missing data. We outline some applications of robust optimization to these three settings.

14.3.1   Corrupted Location

Given observed points $\{\mathbf{x}_i\}$, the additive uncertainty model assumes that $\mathbf{x}_i^{true} = \mathbf{x}_i + \mathbf{u}_i$. Robust optimization protects against the uncertainty $\mathbf{u}_i$ by minimizing the regularized training loss on all possible locations of the $\mathbf{u}_i$ in some uncertainty set, $\mathcal{U}_i$.

Trafalis and Gilbert (2007) consider the ellipsoidal uncertainty set given by

$$
\mathcal{U}_i = \left\{ \mathbf{u}_i : \mathbf{u}_i^\top \Sigma_i \mathbf{u}_i \leq 1 \right\}, \quad i = 1, \dots, m,
$$

so that each constraint becomes $\xi_i \geq \left[1 - y_i(\langle \mathbf{w}, \mathbf{x}_i + \mathbf{u}_i \rangle + b)\right]$, $\forall \mathbf{u}_i \in \mathcal{U}_i$. By duality, this is equivalent to $y_i(\mathbf{w}^\top \mathbf{x}_i + b) \geq 1 + \|\Sigma_i^{1/2} \mathbf{w}\|_2 - \xi_i$, and hence their version of robust SVM reduces to

$$\min_{\mathbf{w}, b, \xi} : \quad r(\mathbf{w}, b) + C \sum_{i=1}^{m} \xi_i$$

$$\text{s.t.} \quad y_i(\mathbf{w}^\top \mathbf{x}_i + b) \geq 1 - \xi_i + \|\Sigma_i^{1/2} \mathbf{w}\|_2; \quad i = 1, \ldots, m; \quad (14.3)$$

$$\xi_i \geq 0; \quad i = 1, \ldots, m.$$

Trafalis and Gilbert (2007) use $r(\mathbf{w}, b) = \frac{1}{2}\|\mathbf{w}\|_2$, while Bhattacharyya et al. (2004) use the sparsity-inducing regularizer $r(\mathbf{w}, b) = \|\mathbf{w}\|_1$. In both settings, the robust problem is an instance of a second-order cone program (SOCP). Available solvers can solve SOCPs with hundreds of thousands of variables and more.

If the uncertainty $\mathbf{u}_i$ is stochastic, one can use this robust formulation to find a classifier that satisfies constraints on the probability (w.r.t. the distribution of $\mathbf{u}_i$) that each constraint is violated. In Shivaswamy et al. (2006), the authors consider two varieties of such chance constraints for $i = 1, \ldots, m$:

$$(a) \quad \Pr_{\mathbf{u}_i \sim \mathcal{N}(\tilde{\mathbf{0}}, \Sigma_i)} \left(y_i(\mathbf{w}^\top (\mathbf{x}_i + \mathbf{u}_i) + b) \geq 1 - \xi_i \right) \geq 1 - \kappa_i; \quad (14.4)$$

$$(b) \quad \inf_{\mathbf{u}_i \sim (\tilde{\mathbf{0}}, \Sigma_i)} \Pr_{\mathbf{u}_i} \left(y_i(\mathbf{w}^\top (\mathbf{x}_i + \mathbf{u}_i) + b) \geq 1 - \xi_i \right) \geq 1 - \kappa_i.$$

Constraint (a) controls the probability of constraint violation when the uncertainty follows a known Gaussian distribution. Constraint (b) is more conservative: it controls the worst-case probability of constraint violation, over all centered distributions with variance Σ_i. Theorem 14.1 says that the robust formulation with ellipsoidal uncertainty sets as above can be used to control both of these quantities.

Theorem 14.1. *For $i = 1, \ldots, m$ consider the robust constraint as given above:*

$$y_i(\mathbf{w}^\top \mathbf{x}_i + b) \geq 1 - \xi_i + \gamma_i \|\Sigma^{1/2} \mathbf{w}\|_2.$$

If we take $\gamma_i = \Phi^{-1}(\kappa_i)$, for Φ the Gaussian c.d.f., this constraint is equivalent to constraint (a) of (14.4), while taking $\gamma_i = \sqrt{\kappa_i/(1 - \kappa_i)}$ yields constraint (b).

14.3.2 Missing Data

Globerson and Roweis (2006) use robust optimization with polyhedral uncertainty set to address the problem where some of the features of the testing samples may be deleted (possibly in an adversarial fashion). Using a dummy feature to remove the bias term b if necessary, we can rewrite the nominal problem as

$$\min_{\mathbf{w}} : \frac{1}{2}\|\mathbf{w}\|_2^2 + C\sum_{i=1}^{m}[1 - y_i\mathbf{w}^\top\mathbf{x}_i]_+.$$

For a given choice of $\mathbf{w}$, the value of the term $[1 - y_i\mathbf{w}^\top\mathbf{x}_i]_+$ in the objective, under an adversarial deletion of K features, becomes

$$\begin{aligned}
\max_{\boldsymbol{\alpha}_i} \quad & [1 - y_i\mathbf{w}^\top(\mathbf{x}_i \circ (1 - \boldsymbol{\alpha}_i))]_+ \\
\text{s.t:} \quad & \alpha_{ij} \in \{0,1\}; \quad j = 1,\ldots,n; \\
& \sum_{j=1}^{n}\alpha_{ij} = K,
\end{aligned}$$

where $\circ$ denotes pointwise vector multiplication. While this optimization problem is combinatorial, relaxing the integer constraint $\alpha_{ij} \in \{0,1\}$ to be $0 \le \alpha_{ij} \le 1$ does not change the objective value. Thus, taking the dual of the maximization and substituting into the original problem, one obtains the classifier that is maximally resistant to up to K missing features:

$$\begin{aligned}
\min_{\mathbf{w},\mathbf{v}_i,z_i,t_i,\boldsymbol{\xi}} \quad & \frac{1}{2}\|\mathbf{w}\|_2^2 + C\sum_{i=1}^{m}\xi_i \\
\text{s.t.} \quad & y_i\mathbf{w}^\top\mathbf{x}_i - t_i \ge 1 - \xi_i; \quad i = 1,\ldots,m; \\
& \xi_i \ge 0; \quad i = 1,\ldots,m; \\
& t_i \ge Kz_i + \sum_{j=1}^{n}v_{ij}; \quad i = 1,\ldots,m; \\
& \mathbf{v}_i \ge \mathbf{0}; \quad i = 1,\ldots,m; \\
& z_i + v_{ij} \ge y_i x_{ij} w_{ij}; \quad i = 1,\ldots,m; \quad j = 1,\ldots n.
\end{aligned}$$

This is again an SOCP, and hence fairly large instances can be solved with specialized software.

14.3.3   Corrupted Labels

When the labels are corrupted, the problem becomes more difficult to address due to its combinatorial nature. However, it too has been recently addressed using robust optimization (Caramanis and Mannor, 2008). While there is still a combinatorial price to pay in the complexity of the classifier class, robust optimization can be used to find the optimal classifier; see Caramanis and Mannor (2008) for details.

14.4   Robust Optimization and Regularization

In this section and sections 14.5 and 14.6, we demonstrate that robustness
can provide a unified explanation for many desirable properties of a learning
algorithm, from regularity and sparsity, to consistency and generalization.
A main message of this chapter is that many regularized problems exhibit
a "hidden robustness"—they are in fact equivalent to a robust optimization problem—which can then be used to directly prove properties such as
consistency and sparsity, and also to design new algorithms. The main problems that highlight this equivalence are regularized support vector machines,
ℓ_2-regularized regression, and ℓ_1-regularized regression, also known as Lasso.

14.4.1   Support Vector Machines

We consider regularized SVMs, and show that they are algebraically equivalent to a robust optimization problem. We use this equivalence to provide a
probabilistic interpretation of SVMs, which allows us to propose new probabilistic SVM-type formulations. This section is based on Xu et al. (2009).

At a high level it is known that regularization and robust optimization
are related; see for instance, El Ghaoui and Lebret (1997), Anthony and
Bartlett (1999), and Section 14.3. Yet, the precise connection between
robustness and regularized SVMs did not appear until Xu et al. (2009).
One of the mottos of robust optimization is to harness the consequences of
probability theory without paying the computational cost of having to use
its axioms. Consider the additive uncertainty model from Section 14.3.1:
$\mathbf{x}_i + \mathbf{u}_i$. If the uncertainties $\mathbf{u}_i$ are stochastic, various limit results (LLN, CLT,
etc.) promise that even independent variables will exhibit strong aggregate
coupling behavior. For instance, the set $\{(\mathbf{u}_1, \ldots, \mathbf{u}_m) : \sum_{i=1}^m \|\mathbf{u}_i\| \le c\}$ will
have increasing probability as m grows. This motivates designing uncertainty
sets with this kind of coupling across uncertainty parameters. We leave it
to the reader to check that the *constraint-wise* robustness formulations of
Section 14.3.1 cannot be made to capture such coupling constraints across
the disturbances $\{\mathbf{u}_i\}$.

We rewrite SVM without slack variables, as an unconstrained optimization. The natural robust formulation now becomes

$$\min_{\mathbf{w},b} \max_{\boldsymbol{u} \in \mathcal{U}} \{r(\mathbf{w}, b) + \sum_{i=1}^m \max\left[1 - y_i(\langle \mathbf{w}, \mathbf{x}_i - \boldsymbol{u}_i \rangle + b), 0\right]\}, \qquad (14.5)$$

where $\boldsymbol{u}$ denotes the collection of uncertainty vectors, $\{\mathbf{u}_i\}$. Describing our
coupled uncertainty set requires a few definitions. Definition 14.1 character-
izes the effect of different uncertainty sets, and captures the coupling that

they exhibit. As an immediate consequence we obtain an equivalent robust optimization formulation for regularized SVMs.

Definition 14.1. *A set $\mathcal{U}_0 \subseteq \mathbb{R}^n$ is called an* atomic uncertainty set *if*

    (I) $\mathbf{0} \in \mathcal{U}_0$;

    (II)   *For any* $\mathbf{w}_0 \in \mathbb{R}^n$: $\sup_{\boldsymbol{u} \in \mathcal{U}_0} [\mathbf{w}_0^\top \boldsymbol{u}] = \sup_{\boldsymbol{u}' \in \mathcal{U}_0} [-\mathbf{w}_0^\top \boldsymbol{u}'] < +\infty.$

Definition 14.2. *Let $\mathcal{U}_0$ be an atomic uncertainty set. A set $\mathcal{U} \subseteq \mathbb{R}^{n \times m}$ is called a* sublinear aggregated uncertainty set *of $\mathcal{U}_0$, if*

$$\mathcal{U}^- \subseteq \mathcal{U} \subseteq \mathcal{U}^+,$$

where $\mathcal{U}^- \triangleq \bigcup_{t=1}^{m} \mathcal{U}_t^-; \qquad \mathcal{U}_t^- \triangleq \{(\boldsymbol{u}_1, \ldots, \boldsymbol{u}_m) | \boldsymbol{u}_t \in \mathcal{U}_0; \ \boldsymbol{u}_{i \neq t} = \mathbf{0}\}.$

$$\mathcal{U}^+ \triangleq \{(\alpha_1 \boldsymbol{u}_1, \ldots, \alpha_m \boldsymbol{u}_m) | \sum_{i=1}^{m} \alpha_i = 1; \ \alpha_i \geq 0, \ \boldsymbol{u}_i \in \mathcal{U}_0, \ i = 1, \ldots, m\}.$$

Sublinear aggregated uncertainty models the case where the disturbances on each sample are treated identically, but their aggregate behavior across multiple samples is controlled. Some interesting examples include

    (1) $\mathcal{U} = \{(\boldsymbol{u}_1, \ldots, \boldsymbol{u}_m) | \sum_{i=1}^{m} \|\boldsymbol{u}_i\| \leq c\};$

    (2) $\mathcal{U} = \{(\boldsymbol{u}_1, \ldots, \boldsymbol{u}_m) | \exists t \in [1:m]; \ \|\boldsymbol{u}_t\| \leq c; \ \boldsymbol{u}_i = \mathbf{0}, \forall i \neq t\};$ and

    (3) $\mathcal{U} = \{(\boldsymbol{u}_1, \ldots, \boldsymbol{u}_m) | \sum_{i=1}^{m} \sqrt{c\|\boldsymbol{u}_i\|} \leq c\}.$

All these examples share the same atomic uncertainty set $\mathcal{U}_0 = \{\boldsymbol{u} | \|\boldsymbol{u}\| \leq c\}$. Figure 14.1 illustrates a sublinear aggregated uncertainty set for $n = 1$ and $m = 2$, that is, the training set consists of two univariate samples.

Theorem 14.2. *Assume $\{\mathbf{x}_i, y_i\}_{i=1}^{m}$ are nonseparable, $r(\cdot, \cdot) : \mathbb{R}^{n+1} \to \mathbb{R}$ is an arbitrary function, and $\mathcal{U}$ is a sublinear aggregated uncertainty set with corresponding atomic uncertainty set $\mathcal{U}_0$. Then the min-max problem*

$$\min_{\mathbf{w}, b} \sup_{(\boldsymbol{u}_1, \ldots, \boldsymbol{u}_m) \in \mathcal{U}} \left\{ r(\mathbf{w}, b) + \sum_{i=1}^{m} \max \left[1 - y_i(\langle \mathbf{w}, \mathbf{x}_i - \boldsymbol{u}_i \rangle + b), 0 \right] \right\} \quad (14.6)$$

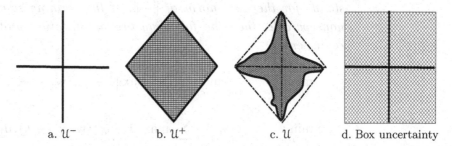

a. $\mathcal{U}^-$ b. $\mathcal{U}^+$ c. $\mathcal{U}$ d. Box uncertainty

Figure 14.1: A sublinear aggregated uncertainty set $\mathcal{U}$, and its contrast with the box uncertainty set.

is equivalent to the following optimization problem on $\mathbf{w}, b, \boldsymbol{\xi}$:

$$
\begin{aligned}
\min_{\mathbf{w}, b, \boldsymbol{\xi}} : \quad & r(\mathbf{w}, b) + \sup_{\boldsymbol{u} \in \mathcal{U}_0} (\mathbf{w}^\top \boldsymbol{u}) + \sum_{i=1}^{m} \xi_i, \\
\text{s.t.} : \quad & \xi_i \geq 1 - [y_i(\langle \mathbf{w}, \mathbf{x}_i \rangle + b)], \quad i = 1, \ldots, m; \\
& \xi_i \geq 0, \quad i = 1, \ldots, m.
\end{aligned}
\tag{14.7}
$$

The minimization of (14.7) is attainable when $r(\cdot, \cdot)$ *is lower semi-continuous.*

Proof. We give only the proof idea. The details can be found in Xu et al. (2009). Define

$$
v(\mathbf{w}, b) \triangleq \sup_{\boldsymbol{u} \in \mathcal{U}_0} (\mathbf{w}^\top \boldsymbol{u}) + \sum_{i=1}^{m} \max \left[1 - y_i(\langle \mathbf{w}, \mathbf{x}_i \rangle + b), 0 \right].
$$

In the first step, we show

$$
v(\hat{\mathbf{w}}, \hat{b}) \leq \sup_{(\boldsymbol{u}_1, \ldots, \boldsymbol{u}_m) \in \mathcal{U}^-} \sum_{i=1}^{m} \max \left[1 - y_i(\langle \hat{\mathbf{w}}, \mathbf{x}_i - \boldsymbol{u}_i \rangle + \hat{b}), 0 \right].
\tag{14.8}
$$

This follows because the samples are nonseparable. In the second step, we prove the reverse inequality:

$$
\sup_{(\boldsymbol{u}_1, \ldots, \boldsymbol{u}_m) \in \mathcal{U}^+} \sum_{i=1}^{m} \max \left[1 - y_i(\langle \hat{\mathbf{w}}, \mathbf{x}_i - \boldsymbol{u}_i \rangle + \hat{b}), 0 \right] \leq v(\hat{\mathbf{w}}, \hat{b}).
\tag{14.9}
$$

This holds regardless of separability. Combining the two, adding the regularizer, and then infimizing both sides concludes the proof. $\square$

An immediate corollary is that a special case of our robust formulation is equivalent to the norm-regularized SVM setup:

Corollary 14.3. *Let* $\mathcal{T} \triangleq \left\{ (\boldsymbol{u}_1, \ldots \boldsymbol{u}_m) \mid \sum_{i=1}^{m} \|\boldsymbol{u}_i\|^* \leq c \right\}$, *where* $\| \cdot \|^*$

stands for the dual norm of $\| \cdot \|$. *If the training samples* $\{\mathbf{x}_i, y_i\}_{i=1}^{m}$ *are nonseparable, then the following two optimization problems on* $(\mathbf{w}, b)$ *are equivalent.*

$$\min_{\mathbf{w}, b} : \quad \max_{(\boldsymbol{u}_1, \ldots, \boldsymbol{u}_m) \in \mathcal{J}} \sum_{i=1}^{m} \max \left[1 - y_i (\langle \mathbf{w}, \mathbf{x}_i - \boldsymbol{u}_i \rangle + b), 0 \right], \quad (14.10)$$

$$\min_{\mathbf{w}, b} : \quad c\|\mathbf{w}\| + \sum_{i=1}^{m} \max \left[1 - y_i (\langle \mathbf{w}, \mathbf{x}_i \rangle + b), 0 \right]. \quad (14.11)$$

Proof. Let $\mathcal{U}_0$ be the dual-norm ball $\{\boldsymbol{u} | \|\boldsymbol{u}\|^* \leq c\}$ and $r(\mathbf{w}, b) \equiv 0$. Then $\sup_{\|\boldsymbol{u}\|^* \leq c} (\mathbf{w}^{\top} \boldsymbol{u}) = c\|\mathbf{w}\|$. The corollary follows from Theorem 14.2. Notice that the equivalence holds for any $\mathbf{w}$ and b. $\qquad \square$

This corollary explains the common belief that regularized classifiers tend to be more robust. Specifically, it explains the observation that when the disturbance is noise like and neutral rather than adversarial, a norm-regularized classifier (without explicit robustness) has a performance often superior to a *box-type* robust classifier (see Trafalis and Gilbert, 2007). One take-away message is that while robust optimization is adversarial in its formulation, it can be quite flexible, and can be designed to yield solutions, such as the regularized solution above, that are appropriate for a non-adversarial setting.

One interesting research direction is to use this equivalence to find good regularizers without the need for cross validation. This could be done by mapping a measure of the variation in the training data to an appropriate uncertainty set, and then using the above equivalence to map back to a regularizer.

14.4.1.1 *Kernelization*

The previous results can easily be generalized to the kernelized setting. The kernelized SVM formulation considers a linear classifier in the feature space $\mathcal{H}$, a Hilbert space containing the range of some feature mapping $\Phi(\cdot)$. The standard formulation is as follows,

$$\min_{\mathbf{w}, b, \boldsymbol{\xi}} : \quad r(\mathbf{w}, b) + \sum_{i=1}^{m} \xi_i$$
$$\text{s.t.} : \quad \xi_i \geq \left[1 - y_i (\langle \mathbf{w}, \Phi(\mathbf{x}_i) \rangle + b) \right], \quad i = 1, \ldots, m;$$
$$\xi_i \geq 0, \quad i = 1, \ldots, m;$$

where we use the representer theorem (see Schölkopf and Smola (2001)).

The definitions of an atomic uncertainty set and a sublinear aggregated uncertainty set in the feature space are identical to Definitions 14.1 and 14.2, with $\mathbb{R}^n$ replaced by $\mathcal{H}$. Theorem 14.4 is a feature space counterpart of Theorem 14.2, and the proof follows from a similar argument.

Theorem 14.4. *Assume* $\{\Phi(\mathbf{x}_i), y_i\}_{i=1}^m$ *are not linearly separable,* $r(\cdot)$: $\mathcal{H} \times \mathbb{R} \to \mathbb{R}$ *is an arbitrary function,* $\mathcal{U} \subseteq \mathcal{H}^m$ *is a sublinear aggregated uncertainty set with corresponding atomic uncertainty set* $\mathcal{U}_0 \subseteq \mathcal{H}$. *Then the min-max problem*

$$\min_{\mathbf{w},b} \quad \sup_{(\boldsymbol{u}_1,\ldots,\boldsymbol{u}_m)\in\mathcal{U}} \left\{ r(\mathbf{w},b) + \sum_{i=1}^m \max\left[1 - y_i(\langle \mathbf{w}, \Phi(\mathbf{x}_i) - \boldsymbol{u}_i \rangle + b), 0\right] \right\}$$

is equivalent to

$$\begin{aligned}
\min_{\mathbf{w},b,\boldsymbol{\xi}} : \quad & r(\mathbf{w},b) + \sup_{\boldsymbol{u}\in\mathcal{U}_0} (\langle \mathbf{w}, \boldsymbol{u} \rangle) + \sum_{i=1}^m \xi_i, \\
\text{s.t.} : \quad & \xi_i \geq 1 - y_i(\langle \mathbf{w}, \Phi(\mathbf{x}_i) \rangle + b), \quad i = 1,\ldots,m; \\
& \xi_i \geq 0, \quad i = 1,\ldots,m.
\end{aligned} \tag{14.12}$$

The minimization of (14.12) is attainable when $r(\cdot,\cdot)$ *is lower semi-continuous.*

For some widely used feature mappings (e.g., RKHS of a Gaussian kernel), $\{\Phi(\mathbf{x}_i), y_i\}_{i=1}^m$ are always separable. In this case, the equivalence reduces to a bound.

Corollary 14.5 is the feature space counterpart of Corollary 14.3, where $\|\cdot\|_{\mathcal{H}}$ stands for the RKHS norm, that is, for $\mathbf{z} \in \mathcal{H}$, $\|\mathbf{z}\|_{\mathcal{H}} = \sqrt{\langle \mathbf{z}, \mathbf{z} \rangle}$.

Corollary 14.5. *Let* $\mathcal{T}_{\mathcal{H}} \triangleq \left\{ (\boldsymbol{u}_1, \ldots \boldsymbol{u}_m) | \sum_{i=1}^m \|\boldsymbol{u}_i\|_{\mathcal{H}} \leq c \right\}$. *If* $\{\Phi(\mathbf{x}_i), y_i\}_{i=1}^m$ *are non-separable, then the following two optimization problems on* $(\mathbf{w}, b)$ *are equivalent:*

$$\min_{\mathbf{w},b} : \quad \max_{(\boldsymbol{u}_1,\ldots,\boldsymbol{u}_m)\in\mathcal{T}_{\mathcal{H}}} \sum_{i=1}^m \max\left[1 - y_i(\langle \mathbf{w}, \Phi(\mathbf{x}_i) - \boldsymbol{u}_i \rangle + b), 0\right],$$

$$\min_{\mathbf{w},b} : \quad c\|\mathbf{w}\|_{\mathcal{H}} + \sum_{i=1}^m \max\left[1 - y_i(\langle \mathbf{w}, \Phi(\mathbf{x}_i) \rangle + b), 0\right]. \tag{14.13}$$

Equation (14.13) is a variant form of the standard SVM that has a squared RKHS norm regularization term, and by convexity arguments the two formulations are equivalent up to a change of tradeoff parameter c. Therefore, Corollary 14.5 essentially means that the standard kernelized SVM is implicitly a robust classifier (without regularization) with disturbance in the featurespace, and the sum of the magnitudes of the disturbance is bounded.

Disturbance in the feature space is less intuitive than disturbance in the sample space, and Lemma 14.6 relates these two different notions.

Lemma 14.6. *Suppose there exists $\mathfrak{X} \subseteq \mathbb{R}^n$, $\rho > 0$, and a continuous non-decreasing function $f : \mathbb{R}^+ \to \mathbb{R}^+$ satisfying $f(0) = 0$, such that*

$$k(\mathbf{x}, \mathbf{x}) + k(\mathbf{x}', \mathbf{x}') - 2k(\mathbf{x}, \mathbf{x}') \leq f(\|\mathbf{x} - \mathbf{x}'\|_2^2), \quad \forall \mathbf{x}, \mathbf{x}' \in \mathfrak{X}, \|\mathbf{x} - \mathbf{x}'\|_2 \leq \rho.$$

Then,

$$\|\Phi(\hat{\mathbf{x}} + \boldsymbol{u}) - \Phi(\hat{\mathbf{x}})\|_{\mathcal{H}} \leq \sqrt{f(\|\boldsymbol{u}\|_2^2)}, \quad \forall \|\boldsymbol{u}\|_2 \leq \rho, \ \hat{\mathbf{x}}, \hat{\mathbf{x}} + \boldsymbol{\delta} \in \mathfrak{X}.$$

Lemma 14.6 essentially says that under certain conditions, robustness in the feature space is a stronger requirement than robustness in the sample space. Therefore, a classifier that achieves robustness in the feature space also achieves robustness in the sample space. Notice that the condition of Lemma 14.6 is rather weak. In particular, it holds for any continuous $k(\cdot, \cdot)$ and bounded domain $\mathfrak{X}$.

14.4.1.2 *Probabilistic Interpretations*

As discussed and demonstrated above, robust optimization can often be used for *probabilistic analysis*. In this section, we show that robust optimization and the equivalence theorem can be used to construct a classifier with *probabilistic margin protection*, that is, a classifier with probabilistic constraints on the chance of violation beyond a given threshold. Second, we show that in the Bayesian setup, if one has a prior only on the total magnitude of the disturbance vector, robust optimization can be used to tune the regularizer.

Probabilistic Protection. We can use Problem (14.6) to obtain an upper bound for a chance-constrained classifier. Suppose the disturbance is stochastic with known distribution. We denote the disturbance vector by $(\boldsymbol{u}_1^r, \dots \boldsymbol{u}_m^r)$ to emphasize that it is now a random variable. The chance-constrained classifier minimizes the hinge loss that occurs with probability above some given confidence level $\eta \in [0, 1]$. The classifier is given by the optimization problem

$$\min_{\mathbf{w}, b, l} : \quad l \tag{14.14}$$

$$\text{s.t.} : \quad \mathbb{P}\Big\{ \sum_{i=1}^{m} \max\big[1 - y_i(\langle \mathbf{w}, \mathbf{x}_i - \boldsymbol{u}_i^r \rangle + b), 0\big] \leq l \Big\} \geq 1 - \eta.$$

The constraint controls the η-quantile of the average (or equivalently the sum of) empirical errors. In Shivaswamy et al. (2006), Lanckriet et al. (2003), and Bhattacharyya et al. (2004), the authors explore a different direction; starting from the constraint formulation of SVM as in (14.2), they impose probabilistic constraints on each random variable individually. This formulation requires all constraints to be satisfied with high probability *simultaneously*. Thus, instead of controlling the η-quantile of the average loss, they control the η-quantile of the hinge loss for each sample. For the same reason that box uncertainty in the robust setting may be too conservative, this constraint-wise formulation may also be too conservative.

Problem (14.14) is generally intractable. However, we can approximate it as follows. Let

$$\hat{c} \triangleq \inf\{\alpha \mid \mathbb{P}(\sum_{i=1}^{m}\|\boldsymbol{u}_i\|^* \leq \alpha) \geq 1-\eta\}.$$

Notice that $\hat{c}$ is easily simulated, given μ. Then for any $(\mathbf{w}, b)$, with probability no less than $1-\eta$, the following holds:

$$\sum_{i=1}^{m}\max\left[1 - y_i(\langle \mathbf{w}, \mathbf{x}_i - \boldsymbol{u}_i^r\rangle + b), 0\right]$$

$$\leq \max_{\sum_i \|\boldsymbol{u}_i\|^* \leq \hat{c}} \sum_{i=1}^{m}\max\left[1 - y_i(\langle \mathbf{w}, \mathbf{x}_i - \boldsymbol{u}_i\rangle + b), 0\right].$$

Thus (14.14) is upper-bounded by (14.11) with $c = \hat{c}$. This gives an additional probabilistic robustness property of the standard regularized classifier. We observe that we can follow a similar approach using the constraint-wise robust setup, that is, the box uncertainty set. The interested reader can check that this would lead to considerably more pessimistic approximations of the chance constraint.

A Bayesian Regularizer. Next, we show how the above can be used in a Bayesian setup, to obtain an appropriate regularization coefficient. Suppose the total disturbance $c^r \triangleq \sum_{i=1}^{m}\|\boldsymbol{u}_i^r\|^*$ is a random variable and follows a prior distribution $\rho(\cdot)$. This can model, for example, the case where the training sample set is a mixture of several data sets in which the disturbance magnitude of each set is known. Such a setup leads to the following classifier which minimizes the Bayesian (robust) error:

$$\min_{\mathbf{w},b}: \quad \int\left\{\max_{\sum\|\boldsymbol{\delta}_i\|^*\leq c}\sum_{i=1}^{m}\max\left[1-y_i(\langle\mathbf{w},\mathbf{x}_i-\boldsymbol{u}_i\rangle+b),0\right]\right\}d\rho(c) \quad (14.15)$$

By Corollary 14.3, the Bayesian classifier (14.15) is equivalent to

$$\min_{\mathbf{w},b}: \quad \int\left\{c\|\mathbf{w}\| + \sum_{i=1}^{m}\max\left[1-y_i(\langle\mathbf{w},\mathbf{x}_i\rangle+b),0\right]\right\}d\rho(c),$$

which can be further simplified as

$$\min_{\mathbf{w},b} : \quad \bar{c}\|\mathbf{w}\| + \sum_{i=1}^{m} \max\left[1 - y_i\big(\langle\mathbf{w}, \mathbf{x}_i\rangle + b\big), 0\right],$$

where $\bar{c} \triangleq \int c\, d\rho(c)$. This provides a justifiable parameter tuning method different from cross validation: simply using the expected value of c^r.

14.4.2 Tikhonov Regularized ℓ_2-Regression

We now move from classification and SVMs to regression, and show that ℓ_2-regularized regression, like SVM, is equivalent to a robust optimization problem. This equivalence is then used to define new regularization-like algorithms, and also to prove properties of the regularized solution.

Given input-output pairs $\mathbf{x}_i, y_i$ which form the rows of X and the elements of vector $\mathbf{y}$, respectively, the goal is to find a predictor $\boldsymbol{\beta}$ that minimizes the squared loss $\|\mathbf{y} - X\boldsymbol{\beta}\|_2^2$. As is well known, this problem is often notoriously ill-conditioned, and may not have a unique solution. The classical and much-explored remedy has been, as in the SVM case, regularization. Regularizing with an ℓ_2-norm, known in statistics as ridge regression (Hoerl, 1962) and in analysis as Tikhonov regularization (Tikhonov and Arsenin, 1977), solves the problem[1]

$$\min_{\boldsymbol{\beta}} : \quad \|\mathbf{y} - X\boldsymbol{\beta}\|_2 + \lambda\|\boldsymbol{\beta}\|_2. \tag{14.16}$$

The main result of this section states that Tikhonov-regularized regression is the solution to a robust optimization, where X is subject to matrix-disturbance U with a bounded Frobenius norm.

Theorem 14.7. *The robust optimization formulation*

$$\min_{\boldsymbol{\beta}} : \quad \max_{U:\|U\|_F \leq \lambda} \|\mathbf{y} - (X + U)\boldsymbol{\beta}\|_2$$

is equivalent to Tikhonov-regularized regression (14.16).

Proof. For any perturbation U, we have $\|\mathbf{y} - (X+U)\boldsymbol{\beta}\|_2 = \|\mathbf{y} - X\boldsymbol{\beta} - U\boldsymbol{\beta}\|_2$. By the triangle inequality and because $\|U\|_F \leq \lambda$, we thus have $\|\mathbf{y} - (X + U)\boldsymbol{\beta}\|_2 \leq \|\mathbf{y} - X\boldsymbol{\beta}\| + \lambda\|\boldsymbol{\beta}\|_2$. On the other hand, for any given $\boldsymbol{\beta}$, we can choose a rank-1 U so that $U\boldsymbol{\beta}$ is aligned with $(\mathbf{y} - X\boldsymbol{\beta})$, and thus equality is attained. □

---

1. This problem is equivalent to one where we square the norm, up to a change in the regularization coefficient, λ.

This connection was first explored in the seminal work of El Ghaoui and Lebret (1997). There, they further show that the solution to the robust counterpart is almost as easily determined as that to the nominal problem: one need perform only a line search, in the case where the SVD of A is available. Thus, the computational cost of the robust regression is comparable to the original formulation.

As with SVMs, the "hidden robustness" has several consequences. By changing the uncertainty set, robust optimization allows for a rich class of regularization-like algorithms. Motivated by problems from robust control, El Ghaoui and Lebret (1997) then consider perturbations that have structure, leading to structured robust least-squares problems. They then analyze tractability and approximations to these structured least squares.[2] Finally, they use the robustness equivalence to prove regularity properties of the solution. Refer to El Ghaoui and Lebret (1997) for further details about structured robustness, tractability, and regularity.

14.4.3 Lasso

In this section, we consider a similar problem: ℓ_1-regularized regression, also known as Lasso (Tibshirani, 1996). Lasso has been explored extensively for its remarkable sparsity properties (e.g., Tibshirani, 1996; Bickel et al., 2009; Wainwright, 2009), most recently under the banner of compressed sensing (e.g., Chen et al. (1999); Candès et al. (2006), Candès and Tao (2006); Candès and Tao (2007); Candès and Tao (2008), Donoho (2006), for an incomplete list). Following the theme of this section, we show that the solution to Lasso is the solution to a robust optimization problem. As with Tikhonov regularization, robustness provides a connection of the regularizer to a physical property: protection from noise. This allows a principled selection of the regularizer. Moreover, by considering different uncertainty sets, we obtain generalizations of Lasso. Next, we go on to show that robustness can itself be used as an avenue for exploring properties of the solution. In particular, we show that robustness explains why the solution is sparse—that is, *Lasso is sparse because it is robust*. The analysis and the specific results obtained differ from standard sparsity results, providing different geometric intuition. This section is based on results reported in Xu et al. (2010a), where full proofs to all stated results can be found.

Lasso, or ℓ_1-regularized regression, has a form similar to ridge regression,

---

2. Note that arbitrary uncertainty sets may lead to intractable problems. This is because the inner maximization in the robust formulation is of a convex function, and hence is nonconvex.

differing only in the regularizer: [3]

$$\min : \|\mathbf{y} - X\boldsymbol{\beta}\|_2 + \lambda\|\boldsymbol{\beta}\|_1.$$

For a general uncertainty set $\mathcal{U}$, using the same notation as in Section 14.4.2, the robust regression formulation becomes

$$\min_{\boldsymbol{\beta} \in \mathbb{R}^m} \quad \max_{U \in \mathcal{U}} \|\mathbf{y} - (X + U)\boldsymbol{\beta}\|_2, \tag{14.17}$$

In the previous section, the uncertainty set was $\mathcal{U} = \{U : \|U\|_F \leq \lambda\}$. We consider a different uncertainty set here. Writing

$$U = \begin{bmatrix} | & | & \cdots & | \\ \mathbf{u}_1 & \mathbf{u}_2 & \cdots & \mathbf{u}_m \\ | & | & \cdots & | \end{bmatrix}, \qquad \text{where } (\mathbf{u}_1, \ldots, \mathbf{u}_m) \in \mathcal{U},$$

let the uncertainty set $\mathcal{U}$ have the form

$$\mathcal{U} \triangleq \left\{ (\boldsymbol{u}_1, \cdots, \boldsymbol{u}_m) \Big| \|\boldsymbol{u}_i\|_2 \leq c_i, \ i = 1, \cdots, m \right\}. \tag{14.18}$$

This is a *featurewise uncoupled* uncertainty set: the uncertainty in different features need not satisfy any joint constraints. In contrast, the constraint $\|U\|_F \leq 1$ used in Section 14.4.2 is featurewise coupled. We revisit coupled uncertainty sets below.

Theorem 14.8. *The robust regression problem (14.17) with an uncertainty set of the form (14.18) is equivalent to the following ℓ_1-regularized regression problem:*

$$\min_{\boldsymbol{\beta} \in \mathbb{R}^m} \quad \left\{ \|\mathbf{y} - X\boldsymbol{\beta}\|_2 + \sum_{i=1}^m c_i|\beta_i| \right\}. \tag{14.19}$$

Proof. Fix $\boldsymbol{\beta}^*$. We prove that $\max_{U \in \mathcal{U}} \|\mathbf{y} - (X + U)\boldsymbol{\beta}^*\|_2 = \|\mathbf{y} - X\boldsymbol{\beta}^*\|_2 + \sum_{i=1}^m c_i|\beta_i^*|$.

The inequality

$$\max_{U \in \mathcal{U}} \|\mathbf{y} - (X + U)\boldsymbol{\beta}^*\|_2 \leq \|\mathbf{y} - X\boldsymbol{\beta}^*\|_2 + \sum_{i=1}^m |\beta_i^*|c_i$$

follows from the triangle inequality, as in our proof in Section 14.4.2. The

---

3. Again we remark that with a change of regularization parameter, this is equivalent to the more common form appearing with a square outside the norm.

other inequality follows, if we take

$$\mathbf{u} \triangleq \begin{cases} \frac{\mathbf{y} - X\boldsymbol{\beta}^*}{\|\mathbf{y} - X\boldsymbol{\beta}^*\|_2} & \text{if } X\boldsymbol{\beta}^* \neq \mathbf{y}, \\ \text{any vector with unit } \ell_2\text{-norm} & \text{otherwise;} \end{cases}$$

and let

$$\boldsymbol{u}_i^* \triangleq \begin{cases} -c_i \mathrm{sgn}(\beta_i^*)\mathbf{u} & \text{if } x_i^* \neq 0; \\ -c_i \mathbf{u} & \text{otherwise.} \end{cases}$$

$\square$

Taking $c_i = c$ and normalizing $\mathbf{x}_i$ for all i, Problem (14.19) recovers the well-known Lasso (Tibshirani, 1996; Efron et al., 2004).

14.4.3.1   *General Uncertainty Sets*

Using this equivalence, we generalize to Lasso-like regularization algorithms in two ways: (a) to the case of an arbitrary norm and (b) to the case of coupled uncertainty sets.

Theorem 14.9. *For $\|\cdot\|_a$, an arbitrary norm in the Euclidean space, the robust regression problem*

$$\min_{\boldsymbol{\beta} \in \mathbb{R}^m} \left\{ \max_{U \in \mathcal{U}_a} \|\mathbf{y} - (X + U)\boldsymbol{\beta}\|_a \right\},$$

where

$$\mathcal{U}_a \triangleq \left\{ (\boldsymbol{u}_1, \cdots, \boldsymbol{u}_m) \,\middle|\, \|\boldsymbol{u}_i\|_a \leq c_i, \ i = 1, \cdots, m \right\},$$

is equivalent to the following regularized regression problem

$$\min_{\boldsymbol{\beta} \in \mathbb{R}^m} \left\{ \|\mathbf{y} - X\boldsymbol{\beta}\|_a + \sum_{i=1}^m c_i |\beta_i| \right\}.$$

We next consider featurewise coupled uncertainty sets. They can be used to incorporate additional information about potential noise in the problem, when available, to limit the conservativeness of the worst-case formulation. Consider the uncertainty set

$$\mathcal{U}' \triangleq \left\{ (\boldsymbol{u}_1, \cdots, \boldsymbol{u}_m) \,\middle|\, f_j(\|\boldsymbol{u}_1\|_a, \cdots, \|\boldsymbol{u}_m\|_a) \leq 0; \ j = 1, \cdots, k \right\},$$

where each $f_j(\cdot)$ is a convex function. The resulting robust formulation is equivalent to a more general regularization type of problem and, moreover, it is tractable.

Theorem 14.10. *Let $\mathcal{U}'$ be as above, and assume that the set*

$$\mathcal{Z} \triangleq \{\mathbf{z} \in \mathbb{R}^m | f_j(\mathbf{z}) \leq 0, \ j = 1, \cdots, k; \ \mathbf{z} \geq \mathbf{0}\}$$

has a nonempty relative interior. Then the robust regression problem

$$\min_{\boldsymbol{\beta} \in \mathbb{R}^m} \left\{ \max_{U \in \mathcal{U}'} \|\mathbf{y} - (X + U)\boldsymbol{\beta}\|_a \right\}$$

is equivalent to the following regularized regression problem:

$$\min_{\boldsymbol{\lambda} \in \mathbb{R}_+^k, \boldsymbol{\kappa} \in \mathbb{R}_+^m, \boldsymbol{\beta} \in \mathbb{R}^m} \left\{ \|\mathbf{y} - X\boldsymbol{\beta}\|_a + v(\boldsymbol{\lambda}, \boldsymbol{\kappa}, \boldsymbol{\beta}) \right\}, \tag{14.20}$$

where $v(\boldsymbol{\lambda}, \boldsymbol{\kappa}, \boldsymbol{\beta}) \triangleq \max_{\mathbf{c} \in \mathbf{R}^m} \left[(\boldsymbol{\kappa} + |\boldsymbol{\beta}|)^\top \mathbf{c} - \sum_{j=1}^{k} \lambda_j f_j(\mathbf{c}) \right]$

and, in particular, is efficiently solvable.

The next two corollaries are a direct application of Theorem 14.10.

Corollary 14.11. *Suppose*

$$\mathcal{U}' = \left\{ (\boldsymbol{\delta}_1, \cdots, \boldsymbol{\delta}_m) \, \Big| \, \big\| \|\boldsymbol{\delta}_1\|_a, \cdots, \|\boldsymbol{\delta}_m\|_a \big\|_s \leq l \right\},$$

for arbitrary norms $\| \cdot \|_a$ and $\| \cdot \|_s$. Then the robust problem is equivalent to the regularized regression problem

$$\min_{\boldsymbol{\beta} \in \mathbb{R}^m} \left\{ \|\mathbf{y} - X\boldsymbol{\beta}\|_a + l\|\boldsymbol{\beta}\|_s^* \right\},$$

where $\| \cdot \|_s^*$ is the dual norm of $\| \cdot \|_s$.

This corollary interprets *arbitrary* norm-based regularizers from a robust regression perspective. For example, taking both $\| \cdot \|_a$ and $\| \cdot \|_s$ to be the Euclidean norm, $\mathcal{U}'$ is the set of matrices with bounded Frobenius norm, and Corollary 14.11 recovers Theorem 14.7.

The next corollary considers general polytope uncertainty sets, where the columnwise norm vector of the realizable uncertainty belongs to a polytope. To illustrate the flexibility and potential use of such an uncertainty set, take $\| \cdot \|_a$ to be the ℓ_1-norm and the polytope to be the standard simplex, the resulting uncertainty set consists of matrices with bounded $\|\cdot\|_{2,1}$-norm. This is the ℓ_1-norm of the ℓ_2-norm of the columns, and has numerous applications, including outlier removal (Xu et al., 2010c).

Corollary 14.12. *Suppose*

$$\mathcal{U}' = \left\{ (\boldsymbol{u}_1, \cdots, \boldsymbol{u}_m) \, \Big| \, T\mathbf{c} \leq \mathbf{s}; \ \text{where:} \ c_j = \|\boldsymbol{u}_j\|_a \right\},$$

for a given matrix T, vector $\mathbf{s}$, and arbitrary norm $\|\cdot\|_a$. Then the robust regression is equivalent to the following regularized regression problem with variables $\boldsymbol{\beta}$ and $\boldsymbol{\lambda}$:

$$\min_{\boldsymbol{\beta},\boldsymbol{\lambda}}: \quad \|\mathbf{y} - X\boldsymbol{\beta}\|_a + \mathbf{s}^\top \boldsymbol{\lambda}$$

$$\text{s.t.} \quad \boldsymbol{\beta} \leq T^\top \boldsymbol{\lambda};$$

$$-\boldsymbol{\beta} \leq T^\top \boldsymbol{\lambda};$$

$$\boldsymbol{\lambda} \geq \mathbf{0}.$$

14.4.3.2   Sparsity

In this section, we investigate the sparsity properties of robust regression, and show in particular that Lasso is sparse *because it is robust*. This new connection between robustness and sparsity suggests that robustifying with respect to a featurewise independent uncertainty set might be a plausible way to achieve sparsity for other problems.

We show that if there is any perturbation in the uncertainty set that makes some feature *irrelevant*, that is, not contributing to the regression error, then the optimal robust solution puts no weight there. Thus, if the features in an index set $I \subset \{1, \ldots, n\}$ can be perturbed so as to be made irrelevant, then the solution will be supported on the complement, I^c.

To state the main theorem of this section, we introduce some notation. Given an index subset $I \subseteq \{1, \ldots, n\}$, and a matrix U, let U^I denote the restriction of U to feature set I, that is, U^I equals U on each feature indexed by $i \in I$, and is zero elsewhere. Similarly, given a featurewise uncoupled uncertainty set $\mathcal{U}$, let $\mathcal{U}^I$ be the restriction of $\mathcal{U}$ to the feature set I, that is, $\mathcal{U}^I \triangleq \{U^I \,|\, U \in \mathcal{U}\}$. Any element $U \in \mathcal{U}$ can be written as $U^I + U^{I^c}$ (here $I^c \triangleq \{1, \ldots, n\} \setminus I$) with $U^I \in \mathcal{U}^I$ and $U^{I^c} \in \mathcal{U}^{I^c}$.

Theorem 14.13. *The robust regression problem*

$$\min_{\boldsymbol{\beta} \in \mathbb{R}^m} \left\{ \max_{\Delta A \in \mathcal{U}} \|\mathbf{y} - (X + U)\boldsymbol{\beta}\|_2 \right\} \tag{14.21}$$

has a solution supported on an index set I if there exists some perturbation $\tilde{U} \in \mathcal{U}^{I^c}$, such that the robust regression problem

$$\min_{\boldsymbol{\beta} \in \mathbb{R}^m} \left\{ \max_{U \in \mathcal{U}^I} \|\mathbf{y} - (X + \tilde{U} + U)\boldsymbol{\beta}\|_2 \right\} \tag{14.22}$$

has a solution supported on the set I.

Theorem 14.13 is a special case of Theorem 14.13' with $c_j = 0$ for all $j \notin I$.
Theorem 14.13': Let $\boldsymbol{\beta}^*$ be an optimal solution of the robust regression

problem

$$\min_{\boldsymbol{\beta}\in\mathbb{R}^m}\left\{\max_{U\in\mathcal{U}}\|\mathbf{y}-(X+U)\boldsymbol{\beta}\|_2\right\},\tag{14.23}$$

and let $I\subseteq\{1,\cdots,m\}$ be such that $\beta_j^*=0\ \forall j\notin I$. Let

$$\tilde{\mathcal{U}}\triangleq\Big\{(\boldsymbol{u}_1,\cdots,\boldsymbol{u}_m)\Big|\|\boldsymbol{u}_i\|_2\le c_i,\ i\in I;\ \|\boldsymbol{u}_j\|_2\le c_j+l_j,\ j\notin I\Big\}.$$

Then, $\boldsymbol{\beta}^*$ is an optimal solution of

$$\min_{\boldsymbol{\beta}\in\mathbb{R}^m}\left\{\max_{U\in\tilde{\mathcal{U}}}\|\mathbf{y}-(\tilde{X}+U)\boldsymbol{\beta}\|_2\right\}\tag{14.24}$$

for any $\tilde{X}$ that satisfies $\|\tilde{\mathbf{x}}_j-\mathbf{x}_j\|\le l_j$ for $j\notin I$, and $\tilde{\mathbf{x}}_i=\mathbf{x}_i$ for $i\in I$.

In fact, we can replace the ℓ_2-norm loss with any loss function $f(\cdot)$ which satisfies the condition that if $\beta_j=0$, X and X' differ only in the jth column, and then $f(\mathbf{y},X,\boldsymbol{\beta})=f(\mathbf{y},X',\boldsymbol{\beta})$. This theorem thus suggests a methodology for constructing sparse algorithms by solving a robust optimization with respect to columnwise uncoupled uncertainty sets.

When we consider ℓ_2-loss, we can translate the condition of a feature being "irrelevant" to a geometric condition: orthogonality. We now use the result of Theorem 14.13 to show that robust regression has a sparse solution as long as an incoherence-type property is satisfied. This result is more in line with the traditional sparsity results, but we note that the geometric reasoning is different, now based on robustness. Specifically, we show that a feature receives zero weight if it is "nearly" (i.e., within an allowable perturbation) orthogonal to the signal and all relevant features.

Theorem 14.14. *Let $c_i=c$ for all i and consider ℓ_2-loss. Suppose that there exists $I\subset\{1,\cdots,m\}$ such that for all $\mathbf{v}\in\mathrm{span}\big(\{\mathbf{x}_i,i\in I\}\bigcup\{\mathbf{y}\}\big)$, $\|\mathbf{v}\|=1$, we have $\mathbf{v}^\top\mathbf{x}_j\le c,\ \forall j\notin I$. Then there exists an optimal solution $\boldsymbol{\beta}^*$ that satisfies $\beta_j^*=0,\ \forall j\notin I$.*

The proof proceeds as Theorem 14.13' would suggest: the columns in I^c can be perturbed so that they are made irrelevant, and thus the optimal solution will not be supported there; see Xu et al. (2010a) for details..

14.5   Robustness and Consistency

In this section we explore a fundamental connection between learning and robustness by using robustness properties to re-prove the consistency of kernelized SVM, and then of Lasso. The key difference from the proofs here and those seen elsewhere (e.g., Steinwart, 2005; Steinwart and Christmann, 2008;

Wainwright, 2009) is that we replace the metric entropy, VC-dimension, and stability conditions typically used, with a robustness condition. Thus we conclude that *SVM and Lasso are consistent because they are robust.*

14.5.1 Consistency of SVM

Let $\mathcal{X} \subseteq \mathbb{R}^n$ be bounded, and suppose the training samples $(\mathbf{x}_i, y_i)_{i=1}^{\infty}$ are generated according to an unknown i.i.d. distribution $\mathbb{P}$ supported on $\mathcal{X} \times \{-1, +1\}$. Theorem 14.15 shows that our robust classifier, and thus regularized SVM, asymptotically minimizes an upper bound of the expected classification error and hinge loss as the number of samples increases.

Theorem 14.15. *Let* $K \triangleq \max_{x \in \mathcal{X}} \|x\|_2$. *Then there exists a random sequence* $\{\gamma_{m,c}\}$ *such that*

1. *The following bounds on the Bayes loss and the hinge loss hold uniformly for all* $(\mathbf{w}, b)$:

$$\mathbb{E}_{(\mathbf{x},y) \sim \mathbb{P}}(\mathbf{1}_{y \neq sgn(\langle \mathbf{w}, \mathbf{x} \rangle + b)}) \leq \gamma_{m,c} + c\|\mathbf{w}\|_2 + \frac{1}{m}\sum_{i=1}^{m} \max\left[1 - y_i(\langle \mathbf{w}, \mathbf{x}_i \rangle + b), 0\right];$$

$$\mathbb{E}_{(\mathbf{x},y) \sim \mathbb{P}}\left(\max(1 - y(\langle \mathbf{w}, \mathbf{x} \rangle + b), 0)\right) \leq$$
$$\gamma_{m,c}(1 + K\|\mathbf{w}\|_2 + |b|) + c\|\mathbf{w}\|_2 + \frac{1}{m}\sum_{i=1}^{m} \max\left[1 - y_i(\langle \mathbf{w}, \mathbf{x}_i \rangle + b), 0\right].$$

2. *For every* $c > 0$, $\lim_{m \to \infty} \gamma_{m,c} = 0$ *almost surely, and the convergence is uniform in* $\mathbb{P}$.

Proof. We outline the basic idea of the proof here; refer to Xu et al. (2009) for the technical details. We consider the testing sample set as a perturbed copy of the training sample set, and measure the magnitude of the perturbation. For testing samples that have "small" perturbations, Corollary 14.3 guarantees that the quantity $c\|\mathbf{w}\|_2 + \frac{1}{m}\sum_{i=1}^{m} \max\left[1 - y_i(\langle \mathbf{w}, \mathbf{x}_i \rangle + b), 0\right]$ upper-bounds their total loss. Therefore, we only need to show that the fraction of testing samples having "large" perturbations diminishes to prove the theorem. We show this using a balls and bins argument. Partitioning $\mathcal{X} \times \{-1, +1\}$, we match testing and training samples that fall in the same partition. We then use the Bretagnolle-Huber-Carol inequality for multinomial distributions to conclude that the fraction of unmatched points diminishes to zero. $\qquad\square$

Based on Theorem 14.15, it can be further shown that the expected classification error of the solutions of SVM converges to the Bayes risk, that is, SVM is consistent.

14.5.2 Consistency of Lasso

In this section, we re-prove the asymptotic consistency of Lasso by using robustness. The basic idea of the consistency proof is as follows. We show that the robust optimization formulation can be seen to have the maximum *expected error* w.r.t. a class of probability measures. This class includes a kernel density estimator, and using this, we show that Lasso is consistent.

14.5.2.1 *Robust Optimization and Kernel Density Estimation*

En route to proving the consistency of Lasso based on robust optimization, we discuss another result of independent interest. We link robust optimization to worst-case expected utility, that is, the worst-case expectation over a set of measures. For the proofs, and more along this direction, see Xu et al. (2010a,b). Throughout this section, we use $\mathcal{P}$ to represent the set of all probability measures (on Borel σ-algebra) of $\mathbb{R}^{m+1}$.

We first establish a general result on the equivalence between a robust optimization formulation and a worst-case expected utility:

Proposition 14.16. *Given a function* $f : \mathbb{R}^{m+1} \to \mathbb{R}$ *and Borel sets* $\mathcal{Z}_1, \cdots, \mathcal{Z}_n \subseteq \mathbb{R}^{m+1}$, *let*

$$\mathcal{P}_n \triangleq \{\mu \in \mathcal{P} | \forall S \subseteq \{1, \cdots, n\} : \mu(\bigcup_{i \in S} \mathcal{Z}_i) \geq |S|/n\}.$$

The following holds:

$$\frac{1}{n} \sum_{i=1}^{n} \sup_{(\mathbf{x}_i, y_i) \in \mathcal{Z}_i} f(\mathbf{x}_i, y_i) = \sup_{\mu \in \mathcal{P}_n} \int_{\mathbb{R}^{m+1}} f(\mathbf{x}, y) d\mu(\mathbf{x}, y).$$

This leads to the Corollary 14.17 for Lasso, which states that for a given solution $\boldsymbol{\beta}$, the robust regression loss over the training data is equal to the worst-case expected *generalization error*.

Corollary 14.17. *Given* $\mathbf{y} \in \mathbb{R}^n$, $X \in \mathbb{R}^{n \times m}$, *the following equation holds for any* $\boldsymbol{\beta} \in \mathbb{R}^m$,

$$\|\mathbf{y} - X\boldsymbol{\beta}\|_2 + \sqrt{n} c_n(\|\boldsymbol{\beta}\|_1 + 1) = \sup_{\mu \in \hat{\mathcal{P}}(n)} \sqrt{n \int_{\mathbb{R}^{m+1}} (y' - \mathbf{x}'^\top \boldsymbol{\beta})^2 d\mu(\mathbf{x}', y')}. \quad (14.25)$$

Where we let x_{ij} and u_{ij} be the (i,j)-entries of X and U, respectively, and

$$\hat{\mathcal{P}}(n) \triangleq \bigcup_{\|\boldsymbol{\sigma}\|_2 \le \sqrt{n}c_n; \forall i: \|\boldsymbol{u}_i\|_2 \le \sqrt{n}c_n} \mathcal{P}_n(X, U, \mathbf{y}, \boldsymbol{\sigma});$$

$$\mathcal{P}_n(X, U, \mathbf{y}, \boldsymbol{\sigma}) \triangleq \{\mu \in \mathcal{P} | \mathcal{Z}_i = [y_i - \sigma_i, y_i + \sigma_i] \times \prod_{j=1}^{m} [x_{ij} - u_{ij}, x_{ij} + u_{ij}];$$

$$\forall S \subseteq \{1, \cdots, n\} : \mu(\bigcup_{i \in S} \mathcal{Z}_i) \ge |S|/n\}.$$

The proof of consistency relies on showing that the set $\hat{\mathcal{P}}(n)$ of distributions contains a kernel density estimator. Recall the basic definition: the *kernel density estimator* for a density h in $\mathbb{R}^d$, originally proposed in Rosenblatt (1956) and Parzen (1962), is defined by

$$h_n(\mathbf{x}) = (nc_n^d)^{-1} \sum_{i=1}^{n} K\left(\frac{\mathbf{x} - \hat{\mathbf{x}}_i}{c_n}\right),$$

where $\{c_n\}$ is a sequence of positive numbers, $\hat{\mathbf{x}}_i$ are i.i.d. samples generated according to h, and K is a Borel measurable function (kernel) satisfying $K \ge 0$, $\int K = 1$. See Devroye and Györfi (1985), Scott (1992), and references therein for detailed discussions. A celebrated property of a kernel density estimator is that it converges in $\mathcal{L}^1$ to h when $c_n \downarrow 0$ and $nc_n^d \uparrow \infty$ (Devroye and Györfi, 1985).

14.5.2.2 *Density Estimation and Consistency of Lasso*

We now use robustness of Lasso to prove its consistency. Throughout, we use c_n to represent the robustness level c where there are n samples. We take c_n to zero as n grows.

Recall the standard generative model in statistical learning: let $\mathbb{P}$ be a probability measure with bounded support that generates i.i.d. samples $(y_i, \mathbf{x}_i)$, and has a density $f^*(\cdot)$. Denote the set of the first n samples by $\mathcal{S}_n$. Define

$$\boldsymbol{\beta}(c_n, \mathcal{S}_n) \triangleq \arg\min_{\boldsymbol{\beta}} \left\{ \sqrt{\frac{1}{n} \sum_{i=1}^{n} (y_i - \mathbf{x}_i^{\top}\boldsymbol{\beta})^2} + c_n\|\beta\|_1 \right\}$$

$$= \arg\min_{\boldsymbol{\beta}} \left\{ \frac{\sqrt{n}}{n} \sqrt{\sum_{i=1}^{n} (y_i - \mathbf{x}_i^{\top}\boldsymbol{\beta})^2} + c_n\|\beta\|_1 \right\};$$

$$\boldsymbol{\beta}(\mathbb{P}) \triangleq \arg\min_{\boldsymbol{\beta}} \left\{ \sqrt{\int_{y,\mathbf{x}} (y - \mathbf{x}^{\top}\boldsymbol{\beta})^2 d\mathbb{P}(y, \mathbf{x})} \right\}.$$

In words, $\boldsymbol{\beta}(c_n, \mathcal{S}_n)$ is the solution to Lasso with the tradeoff parameter set to $c_n\sqrt{n}$, and $\boldsymbol{\beta}(\mathbb{P})$ is the "true" optimal solution. We establish that $\boldsymbol{\beta}(c_n, \mathcal{S}_n) \to \boldsymbol{\beta}(\mathbb{P})$ using robustness.

Theorem 14.18. *Let $\{c_n\}$ be such that $c_n \downarrow 0$ and $\lim_{n\to\infty} n(c_n)^{m+1} = \infty$. Suppose there exists a constant H such that $\|\boldsymbol{\beta}(c_n, \mathcal{S}_n)\|_2 \le H$ for all n. Then,*

$$\lim_{n\to\infty} \sqrt{\int_{y,\mathbf{x}} (y - \mathbf{x}^\top \boldsymbol{\beta}(c_n, \mathcal{S}_n))^2 d\mathbb{P}(y, \mathbf{x})} = \sqrt{\int_{y,\mathbf{x}} (y - \mathbf{x}^\top \boldsymbol{\beta}(\mathbb{P}))^2 d\mathbb{P}(y, \mathbf{x})},$$

almost surely.

We give an outline of the proof; refer to Xu et al. (2010a) for the details. In Section 14.4.3 we showed that Lasso is a special case of robust optimization. Then, in Section 14.5.2.1, we proved that robust optimization is equivalent to a worst-case expectation. The proof follows by showing that the sets $\mathcal{P}_n$, in the worst-case expectation equivalent to Lasso, contain a kernel density estimator. Since these sets shrink, consistency follows.

The assumption that $\|\mathbf{x}(c_n, \mathcal{S}_n)\|_2 \le H$ can be removed. As in Theorem 14.18, the proof technique rather than the result itself is of interest. We refer the interested reader to Xu et al. (2010a).

14.6   Robustness and Generalization

We have already seen that regularized regression and regularized SVMs are special cases of robust optimization, and hence exhibit robustness to perturbed data. This robustness was used above to show that ridge regression has a Lipschitz solution, that Lasso is sparse, and that SVM and Lasso are consistent. In this section, we show that robustness can be used to control the estimation of the risk (i.e., generalization error) of learning algorithms. The results we describe are based on Xu and Mannor (2010b).

Several approaches have been proposed to bound the deviation of the risk from its empirical measurement, and among these methods, those based on uniform convergence and stability are the most widely used (e.g., Vapnik and Chervonenkis, 1991; Evgeniou et al., 2000; Alon et al., 1997; Bartlett, 1998; Bartlett and Mendelson, 2002; Bartlett et al., 2005; Bousquet and Elisseeff, 2002; Poggio et al., 2004; Mukherjee et al., 2006, and many others). We provide a new, robustness-driven approach to proving generalization bounds.

Whereas in previous sections "robustness" was defined directly in terms of robust optimization, here we abstract this definition. Because we consider abstract algorithms in this section, we introduce some necessary notations, that differ from those in previous sections. We use $\mathcal{Z}$ to denote the set from which each sample is drawn, and $\mathcal{H}$ to denote the hypothesis set. Throughout this section we use $\mathbf{s} \in \mathcal{Z}^m$ to denote the training sample set consisting of m training samples $(s_1, \cdots, s_m)$. A learning algorithm $\mathcal{A}$ is thus a mapping from $\mathcal{Z}^m$ to $\mathcal{H}$. We use $\mathcal{A}_{\mathbf{s}}$ to represent the hypothesis learned, given training set $\mathbf{s}$. For each hypothesis $h \in \mathcal{H}$ and each point $z \in \mathcal{Z}$, there is an associated loss $l(h, z)$, which is nonnegative and upper-bounded uniformly by a scalar M. In the special case of supervised learning, the sample space can be decomposed as $\mathcal{Z} = \mathcal{Y} \times \mathcal{X}$, and the goal is to learn a mapping from $\mathcal{X}$ to $\mathcal{Y}$, that is, to predict the y-component given the x-component. Hence we use $\mathcal{A}_{\mathbf{s}}(x)$ to represent the predicted y-component (label) of $x \in \mathcal{X}$ when $\mathcal{A}$ is trained on $\mathbf{s}$. We call $\mathcal{X}$ the input space and $\mathcal{Y}$ the output space. We use $|_x$ and $|_y$ to denote the x-component and y-component of a point. For example, $s_{i|x}$ is the x-component of s_i. Finally, we use $\mathcal{N}(\varepsilon, T, \rho)$ to denote the ε-covering number of a space T equipped with a metric ρ (see van der Vaart and Wellner, 2000, for a precise definition).

Definition 14.3 says that an algorithm is robust, if we can partition the sample set into finite subsets, such that if a new sample falls into the same subset as a training sample, then the loss of the former is close to the loss of the latter.

Definition 14.3. *Algorithm $\mathcal{A}$ is $(K, \varepsilon(\mathbf{s}))$ robust if $\mathcal{Z}$ can be partitioned into K disjoint sets, denoted by $\{C_i\}_{i=1}^K$, such that $\forall s \in \mathbf{s}$,*

$$s, z \in C_i, \quad \Longrightarrow \quad |l(\mathcal{A}_{\mathbf{s}}, s) - l(\mathcal{A}_{\mathbf{s}}, z)| \leq \varepsilon(\mathbf{s}). \tag{14.26}$$

14.6.1   Generalization Properties of Robust Algorithms

In this section we use Definition 14.3 to derive PAC bounds for robust algorithms. Let the sample set $\mathbf{s}$ consist of m i.i.d. samples generated by an unknown distribution μ. Let $\hat{l}(\cdot)$ and $l_{\mathrm{emp}}(\cdot)$ denote the expected error and the training error, respectively. That is,

$$\hat{l}(\mathcal{A}_{\mathbf{s}}) \triangleq \mathbb{E}_{z \sim \mu} l(\mathcal{A}_{\mathbf{s}}, z); \quad l_{\mathrm{emp}}(\mathcal{A}_{\mathbf{s}}) \triangleq \frac{1}{m} \sum_{s_i \in \mathbf{s}} l(\mathcal{A}_{\mathbf{s}}, s_i).$$

Theorem 14.19. *If $\mathbf{s}$ consists of m i.i.d. samples, the loss function $l(\cdot, \cdot)$ is upper-bounded by M, and $\mathcal{A}$ is $(K, \varepsilon(\mathbf{s}))$-robust, then for any $\delta > 0$, with probability at least $1 - \delta$,*

$$\left| \hat{l}(\mathcal{A}_{\mathbf{s}}) - l_{\mathrm{emp}}(\mathcal{A}_{\mathbf{s}}) \right| \leq \varepsilon(s) + M \sqrt{\frac{2K \ln 2 + 2 \ln(1/\delta)}{m}}.$$

Proof. The proof follows by partitioning the set and using inequalities for multinomial random variables, à la the Bretagnolle-Huber-Carol inequality.

□

Theorem 14.19 requires that we fix a K a priori. However, it is often worthwhile to consider adaptive K. For example, in the large-margin classification case, typically the margin is known only after $\mathbf{s}$ is realized. That is, the value of K depends on $\mathbf{s}$. Because of this dependency, we need a generalization bound that holds uniformly for all K.

Corollary 14.20. *If* $\mathbf{s}$ *consists of* m *i.i.d. samples, and* $\mathcal{A}$ *is* $(K, \varepsilon_K(\mathbf{s}))$-*robust for all* $K \geq 1$*, then for any* $\delta > 0$*, with probability at least* $1 - \delta$*,*

$$\left| \hat{l}(\mathcal{A}_{\mathbf{s}}) - l_{\mathrm{emp}}(\mathcal{A}_{\mathbf{s}}) \right| \leq \inf_{K \geq 1} \left[\varepsilon_K(s) + M \sqrt{\frac{2K \ln 2 + 2 \ln \frac{K(K+1)}{\delta}}{m}} \right].$$

If $\varepsilon(s)$ does not depend on $\mathbf{s}$, we can sharpen the bound given in Corollary 14.20.

Corollary 14.21. *If* $\mathbf{s}$ *consists of* m *i.i.d. samples, and* $\mathcal{A}$ *is* (K, ε_K)-*robust for all* $K \geq 1$*, then for any* $\delta > 0$*, with probability at least* $1 - \delta$*,*

$$\left| \hat{l}(\mathcal{A}_{\mathbf{s}}) - l_{\mathrm{emp}}(\mathcal{A}_{\mathbf{s}}) \right| \leq \inf_{K \geq 1} \left[\varepsilon_K + M \sqrt{\frac{2K \ln 2 + 2 \ln \frac{1}{\delta}}{m}} \right].$$

14.6.2   Examples of Robust Algorithms

In this section we provide some examples of robust algorithms. For the proofs of these examples, refer to Xu and Mannor (2010b,a). Our first example is majority voting (MV) classification (e.g., Devroye et al., 1996, Section 6.3), which partitions the input space $\mathcal{X}$ and labels each partition set according to a majority vote of the training samples belonging to it.

Example 14.1 (majority voting). *Let* $\mathcal{Y} = \{-1, +1\}$*. Partition* $\mathcal{X}$ *to* $\mathcal{C}_1, \cdots, \mathcal{C}_K$*, and use* $\mathcal{C}(x)$ *to denote the set to which* x *belongs. A new sample* $x_a \in \mathcal{X}$ *is labeled*

$$\mathcal{A}_{\mathbf{s}}(x_a) \triangleq \begin{cases} 1, & \text{if } \sum_{s_i \in \mathcal{C}(x_a)} \mathbf{1}(s_{i|y} = 1) \geq \sum_{s_i \in \mathcal{C}(x_a)} \mathbf{1}(s_{i|y} = -1); \\ -1, & \text{otherwise.} \end{cases}$$

If the loss function is the prediction error $l(\mathcal{A}_s, z) = \mathbf{1}_{z_{|y} \neq \mathcal{A}_{\mathbf{s}}(z_{|x})}$*, then MV is* $(2K, 0)$-*robust.*

The MV algorithm has a natural partition of the sample space that makes it robust. Another class of robust algorithms is those that have approximately the same testing loss for testing samples that are close (in the sense of geometric distance) to each other, since we can partition the sample space with norm balls, as in the standard definition of covering numbers (van der Vaart and Wellner, 2000). Theorem 14.22 states that an algorithm is robust if two samples being close implies that they have a similar testing error. Thus, in particular, this means that robustness is weaker than uniform stability (Bousquet and Elisseeff, 2002).

Theorem 14.22. *Fix $\gamma > 0$ and metric ρ of $\mathcal{Z}$. Suppose $\mathcal{A}$ satisfies*

$$|l(\mathcal{A}_{\mathbf{s}}, z_1) - l(\mathcal{A}_{\mathbf{s}}, z_2)| \le \varepsilon(\mathbf{s}), \quad \forall z_1, z_2 : z_1 \in \mathbf{s}, \rho(z_1, z_2) \le \gamma,$$

and $\mathcal{N}(\gamma/2, \mathcal{Z}, \rho) < \infty$. Then $\mathcal{A}$ is $\big(\mathcal{N}(\gamma/2, \mathcal{Z}, \rho), \varepsilon(\mathbf{s})\big)$-robust.

Theorem 14.22 leads Example 14.3: if the testing error, given the output of an algorithm, is Lipschitz continuous, then the algorithm is robust.

Example 14.2 (Lipschitz continuous functions). *If $\mathcal{Z}$ is compact w.r.t. metric ρ, and $l(\mathcal{A}_{\mathbf{s}}, \cdot)$ is Lipschitz continuous with Lipschitz constant $c(\mathbf{s})$, that is,*

$$|l(\mathcal{A}_{\mathbf{s}}, z_1) - l(\mathcal{A}_{\mathbf{s}}, z_2)| \le c(\mathbf{s})\rho(z_1, z_2), \quad \forall z_1, z_2 \in \mathcal{Z},$$

then $\mathcal{A}$ is $\big(\mathcal{N}(\gamma/2, \mathcal{Z}, \rho), c(\mathbf{s})\gamma\big)$-robust for all $\gamma > 0$.

Theorem 14.22 also implies that SVM, Lasso, feed-forward neural networks, and PCA are robust, as stated in Examples 14.3–14.6.

Example 14.3 (support vector machines). *Let $\mathcal{X}$ be compact. Consider the standard SVM formulation (Cortes and Vapnik, 1995; Schölkopf and Smola, 2001), as discussed in Sections 14.3 and 14.4.*

$$\min_{\mathbf{w}, d} \quad c\|\mathbf{w}\|_{\mathcal{H}}^2 + \sum_{i=1}^{m} \xi_i$$
$$\text{s.t.} \quad 1 - s_{i|y}[\langle \mathbf{w}, \phi(s_{i|x})\rangle + d] \le \xi_i, \quad i = 1, \cdots, m;$$
$$\xi_i \ge 0, \quad i = 1, \cdots, m.$$

Here $\phi(\cdot)$ is a feature mapping, $\|\cdot\|_{\mathcal{H}}$ is its RKHS kernel, and $k(\cdot, \cdot)$ is the kernel function. Let $l(\cdot, \cdot)$ be the hinge loss, that is, $l((w, d), z) = [1 - z_{|y}(\langle w, \phi(z_{|x})\rangle + d)]^+$, and define $f_{\mathcal{H}}(\gamma) \triangleq \max_{\mathbf{a}, \mathbf{b} \in \mathcal{X}, \|\mathbf{a} - \mathbf{b}\|_2 \le \gamma} (k(\mathbf{a}, \mathbf{a}) + k(\mathbf{b}, \mathbf{b}) - 2k(\mathbf{a}, \mathbf{b}))$. If $k(\cdot, \cdot)$ is continuous, then for any $\gamma > 0$, $f_{\mathcal{H}}(\gamma)$ is finite, and SVM is $(2\mathcal{N}(\gamma/2, \mathcal{X}, \|\cdot\|_2), \sqrt{f_{\mathcal{H}}(\gamma)/c})$ robust.

Example 14.4 (Lasso). *Let $\mathcal{Z}$ be compact and the loss function be $l(\mathcal{A}_{\mathbf{s}}, z) = |z_{|y} - \mathcal{A}_{\mathbf{s}}(z_{|x})|$. Lasso (Tibshirani, 1996), which is the following regression* *formulation:*

$$\min_{\mathbf{w}} : \quad \frac{1}{m}\sum_{i=1}^{m}(s_{i|y} - \mathbf{w}^{\top}s_{i|x})^2 + c\|\mathbf{w}\|_1,$$

is $(\mathcal{N}(\gamma/2, \mathcal{Z}, \|\cdot\|_{\infty}), (Y(\mathbf{s})/c + 1)\gamma)$-*robust for all* $\gamma > 0$, *where* $Y(\mathbf{s}) \triangleq \frac{1}{n}\sum_{i=1}^{n}s_{i|y}^{2}$.

Example 14.5 (Feed-forward neural networks). *Let* $\mathcal{Z}$ *be compact and the loss function be* $l(\mathcal{A}_{\mathbf{s}}, z) = |z_{|y} - \mathcal{A}_{\mathbf{s}}(z_{|x})|$. *Consider the d-layer neural network (trained on* $\mathbf{s}$), *which is the following predicting rule, given an input* $x \in \mathcal{X}$

$$x^0 \quad := \quad z_{|x}$$

$$\forall v = 1, \cdots, d-1: \quad x_i^v \quad := \quad \sigma\Big(\sum_{j=1}^{N_{v-1}} w_{ij}^{v-1} x_j^{v-1}\Big); \quad i = 1, \cdots, N_v;$$

$$\mathcal{A}_{\mathbf{s}}(x) \quad := \quad \sigma\Big(\sum_{j=1}^{N_{d-1}} w_j^{d-1} x_j^{d-1}\Big);$$

If there exist α *and* β *such that the d-layer neural network satisfying that* $|\sigma(a) - \sigma(b)| \leq \beta|a - b|$, *and* $\sum_{j=1}^{N_v}|w_{ij}^v| \leq \alpha$ *for all* v, i, *then it is* $(\mathcal{N}(\gamma/2, \mathcal{Z}, \|\cdot\|_{\infty}), \alpha^d\beta^d\gamma)$-*robust, for all* $\gamma > 0$.

In Example 14.5, the number of hidden units in each layer has no effect on the robustness of the algorithm and, consequently, on the bound on the testing error. This indeed agrees with Bartlett (1998), where the author showed (using a different approach based on fat-shattering dimension) that for neural networks, the weight plays a more important role than the number of hidden units.

Example 14.6 considers an unsupervised learning algorithm, namely, the principal component analysis algorithm. We show that it is robust if the sample space is *bounded*. This does not contradict the well-known fact that the principal component analysis is sensitive to outliers which are far from the origin.

Example 14.6 (Principal component analysis (PCA)). *Let* $\mathcal{Z} \subset \mathbb{R}^m$ *be such that* $\max_{z \in \mathcal{Z}} \|z\|_2 \leq B$. *If the loss function is* $l((w_1, \cdots, w_d), z) = \sum_{k=1}^{d}(w_k^{\top}z)^2$, *then finding the first d principal components, which solves the* *optimization problem over d vectors $w_1, \cdots, w_d \in \mathbb{R}^m$,*

$$\max_{\mathbf{w}_1, \cdots, \mathbf{w}_k} \quad \sum_{i=1}^{m} \sum_{k=1}^{d} (\mathbf{w}_k^\top s_i)^2$$
$$\text{s.t.} \quad \|\mathbf{w}_k\|_2 = 1, \quad k = 1, \cdots, d;$$
$$\mathbf{w}_i^\top \mathbf{w}_j = 0, \quad i \neq j.$$

is $(\mathcal{N}(\gamma/2, \mathcal{Z}, \|\cdot\|_2), 2d\gamma B)$-robust.

14.7   Conclusion

The purpose of this chapter has been to hint at the wealth of applications and uses of robust optimization in machine learning. Broadly speaking, there are two main methodological frameworks developed here: robust optimization used as a way to make an optimization-based machine learning algorithm robust to noise; and robust optimization as a fundamental tool for analyzing properties of machine learning algorithms and for constructing algorithms with special properties. The properties we have discussed here include sparsity, consistency and generalization. There are many directions of interest that future work can pursue. We highlight two that we consider of particular interest and promise. The first is learning in the high-dimensional setting, where the dimensionality of the models (or parameter space) is of the same order of magnitude as the number of training samples available. Hidden structure such as sparsity or low rank has offered ways around the challenges of this regime. Robustness and robust optimization may offer clues as to how to develop new tools and new algorithms for this setting. A second direction of interest is the design from data of uncertainty sets for robust optimization. Constructing uncertainty sets from data is a central problem in robust optimization that has not been adequately addressed, and machine learning methodology may be able to provide a way forward.

14.8   References

N. Alon, S. Ben-David, N. Cesa-Bianchi, and D. Haussler. Scale-sensitive dimension, uniform convergence, and learnability. *Journal of the ACM*, 44(4):615–631, 1997.

M. Anthony and P. L. Bartlett. *Neural Network Learning: Theoretical Foundations.* Cambridge University Press, 1999.

P. L. Bartlett. The sample complexity of pattern classification with neural networks: The size of the weight is more important than the size of the network. *IEEE Transactions on Information Theory*, 44(2):525–536, 1998.

P. L. Bartlett and S. Mendelson. Rademacher and Gaussian complexities: Risk bounds and structural results. *Journal of Machine Learning Research*, 3:463–482, November 2002.

P. L. Bartlett, O. Bousquet, and S. Mendelson. Local Rademacher complexities. *Annals of Statistics*, 33(4):1497–1537, 2005.

A. Ben-Tal and A. Nemirovski. Robust solutions of linear programming problems contaminated with uncertain data. *Mathematical Programming, Series A*, 88(3): 411–424, 2000.

A. Ben-Tal, L. El Ghaoui, and A. Nemirovski. *Robust Optimization*. Princeton University Press, 2009.

D. Bertsimas and M. Sim. The price of robustness. *Operations Research*, 52(1): 35–53, January-February 2004.

D. Bertsimas, D. B. Brown, and C. Caramanis. Theory and applications of robust optimization. To appear in SIAM Review, 2010.

C. Bhattacharyya, L. R. Grate, M. I. Jordan, L. El Ghaoui, and I. S. Mian. Robust sparse hyperplane classifiers: Application to uncertain molecular profiling data. *Journal of Computational Biology*, 11(6):1073–1089, 2004.

P. Bickel, Y. Ritov, and A. Tsybakov. Simultaneous analysis of Lasso and Dantzig selector. *Annals of Statistics*, 37(4):1705–1732, 2009.

B. E. Boser, I. M. Guyon, and V. N. Vapnik. A training algorithm for optimal margin classifiers. In *Proceedings of the 5th Annual ACM Workshop on Computational Learning Theory*, pages 144–152, New York, 1992. ACM Press.

O. Bousquet and A. Elisseeff. Stability and generalization. *Journal of Machine Learning Research*, 2:499–526, 2002.

E. J. Candès and T. Tao. Near-optimal signal recovery from random projections: Universal encoding strategies? *IEEE Transactions on Information Theory*, 52 (12):5406–5425, 2006.

E. J. Candès and T. Tao. The Dantzig selector: Statistical estimation when p is much larger than n. *Annals of Statistics*, 35(6):2313–2351, 2007.

E. J. Candès and T. Tao. Reflections on compressed sensing. *IEEE Information Theory Society Newsletter*, 58(4):20–23, 2008.

E. J. Candès, J. Romberg, and T. Tao. Robust uncertainty principles: Exact signal reconstruction from highly incomplete frequency information. *IEEE Transactions on Information Theory*, 52(2):489–509, 2006.

C. Caramanis and S. Mannor. Learning in the limit with adversarial disturbances. In *Proceedings of the 21st Annual Conference on Learning Theory*, pages 467–478, 2008.

S. S. Chen, D. L. Donoho, and M. A. Saunders. Atomic decomposition by basis pursuit. *SIAM Journal on Scientific Computing*, 20(1):33–61, 1999.

C. Cortes and V. N. Vapnik. Support vector networks. *Machine Learning*, 20:1–25, 1995.

A. d'Aspremont, L. El Ghaoui, M. I. Jordan, and G. R. Lanckriet. A direct formulation for sparse PCA using semidefinite programming. *SIAM Review*, 49 (3):434–448, 2007.

L. Devroye and L. Györfi. *Nonparametric Density Estimation: the l_1 View*. John Wiley & Sons, 1985.

L. Devroye, L. Györfi, and G. Lugosi. *A Probabilistic Theory of Pattern Recognition*.

Springer, New York, 1996.

D. L. Donoho. Compressed sensing. *IEEE Transactions on Information Theory*, 52(4):1289–1306, 2006.

G. E. Dullerud and F. Paganini. *A Course in Robust Control Theory: A Convex Approach*, volume 36 of *Texts in Applied Mathematics*. Springer-Verlag, New York, 2000.

B. Efron, T. Hastie, I. Johnstone, and R. Tibshirani. Least angle regression. *Annals of Statistics*, 32(2):407–499, 2004.

L. El Ghaoui and H. Lebret. Robust solutions to least-squares problems with uncertain data. *SIAM Journal on Matrix Analysis and Applications*, 18(4):1035–1064, 1997.

T. Evgeniou, M. Pontil, and T. Poggio. Regularization networks and support vector machines. In A. J. Smola, P. L. Bartlett, B. Schölkopf, and D. Schuurmans, editors, *Advances in Large Margin Classifiers*, pages 171–203, Cambridge, MA, 2000. MIT Press.

A. Globerson and S. Roweis. Nightmare at test time: Robust learning by feature deletion. In *Proceedings of the 23rd International Conference on Machine Learning*, pages 353–360, New York, 2006. ACM Press.

A. Hoerl. Application of ridge analysis to regression problems. *Chemical Engineering Progress*, 58:54–59, 1962.

P. Kall and S. W. Wallace. *Stochastic Programming*. John Wiley & Sons, 1994.

S.-J. Kim, A. Magnani, and S. Boyd. Robust Fisher discriminant analysis. In *Advances in Neural Information Processing Systems*, pages 659–666, 2005.

G. R. Lanckriet, L. El Ghaoui, C. Bhattacharyya, and M. I. Jordan. A robust minimax approach to classification. *Journal of Machine Learning Research*, 3: 555–582, March 2003.

S. Mukherjee, P. Niyogi, T. Poggio, and R. Rifkin. Learning theory: Stability is sufficient for generalization and necessary and sufficient for consistency of empirical risk minimization. *Advances in Computational Mathematics*, 25(1-3): 161–193, 2006.

A. Nilim and L. El Ghaoui. Robust control of Markov decision processes with uncertain transition matrices. *Operations Research*, 53(5):780–798, September 2005.

E. Parzen. On the estimation of a probability density function and the mode. *Annals of Mathematical Statistics*, 33(3):1065–1076, 1962.

T. Poggio, R. Rifkin, S. Mukherjee, and P. Niyogi. General conditions for predictivity in learning theory. *Nature*, 428(6981):419–422, 2004.

A. Prékopa. *Stochastic Programming*. Kluwer Academic Publishers, 1995.

M. Rosenblatt. Remarks on some nonparametric estimates of a density function. *Annals of Mathematical Statistics*, 27(3):832–837, 1956.

B. Schölkopf and A. J. Smola. *Learning with Kernels*. MIT Press, 2001.

D. W. Scott. *Multivariate Density Estimation: Theory, Practice, and Visualization*. John Wiley & Sons, New York, 1992.

P. K. Shivaswamy, C. Bhattacharyya, and A. J. Smola. Second order cone programming approaches for handling missing and uncertain data. *Journal of Machine Learning Research*, 7:1283–1314, July 2006.

A. L. Soyster. Convex programming with set-inclusive constraints and applications to inexact linear programming. *Operations Research*, 21(5):1154–1157, 1973.

I. Steinwart. Consistency of support vector machines and other regularized kernel classifiers. *IEEE Transactions on Information Theory*, 51(1):128–142, 2005.

I. Steinwart and A. Christmann. *Support Vector Machines*. Springer, New York, 2008.

R. Tibshirani. Regression shrinkage and selection via the Lasso. *Journal of the Royal Statistical Society, Series B*, 58(1):267–288, 1996.

A. N. Tikhonov and V. Arsenin. *Solutions of Ill-Posed Problems*. Winston, New York, 1977.

T. Trafalis and R. Gilbert. Robust support vector machines for classification and computational issues. *Optimization Methods and Software*, 22(1):187–198, February 2007.

A. W. van der Vaart and J. A. Wellner. *Weak Convergence and Empirical Processes*. Springer-Verlag, New York, 2000.

V. N. Vapnik and A. Chervonenkis. The necessary and sufficient conditions for consistency in the empirical risk minimization method. *Pattern Recognition and Image Analysis*, 1(3):284–305, 1991.

V. N. Vapnik and A. Lerner. Pattern recognition using generalized portrait method. *Automation and Remote Control*, 24:744–780, 1963.

M. Wainwright. Sharp thresholds for noisy and high-dimensional recovery of sparsity using ℓ_1-constrained quadratic programming (Lasso). *IEEE Transactions on Information Theory*, 55:2183–2202, 2009.

H. Xu and S. Mannor. Robustness and generalization. ArXiv 1005.2243, 2010a.

H. Xu and S. Mannor. Robustness and generalizability. In *Proceeding of the 23rd Annual Conference on Learning Theory*, pages 503–515, 2010b.

H. Xu, C. Caramanis, and S. Mannor. Robustness and regularization of support vector machines. *Journal of Machine Learning Research*, 10(July):1485–1510, 2009.

H. Xu, C. Caramanis, and S. Mannor. Robust regression and Lasso. *IEEE Transactions on Information Theory*, 56(7):3561–3574, 2010a.

H. Xu, C. Caramanis, and S. Mannor. A distributional interpretation to robust optimization. submitted, 2010b.

H. Xu, C. Caramanis, and S. Sanghavi. Robust PCA via outlier pursuit. To appear *Advances in Neural Information Processing Systems*, 2010c.

K. Zhou, J. Doyle, and K. Glover. *Robust and Optimal Control*. Prentice-Hall, 1996.

15   Improving First and Second-Order Methods by Modeling Uncertainty

Nicolas Le Roux   nicolas.le.roux@gmail.com
Microsoft Research Cambridge

Yoshua Bengio   yoshua.bengio@umontreal.ca
University of Montreal

Andrew Fitzgibbon   awf@microsoft.com
Microsoft Research Cambridge

Machine learning's goal is to provide algorithms able to deal with new situations or data. Thus, what matters is their performance on unseen test data. For that purpose, we have at our disposal data on which we can train our model. Much previous work aimed at taking account of the fact that the training data are only a sample from the distribution of interest, for instance, by optimizing training error plus a regularization term. We present here a new way to take that information into account, based on an estimator of the gradient of generalization error that takes this uncertainty into account through a weak prior. We show how taking into account the uncertainty across training data can yield faster and more stable convergence, even when using a first-order method. We then show that in spite of apparent similarities with second-order methods, taking this uncertainty into account is different and can be used in conjunction with an approximate Newton method to yield even faster convergence.

15.1   Introduction

Machine learning often looks like optimization: write down the likelihood of some training data under some model and find the model parameters which maximize that likelihood, or which minimize some divergence between the model and the data. In this context, conventional wisdom is that one should find in the optimization literature the state-of-the-art optimizer for one's problem, and use it.

However, this should not hide the fundamental difference between these two concepts: while optimization is about minimizing some error on the training data, it is the performance on the test data we care about in machine learning. Of course, by their very definition, test data are not available to us at training time, and thus we must use alternative techniques to prevent the model from focusing too much on the training data at the expense of generalization performance (a phenomenon known as overfitting), such as weight decay or limited model capacity.

The goal of this chapter is to prove that this misfit between training and test error can be dealt with by modifying the optimization procedure rather than the objective function itself, using a technique similar to the natural gradient. It is organized as follows: we start by exploring the differences between the optimization and the learning frameworks in section 15.2, before introducing a model of the gradients which modifies the search direction in section 15.3. Then, after exploring the similarities and differences between the covariance and the Hessian in section 15.4, we propose a modification of our model of the gradients, enabling us to use second-order information to find the optimal search direction, in section 15.5. Section 15.6 presents TONGA, an efficient algorithm to obtain these new search directions, which is tested in section 15.7.

15.2   Optimization Versus Learning

15.2.1   Optimization Methods

The goal of optimization is to minimize a function f, which we will assume to be twice differentiable and defined from a space E to $\mathbb{R}$, over E. This is a problem with a considerable literature (see Nocedal and Wright (2006), for instance). It is well known that second-order descent methods, which rely on the Hessian of f (or approximations thereof), enjoy much faster theoretical convergence than first-order methods (quadratic versus linear), in terms of number of updates. Such methods include the following: Newton, Gauss-Newton, Levenberg-Marquardt, and quasi-Newton (such as BFGS).

15.2.2   Online Learning

The learning framework for online learning differs slightly from the optimization one. The function f we wish to minimize (which we call the cost function) is defined as the expected value of a function $\mathcal{L}$ under a distribution p over the space E of possible inputs, that is,

$$f(\theta) = \int_{x \in E} \mathcal{L}(\theta, x) p(x) \, dx, \tag{15.1}$$

and we have access only to samples x_i drawn from p. If we have n samples, we can define a new function,

$$\widehat{f}(\theta) = \frac{1}{n} \sum_i \mathcal{L}(\theta, x_i) . \tag{15.2}$$

Let us call f the *test cost*, and $\widehat{f}$ the *training cost*. The x_i are the *training data*. As n goes to infinity, the difference between f and $\widehat{f}$ vanishes.

Bottou and Bousquet (2008) study the case where one has access to a potentially infinite amount of training data but only a finite amount of time. This setting, which they dub *large-scale learning*, calls for a tradeoff between the quality of the optimization for each data point and the number of data points treated. They show that

1. good optimization algorithms may be poor learning algorithms;

2. stochastic gradient descent enjoys a faster convergence rate than batch gradient descent;

3. introducing second-order information can win us a constant factor (the condition parameter).

Therefore, the choice lies between first- and second-order stochastic gradient descent, depending on the additional cost of taking second-order information into account and the condition parameter. Several authors have developed algorithms allowing for efficient use of this second-order information in a stochastic setting (Schraudolph et al., 2007; Bordes et al., 2009). However, we argue, all of these methods are derived from optimization methods without taking into account the particular nature of the learning problem.

More precisely, the gradient we would like to compute is the one of the true cost defined in (15.1). Differentiating both sides of this equation with respect to θ yields (assuming we can swap the integral and the derivative)

$$g^*(\theta_0) = \frac{\partial f}{\partial \theta}(\theta_0) = \int_{x \in E} \frac{\partial \mathcal{L}}{\partial \theta}(\theta_0, x_i) p(x) \, dx . \tag{15.3}$$

Similarly, differentiating both sides of (15.2) with respect to θ yields

$$g(\theta_0) = \frac{\partial \widehat{f}}{\partial \theta}(\theta_0) = \frac{1}{n} \sum_i \frac{\partial \mathcal{L}}{\partial \theta}(\theta_0, x_i) . \qquad (15.4)$$

Thus, for each parameter value θ_0, the true gradient g^* is the expectation of $\frac{\partial \mathcal{L}}{\partial \theta}(\theta_0, x)$ under p, and we are given only samples $g_i = \frac{\partial \mathcal{L}}{\partial \theta}(\theta_0, x_i)$ from this distribution. This bears a lot of resemblance to the standard setting of machine learning: given a set of samples (here, the g_i's from a distribution, one wishes to estimate interesting properties of that distribution (here, its expectation). We will thus proceed in the standard way, that is, we shall build a model of the gradients g_i and estimate its parameters. At this point, it is important to recall that our model is valid only for one value of θ_0, and thus needs to be reevaluated every time we move in parameter space. We shall discuss this issue further in section 15.3.1.

Also, note that the same reasoning may be applied to stochastic optimization. Modeling the distribution of the gradients will prevent us from focusing too much on the previously seen examples, which should enhance the final performance of the algorithm. This intuition will be proved in section 15.7.

15.3 Building a Model of the Gradients

We shall now describe in detail the model of gradients we use. Since our goal is to achieve fast treatment of incoming data, it must be simple enough to be estimated accurately with little computation. Additionally, a simpler model will regularize our estimate, making it more robust. We will thus use a Gaussian model, which we will see has the extra advantage of having an interpretation as an approximation to the central-limit theorem.

The only quantity we are interested in is the mean of this Gaussian, which is the true gradient g^*.

The likelihood term is

$$g_i | g^* \sim \mathcal{N}(g^*, C^*) \qquad (15.5)$$

where C^* is the true covariance of the gradients, that is,

$$C^* = \int_x \left(\frac{\partial \mathcal{L}(\theta, x)}{\partial \theta} - g^* \right) \left(\frac{\partial \mathcal{L}(\theta, x)}{\partial \theta} - g^* \right)^T p(x) \, dx. \qquad (15.6)$$

Indeed, according to the central-limit theorem, if the g_i's were averages of gradients over a minibatch, then their distribution would converge to a Gaussian as the minibatch size grows to infinity, thus yielding the correct likelihood. Of course, one must bear in mind that this becomes an approximation for finite sizes, and even more so when g_i is a single gradient.

Before receiving any gradient, we know neither the direction nor the amplitude of the true gradient. Hence it is reasonable to take a prior over g^* that is centered on 0 and has an isotropic covariance:

$$g^* \sim \mathcal{N}(0, \sigma^2 I) \,. \tag{15.7}$$

Assuming we receive n gradients $g_1, \ldots, g_n$ with average g, the posterior distribution over g^* is obtained by combining (15.7) and (15.5), yielding

$$g^*|(g_1, \ldots, g_n) \sim \mathcal{N}\left(\left[I + \frac{C^*}{n\sigma^2}\right]^{-1} g, \left[nC^{*-1} + \frac{I}{\sigma^2}\right]^{-1}\right) \,. \tag{15.8}$$

Even though we care about only the mean g^*, its posterior depends on the covariance matrix C^*, which should be estimated using data. Though the proper Bayesian way would be to place a prior on C^* (which could be inverse-Wishart to keep the model conjugate) and estimate the joint posterior over (g, C^*), we will set C^* to the empirical covariance matrix C of the gradients. In doing so, we will lose in robustness but gain in computational efficiency.

Replacing C^* with the empirical covariance C in (15.8), we have the estimator

$$g^*|(g_1, \ldots, g_n) \sim \mathcal{N}\left(\left[I + \frac{C}{n\sigma^2}\right]^{-1} g, \left[nC^{-1} + \frac{I}{\sigma^2}\right]^{-1}\right) , \tag{15.9}$$

with

$$C = \frac{1}{n} \sum_{i=1}^{n} (g_i - g) (g_i - g)^T \,. \tag{15.10}$$

Now that we have estimated the posterior distribution over g^*, we can estimate the expected decrease in $\mathcal{L}$ for a given update $\Delta\theta$, which is simply

$$E[\Delta\mathcal{L}] = (\Delta\theta)^T E[g^*] \,. \tag{15.11}$$

For a given norm of $\Delta\theta$, the optimal decrease is obtained for $\Delta\theta \propto -E[g^*]$, that is,

$$(\Delta\theta)_{\text{opt}} \propto -\left[I + \frac{C}{n\sigma^2}\right]^{-1} g \,. \tag{15.12}$$

This quantity, which we call *consensus gradient*, is reminiscent of the natural gradient of Amari (1998). In his work, Amari showed that in a neural network, the direction of steepest descent in the Riemannian manifold

defined by that network is

$$(\Delta\theta)_{\text{Amari}} \propto - \left[GG^T + \lambda I\right]^{-1} g \,, \qquad (15.13)$$

where G is the matrix containing one gradient per column (λI acts as a regularizer and the direction of steepest descent uses $\lambda = 0$). However, there are some important differences. First, here the covariance matrix C is centered and scaled, and thus not the Fisher information matrix, GG^T, as in Amari's work. Second, and perhaps more important, this formulation makes it obvious that the term $\frac{C}{n\sigma^2}$ acts here as a regularizer of the standard gradient direction, rather than defining a completely new direction based on another metric.

Let us pause a bit and analyze the behavior of such a direction. If there is a strong disagreement between gradients along a direction $\boldsymbol{d}$, the covariance will be large along this direction (that is, the value of $\boldsymbol{d}^T C \boldsymbol{d}$ will be large), which will reduce the update $\Delta\theta$ along $\boldsymbol{d}$. Thus, (15.12) naturally and gracefully deals with incoherent or noisy data. This is in stark contrast with, for example, outlier detectors which discard data points entirely. Moreover, the direction along which to shrink the updates is also learned and does not require any heuristic.

If, on the other hand, there is very little disagreement along a direction, then the step along this direction will be taken as usual. It is worth emphasizing that in this setting, the smallest eigenvalues of C are unimportant, as they will have very little effect on the final direction, as opposed to existing natural gradient methods, where they dominate the final update. As they are often harder to estimate correctly, those methods need to add a regularizer, which is unnecessary here. Once again, in our case the matrix C is the regularizer, not the identity matrix. soit clair. Le style me parat aussi lourd et redondant.

Figure 15.1 shows an example of consensus gradient and one of mean gradient directions, with varying amounts of disagreement among gradients.

Figure 15.1: **Left**: when gradients (solid lines) agree, the consensus gradient direction (dashed line) is indistinguishable from the mean gradient (dashed-dotted line). **Right**: when gradients disagree, the consensus gradient shrinks in direction of high variance while leaving the others untouched.

The dot product of the empirical gradient and the consensus gradient direction of (15.12) is $g^T \left[I + \frac{C}{n\sigma^2} \right]^{-1} g$, which is always positive since $\left(I + \frac{C}{n\sigma^2} \right)$ is a positive definite matrix. Thus, though the consensus gradient direction is a modification of the original direction, they will never be in disagreement (that is, if the magnitude of the update is small enough, the training cost is guaranteed to decrease as well).

Moreover, when the number n of gradients goes to infinity, the optimal direction converges to g: this makes sense, since the modification to standard gradient descent arises from the uncertainty around the particular set of samples chosen, which becomes nonexistent in the case of infinite sample size. However, as we recover a standard optimization problem, we may be disappointed by the use of a first-order method, which, as mentioned earlier, is theoretically slower than second-order ones.

15.3.1 Setting a Zero-Centered Prior at Each Timestep

Before moving on to the second-order version of our consensus gradient algorithm, we will briefly comment on the choice of our prior at each step. Indeed, (15.9) has been obtained using the zero-centered Gaussian prior defined in (15.7). Except for the first update, one may wonder why we would use such a distribution rather than the posterior distribution at the previous timestep as our prior. There are two reasons for that. The first one is that whenever we update the parameters of our model, the distribution over the gradients changes. If the function to optimize were truly quadratic, we could quantify the change in gradient exactly using the Hessian. Unfortunately, this is not the case, and if this path is explored (as we believe it should be), it will involve approximations of the posterior. The second reason is computational. Even if we were able to follow the mean of the posterior exactly, the resulting distributions would become more and more complex over time (while still being Gaussians, their means and covariances would depend on a sum of covariance matrices). Thus, while acknowledging that using the prior of (15.7) at every timestep is a suboptimal strategy that future work might enhance, we believe that it is very appealing because of the simplicity of the algorithm.

15.4 The Relative Roles of the Covariance and the Hessian

In its original formulation, the natural gradient algorithm has often been considered as approximation to the Newton method. Indeed, their updates look very similar ($d = (GG^T)^{-1}g$ for the natural gradient and $d = H^{-1}g$ for the Newton method), and there are several reasons to believe that the covariance (either centered, that is, C, or uncentered, that is, GG^T) and H have analogous properties. However, as we will see, they encode completely different kinds of information. From there, it seems natural to exploit both, yielding an algorithm combining their advantages.

15.4.1 Similarities Between C and H

Let us first focus on the similarities between the covariance matrix and the Hessian.

15.4.1.1 *Maximum Likelihood*

Let us assume that we are training a density model by minimizing the negative log-likelihood. The cost function f_{nll} is defined by

$$f_{\mathrm{nll}}(\theta) = - \int_x \log[L(\theta, x)]p(x)\, dx \; . \tag{15.14}$$

Note that this L is related to the $\mathcal{L}$ used in section 15.2.2 through $\mathcal{L} = -\log L$, but with the constraint that L is a distribution. Let us consider the case where there is a parameter vector θ *such that our model is perfect* (where $p(x) = L(\theta, x)$) and that we are at this θ. Then the covariance matrix of the gradients at that point is equal to the Hessian of f_{nll}. In the general case, this equality does not hold.

15.4.1.2 *Gauss-Newton*

Gauss-Newton is an approximation to the Newton method when f can be written as a sum of squared residuals:

$$f(\theta) = \frac{1}{2} \sum_i f_i(\theta)^2. \tag{15.15}$$

Computing the Hessian of f yields

$$\frac{\partial^2 f(\theta)}{\partial \theta^2} = \sum_i f_i(\theta)\frac{\partial^2 f_i}{\partial \theta^2} + \sum_i \frac{\partial f_i}{\partial \theta}\frac{\partial f_i}{\partial \theta}^T \; . \tag{15.16}$$

If the f_i get close to 0 (relative to their gradient), the first term may be ignored, yielding the following approximation to the Hessian:

$$H \approx \sum_i \frac{\partial f_i}{\partial \theta}\frac{\partial f_i}{\partial \theta}^T \; . \tag{15.17}$$

One, however, must be aware of the following:

- this approximation is interesting only when the f_i are *residuals* (that is, when the approximation is valid close to the optimum);
- the gradients involved are those of f_i and not of f_i^2;
- the term on the right-hand side is the *uncentered* covariance of these gradients.

In order to compare the result of (15.17) to the natural gradient, we will assume that the sum in (15.15) is over data points, that is,

$$f(\theta) = \frac{1}{2} \sum_i f_i(\theta)^2 = \frac{1}{N} \sum_i \mathcal{L}(\theta, x_i) \tag{15.18}$$

with the cost for each data point being

$$\mathcal{L}(\theta, x_i) = \frac{N}{2} f_i(\theta)^2 . \tag{15.19}$$

The gradient of this cost with respect to θ is

$$g_i = \frac{\partial \mathcal{L}(\theta, x_i)}{\partial \theta} = N f_i(\theta) \frac{\partial f_i(\theta)}{\partial \theta} . \tag{15.20}$$

At the optimum (where the average of the gradients is zero and the centered and uncentered covariance matrices are equal), the covariance matrix of the g_i's is

$$C = \sum_i g_i g_i^T = N^2 \sum_i f_i(\theta)^2 \frac{\partial f_i}{\partial \theta} \frac{\partial f_i}{\partial \theta}^T , \tag{15.21}$$

which is a weighted sum of the terms involved in (15.17). Thus the natural gradient and the Gauss-Newton approximation, while related, are different quantities and (as we will show) have very different properties.

15.4.2   Differences Between C and H

Remember what the Hessian is: a measure of the change in gradient when we move in *parameter space*. In other words, the Hessian helps to answer the following question: *If I were at a slightly different position in parameter space, how different would the gradient be?* It is a quantity defined for any (twice differentiable) function.

On the other hand, the covariance matrix of the gradients captures the uncertainty around this particular choice of training data, that is, the change in gradient when we move in *input space*. In other words, the covariance helps us to answer the following question: *If I had slightly different training data, how different would the gradient be?* This quantity makes only sense when there are training data.

Whereas the Hessian seems naturally suited to optimization problems (it allows us to be less shortsighted when minimizing a function), the covariance matrix unleashes its power in the learning setting, where we are given only a subset of the data. We do not claim that there are no numerical similarities between them, and indeed the experiments hint at differing and complementary effects, so we really wish to clarify how they differ.

From this observation, it seems natural to combine these two matrices.

15.5   A Second-Order Model of the Gradients

In our first model of the gradients, in section 15.3, we did not assume any particular form of the function $\mathcal{L}$ to minimize. In the Newton method, however, one assumes that the cost function is locally quadratic, that is,

$$\mathcal{L}(\theta) \approx f(\theta) = \frac{1}{2}(\theta - \theta^*)^T H(\theta - \theta^*) \tag{15.22}$$

for some value of θ^*.

The derivative of this cost is

$$g^*(\theta) = \frac{\partial f(\theta)}{\partial \theta} = H(\theta - \theta^*) \,. \tag{15.23}$$

We can see that in the context of a quadratic function, the isotropic prior over g^* proposed in (15.7) is erroneous, as g^* is clearly influenced by H. We shall, rather, consider an isotropic Gaussian prior on the quantity $\theta - \theta^*$, as we do not have any information about the position of θ relative to θ^*. The resulting prior distribution over g^* is

$$g^* \sim \mathcal{N}\left(0, \sigma^2 H^2\right), \tag{15.24}$$

where we omit the dependence on θ to keep the notation uncluttered. In a fashion similar to section 15.3, we will suppose that we are given only a finite training set composed of n data points x_i with associated gradients g_i. The empirical gradient g is the mean of the g_i's. Using the central limit theorem, we again have

$$g|g^* \sim \mathcal{N}\left(g, \frac{C^*}{n}\right) \tag{15.25}$$

where C^* is the true covariance of the gradients, which we will once again replace with the empirical covariance C. Therefore, the posterior distribution over g is

$$g^*|g \sim \mathcal{N}\left(\left[I + \frac{CH^{-2}}{n\sigma^2}\right]g, \left[\frac{H^{-2}}{\sigma^2} + nC^{-1}\right]^{-1}\right). \tag{15.26}$$

Since the function $\mathcal{L}$ is locally quadratic, we wish to move in the direction $H^{-1}g$. This direction follows the Gaussian distribution

$$H^{-1}g|g \sim \mathcal{N}\left(\left[I + \frac{H^{-1}CH^{-1}}{n\sigma^2}\right]^{-1} H^{-1}g, \left[\frac{I}{\sigma^2} + nHC^{-1}H\right]^{-1}\right). \tag{15.27}$$

Since the mean of the Gaussian in (15.27) appears complicated, we shall explain it. Let us write d_i for the Newton directions:

$$d_i = H^{-1}g_i. \tag{15.28}$$

Since C is the covariance matrix of the gradients g_i, $H^{-1}CH^{-1} = C^H$ is the covariance matrix of the d_i's. We can therefore rewrite

$$H^{-1}g|g \sim \mathcal{N}\left(\left[I + \frac{C^H}{n\sigma^2}\right]^{-1} d, \left[\frac{I}{\sigma^2} + n(C^H)^{-1}\right]^{-1}\right) \tag{15.29}$$

where d is the average of the Newton directions, that is, $d = H^{-1}g$. The direction which maximizes the expected gain is thus

$$\Delta\theta \propto -\left[I + \frac{C^H}{n\sigma^2}\right]^{-1} d. \tag{15.30}$$

This formula is exactly the consensus gradient (15.12), but on the Newton directions. This makes perfect sense, as the Newton method is the standard gradient descent on a space linearly reparameterized by $H^{0.5}$. Here, the direction is the one obtained after having computed the consensus gradient in the same linearly reparameterized space.

From a computational perspective, this simple combination is excellent news. It means that one may choose his or her favorite second-order gradient descent method to compute the Newton directions, and then his or her favorite consensus gradient algorithm to apply to these Newton directions, to yield an algorithm combining the advantages of both methods.

As a side note, one can see that as the number n of data points used to compute the mean increases, the prior vanishes and the posterior distribution concentrates around the empirical Newton direction. This is in contrast with the method of section 15.3, which converged to the first-order gradient descent algorithm.

15.6 An Efficient Implementation of Online Consensus Gradient: TONGA

So far, we have

- provided a justification for the consensus gradient as a means of dealing with the uncertainty arising from having only a finite number of samples in our dataset;

- explored the similarities and differences between the covariance matrix C and the Hessian H;

- shown how the information in these two matrices could be combined to yield an efficient algorithm, both from an optimization and from a learning point of view.

However, these techniques require matrix inversions, which makes them unsuitable for practical cases, where the number of model parameters and of training data may be very large. Also, since our main focus is online learning, we wish to be able to update our parameters after each example (stochastic), or each small group of examples (minibatch), as recommended in Bottou and Bousquet (2008).

Section 15.6.1 will uncover a set of optimizations and approximations which renders possible fast online natural or consensus gradient algorithms: TONGA. This algorithm will provide the basis for the second-order version using the Hessian.

15.6.1 Computing a Low-Rank Approximation of the Covariance Matrix

In a model with P parameters, the covariance C of the gradients over n data points takes $O(nP^2)$ to compute and has an $O(P^2)$ memory storage requirement. Computing its first k eigenvectors is in $O(kP^2)$. When P is large, none of these operations is feasible. This section will thus introduce a way of finding the first k eigenvectors of the covariance matrix without ever storing it.

For the moment, we will assume that the centered covariance matrix may be written in the form $C = GG^T$ for some matrix G. The proof that such a factorization is possible and the explicit formula for G will appear in section 15.6.2. We will assume that G has n columns (and P rows for C to have the correct size).

Writing G in terms of its compact SVD, we get

$$G = U_G \Sigma_G V_G^T, \tag{15.31}$$

where (assuming we have $n < P$) U_G is of size $P \times n$, and Σ_G and V_G are of size $n \times n$. With this notation, the eigenvectors of C associated with non-zero eigenvalues are the columns of U_G and the associated eigenvalues are the diagonal elements of Σ_G^2.

Let us now consider the matrix

$$D = G^T G \, . \tag{15.32}$$

This is an $n \times n$ matrix whose eigenvectors are the columns of V_G and whose eigenvalues are the same as those of C. Left-multiplying those eigenvectors by G and right-multiplying them by Σ_V^{-1}, we get

$$G V_G \Sigma_G^{-1} = U_G \, . \tag{15.33}$$

Thus, we can retrieve the first k eigenvectors and eigenvalues of C by computing D (for a cost of $O(Pn^2)$), extracting its first k eigenvectors (for a cost of $O(kn^2)$), and then performing the matrix multiplications (for a cost of $O(Pn^2)$). Therefore, if n is much smaller than P, this method is much faster than computing C and its eigenvectors directly ($O(Pn^2)$, instead of $O(nP^2)$).

Another advantage is that it is never required to store or even compute C, but only to have access to the matrix G. Section 15.6.2 will show how to get this matrix G efficiently whenever a new data point comes in.

15.6.2 A Fast Update of the Covariance Matrix

Since efficiency is our main goal, we need a fast way to update the covariance matrix of the data points as they arrive. Also, we need to satisfy two constraints. First, the covariance needs to be estimated over many data points. Second, as it will change during the optimization, we need to progressively reduce the contribution of the older data points and replace it with the contribution of the newer ones. For that purpose, we shall use exponentially moving mean μ_n and covariance C_n:

$$
\begin{align}
\mu_1 &= g_1 \tag{15.34}\\
C_1 &= 0 \tag{15.35}\\
\mu_n &= \gamma \mu_{n-1} + (1 - \gamma) g_n \tag{15.36}\\
C_n &= \gamma C_{n-1} + \gamma (1 - \gamma)(g_n - \mu_{n-1})(g_n - \mu_{n-1})^T \tag{15.37}
\end{align}
$$

where g_i is the gradient obtained at time step i and γ is the discount factor. The closer γ is to 1, the longer an example seen at time t will influence the means and covariance estimated at later times.

Thus, since we wish to keep a factorization of C under the form GG^T, whenever a new gradient g_n comes in, we simply have to

1. multiply G by $\sqrt{\gamma}$
2. append the column $(g_n - \mu_{n-1})\sqrt{\gamma(1-\gamma)}$ to G.

15.6.3   Finding the Consensus Gradient Direction Between Two Updates

In section 15.6.1, we have showed that computing the first k eigenvectors of the matrix $C = GG^T$ when G has n columns and is in $O(Pn^2)$. We could thus use the following strategy:

1. compute the first k eigenvectors of C,

2. compute the consensus gradient update using the eigendecomposition of C,

3. write the low-rank approximation under the form UU^T (U then being the matrix of unnormalized eigenvectors),

4. update the matrix U when a new data point arrives, following section 15.6.2, where G plays the role of U,

5. recompute the first k eigenvectors of the new C for a cost of $O(P(k+1)^2)$,

6. iterate from 2.

One can see that the cost of this algorithm is $O(Pk^2)$ for every new gradient, which is approximately k^2 slower than standard gradient descent. The idea will thus be to update this covariance matrix as new data points arrive, but not to recompute the eigendecomposition every time. Instead, we will add data points until there are $k + B$ vectors in the matrix G (with B a hyperparameter), at which point we will recompute the eigendecomposition of this new covariance matrix.

There is a problem, however. While it is easy to compute the consensus gradient direction when one has access to the eigendecomposition of C, this will not be the case when several data points have been added. Luckily, the computation remains tractable, as we will see. b steps after the last eigendecomposition, the matrix G may be written as

$$G = [K_0 U \quad K_1(g_1 - \mu_0) \ldots K_b(g_b - \mu_{b-1})]$$

(the constants $K_0, \ldots, K_b$ stem from the $\sqrt{\gamma}$ and $\sqrt{\gamma(1-\gamma)}$ factors of section 15.6.2). Since $C = GG^T$, and in order to compute the naturalized gradient $d = (I + C/[n\sigma^2])^{-1}g_b$, we wish to find the direction d such that

$$\left(I + \frac{GG^T}{n\sigma^2}\right) d = g_b . \tag{15.38}$$

We will assume that d is of the form $d = Gx + \lambda \mu_{b-1}$ for some vector x and some value of λ. With $y = [0 \ \dots \ 0 \ (1/K_b)]^T$, we have $g_b = Gy + \mu_{b-1}$, and (15.38) thus becomes

$$Gx + \lambda \mu_{b-1} + \frac{GG^T Gx + \lambda GG^T \mu_{b-1}}{n\sigma^2} = Gy + \mu_{b-1} \ .$$

Using $\lambda = 1$ and moving the fraction to the right-hand side, we get

$$Gx \ = \ Gy - \frac{GG^T Gx + \lambda GG^T \mu_{b-1}}{n\sigma^2} \tag{15.39}$$

$$x \ = \ \left(I + \frac{G^T G}{n\sigma^2} \right)^{-1} \left(y - \frac{G^T \mu_{b-1}}{n\sigma^2} \right) \tag{15.40}$$

(assuming G is of full rank), yielding

$$d = G \left(I + \frac{G^T G}{n\sigma^2} \right)^{-1} \left(y - \frac{G^T \mu_{b-1}}{n\sigma^2} \right) + \mu_{b-1} \ . \tag{15.41}$$

Since G is of size $P \times (k + b)$, computing d costs $O((k + b)^3 + P(k + B)) = O(P(k + B))$, since we will limit ourselves to the setting where the rank of the covariance matrix is much less than the square root of the number of parameters.

15.6.4   Analysis of the Computational Cost

We will now briefly analyze the average computational cost of a gradient update where there are P parameters. We will assume that the gradients are computed over minibatches of size m:

1. every B steps, we compute the first k eigenvectors of the covariance matrix of $k + B$ data points for a total cost of $O(P(k + B)^2)$ (see section 15.6.1)

2. every step, we compute the consensus gradient direction for a total cost of $O(P(k + B))$ (see section 15.6.3)

3. computing the average gradient over a minibatch costs $O(Pm)$ at every step.

The average cost per update is thus $O\left(P \left[m + k + B + \frac{(k+B)^2}{B} \right] \right)$ as opposed to $O(Pm)$ for standard minibatch gradient descent. Thus, if we keep $(k + B)$ close to m, the cost of each iteration will be of the same order of magnitude as the standard gradient descent.

15.6.5 Block-Diagonal Online Consensus Gradient for Neural Networks

We now have a strategy to compute the consensus gradient direction using a low-rank approximation of the covariance matrix (with the rank varying between k and $k+B$). The question remains as to which value of k provides a reasonable approximation. Unfortunately, experiments showed that, in general, a high value of k (around 200 for $P = 2000$) was necessary for d to be a meaningful modification of the original gradient direction. In this setting, provided m, the minibatch size, is small (between 5 and 10), each update is at least 20 times slower than the standard gradient descent, and this extra computational cost cannot be made up by better search directions.

One might thus wonder if there are better approximations of the covariance matrix C than computing its first k eigenvectors. Collobert (2004) showed that the Hessian of a neural network with one hidden layer trained with the cross-entropy cost converges to a block-diagonal matrix during optimization. These blocks are composed of the weights linking all the hidden units to one output unit and all the input units to one hidden unit (*fan-in*). Since we listed some of the numerical similarities between the Hessian and the covariance, it may be useful to investigate the use of such a block structure for the covariance estimator. We will thus use a block-diagonal approximation of the covariance matrix. Instead of computing the first k eigenvectors of the entire covariance matrix, we will compute the first k eigenvectors of each block. Some remarks are worth making on that point:

- the rank of the approximation is not k but $k \times$ (number of blocks), which is much higher;

- all the terms outside of these blocks are set to 0. Thus, this approximation will be better only if these elements are actually negligible in the original covariance matrix;

- one may pick a different value of k for each block, depending on its size or the knowledge one has about the problem.

Figure 15.2 shows the correlation between the standard stochastic gradients of the parameters of a $16 - 50 - 26$ neural network. The first blocks represent the weights going from the input units to each hidden unit (thus 50 blocks of size 17, bias included), and the following blocks represent the weights going from the hidden units to each output unit (26 blocks of size 51). One can see that the block-diagonal approximation is reasonable. In the matrices shown in figure 15.2, which are of size 2176, a value of $k = 5$ yields an approximation of rank 380.

Another way of verifying the validity of our block-diagonal assumption

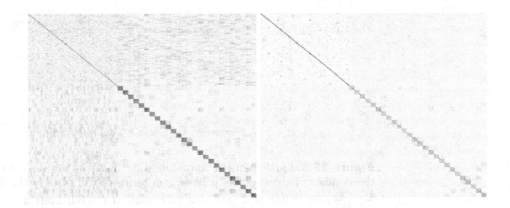

Figure 15.2: Absolute value of correlation between the standard stochastic gradients after one epoch in a neural network with 16 input units, 50 hidden units and 26 output units when following stochastic gradient directions (left) and consensus gradient directions (right). The first blocks in the diagonal are for input to hidden weights (per hidden unit), and the larger ones that follow are for hidden output weights (per output unit), showing a strong within-block correlation. One can see that the off-block terms are not zero, but still are much smaller than the terms in the block. Also, following natural directions helped in making the covariance more block-diagonal, though the reason behind it is unknown.

is to compute the error induced by our low-rank approximations, with or without this assumption. Figure 15.3 shows the relative approximation error of the covariance matrix as a ratio of Frobenius norms $\frac{\|C - \bar{C}\|_F^2}{\|C\|_F^2}$ for different types of approximations $\bar{C}$ (full or block-diagonal). We can first notice that approximating only the blocks yields a ratio of .35 (in comparison, taking only the diagonal of C yields a ratio of .80), even though we considered only $82,076$ out of the $4,734,976$ elements of the matrix (1.73 percent of the total). This ratio is almost obtained with $k = 6$. We can also notice that for $k < 30$, the block-diagonal approximation is much better (in terms of the Frobenius norm) than the full approximation, which proves its effectiveness in the case of neural networks. Yet this approximation also readily applies to any mixture algorithm where we can assume some form of decoupling between the components.

Thus in all our experiments, we used a value of $k = 5$, which allowed us to keep a cost per iteration of the same order of magnitude as standard gradient descent.

15.7 Experiments

In our experiments we wish to validate the two claims we have made so far:

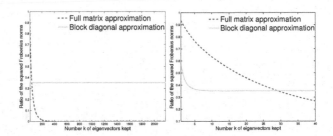

Figure 15.3: Quality of the approximation $\bar{C}$ of the covariance C, depending on the number of eigenvectors kept (k), in terms of the ratio of the Frobenius norms $\frac{\|C-\bar{C}\|_F^2}{\|C\|_F^2}$, for different types of approximation $\bar{C}$ (full matrix or block-diagonal). On the right we zoom on smaller values of k where the full matrix low-rank approximation overtakes the block-diagonal approximation.

1. that taking the uncertainty into account will speed up learning;

2. that C and H encode different pieces of information and that combining them will lead to even faster convergence.

The first set of experiments will thus compare TONGA with standard stochastic gradient descent, whereas the second will compare an approximate Newton method with the second-order TONGA, which we call *Natural-Newton*.

15.7.1   Datasets

Several datasets, architectures, and losses were used in our experiments.

15.7.1.1   *Experiments with TONGA*

We tried TONGA on two different datasets:

1. the MNIST digits dataset consists of $50,000$ training samples, $10,000$ validation samples, and $10,000$ test samples, each one composed of 784 pixels. There are 10 classes (one for every digit)

2. the UCI USPS dataset consists of 9298 samples (broken into 6291 for training, 1000 for validation, and 2007 for the official test set), each one composed of 256 pixels. There are 10 different classes (one for every digit).

In both cases we minimized the negative log-likelihood on the training set, using a neural network with one hidden layer. The block-diagonal approximation of section 15.6.5 was used for TONGA, and no second-order information was used.

15.7.1.2   *Experiments with Natural-Newton (Second-Order TONGA)*

Whereas the goal of the experiments with TONGA was to determine if it was possible to use the information contained in the covariance matrix in an efficient manner, the experiments with Natural-Newton aim at exploring the differences between C and H.

As mentioned in section 15.5, one needs to use a second-order gradient descent method to compute the Newton directions. We chose to use the SGD-QN algorithm (Bordes et al., 2009), since it had recently won the Wild Track competition at the Pascal Large Scale Learning Challenge, on the same datasets it was used on: Alpha, Gamma, Delta, Epsilon, Zeta, and Face.

Labels were available only for the training examples of the challenge. We therefore split these examples into several sets:

- the first 100K (1M for the Face dataset) examples constituted our training set;
- the last 100K (1M for the Face dataset) examples constituted our test set.

The architecture used was a linear SVM. We did not change the hyperparameters of the SGD-QN algorithm; the interested reader may find them in the original paper.

Since this method uses a diagonal approximation to the Hessian, we decided to use a diagonal approximation to the covariance matrix. Though this was not required, and we could have used a low-rank covariance matrix, using a diagonal approximation shows the improvements over the original method that one can obtain with little extra effort. Thus, though (15.30) was used, none of the tricks presented in section 15.6 were necessary, except for the exponentially moving covariance matrix.

15.7.2   Experimental Details for Natural-Newton

15.7.2.1   *Frequency of Updates*

The covariance matrix of the gradients changes very slowly. Therefore, one does not need to update it as often as the Hessian approximation. In the SGD-QN algorithm, the authors introduce a counter *skip* which specifies how many gradient updates are done before the approximation to the Hessian is updated. We introduce an additional variable *skipC*, which specifies how many Hessian approximation updates are done before updating the covariance approximation. The total number of gradient updates between two covariance approximation updates is therefore $skip \cdot skipC$.

Experiments using the validation set showed that using values of *skipC* lower than 8 did not yield any improvement while increasing the cost of each update. We therefore used this value in all our experiments. This allows us to use the information contained in the covariance with very little computation overhead.

15.7.2.2 Limiting the Influence of the Covariance

Equation (15.29) tells us that the direction to follow is

$$\left[I + \frac{C^H}{n\sigma^2}\right]^{-1}\widehat{d}. \tag{15.42}$$

The only unknown in this formula is σ^2, which is the variance of our Gaussian prior on $\theta - \theta^*$. To avoid having to set this quantity by hand at every time step, we will devise a heuristic to find a sensible value of σ^2. While this will lack the purity of a full Bayesian treatment, it will allow us to reduce the number of parameters to be set by hand, which we think is a valuable feature of any gradient descent algorithm.

If we knew the distance from our position in parameter space, θ, to the optimal solution, θ^*, then the optimal value for σ^2 would be $\|\theta - \theta^*\|^2$. Of course, this information is not available to us. However, if the function to be optimized were truly quadratic, the squared norm of the Newton direction would be exactly $\|\theta - \theta^*\|^2$. We shall therefore replace σ^2 with the squared norm of the last-computed Newton direction. Since this estimate may be too noisy, we will replace it with the squared norm of the running average of the Newton directions, that is, $\|\mu_n\|^2$.

Even then, however, we may still get undesirable variations. We shall therefore adopt a conservative strategy: we will set an upper bound on the correction to the Newton method brought by (15.42). More precisely, we will bound the eigenvalues of $\frac{C^H}{n\|\mu_n\|^2}$ by a positive number B_C. The parameter update then becomes

$$\theta_n - \theta_{n-1} = -\left[I + \min\left(B_C, \frac{C^H}{n\|\mu_n\|^2}\right)\right]^{-1}H^{-1}g_n \tag{15.43}$$

where B_C is a scalar hyperparameter and $\min(B_C, M)$ is defined for symmetric matrices M with eigenvectors $u_1, \ldots, u_n$ and eigenvalues $\lambda_1, \ldots, \lambda_n$ as

$$\min(B_C, M) = \sum_{i=1}^{n} \min(B_C, \lambda_i)u_i u_i^T \tag{15.44}$$

(we bound each eigenvalue of M by B_C). If we set $B_C = 0$, we recover the standard Newton method. This modification transforms the algorithm in a conservative way, trading off potential gains brought by the covariance matrix for guarantees that the parameter update will not differ too much from the Newton direction.

In our experiments, the last 50K (500K for the Face dataset) examples of the training set were used as validation examples to tune the bound B_C defined in (15.43).

The pseudocode for the full algorithm, which we call *Natural-Newton*, is shown in algorithm 15.1.

Algorithm 15.1 Pseudocode of the Natural-Newton algorithm

Require: : skip (number of gradient updates between Hessian updates)
Require: : skipC (number of Hessian updates between covariance updates). Default value is skipC = 8.
Require: : θ_0 (the original set of parameters)
Require: : γ (the discount factor). Default value is 0.995.
Require: : T (the total number of epochs)
Require: : t_0
Require: : λ (the weight decay)
Require: : B_C the bound on the eigenvalues of the covariance matrix. Default value is $B_C = 2$.

1:   t = 0, count = skip, countC = skipC
2:   $H^- = \lambda \, \mathbf{I}, \mathbf{D} = \mathbf{I}$
3:   $\mu_0 = 0$ (the running mean vector), $C_0^H = 0$ (the running covariance matrix)
4:   **while** $t \neq T$ **do**
5:      $g_t \leftarrow \frac{\partial \mathcal{L}(\theta_t, x_t, y_t)}{\partial \theta_t}$
6:      $\theta_{t+1} \leftarrow \theta_t - (t + t_0)^{-1}\mathbf{D}H^- g_t$
7:      **if** count == 0 **then**
8:         count $\leftarrow$ skip
9:         Update H^-, the approximate inverse Hessian computed by SGD-QN.
10:        **if** countC == 0 **then**
11:           countC $\leftarrow$ skipC
12:           $\mu_t \leftarrow \gamma\mu_{t-1} + (1 - \gamma)d_t$
13:           $C_t^H \leftarrow \gamma C_{t-1}^H + \gamma(1 - \gamma)(d_t - \mu_{t-1})(d_t - \mu_{t-1})^T$
14:           $\mathbf{D} = \left(I + \frac{min(B_C, C_{t+1}^H)}{N \cdot \|\mu_{t+1}\|^2}\right)^{-1}$
15:        **else**
16:           countC $\leftarrow$ countC - 1
17:        **end if**
18:      **else**
19:         count $\leftarrow$ count - 1
20:      **end if**
21:   **end while**

15.7.2.3   Parameter Tuning

In all the experiments γ has been set to 0.995, as in TONGA. Again, to test the sensitivity of the algorithm to this parameter, we tried other values (0.999, 0.992, 0.99, and 0.9) without noticing any significant difference in validation errors.

We optimized the bound on the covariance (section 15.7.2.2) based on validation set error. The best value was chosen for the test set, but we found that a value of 2 yielded near-optimal results on all datasets, the difference between $B = 1$, $B = 2$, and $B = 5$ being minimal, as shown in figure 15.6 in the case of the Alpha dataset.

15.7.3   Results

15.7.3.1   *TONGA*

We performed a small number of experiments with TONGA's low-rank approximation of the full covariance matrix, keeping the overhead of the consensus gradient small (i.e., limiting the rank of the approximation). Regrettably, TONGA performed only as well as stochastic gradient descent, while being rather sensitive to the hyperparameter values. The following experiments, on the other hand, use TONGA with the block-diagonal approximation and yield impressive results. We believe this is a reflection of the phenomenon illustrated in figure 15.3 (right): the block-diagonal approximation makes for a very cost-effective approximation of the covariance matrix. All the experiments have been done by optimizing hyperparameters on a validation set (not shown here) and selecting the best set of hyperparameters for testing, trying to keep the overhead small due to natural gradient calculations.

One could worry about the number of hyperparameters of TONGA. However, default values of $k = 5$, $B = 50$, and $\gamma = .995$ yielded good results in every experiment.

Figure 15.4 shows that in terms of training CPU time (which includes the overhead due to TONGA), TONGA allows much faster convergence in training NLL, as well as in testing classification error and NLL than ordinary stochastic and minibatch gradient descent on this task. Also note that the minibatch stochastic gradient is able to profit from matrix-matrix multiplications, but this advantage is seen mainly in training classification error.

Note that the gain obtained on the USPS dataset is much slimmer. One possibility is that since the training set is much smaller, the independence

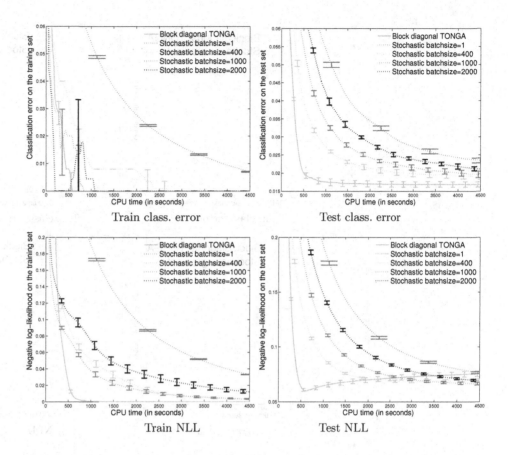

Figure 15.4: Comparison between stochastic gradient (with different minibatch sizes) and TONGA on the MNIST dataset, in terms of training (50,000 examples) and test (10,000 examples) classification error and negative log-likelihood (NLL). The mean and standard error have been computed using nine different initializations.

assumption used to obtain (15.9) becomes invalid.

Finally, though we expected an improvement only on the convergence speed of the test error, the training error decreased faster when using TONGA. This may be due to the stochastic nature of the optimization, where using the covariance prevented disagreeing gradients from having too much influence and ultimately slowing down the optimization.

15.7.3.2   *Natural-Newton*

Natural-Newton exhibited various behaviors on the datasets it was tried on:

- Natural-Newton never performs worse than SGD-QN and always better than TONGA. Using a large value of *skipC* ensures that the overhead of

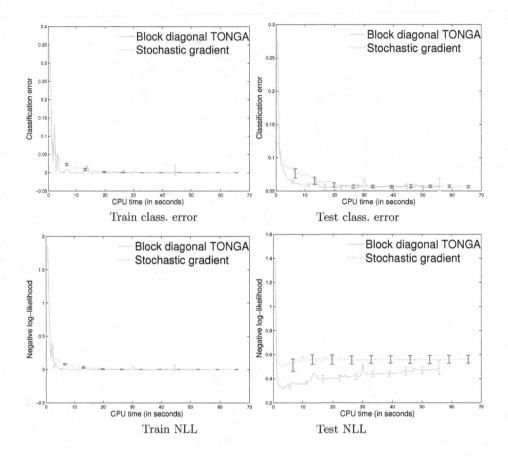

Figure 15.5: Comparison of stochastic gradient and TONGA on the USPS dataset, in terms of training (6291 examples) and test (2007 examples) classification error and negative log-likelihood (NLL). The mean and standard error were computed using nine different initializations.

using the covariance matrix is negligible.

- On the Alpha dataset, using the information contained in the covariance resulted in significantly faster convergence, with or without second-order information.

- On the Epsilon, Zeta, and Face datasets, using the covariance information stabilized the results while yielding the same convergence speed. This is in accordance with the use of the covariance, which reduces the influence of directions where gradients vary wildly.

- On the Gamma and the Delta datasets, using the covariance information helped a lot when the Hessian was not used, and provided no improvement otherwise.

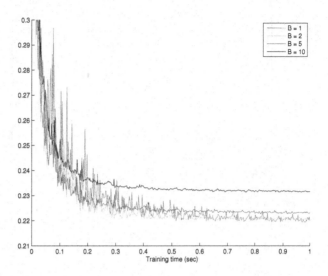

Figure 15.6: Validation error versus time on the Alpha dataset, for various values of B.

15.8 Conclusion

A lot of effort has been put into designing efficient online optimization algorithms, with great results. Most of these algorithms rely on some approximation to the Hessian or to the covariance matrix of the gradients. While the latter is commonly believed to be an approximation of the former, we showed that they encode very different kinds of information. Based on this, we proposed a way of combining information contained in the Hessian and in the covariance matrix of the gradients.

Experiments showed that on most datasets, our method offered either faster convergence or increased robustness, compared with the original algorithm. Furthermore, the second-order version of our algorithm never performed worse than the Newton algorithm it was built upon.

Moreover, our algorithm is able to use any existing second-order algorithm as base method. Therefore, while we used SGD-QN for our experiments, one may pick any algorithm best suited for a given task.

We hope to have shown two things. First, the covariance matrix of the gradients is usefully viewed not as an approximation to the Hessian, but as a source of additional information about the problem, for typical machine learning objective functions. Second, it is possible with little extra effort to use this information in addition to that provided by the Hessian matrix, in some cases yielding faster or more robust convergence.

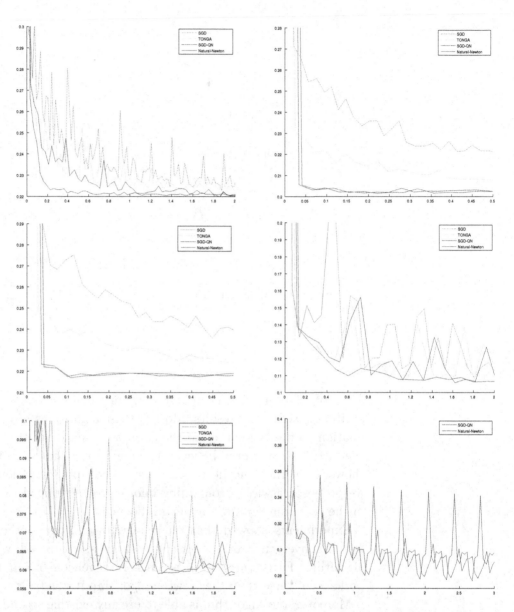

Figure 15.7: Test error versus time on the Alpha (top left), the Gamma (top right), the Delta (middle left), the Epsilon (middle right), the Zeta (bottom left), and the Face (bottom right) datasets.

Despite all these successes, we believe that these algorithms may be improved in several ways, whether it is by retaining some of the information contained in the posterior distribution between timesteps or in the selection of the parameter σ^2.

15.9   References

S. Amari. Natural gradient works efficiently in learning. *Neural Computation*, 10 (2):251–276, 1998.

A. Bordes, L. Bottou, and P. Gallinari. SGD-QN: Careful quasi-newton stochastic gradient descent. *Journal of Machine Learning Research*, 10:1737–1754, July 2009.

L. Bottou and O. Bousquet. The tradeoffs of large scale learning. In J. C. Platt, D. Koller, Y. Singer, and S. Roweis, editors, *Advances in Neural Information Processing Systems 20*, pages 161–168. MIT Press, 2008.

R. Collobert. *Large Scale Machine Learning*. PhD thesis, Université de Paris VI, LIP6, 2004.

J. Nocedal and S. J. Wright. *Numerical Optimization*. Springer Verlag, New York, second edition, 2006.

N. N. Schraudolph, J. Yu, and S. Günter. A stochastic quasi-Newton method for online convex optimization. In M. Meila and X. Shen, editors, *Proceedings of the 11th International Conference on Artificial Intelligence and Statistics*, volume 2 of *Workshop and Conference Proceedings*, pages 433–440. MIT Press, 2007.

16   Bandit View on Noisy Optimization

Jean-Yves Audibert   audibert@certis.enpc.fr
Imagine, Université Paris Est; Willow, CNRS/ENS/INRIA
Paris, France

Sébastien Bubeck   sebastien.bubeck@inria.fr
Sequel Project, INRIA Lille - Nord Europe
Lille, France

Rémi Munos   remi.munos@inria.fr
Sequel Project, INRIA Lille - Nord Europe
Lille, France

This chapter deals with the problem of making the best use of a finite number of noisy evaluations to optimize an unknown function. We are concerned primarily with the case where the function is defined over a finite set. In this discrete setting, we discuss various objectives for the learner, from optimizing the allocation of a given budget of evaluations to optimal stopping time problems with (ε, δ)-PAC guarantees. We also consider the so-called online optimization framework, where the result of an evaluation is associated to a reward, and the goal is to maximize the sum of obtained rewards. In this case, we extend the algorithms to continuous sets and (weakly) Lipschitzian functions (with respect to a prespecified metric).

16.1   Introduction

In this chapter, we investigate the problem of function optimization with a finite number of noisy evaluations. While at first one may think that simple repeated sampling can overcome the difficulty introduced by noisy evaluations, it is far from being an optimal strategy. Indeed, to make the best use of the evaluations, one may want to estimate the seemingly best options more precisely, while for bad options a rough estimate might be enough. This reasoning leads to non-trivial algorithms, which depend on the objective criterion that we set and on how we define the budget constraint on the number of evaluations. The main mathematical tool that we use to build good strategies is a set of concentration inequalities that we briefly recall in section 16.2. Then in section 16.3, we discuss the fundamental case of discrete optimization under various budget constraints. Finally, in section 16.4 we consider the case where the optimization has to be performed online, in the sense that the value of an evaluation can be considered a reward, and the goal of the learner is to maximize his or her cumulative rewards. In this case, we also consider the extension to continuous optimization.

16.1.1 Problem Setup and Notation

Consider a finite set of options $\{1, \ldots, K\}$, also called actions or arms (in reference to the multi-armed bandit terminology). To each option $i \in \{1, \ldots, K\}$ we associate a (reward) distribution ν_i on $[0, 1]$, with mean μ_i. Let i^* denote an optimal arm, that is, $\mu_{i^*} = \max_{1 \leq j \leq K} \mu_j$. We denote the suboptimality gap of option i by $\Delta_i = \mu_{i^*} - \mu_i$, and the minimal positive gap by $\Delta = \min_{i : \Delta_i > 0} \Delta_i$. We assume that when one evaluates an option i, one receives a random variable drawn from the underlying probability distribution ν_i (independently from the previous draws). We investigate strategies that perform sequential evaluations of the options to find the one with the highest mean. More precisely, at each time step $t \in \mathbb{N}$, a strategy chooses an option I_t to evaluate. We denote by $T_i(t)$ the number of times we evaluated option i up to time t, and by $\widehat{X}_{i, T_i(t)}$ the empirical mean estimate of option i at time t (based on $T_i(t)$ i.i.d. random variables). In this chapter, we consider two objectives for the strategy.

1. The learner possesses an evaluation budget, and once this budget is exhausted, he or she has to select an option J as the candidate for being the best option. The performance of the learner is evaluated only through the quality of option J. This setting corresponds to the pure exploration multi-armed bandit setting (Bubeck et al., 2009; Audibert et al., 2010). We study this problem under two different assumptions on the evaluation budget in Section 16.3.

2. The result of an evaluation is associated to a reward, and the learner wants to maximize his or her cumulative rewards. This setting corresponds to the classical multi-armed bandit setting (Robbins, 1952; Lai and Robbins, 1985; Auer et al., 2002). We study this problem in Section 16.4.

16.2   Concentration Inequalities

In this section, we state the fundamental concentration properties of sums of random variables. While we do not directly use the following theorems in this chapter (since we do not provide any proof), this concentration phenomenon is the cornerstone of our reasoning, and a good understanding of it is necessary to get the insights behind our proposed algorithms.

We start with the celebrated Hoeffding-Azuma inequality (Hoeffding, 1963) for the sum of martingale differences. See, for instance, Williams (1991) for an introductory-level textbook on martingales, and Lugosi (1998) and Massart (2007) for lecture notes on concentration inequalities.

Theorem 16.1 (Hoeffding-Azuma inequality for martingales). *Let $\mathcal{F}_1 \subset \cdots \subset \mathcal{F}_n$ be a filtration, and $X_1, \ldots, X_n$ be real random variables such that X_t is $\mathcal{F}_t$-measurable, $\mathbb{E}(X_t | \mathcal{F}_{t-1}) = 0$ and $X_t \in [A_t, A_t + c_t]$ where A_t is a random variable $\mathcal{F}_{t-1}$-measurable and c_t is a positive constant. Then, for any $\varepsilon > 0$, we have*

$$\mathbb{P}\left(\sum_{t=1}^{n} X_t \geq \varepsilon \right) \leq \exp\left(-\frac{2\varepsilon^2}{\sum_{t=1}^{n} c_t^2} \right), \tag{16.1}$$

or equivalently, for any $\delta > 0$, with probability at least $1 - \delta$, we have

$$\sum_{t=1}^{n} X_t \leq \sqrt{\frac{\log(\delta^{-1})}{2} \sum_{t=1}^{n} c_t^2}. \tag{16.2}$$

In particular, when $X_1, \ldots, X_n$ are i.i.d. centered random variables taking their values in $[a, b]$ for some real numbers a and b, with probability at least $1 - \delta$, we have

$$\sum_{t=1}^{n} X_t \leq (b - a)\sqrt{\frac{n \log(\delta^{-1})}{2}}. \tag{16.3}$$

The next result is a refinement of the previous concentration inequality which takes into account the variance of the random variables. More precisely up to a second-order term it replaces the range (squared) of the random variables with their variances.

Theorem 16.2 (Bernstein's inequality for martingales). *Let $\mathcal{F}_1 \subset \cdots \subset \mathcal{F}_n$ be a filtration, and $X_1, \ldots, X_n$ real random variables such that X_t is $\mathcal{F}_t$-measurable, $\mathbb{E}(X_t | \mathcal{F}_{t-1}) = 0$, $|X_t| \leq b$ for some $b > 0$, and $\mathbb{E}(X_t^2 | \mathcal{F}_{t-1}) \leq v$*

for some $v > 0$. Then, for any $\varepsilon > 0$, we have

$$\mathbb{P}\Big(\sum_{t=1}^{n} X_t \geq \varepsilon \Big) \leq \exp\Big(-\frac{\varepsilon^2}{2nv + 2b\varepsilon/3} \Big), \tag{16.4}$$

and for any $\delta > 0$, with probability at least $1 - \delta$, we have

$$\sum_{t=1}^{n} X_t \leq \sqrt{2nv \log(\delta^{-1})} + \frac{b \log(\delta^{-1})}{3}. \tag{16.5}$$

Inequalities (16.4) and (16.5) are two ways of expressing the concentration of the mean of i.i.d. random variables. They are almost equivalent to the extent that up to minor modification of the constants, one can go from (16.4) to (16.5) and conversely by a change of variables.

The next inequality was proved by Audibert et al. (2009). It allows to replace the true variance with its empirical estimate in Bernstein's bound.

Theorem 16.3 (Empirical Bernstein bound). *Let $X_1, \ldots, X_n$ be i.i.d. centered real random variables in $[a, b]$ for some $a, b \in \mathbb{R}$. Then, for any $\delta > 0$ and $s \in \{1, \ldots, n\}$, with probability at least $1 - \delta$, we have*

$$\sum_{t=1}^{s} X_t \leq \sqrt{2nV_s \log(3\delta^{-1})} + 3(b - a) \log(3\delta^{-1}),$$

where $V_s = \frac{1}{s} \sum_{t=1}^{s} \big(X_t - \frac{1}{s} \sum_{\ell=1}^{s} X_\ell \big)^2$.

Variants and refinement of this bound can be found in Maurer and Pontil (2009) and Audibert (2010).

16.3   Discrete Optimization

In this section, we focus on strategies that use a finite budget of evaluations to find the best option. We consider two different (but related) assumptions on this budget.

■ There is a fixed budget of n evaluations (Bubeck et al., 2009; Audibert et al., 2010). The value of n can be known or unknown by the learner. When it is unknown, the learner has thus to design an anytime strategy, that is, a policy with good theoretical guarantees whatever the budget is.

■ The strategy must stop as soon as possible with the guarantee that an ε-optimal option has been found with probability at least $1 - \delta$, where ε and δ are fixed before the procedure starts (Maron and Moore, 1993; Domingo et al., 2002; Dagum et al., 2000; Even-Dar et al., 2006; Mnih et al., 2008).

Let $A_1 = \{1, \ldots, K\}$, $\overline{\log}(K) = \frac{1}{2} + \sum_{i=2}^{K} \frac{1}{i}$, $n_0 = 0$ and for $k \in \{1, \ldots, K-1\}$,

$$n_k = \left\lceil \frac{1}{\overline{\log}(K)} \frac{n - K}{K + 1 - k} \right\rceil.$$

For each phase $k = 1, 2, \ldots, K-1$:

(1) For each $i \in A_k$, select option i for $n_k - n_{k-1}$ evaluations.

(2) Let $A_{k+1} = A_k \setminus \arg\min_{i \in A_k} \hat{X}_{i,n_k}$ (we remove only one element from A_k; if there is a tie, randomly select the option to dismiss among the worst options).

Recommend the unique element of A_K.

Figure 16.1: SR (successive rejects) algorithm.

16.3.1 Fixed Budget

In this section, the number of evaluations is fixed, and the goal is to make the best use of the budget. We propose a strategy, that is simple, yet almost optimal in a strong sense (see theorem 16.4). The algorithm, called SR (successive rejects) is described precisely in figure 16.1. Informally, it proceeds as follows. First the algorithm divides the budget (i.e., the n evaluations) in $K - 1$ phases. At the end of each phase, the algorithm dismisses the option with the lowest empirical mean. During the next phase, it equally often evaluates all the options which have not been dismissed. The recommended arm J is the last surviving option. The lengths of the phases are carefully chosen to obtain an optimal (up to a logarithmic factor) convergence rate. More precisely, one option is evaluated $n_1 = \left\lceil \frac{1}{\overline{\log}(K)} \frac{n-K}{K} \right\rceil$ times, one $n_2 = \left\lceil \frac{1}{\overline{\log}(K)} \frac{n-K}{K-1} \right\rceil$ times, ..., and two options are evaluated $n_{K-1} = \left\lceil \frac{1}{\overline{\log}(K)} \frac{n-K}{2} \right\rceil$ times. SR does not exceed the budget of n evaluations, since, from the definition $\overline{\log}(K) = \frac{1}{2} + \sum_{i=2}^{K} \frac{1}{i}$ we have

$$n_1 + \ldots + n_{K-1} + n_{K-1} \leq K + \frac{n-K}{\overline{\log}(K)} \left(\frac{1}{2} + \sum_{k=1}^{K-1} \frac{1}{K+1-k} \right) = n.$$

Theorem 16.4 (Successive rejects). *Assume that there is a unique arm i^* with maximal mean and let $H = \frac{1}{\Delta} + \sum_{i \neq i^*} \frac{1}{\Delta_i}$. Then the probability of error* *of SR satisfies*

$$\mathbb{P}(J \neq i^*) \leq \frac{K(K-1)}{2} \exp\left(-\frac{n-K}{\log(2K)H}\right). \qquad (16.6)$$

Moreover, if $\nu_1, \ldots, \nu_K$ are Bernoulli distributions with parameters in $[p, 1-p]$, $p \in (0, 1/2)$, then for any strategy there exists a permutation $\sigma : \{1, \ldots, K\} \to \{1, \ldots, K\}$ such that the probability of error of the strategy on the problem defined by $\tilde{\nu}_1 = \nu_{\sigma(1)}, \ldots, \tilde{\nu}_K = \nu_{\sigma(K)}$ satisfies

$$\mathbb{P}(J \neq i^*) \geq \exp\left(-\frac{(5 + o(1))n\log(2K)}{p(1-p)H}\right), \qquad (16.7)$$

where the $o(1)$ term depends only on K and n, and goes to 0 when n goes to infinity.

16.3.1.1   *Interpretation of Theorem 16.4*

Essentially, equation (16.6) indicates that if the number of evaluations is on the order of $H \log^2 K$, then SR finds the best option with high probability. On the other hand, equation (16.7) shows that it is statistically impossible to find the best option with fewer than (order of) $H/\log K$ evaluations. Thus H is a good measure of the *hardness* of the task; it characterizes the order of magnitude of the number of evaluations required to find the best option with a reasonable probability.

Closing the logarithmic gap between the upper and lower bounds in theorem 16.4 is an open problem. Audibert et al. (2010) exhibit an algorithm which requires only (on the order of) $H \log n$ evaluations to find the best option with high probability. However, this algorithm needs to know the value of H to tune its parameters. One can overcome this difficulty by trying to estimate H online, which leads to the algorithm Adaptive UCB-E that is described precisely in figure 16.2. We do not give any further details about this algorithm and refer the interested reader to Audibert et al. (2010); we simply point out that in our numerical simulations, Adaptive UCB-E outperformed SR.

16.3.1.2   *Anytime Versions of SR and Adaptive UCB-E.*

Both algorithms that we propose depend heavily on the knowledge of the number of evaluations n. However in many natural cases this number is only implicitly defined (for instance through CPU time). Thus, it is important to have strategies which do not need to know the time horizon in advance.

Parameter: exploration rate $c > 0$.

Definitions: For $k \in \{1, \ldots, K-1\}$, let $n_k = \left\lceil \frac{1}{\log(K)} \frac{n-K}{K+1-k} \right\rceil$, $t_0 = 0$, $t_1 = Kn_1$, and for $k > 1$, $t_k = n_1 + \ldots n_{k-1} + (K - k + 1)n_k$.

For $i \in \{1, \ldots, K\}$ and $a > 0$, let $B_{i,s}(a) = \widehat{X}_{i,s} + \sqrt{\frac{a}{s}}$ for $s \geq 1$ and $B_{i,0} = +\infty$.

Algorithm: For each phase $k = 0, 1, \ldots, K-1$:

Let $\widehat{H}_k = K$ if $k = 0$, and otherwise

$$\widehat{H}_k = \max_{K-k+1 \leq i \leq K} i \widehat{\Delta}_{<i>}^{-2},$$

where $\widehat{\Delta}_i = \left(\max_{1 \leq j \leq K} \widehat{X}_{j,T_j(t_k)} \right) - \widehat{X}_{i,T_i(t_k)}$ and $< i >$ is an ordering such that $\widehat{\Delta}_{<1>} \leq \ldots \leq \widehat{\Delta}_{<K>}$.

For $t = t_k + 1, \ldots, t_{k+1}$:

   Evaluate $I_t \in \arg\max_{i \in \{1, \ldots, K\}} B_{i, T_i(t-1)}(cn/\widehat{H}_k)$.

Recommendation: Let $J \in \arg\max_{i \in \{1, \ldots, K\}} \widehat{X}_{i, T_i(n)}$.

Figure 16.2: Adaptive UCB-E (Upper Confidence Bound Exploration).

One simple and famous trick for this purpose is the doubling trick. The idea is to introduce metaphases, $s = 1, 2, \ldots$, such that from the evaluations $t = 2^{s-1} + 1$ to $t = 2^s$, one runs a new instance of the algorithm with n replaced by 2^{s-1}. While it is often assumed that the new instance of the algorithm does not use the samples obtained in the previous phases, here we do not need to make this assumption. For instance, the anytime version of SR would work as follows. At time 2^s there is only one surviving option. Then at time $2^s + 1$ we "revive" all the options and run SR with n replaced by 2^{s+1} (to define the length of the phases of SR). However, the empirical mean of each option is computed over the whole run of the algorithm, starting with $t = 1$.

16.3.2 Hoeffding and Bernstein Races

Racing algorithms aim to reduce the computational burden of performing tasks such as model selection using a holdout set by discarding poor models quickly (Maron and Moore, 1993; Ortiz and Kaelbling, 2000). A racing algorithm terminates either when it runs out of time (i.e., at the end of the n-th round) or when it can say that with probability at least $1 - \delta$, it has found the best option, that is, an option $i^* \in \arg\max_{i \in \{1, \ldots, K\}} \mu_i$. The goal is to stop as soon as possible, and the time constraint n is here to stop the algorithm when the two best options have (almost) equal mean rewards.

Parameter: the confidence level δ.

Let $A = \{1, \ldots, K\}$ and $t = 1$

While $|A| > 1$

(1) sample every option in A for the t-th time.

(2) remove from A all the options having an empirical mean differing from the highest empirical mean by more than $\sqrt{2 \log(nK/\delta)/t}$, that is,

$$A \leftarrow A \setminus \left\{ j \in A : \widehat{X}_{j,t} \leq \max_{1 \leq i \leq K} \widehat{X}_{i,t} - \sqrt{\frac{2 \log(nK/\delta)}{t}} \right\}.$$

(3) $t \leftarrow t + 1$.

Output the unique element of A.

Figure 16.3: Hoeffding race.

The Hoeffding race introduced by Maron and Moore (1993) is an algorithm based on discarding options which are likely to have a smaller mean than the optimal one until only one option remains. Precisely, for each time step and each option i, $\delta/(nK)$-confidence intervals are constructed for the mean μ_i. Options with an upper confidence bound smaller than the lower confidence bound of another option are discarded. The algorithm samples one by one all the options that have not been discarded. The process is detailed in Figure 16.3. The correctness of this algorithm is proved by Maron and Moore (1993), and its sample complexity is given by the following theorem (Even-Dar et al., 2006; Mnih et al., 2008).

Theorem 16.5 (Hoeffding race). *With probability at least $1 - \delta$, the optimal option is not discarded, and the non-discarded option(s) (which can be multiple when the algorithm runs out of time) satisfy(ies)*

$$\Delta_i = O\left(\sqrt{\frac{\log(nK/\delta)}{n/K}} \right).$$

Besides, if there is a unique optimal arm i^*, with probability at least $1 - \delta$, the Hoeffding race stops after at most $O\left(\sum_{i \neq i^*} \frac{1}{\Delta_i^2} \log\left(\frac{nK}{\delta}\right) \right)$ time steps.

Empirical and theoretical studies show that replacing the Hoeffding inequality with the empirical Bernstein bound to build the confidence intervals generally leads to significant improvements. The algorithm based on the empirical Bernstein bound is described in Figure 16.4. Theorem 16.6 provides its theoretical guarantee, and table 16.1 shows the percentage of work saved by each method (1− number of samples taken by method divided by nK), as well as the number of options remaining after termination (see Mnih et al. (2008) for a more detailed description of the experiments).

Theorem 16.6 (Bernstein race). *Let σ_i denote the standard deviation of ν_i. With probability at least $1 - \delta$, the optimal option is not discarded, and the non-discarded option(s) (which can be multiple when the algorithm runs out of time) satisfy(ies)*

$$\Delta_i = O\left((\sigma_i + \sigma_{i^*})\sqrt{\frac{\log(nK/\delta)}{n/K}} + \frac{\log(nK/\delta)}{n/K}\right).$$

Besides, if there is a unique optimal arm i^*, with probability at least $1 - \delta$, the Bernstein race stops after at most $O\left(\sum_{i\neq i^*} \frac{\sigma_i^2 + \sigma_{i^*}^2 + \Delta_i}{\Delta_i^2} \log\left(\frac{nK}{\delta}\right)\right)$ time steps.

Parameter: the confidence level δ.

Let $A = \{1, \ldots, K\}$ and $t = 1$

While $|A| > 1$

(1) sample every option in A for the t-th time.

(2) remove suboptimal options from A:

$$A \leftarrow A \setminus \left\{j \in A : \widehat{X}_{j,t} + \sqrt{\frac{2V_{j,t}\log(nK/\delta)}{t}} + 6\frac{\log(nK/\delta)}{t}\right.$$

$$\left. \leq \max_{1\leq i\leq K}\left(\widehat{X}_{i,t} - \sqrt{\frac{2V_{i,t}\log(nK/\delta)}{t}}\right)\right\},$$

where $V_{i,t} = \frac{1}{t}\sum_{s=1}^{t}\left(X_{i,s} - \widehat{X}_{i,t}\right)^2$ is the empirical variance of option i.

(3) $t \leftarrow t + 1$.

Output the unique element of A.

Figure 16.4: Bernstein race.

Data set	Hoeffding	Empirical Bernstein
SARCOS	0.0% / 11	44.9% / 4
Covertype2	14.9% / 8	29.3% / 5
Local	6.0% / 9	33.1% / 6

Table 16.1: Percentage of work saved/number of options left after termination

16.3.3   Optimal Stopping Times

Section 16.3.3.1 takes a step back since it considers the single option case (that is, when $K = 1$). The additive and multiplicative stopping time problems are tackled there. Section 16.3.3.2 then deals with the multiple options case for the additive stopping time problem.

16.3.3.1   For a Single Option

Algorithms described in section 16.3 rely on either the Hoeffding or the (empirical) Bernstein inequality, and on a probabilistic union bound corresponding to both the different options and the different time steps. Maximal inequalities based on a martingale argument due to Doob (1953) (see also Freedman (1975) for maximal inequalities more similar to the one below) allow one to reduce the impact on the confidence levels of the union bound across time steps. Precisely, one can write the following version of the empirical Bernstein inequality, which holds uniformly over time.

Theorem 16.7. *Let $X_1, \ldots, X_n$ be $n \geq 1$ i.i.d. random variables taking their values in $[a, b]$. Let $\mu = \mathbb{E}X_1$ be their common expected value. For any $1 \leq t \leq n$, introduce the empirical mean $\hat{X}_t$ and variance V_t, defined respectively by*

$$\hat{X}_t = \frac{\sum_{i=1}^t X_i}{t} \quad and \quad V_t = \frac{\sum_{i=1}^t (X_i - \hat{X}_t)^2}{t}.$$

For any $x > 0$, with probability at least

$$1 - 3 \inf_{1 < \alpha \leq 3} \min\left(\frac{\log n}{\log \alpha}, n\right) e^{-x/\alpha}, \tag{16.8}$$

the following inequality holds simultaneously for any $t \in \{1, 2, \ldots, n\}$:

$$|\hat{X}_t - \mu| \leq \sqrt{\frac{2V_t x}{t}} + \frac{3(b-a)x}{t}. \tag{16.9}$$

This theorem allows one to address the additive stopping time problem in which the learner stops sampling an unknown distribution ν supported in

$[a, b]$ as soon as it can output an estimate $\hat{\mu}$ of the mean μ of ν with additive error at most ε with probability at least $1 - \delta$, that is,

$$\mathbb{P}\big(|\hat{\mu} - \mu| \leq \varepsilon\big) \geq 1 - \delta, \tag{16.10}$$

with the time constraint that the learner is not allowed to sample more than n times. Indeed, from Theorem 16.7, it suffices to stop sampling at time t such that the right-hand side of (16.9) is below ε where x is set such that (16.8) equals $1 - \delta$. Besides, it can be shown that the sampling complexity is in expectation

$$O\left(\Big(\log(\delta^{-1}) + \log\big(\log(3n)\big)\Big) \max\left(\frac{\sigma^2}{\varepsilon^2}, \frac{b - a}{\varepsilon}\right)\right),$$

where σ^2 is the variance of the sampling distribution. This is optimal up to the log-log term.

In the multiplicative stopping time problem, the learner stops sampling an unknown distribution ν supported in $[a, b]$ as soon as it can output an estimate $\hat{\mu}$ of the mean μ of ν with relative error at most ε with probability at least $1 - \delta$, that is,

$$\mathbb{P}\big(|\hat{\mu} - \mu| \leq \varepsilon|\mu|\big) \geq 1 - \delta, \tag{16.11}$$

with the time constraint that the learner is not allowed to sample more than n times. The multiplicative stopping time problem is similar to the additive one, except when μ is close to 0 (but nonzero). Considering relative errors introduces an asymmetry between the left and right bounds of the confidence intervals, which requires more involved algorithms to get better practical performances. The state-of-the-art method to handle the task is the geometric empirical Bernstein stopping proposed by Mnih et al. (2008) and detailed in Figure 16.5. A slightly refined version is given in Audibert (2010).

It uses a geometric grid and parameters ensuring that the event $\mathcal{E} = \{|\hat{X}_t - \mu| \leq c_t, t \geq t_1\}$ occurs with probability at least $1 - \delta$. It operates by maintaining a lower bound, LB, and an upper bound, UB, on the absolute value of the mean of the random variable being sampled, terminates when $(1 + \varepsilon)\text{LB} < (1 - \varepsilon)\text{UB}$, and returns the mean estimate $\hat{\mu} = \text{sign}(\hat{X}_t)\frac{(1+\varepsilon)\text{LB}+(1-\varepsilon)\text{UB}}{2}$. Mnih et al. (2008) proved that the output satisfies (16.11) and that the expected stopping time of the policy is

$$O\left(\left(\log\left(\frac{1}{\delta}\right) + \log\left(\log\frac{3}{\varepsilon|\mu|}\right)\right) \max\left(\frac{\sigma^2}{\varepsilon^2\mu^2}, \frac{b - a}{\varepsilon|\mu|}\right)\right).$$

Parameters: $q > 0$, $t_1 \geq 1$, and $\alpha > 1$ defining the geometric grid $t_k = \lceil \alpha t_{k-1} \rceil$.
(Good default choice: $q = 0.1$, $t_1 = 20$, and $\alpha = 1.1$.)

Initialization:
$c = \frac{3}{\delta t_1^q (1 - \alpha^{-q})}$
LB $\leftarrow 0$
UB $\leftarrow \infty$

For $t = 1, \ldots, t_1 - 1$,
    sample X_t from ν
End For

For $k = 1, 2, \ldots,$
    For $t = t_k, \ldots, t_{k+1} - 1$,
        sample X_t from ν
        compute $\ell_t = \frac{t_{k+1}}{t^2} \log(ct_k^q)$ and $c_t = \sqrt{2\ell_t V_t} + 3(b - a)\ell_t$
        LB $\leftarrow \max(\text{LB}, |\hat{X}_t| - c_t)$
        UB $\leftarrow \min(\text{UB}, |\hat{X}_t| + c_t)$
        If $(1 + \varepsilon)\text{LB} < (1 - \varepsilon)\text{UB}$, Then
                stop simulating X and return the mean estimate
$\text{sign}(\hat{X}_t)\frac{(1+\varepsilon)\text{LB} + (1-\varepsilon)\text{UB}}{2}$ End If
    End For
End For

Figure 16.5: Geometric empirical Bernstein stopping rule.

Up to the log-log term, this is optimal from the work of Dagum et al. (2000).

16.3.3.2   *For Multiple Options*

Let us go back to the case where we consider $K > 1$ options. A natural variant of the best option identification problems addressed in sections 16.3.1 and 16.3.2 is to find, with high probability, a near-optimal option while not sampling for too long a time. Precisely, the learner wants to stop sampling as soon as he or she can say that with probability at least $1 - \delta$, he or she has identified an option i with $\mu_i \geq \max_{1 \leq j \leq K} \mu_j - \varepsilon$. An algorithm solving this problem will be called an (ε, δ)-correct policy. A simple way to get such a policy is to adapt the Hoeffding or Bernstein race (figures 16.3 and 16.4) by adding an ε in the right-hand side of the inequality defining the removal step. It can easily be shown that this strategy is (ε, δ)-correct and has an expected sampling time of $O\left(\frac{K}{\varepsilon^2} \log\left(\frac{nK}{\delta}\right)\right)$. This is minimax optimal up to the $\log(nK)$ term in view of the following lower bound due to Mannor and Tsitsiklis (2004).

Theorem 16.8 (Additive optimal sampling lower bound). *There exist* *positive constants c_1, c_2 such that for any $K \geq 2$, $0 < \delta < 1/250$, $0 < \varepsilon < 1/8$, and any (ε, δ)-correct policy, there exist distributions $\nu_1, \ldots, \nu_K$ on $[0,1]$ such that the average stopping time of the policy is greater than $c_1 \frac{K}{\varepsilon^2} \log\left(\frac{c_2}{\delta}\right)$.*

Parameters: $\varepsilon > 0$, $\delta > 0$.

Let $A = \{1, \ldots, K\}$, $\tilde{\varepsilon} = \varepsilon/4$ and $\tilde{\delta} = \delta/2$.

While $|A| > 1$

(1) sample every option in A for $\left\lfloor \frac{4}{\tilde{\varepsilon}^2} \log(3/\tilde{\delta}) \right\rfloor$ times.

(2) remove from A suboptimal options:

$$A \leftarrow A \setminus \left\{ j \in A : \widehat{X}_{j,t} \text{ is smaller than the median of } (\widehat{X}_{i,t})_{i \in A} \right\},$$

(3) $\tilde{\varepsilon} \leftarrow \frac{3}{4}\tilde{\varepsilon}$ and $\tilde{\delta} \leftarrow \frac{1}{2}\tilde{\delta}$.

Output the unique element of A.

Figure 16.6: Median elimination.

Even-Dar et al. (2006) propose a policy, called median elimination (detailed in figure 16.6), with a sampling complexity matching the previous lower bound according to the following sampling complexity result.

Theorem 16.9 (Median elimination). *The median elimination algorithm is (ε, δ)-correct and stops after at most $O\left(\frac{K}{\varepsilon^2} \log\left(\frac{2}{\delta}\right)\right)$.*

16.4   Online Optimization

In this section we consider a setting different from the one presented in section 16.3. We assume that the result of an evaluation is associated to a reward, and the objective is to maximize the sum of obtained rewards. This notion induces an explicit trade-off between exploration and exploitation: at each time step the strategy has to balance between trying to obtain more information about the options and selecting the option which seems to yield (in expectation) the highest rewards. As we shall see in section 16.4.1, good strategies perform both exploration and exploitation at the same time.

This framework is known as the multi-armed bandit problem. It was introduced by Robbins (1952). Since about 2000 there has been a flurry of activity around this type of problem, with many different extensions. In this section we concentrate on the basic version where there is a finite number of options, as well as on the extension to an arbitrary set of options with a Lipschitz assumption on the mapping from options to expected rewards. A more extensive review of the existing literature (as well as the proofs of the results of section 16.4.1) can be found in Bubeck (2010, chapter 2).

16.4.1   Discrete Case

We propose three strategies for the case of a finite number of options. We describe these algorithms in figure 16.7. They are all based on the same underlying principle: optimism in face of uncertainty. More precisely, these methods assign an upper confidence bound on the mean reward of each option (which holds with high probability), and then select the option with the highest bound.

We now review the theoretical performances of the proposed strategies, and briefly discuss the implications of the different results. In particular, as we shall see, none of these strategies is uniformly (over all possible K-tuple of distributions) better (in the sense that it would have a larger expected sum of rewards) than the others.

To assess a strategy, we use the expected cumulative regret, defined as

$$R_n = n \max_{1 \leq i \leq K} \mu_i - \sum_{t=1}^{n} \mathbb{E}\mu_{I_t}.$$

That is, R_n represents the difference in expected reward between the optimal strategy (which always selects the best option) and the strategy we used.

16.4.1.1   UCB (Auer et al., 2002).

This strategy relies on the basic Hoeffding's inequality (16.3) to build the upper confidence bound. This leads to a simple and natural algorithm, yet one that is almost optimal. More precisely, the distribution-dependent upper bound (16.12) has the optimal logarithmic rate in n, but not the optimal distribution-dependent constant (see theorem 16.13 for the corresponding lower bound). On the other hand, the distribution-free upper bound (16.13) is optimal up to a logarithmic term (see theorem 16.14 for the corresponding lower bound). The two other strategies, UCB-V and MOSS, are designed to improve on these weaknesses.

Theorem 16.10 (Upper Confidence Bound algorithm). *UCB with $\alpha > 1/2$*

UCB (Upper Confidence Bound), UCB-V (Upper Confidence Bound with Variance), and MOSS (Minimax Optimal Strategy for the Stochastic case):

Parameter: exploration rate $\alpha > 0$.

For an arm i, define its index $B_{i,s,t}$ by

$$\text{UCB index:} \quad B_{i,s,t} = \widehat{X}_{i,s} + \sqrt{\frac{\alpha \log(t)}{s}},$$

$$\text{UCB-V index:} \quad B_{i,s,t} = \widehat{X}_{i,s} + \sqrt{\frac{2\alpha V_{i,s} \log(t)}{s}} + 3\alpha \frac{\log(t)}{s},$$

$$\text{MOSS index:} \quad B_{i,s,t} = \widehat{X}_{i,s} + \sqrt{\frac{\max\left(\log(\frac{n}{Ks}), 0\right)}{s}},$$

for $s, t \geq 1$, and $B_{i,0,t} = +\infty$.

At time t, evaluate an option I_t maximizing $B_{i,T_i(t-1),t}$, where $T_i(t-1)$ denotes the number of times we evaluated option i during the $t-1$ first steps.

Figure 16.7: Upper confidence bound-based policies.

satisfies

$$R_n \leq \sum_{i:\Delta_i>0} \frac{4\alpha}{\Delta_i} \log(n) + \Delta_i \left(1 + \frac{4}{\log(\alpha+1/2)}\left(\frac{\alpha+1/2}{\alpha-1/2}\right)^2\right), \quad (16.12)$$

and

$$R_n \leq \sqrt{nK\left(4\alpha \log n + 1 + \frac{4}{\log(\alpha+1/2)}\left(\frac{\alpha+1/2}{\alpha-1/2}\right)^2\right)}. \quad (16.13)$$

16.4.1.2   *UCB-V (Audibert et al., 2009).*

Here the confidence intervals are derived from an empirical version of Bernstein's inequality (see theorem 16.3). This leads to an improvement in the distribution-dependent rate, where basically one can replace the range of the distributions with their variances.

Theorem 16.11 (Upper Confidence Bound with Variance algorithm).

UCB-V with $\alpha > 1$ satisfies[1]

$$R_n \leq 8\alpha \sum_{i:\Delta_i>0} \left(\frac{\sigma_i^2}{\Delta_i} + 2\right) \log(n) + \Delta_i \left(2 + \frac{12}{\log(\alpha+1)} \left(\frac{\alpha+1}{\alpha-1}\right)^2\right). \quad (16.14)$$

16.4.1.3 MOSS (Audibert and Bubeck, 2009).

In this second modification of UCB, one combines the Hoeffding-type confidence intervals by using a tight peeling device. This leads to a minimax strategy, in the sense that the distribution-free upper bound (16.16) is optimal up to a numerical constant. On the other hand, the distribution-dependent bound (16.15) can be slightly worse than the one for UCB. Note also that, contrary to UCB and UCB-V, MOSS needs to know in advance the number of evaluations. Again, one can overcome this difficulty with the doubling trick.

Theorem 16.12 (Minimax Optimal Strategy for the Stochastic case). *MOSS satisfies*

$$R_n \leq \frac{23K}{\Delta} \log\left(\max\left(\frac{110n\Delta^2}{K}, 10^4\right)\right) \quad (16.15)$$

and

$$R_n \leq 25\sqrt{nK}. \quad (16.16)$$

16.4.1.4 Lower Bounds (Lai and Robbins, 1985; Auer et al., 2003).

For the sake of completeness, we state here the two main lower bounds for multi-armed bandits. In theorem 16.13, we use the Kullback-Leibler divergence between two Bernoulli distributions of parameters $p, q \in (0,1)$, defined as

$$\text{KL}(p,q) = p\log\left(\frac{p}{q}\right) + (1-p)\log\left(\frac{1-p}{1-q}\right).$$

---

1. In the context of UCB-V it is interesting to see the influence of the range of the distributions. Precisely, if the support of all distributions ν_i are included in $[0,b]$, and if one uses the upper confidence bound sequence $B_{i,s,t} = \widehat{X}_{i,s} + \sqrt{2\alpha V_{i,s}\log(t)/s} + 3b\alpha\frac{\log(t)}{s}$, then one can easily prove that the leading constant in the bound becomes $\frac{\sigma_i^2}{\Delta_i} + 2b$, which can be much smaller than the b^2/Δ_i factor characterizing the regret bound of UCB.

A useful inequality to compare the lower bound of theorem 16.13 with (16.12) and (16.14) is the following:

$$2(p-q)^2 \leq \text{KL}(p,q) \leq \frac{(p-q)^2}{q(1-q)}.$$

Theorem 16.13 (Distribution-dependent lower bound). *Let us consider a strategy such that for any set of K distributions, any arm i such that $\Delta_i > 0$ and any $a > 0$, we have $\mathbb{E}T_i(n) = o(n^a)$. Then, if $\nu_1, \ldots, \nu_K$ are Bernoulli distributions, all different from a Dirac distribution at 1, the following holds true:*

$$\liminf_{n \to +\infty} \frac{R_n}{\log n} \geq \sum_{i: \Delta_i > 0} \frac{\Delta_i}{\text{KL}(\mu_i, \max_{1 \leq j \leq K} \mu_j)}. \tag{16.17}$$

An extension of Theorem 16.13 can be found in Burnetas and Katehakis (1996).

Theorem 16.14 (Distribution-free lower bound). *Let* sup *represent the supremum taken over all sets of K distributions on $[0,1]$ and* inf *the infimum taken over all strategies. Then the following holds true:*

$$\inf \sup R_n \geq \frac{1}{20}\sqrt{nK}. \tag{16.18}$$

16.4.2   Continuous Case

In many natural examples, the number of options is extremely large, potentially infinite. One particularly important and ubiquitous case is when the set of options is identified by a finite number of continuous-valued parameters. Unfortunately, this type of problem can be arbitrarily difficult without further assumptions. One standard way to constrain the problem is to make a smoothness assumption on the mapping from options to expected reward (the mean payoff function). In this section we present the approach proposed in Bubeck et al. (2008), where there is essentially a weak compactness assumption on the set of options, and a weak Lipschitz assumption on the mean payoff. We make these assumptions more precise in section 16.4.3. Then section 16.4.4 details the algorithm called HOO (Hierarchical Optimistic Optimization), which is based on the recent successful tree optimization algorithms (Kocsis and Szepesvári, 2006; Coquelin and Munos, 2007). Finally, section 16.4.5 provides the theoretical guarantees that one can derive for HOO. The latter can be informally summed up as follows: if one knows the local smoothness of the mean payoff function around its maximum, then with n evaluations it is possible to find an option which is (on the order of) $1/\sqrt{n}$-optimal (no matter what the ambient dimension is).

16.4.3 Assumptions and Notation

Let $\mathcal{X}$ denote the set of options, f the mean payoff function, and $f^* = \sup_{x \in \mathcal{X}} f(x)$ the supremum of f over $\mathcal{X}$. Recall that when one evaluates a point $x \in \mathcal{X}$, one receives an independent random variable in $[0, 1]$ with expectation $f(x)$. Let X_t be the tth point that one chooses to evaluate.

As we said, one needs to place some restriction on the set of possible mean payoff functions. We shall do this by resorting to some (weakly) Lipschitz condition. However, somewhat unconventionally, we shall use dissimilarity functions rather than metric distances, which allows us to deal with function classes of highly different smoothness orders in a unified manner. Formally, a *dissimilarity* ℓ over $\mathcal{X}$ is a non-negative mapping $\ell : \mathcal{X}^2 \to \mathbb{R}$ satisfying $\ell(x, x) = 0$ for all $x \in \mathcal{X}$. The weakly Lipschitz assumption on the mean payoff requires that for all $x, y \in \mathcal{X}$,

$$f^* - f(y) \leq f^* - f(x) + \max\{f^* - f(x), \ell(x, y)\}. \tag{16.19}$$

The choice of this terminology follows from the fact that if f is 1–Lipschitz w.r.t. ℓ, so that for all $x, y \in \mathcal{X}$, one has $|f(x) - f(y)| \leq \ell(x, y)$, then it is also weakly Lipschitz w.r.t. ℓ. On the other hand, weak Lipschitzness is a milder requirement. It implies local (one-sided) 1–Lipschitzness at any global maximum (if one exists) x^* (i.e., such that $f(x^*) = f^*$), since in that case the criterion (16.19) rewrites to $f(x^*) - f(y) \leq \ell(x^*, y)$. In the vicinity of other options x, the constraint is milder as the option x gets worse (as $f^* - f(x)$ increases) since the condition (16.19) rewrites to

$$\forall y \in \mathcal{X}, \qquad f(x) - f(y) \leq \max\{f^* - f(x), \ell(x, y)\}.$$

In fact, it is possible to relax (16.19) and require it only to hold locally at the global maximum (or the set of maxima if there are several). We refer the interested reader to Bubeck et al. (2010) for further details.

We also make a mild assumption on the set $\mathcal{X}$ which can be viewed as some sort of compacity w.r.t. ℓ. More precisely, we assume that there exists a sequence $(\mathcal{P}_{h,i})_{h \geq 0, 1 \leq i \leq 2^h}$ of subsets of $\mathcal{X}$ satisfying

- $\mathcal{P}_{0,1} = \mathcal{X}$, and for all $h \geq 0, 1 \leq i \leq 2^h$, $\mathcal{P}_{h,i} = \mathcal{P}_{h+1,2i-1} \cup \mathcal{P}_{h,2i}$.
- There exist $\nu_1, \nu_2 > 0$ and $\rho \in (0, 1)$ such that each $\mathcal{P}_{h,i}$ is included in a ball of radius $\nu_1 \rho^h$ (w.r.t. ℓ) and contains a ball of radius $\nu_2 \rho^h$. Moreover, for a given h, the balls of radius $\nu_2 \rho^h$ are all disjoint.

Intuitively, for a given h, the sets $(\mathcal{P}_{h,i})_{1 \leq i \leq 2^h}$ represent a covering of $\mathcal{X}$ at "scale" h.

The proposed algorithm takes this sequence of subsets and the real numbers ν_1, ρ as inputs. Moreover, the sequence $(\mathcal{P}_{h,i})$ will be represented as an infinite binary tree, where the nodes are indexed by pairs of integers (h, i), such that the nodes $(h + 1, 2i - 1)$ and $(h + 1, 2i)$ are the children of (h, i). The subset $\mathcal{P}_{h,i}$ is associated with node (h, i).

16.4.4   The Hierarchical Optimistic Optimization (HOO) Strategy

The HOO strategy (see algorithm 16.1) incrementally builds an estimate of the mean payoff function f over $\mathcal{X}$. The core idea is to estimate f precisely around its maxima, while estimating it loosely in other parts of the space $\mathcal{X}$. To implement this idea, HOO maintains the binary tree described in section 16.4.3, whose nodes are associated with subsets of $\mathcal{X}$ such that the regions associated with nodes deeper in the tree (farther from the root) represent increasingly smaller subsets of $\mathcal{X}$. The tree is built in an incremental manner. At each node of the tree, HOO stores some statistics based on the information received in previous evaluations. In particular, HOO keeps track of the number of times a node was traversed up to round n and the corresponding empirical average of the rewards received so far. Based on these, HOO assigns an optimistic estimate (denoted by B) to the maximum mean payoff associated with each node. These estimates are then used to select the next node to "play". This is done by traversing the tree, beginning from the root and always following the node with the highest B–value (see lines 4–14 of algorithm 16.1). Once a node is selected, a point in the region associated with it is chosen (line 16) and is evaluated. Based on the point selected and the reward received, the tree is updated (lines 18–33).

Note that the total running time up to the nth evaluation is quadratic in n. However, it is possible to modify the algorithm slightly to obtain a running time of order $O(n \log n)$. The details can be found in Bubeck et al. (2010).

16.4.5   Regret Bound for HOO

In this section, we show that the regret of HOO depends on how fast the volumes of the set $\mathcal{X}_\varepsilon$ of ε–optimal options shrink as $\varepsilon \to 0$. We formalize this notion with the near-optimality dimension of the mean payoff function. We start by recalling the definition of packing numbers.

Definition 16.1 (Packing number). *The ε–packing number $\mathcal{N}(\mathcal{X}, \ell, \varepsilon)$ of $\mathcal{X}$ w.r.t. the dissimilarity ℓ is the largest integer k such that there exist k disjoint ℓ–open balls with radius ε contained in $\mathcal{X}$.*

Algorithm 16.1   The HOO strategy

Parameters: Two real numbers $\nu_1 > 0$ and $\rho \in (0,1)$, a sequence $(\mathcal{P}_{h,i})_{h \geq 0, 1 \leq i \leq 2^h}$ of subsets of $\mathcal{X}$.

Auxiliary function $\text{LEAF}(\mathcal{T})$: outputs a leaf of $\mathcal{T}$.

Initialization: $\mathcal{T} = \{(0,1)\}$ and $B_{1,2} = B_{2,2} = +\infty$.

```
 1:  for n = 1, 2, ... do                                    ▷ Strategy HOO in round n ≥ 1
 2:      (h, i) ← (0, 1)                                                  ▷ Start at the root
 3:      P ← {(h, i)}                                 ▷ P stores the path traversed in the tree
 4:      while (h, i) ∈ T do                                           ▷ Search the tree T
 5:          if B_{h+1,2i−1} > B_{h+1,2i} then            ▷ Select the "more promising" child
 6:              (h, i) ← (h + 1, 2i − 1)
 7:          else if B_{h+1,2i−1} < B_{h+1,2i} then
 8:              (h, i) ← (h + 1, 2i)
 9:          else                                                      ▷ Tie-breaking rule
10:              Z ∼ Ber(0.5)                              ▷ e.g., choose a child at random
11:              (h, i) ← (h + 1, 2i − Z)
12:          end if
13:          P ← P ∪ {(h, i)}
14:      end while
15:      (H, I) ← (h, i)                                              ▷ The selected node
16:      Choose option x in P_{H,I} and evaluate it         ▷ Arbitrary selection of an option
17:      Receive corresponding reward Y
18:      T ← T ∪ {(H, I)}                                              ▷ Extend the tree
19:      for all (h, i) ∈ P do            ▷ Update the statistics T and μ̂ stored in the path
20:          T_{h,i} ← T_{h,i} + 1                       ▷ Increment the counter of node (h, i)
21:          μ̂_{h,i} ← (1 − 1/T_{h,i}) μ̂_{h,i} + Y/T_{h,i}      ▷ Update the mean μ̂_{h,i} of node (h, i)
22:      end for
23:      for all (h, i) ∈ T do                 ▷ Update the statistics U stored in the tree
24:          U_{h,i} ← μ̂_{h,i} + √((2 log n)/T_{h,i}) + ν₁ρʰ     ▷ Update the U−value of node (h, i)
25:      end for
26:      B_{H+1,2I−1} ← +∞                        ▷ B−values of the children of the new leaf
27:      B_{H+1,2I} ← +∞
28:      T′ ← T                                               ▷ Local copy of the current tree T
29:      while T′ ≠ {(0, 1)} do                       ▷ Backward computation of the B−values
30:          (h, i) ← LEAF(T′)                                    ▷ Take any remaining leaf
31:          B_{h,i} ← min{U_{h,i}, max{B_{h+1,2i−1}, B_{h+1,2i}}}       ▷ Backward computation
32:          T′ ← T′ \ {(h, i)}                                 ▷ Drop updated leaf (h, i)
33:      end while
34:  end for
```

We now define the c–near-optimality dimension, which characterizes the size of the sets $\mathcal{X}_{c\varepsilon}$ as a function of ε. It can be seen as some growth rate in ε of the metric entropy (measured in terms of ℓ and with packing numbers rather than covering numbers) of the set of $c\varepsilon$–optimal options.

Definition 16.2 (Near-optimality dimension). *For $c > 0$, the c–near-optimality dimension of f w.r.t. ℓ equals*

$$\max\left\{0, \limsup_{\varepsilon \to 0} \frac{\log \mathcal{N}(\mathcal{X}_{c\varepsilon}, \ell, \varepsilon)}{\log(\varepsilon^{-1})}\right\}.$$

Theorem 16.15 (Hierarchical Optimistic Optimization). *Let d be the $4\nu_1/\nu_2$–near-optimality dimension of the mean payoff function f w.r.t. ℓ. Then, for all $d' > d$, there exists a constant γ such that for all $n \geq 1$, HOO satisfies*

$$R_n = nf^* - \mathbb{E}\sum_{t=1}^{n} f(X_t) \leq \gamma\, n^{(d'+1)/(d'+2)} \left(\log n\right)^{1/(d'+2)}.$$

To put this result in perspective, we present the following example. Equip $\mathcal{X} = [0,1]^D$ with a norm $\|\cdot\|$ and assume that the mean payoff function f satisfies the Hölder-type property at any global maximum x^* of f (these maxima being additionally assumed to be in finite number):

$$f(x^*) - f(x) = \Theta\big(\|x - x^*\|^\alpha\big) \qquad \text{as} \quad x \to x^*,$$

for some smoothness order $\alpha \in [0, \infty)$. This means that there exist $c_1, c_2, \delta > 0$ such that for all x satisfying $\|x - x^*\| \leq \delta$,

$$c_2\|x - x^*\|^\alpha \leq f(x^*) - f(x) \leq c_1\|x - x^*\|^\alpha.$$

In particular, one can check that f is locally weakly Lipschitz for the dissimilarity defined by $\ell_{c,\beta}(x, y) = c\|x - y\|^\beta$, where $\beta \leq \alpha$ (and $c \geq c_1$ when $\beta = \alpha$) (see Bubeck et al. (2010) for a precise definition). We further assume that HOO is run with parameters ν_1 and ρ and a tree of dyadic partitions such that the assumptions of Section 16.4.3 are satisfied. The following statements can then be formulated on the regret of HOO:

■ **Known smoothness:** If we know the true smoothness of f around its maxima, then we set $\beta = \alpha$ and $c \geq c_1$. This choice $\ell_{c_1,\alpha}$ of a dissimilarity is such that f is locally weakly Lipschitz with respect to it and the near-optimality dimension is $d = 0$. Theorem 16.15 thus implies that the expected regret of HOO is $\tilde{O}(\sqrt{n})$, that is, *the rate of the bound is independent of the dimension D.*

- **Smoothness underestimated:** Here, we assume that the true smoothness of f around its maxima is unknown and that it is underestimated by choosing $\beta < \alpha$ (and some c). Then f is still locally weakly Lipschitz with respect to the dissimilarity $\ell_{c,\beta}$ and the near-optimality dimension is $d = D(1/\beta - 1/\alpha)$; the regret of HOO is $\tilde{O}\big(n^{(d+1)/(d+2)}\big)$.

- **Smoothness overestimated:** Now, if the true smoothness is overestimated by choosing $\beta > \alpha$ or $\alpha = \beta$ and $c < c_1$, then the assumption of weak Lipschitzness is violated and we are unable to provide any guarantee on the behavior of HOO. The latter, when used with an overestimated smoothness parameter, may lack exploration and exploit too heavily from the beginning. As a consequence, it may get stuck in some local optimum of f, missing the global one(s) for a very long time (possibly indefinitely). Such a behavior is illustrated in the example provided in Coquelin and Munos (2007) and shows the possible problematic behavior of the closely related algorithm UCT of Kocsis and Szepesvári (2006). UCT is an example of an algorithm overestimating the smoothness of the function; this is because the B–values of UCT are defined similarly to the ones of the HOO algorithm but without the additional third term in the definition of the U–values. In such cases, the corresponding B–values do not provide high-probability upper bounds on the supremum of f over the corresponding domains, and the resulting algorithms no longer implement the idea of "optimistism in the face of uncertainty".

16.5   References

J.-Y. Audibert. PAC-Bayesian aggregation and multi-armed bandits, 2010. Habilitation thesis, Université Paris Est, arXiv:1011.3396.

J.-Y. Audibert and S. Bubeck. Minimax policies for adversarial and stochastic bandits. In *Proceedings of the 22nd Annual Conference on Learning Theory.* Omnipress, 2009.

J.-Y. Audibert, R. Munos, and C. Szepesvári. Exploration-exploitation trade-off using variance estimates in multi-armed bandits. *Theoretical Computer Science*, 410(19):1876–1902, 2009.

J.-Y. Audibert, S. Bubeck, and R. Munos. Best arm identification in multi-armed bandits. In *Proceedings of the 23rd Annual Conference on Learning Theory*, 2010.

P. Auer, N. Cesa-Bianchi, and P. Fischer. Finite-time analysis of the multiarmed bandit problem. *Machine Learning Journal*, 47(2-3):235–256, 2002.

P. Auer, N. Cesa-Bianchi, Y. Freund, and R. Schapire. The non-stochastic multi-armed bandit problem. *SIAM Journal on Computing*, 32(1):48–77, 2003.

S. Bubeck. *Bandits Games and Clustering Foundations.* PhD thesis, Université Lille 1, 2010.

S. Bubeck, R. Munos, G. Stoltz, and C. Szepesvári. Online optimization in $\mathcal{X}$-

armed bandits. In D. Koller, D. Schuurmans, Y. Bengio, and L. Bottou, editors, *Advances in Neural Information Processing Systems 22*, pages 201–208, 2008.

S. Bubeck, R. Munos, and G. Stoltz. Pure exploration in multi-armed bandits problems. In *Proceedings of the 20th International Conference on Algorithmic Learning Theory*, pages 29–37, 2009.

S. Bubeck, R. Munos, G. Stoltz, and C. Szepesvári. X-armed bandits. arXiv preprint 1001.4475, 2010.

A. Burnetas and M. Katehakis. Optimal adaptive policies for sequential allocation problems. *Advances in Applied Mathematics*, 17(2):122–142, 1996.

P.-A. Coquelin and R. Munos. Bandit algorithms for tree search. In *Proceedings of the 23rd Conference on Uncertainty in Artificial Intelligence*, pages 67–74, 2007.

P. Dagum, R. Karp, M. Luby, and S. Ross. An optimal algorithm for Monte Carlo estimation. *SIAM Journal on Computing*, 29(5):1484–1496, 2000.

C. Domingo, R. Gavaldà, and O. Watanabe. Adaptive sampling methods for scaling up knowledge discovery algorithms. *Data Mining and Knowledge Discovery*, 6 (2):131–152, 2002.

J. Doob. *Stochastic processes*. John Wiley, New York, 1953.

E. Even-Dar, S. Mannor, and Y. Mansour. Action elimination and stopping conditions for the multi-armed bandit and reinforcement learning problems. *Journal of Machine Learning Research*, 7:1079–1105, 2006.

D. Freedman. On tail probabilities for martingales. *The Annals of Probability*, 3 (1):100–118, 1975.

W. Hoeffding. Probability inequalities for sums of bounded random variables. *Journal of the American Statistical Association*, 58(301):13–30, 1963.

L. Kocsis and C. Szepesvári. Bandit based Monte-carlo planning. In *Proceedings of the 15th European Conference on Machine Learning*, pages 282–293, 2006.

T. L. Lai and H. Robbins. Asymptotically efficient adaptive allocation rules. *Advances in Applied Mathematics*, 6(1):4–22, 1985.

G. Lugosi. Concentration-of-measure inequalities. *Lecture notes*, 1998.

S. Mannor and J. N. Tsitsiklis. The sample complexity of exploration in the multi-armed bandit problem. *Journal of Machine Learning Research*, 5:623–648, 2004.

O. Maron and A. W. Moore. Hoeffding races: Accelerating model selection search for classification and function approximation. In *Advances in Neural Information Processing Systems*, pages 59–66, 1993.

P. Massart. *Concentration inequalities and model selection: Ecole d'Eté de Probabilités de Saint-Flour XXXIII-2003*. Springer, 2007.

A. Maurer and M. Pontil. Empirical bernstein bounds and sample-variance penalization. In *Proceedings of the 22th Annual Conference on Learning Theory*, 2009.

V. Mnih, C. Szepesvári, and J.-Y. Audibert. Empirical bernstein stopping. In *Proceedings of the 25th International Conference on Machine Learning*, pages 672–679, 2008.

L. E. Ortiz and L. P. Kaelbling. Sampling methods for action selection in influence diagrams. In *Proceedings of the National Conference on Artificial Intelligence*, pages 378–385, 2000.

H. Robbins. Some aspects of the sequential design of experiments. *Bulletin of the*

American Mathematics Society, 58:527–535, 1952.

D. Williams. *Probability with martingales*. Cambridge University Press, 1991.

17   Optimization Methods for Sparse Inverse Covariance Selection

Katya Scheinberg   katyas@lehigh.edu
Department of Industrial and Systems Engineering
Lehigh University
Bethlehem, PA 18015-1582

Shiqian Ma   sm2756@columbia.edu
Department of Industrial Engineering and Operations Research
Columbia University
New York, NY 10027

17.1   Introduction

In many practical applications of statistical learning the objective is not simply to construct an accurate predictive model, but rather to discover meaningful interactions among the variables. For example, in applications such as reverse engineering of gene networks, discovery of functional brain connectivity patterns from brain-imaging data, and analysis of social interactions, the main focus is on reconstructing the network structure representing dependencies among multiple variables, such as genes, brain areas, and individuals. Probabilistic graphical models, such as Markov networks (or Markov random fields), provide a statistical tool for multivariate data analysis that allows the capture of interactions such as conditional independence relationships between variables. We focus on the task of learning the structure of a Markov network over Gaussian random variables, which is equivalent to learning the zero pattern of the inverse covariance matrix. A standard approach is to choose the sparsest network (inverse covariance matrix) that adequately explains the data. This can be achieved by solving a regularized maximum likelihood problem with the regularization term involving the number of nonzeros (ℓ_0-norm) in the inverse covariance matrix—

a generally intractable problem that is often solved approximately by greedy search (Heckerman, 1995). Recently, however, novel tractable approximations have been suggested that exploit the sparsity-enforcing property of ℓ_1-norm regularization and yield convex optimization problems (Meinshausen and Buhlmann, 2006; Wainwright et al., 2007; Yuan and Lin, 2007; Banerjee et al., 2008; Friedman et al., 2007). In this chapter we focus on one such convex formulation, often referred to as sparse inverse covariance selection (SICS), and describe several optimization approaches to this problem.

17.1.1 Problem Formulation

Let $\mathcal{S}$ be a set of p random variables with joint distribution $P(\mathcal{S})$. It is common to assume a multivariate Gaussian probability density function over $\mathcal{S}$, hence, if $\mathbf{s}$ is a p-dimensional vector which is a realization of p random variables, then

$$p(\mathbf{s}) = (2\pi)^{-p/2} \det(\Sigma)^{-\frac{1}{2}} e^{-\frac{1}{2}(\mathbf{s}-\mu)^T \Sigma^{-1}(\mathbf{s}-\mu)}, \tag{17.1}$$

where μ is the mean and Σ is the covariance matrix of the distribution, respectively. Without loss of generality we assume that the data are centered so that $\mu = 0$; hence the purpose is to estimate Σ. Introducing $X := \det(\Sigma)^{-1}$, we can rewrite (17.1) as

$$p(\mathbf{s}) = (2\pi)^{-p/2} \det(X)^{\frac{1}{2}} e^{-\frac{1}{2}\mathbf{s}^\top X \mathbf{s}}. \tag{17.2}$$

Missing edges in the above graphical model correspond to zero entries in the inverse covariance matrix X, and vice versa (Lauritzen, 1996), and thus the problem of structure learning for the above probabilistic graphical model is equivalent to the problem of learning the zero pattern of the inverse covariance matrix. Note that the maximum likelihood estimate of the covariance matrix Σ is the empirical covariance matrix $S = \frac{1}{n}\sum_{i=1}^{n} \mathbf{s}_i \mathbf{s}_i^\top$ where $\mathbf{s}_i$ is the ith sample, $i = 1, ..., n$. The inverse of S, even if it exists, does not typically contain any elements that are exactly zero. Therefore an explicit sparsity-enforcing constraint needs to be added to the estimation process.

A common approach is to include the (vector) ℓ_1-norm of X as a penalty term in the objective function, which is equivalent to imposing a Laplace prior on Σ^{-1} in a maximum likelihood framework (Friedman et al., 2007; Banerjee et al., 2008; Yuan and Lin, 2007). Formally, the entries X_{ij} of the inverse covariance matrix are assumed to be independent random variables, each following a Laplace distribution

$$p(X_{ij}) = \frac{\lambda}{2} e^{-\lambda |X_{ij} - \alpha_{ij}|} \tag{17.3}$$

with zero location parameter (mean) α_{ij}, yielding

$$p(X) = \prod_{i=1}^{p}\prod_{j=1}^{p} p(X_{ij}) = (\lambda/2)^{p^2} e^{-\lambda\|X\|_1}, \qquad (17.4)$$

where $\|X\|_1 = \sum_{ij}|X_{ij}|$ is the (vector) ℓ_1-norm of X. Then the objective is to find the maximum log-likelihood solution $\arg\max_{X\succ0}\log p(X|\mathbf{S})$, where $\mathbf{S}$ is the $n \times p$ data matrix whose rows are given by $\mathbf{s}^\top$, and $X \succ 0$ denotes that X is positive definite. Invoking Bayes's rule, $p(X|\mathbf{S}) = P(\mathbf{S}|X)P(X)/p(\mathbf{S})$, this max-likelihood estimate can be obtained by

$$\arg\max_{X\succ0} \quad \log\prod_{i=1}^{n}\left[\frac{\det(X)^{\frac{1}{2}}}{(2\pi)^{p/2}}e^{-\frac{1}{2}\mathbf{s}_i^\top X\mathbf{s}_i}\right] + \log[(\lambda/2)^{p^2}e^{-\lambda\|X\|_1}]. \qquad (17.5)$$

We write $\frac{1}{n}\sum_{i=1}^{n}\mathbf{s}_i^\top X\mathbf{s}_i = \langle S, X\rangle$. This yields the following optimization problem (also see Friedman et al. (2007); Banerjee et al. (2008); Yuan and Lin (2007)):

$$\max_{X\succ0} \quad \log\det(X) - \langle S, X\rangle - \rho\|X\|_1, \qquad (17.6)$$

where $\rho = \frac{2}{n}\lambda$.

More generally, one can consider the following formulation:

$$\max_{X\succ0} \quad \log\det(X) - \langle S, X\rangle - \sum_{ij}M_{ij}|X_{ij}|. \qquad (17.7)$$

If M is a product of $\rho = \frac{2}{n}\lambda$ and E is the matrix of all ones, then problem formulation (17.7) reduces to (17.6).

Note that by allowing the matrix M to have arbitrary nonnegative entries, we automatically include in the formulation the case where the diagonal elements of X are not penalized or the case when the absolute values of the entries of X are scaled by their estimated value, as was considered in Yuan and Lin (2007).

We will refer to (17.6) and (17.7) as the SICS problem. The two formulations are very similar in terms of optimization effort. In some cases, for brevity we present methods for (17.6) and explain its extension for (17.7) afterward.

The dual of (17.7) can be written as (Banerjee et al., 2008)

$$\max_{W\succ0}\{\log\det(W) - p : \text{s.t.} \ -M \le W - S \le M\}, \qquad (17.8)$$

where the inequalities involving matrices W, S, and M are element-wise.

Problems (17.6) and (17.7) are both strictly convex, and as long as the off-diagonal elements of M are positive, the optimal solution is always attained.

The optimality conditions for this pair of primal and dual problems imply that $W = X^{-1}$ and that $W_{ij} - S_{ij} = M_{ij}$ if $X_{ij} > 0$ and $W_{ij} - S_{ij} = -M_{ij}$ if $X_{ij} < 0$. These optimality conditions are imperative when the primal sparsity structure needs to be recovered from the dual solution. Several existing methods solve (17.8) and obtain an approximate solution to (17.7) by inverting the solution of (17.8). Since the obtained dual solution is also approximate, the resulting primal approximate solution is not sparse. However, it is often the case that the dual solution is an accurate projection onto the dual feasible set. Hence, for the "true" nonzero elements of X_{ij} the appropriate complementarity conditions $W_{ij} - S_{ij} = M_{ij}$ and $W_{ij} - S_{ij} = -M_{ij}$ are observed from the solution obtained for (17.8).

17.1.2   Overview of Optimization Approaches

Problems (17.6) and (17.7) are special cases of a semidefinite programming problem (SDP) (Wolkowicz et al., 2000) which can be solved in polynomial time by interior-point methods (IPM). However, as is well known, each iteration of an IPM applied to a semidefinite programming problem of size p requires up to $O(p^6)$ operations and $O(p^4)$ memory space, which is very costly. Although an approximate IPM has recently been proposed for the SICS problem (Li and Toh, 2010), another reason that using an IPM is undesirable for our problem is that an IPM does not produce the sparsity pattern of the solution. The sparsity is, in theory, recovered in the limit. In practice it is recovered by thresholding the elements of an approximate solution; hence, numerical inaccuracy can interfere with the structure recovery.

As an alternative to IPMs, more efficient approaches, COVSEL and glasso, were developed for problem (17.6) in Banerjee et al. (2008) and (Friedman et al., 2007). The methods are similar in that they are based on applying a block coordinate descent (BCD) method to the dual of (17.6). At each iteration only one row (and the corresponding symmetric column) of the dual matrix is optimized while the rest of that matrix remains fixed. The resulting subproblem is a convex quadratic problem. The difference between the COVSEL method, described in Banerjee et al. (2008), and the glasso method in Friedman et al. (2007), is that COVSEL solves the subproblems via an interior-point approach, while glasso poses the subproblem as a dual of the Lasso problem (Tibshirani, 1996) and utilizes a coordinate descent (CD) approach (Tibshirani, 1996) to solve the resulting Lasso subproblem. Due to the use of the coordinate descent, the sparsity of the primal matrix is recovered more accurately than with the interior-point approach, and the glasso method (Friedman et al., 2007) is faster than COVSEL because it takes advantage of that sparsity. On the other hand, subproblems solved by glasso are solved by coordinate descent up to an unknown accuracy; hence, the current implementation of this method lacks rigor. A recent row-by-row (RBR) method for general SDP (Wen et al., 2009) is based on the same idea of updating one row and column at a time, like glasso and COVSEL, but can also be applied directly to the primal matrix. The resulting subproblem is in general a second-order cone problem (SOCP), while in the dual case it reduces to a convex quadratic problem (QP). It follows from the result in Wen et al. (2009) that the BCD approach converges to the optimal solution when applied to the primal and dual SICS formulations. The SINCO method proposed by Scheinberg and Rish (2009) is a greedy coordinate descent method applied to the primal problem. We will describe the approaches of glasso and SINCO in more detail below. Sun et al. (2009) propose solving the primal problem (17.6) by using a BCD method. They formulate the subproblem as a min-max problem and solve it using a prox-method proposed by Nemirovski (2005). All of these BCD and CD approaches lack iteration complexity bounds as of now.

As alternatives to block coordinate based approaches several gradient based approaches, for problem (17.6) have been suggested. A projected gradient method for solving the dual problem (17.8) was proposed by Duchi et al. (2008). Variants of Nesterov's method (Nesterov, 2005, 2004) were applied to solving the SICS problem by d'Aspremont et al. (2008) and Lu (2009, 2010). d'Aspremont et al. (2008) apply Nesterov's optimal first-order method to solve the primal problem (17.6) after smoothing the nonsmooth ℓ_1-term, obtaining an iteration complexity bound of $O(1/\varepsilon)$ for an ε-optimal solution. Lu solves the dual problem (17.8), which is a smooth problem, by Nesterov's algorithm, and improves the iteration complexity to $O(1/\sqrt{\varepsilon})$. However, since the practical performance of this algorithm is not attractive, Lu gives a variant of it (VSM) with unknown complexity that exhibits better performance. Yuan (2009) proposes an alternating direction method based on an augmented Lagrangian framework. Goldfarb et al. (2009) and Scheinberg et al. (2010) have developed an alternating linearization method with $O(1/\sqrt{\varepsilon})$ complexity which is similar to the augmented Lagrangian approach. A proximal point algorithm was proposed by Wang et al. (2009) which requires a reformulation which increases the size of the problem. The IPM in Li and Toh (2010) also requires such a reformulation.

In this chapter we give details of BCD approaches developed in Friedman et al. (2007) and Scheinberg and Rish (2009). The fundamental difference between the two approaches is in the choice of the next coordinates that are updated. From the class of the first-order methods we choose to present the alternating linearization method from Goldfarb et al. (2009) and Scheinberg et al. (2010), which currently appears to be the most efficient approach for SICS (see Scheinberg et al. (2010) for computational comparison).

17.1.3　Preliminaries

17.1.3.1　*Properties of Matrix Determinant and Inverse*

Let $X \in S_{++}^n$ be a positive definite matrix. We will list here a few useful properties from linear algebra.

Let X be partitioned as

$$X = \begin{pmatrix} \xi & y^\top \\ y & B \end{pmatrix}$$

where $\xi \in \mathbb{R}$, $y \in \mathbb{R}^{n-1}$, and $B \in S_{++}^{n-1}$. Then

$$\det X = (\xi - y^\top B^{-1} y) \det B. \tag{17.9}$$

Consider a matrix $X + uv^\top$, where $u, v \in \mathbb{R}^n$, then

$$\det(X + uv^\top) = \det(X)(1 + v^\top X^{-1} u) \tag{17.10}$$

and

$$(X + uv^\top)^{-1} = X^{-1} - X^{-1} uv^\top X^{-1} / (1 + v^\top X^{-1} u). \tag{17.11}$$

The last expression is well known as the Sherman-Morrison-Woodbury formula.

17.1.3.2　*Soft Thresholding and Shrinkage*

We will use the well-known fact that the solution to the following optimization problem,

$$\min_x \frac{1}{2} \|x - r\|_2^2 + \lambda \|x\|_1, \tag{17.12}$$

where x and r are vectors in $\mathbb{R}^n$, can be obtained in closed form by a shrinkage operator $x^* = \mathrm{shrink}(r, \lambda)$, where

$$x_i^* = \mathrm{shrink}(r_i, \lambda) = \begin{cases} r_i - \lambda & \text{if } r_i \geq \lambda \\ 0 & \text{if } -\lambda < r_i < \lambda \\ r_i + \lambda & \text{if } r_i \leq -\lambda. \end{cases} \tag{17.13}$$

17.1.3.3   *Coordinate Descent for Lasso*

Let us consider a general Lasso problem (Tibshirani, 1996),

$$\min_x \ \frac{1}{2}\|Ax - b\|_2^2 + \rho\|x\|_1, \tag{17.14}$$

where $A \in \mathbb{R}^{m \times n}$, $b \in \mathbb{R}^m$ and $x \in \mathbb{R}^n$. The idea of a coordinate descent approach is at each step to fix all elements of x except for the ith element and to optimize the objective function for only one variable x_i. Let $\bar{x}$ denote the fixed part of vector x, and $\bar{A}$ the part of matrix A that corresponds to $\bar{x}$. Writing the reduced problem omitting the terms that do not depend on x_i, we get

$$\min_{x_i} \ \frac{1}{2}x_i^\top A_i^\top A_i x_i + \bar{x}^\top \bar{A}^\top A_i x_i - b^\top A_i x_i + \rho|x_i|, \tag{17.15}$$

where A_i is the ith column of A. Let $r_i = -A_i^\top(\bar{A}\bar{x} - b)/\|A_i\|^2$ and $\eta_i = \rho/\|A_i\|^2$; then problem (17.15) is equivalent to

$$\min_{x_i} \ \frac{1}{2}(x_i - r_i)^2 + \eta_i|x_i|, \tag{17.16}$$

which is solved by the soft thresholding step $x_i^* = \mathrm{shrink}(r_i, \eta_i)$. Hence, each step of coordinate descent for Lasso requires computing r_i and η_i. The cost of this computation depends on the manner in which different parts of the expression for r are stored and updated. For instance, if the elements of $A^\top A$ are precomputed and stored, then at each step all r_i's can be updated in $O(n)$ operations. Alternatively, the residual $Ax - b$ can be stored and updated at the cost of $O(m)$ storage and operations, in which case each r_i can be computed in $O(m)$ operations whenever it is required.

17.2   Block Coordinate Descent Methods

17.2.1   Row-by-Row Method

For simplicity let us consider the case when $M = \rho E$, that is, formulation (17.6). Instead of solving (17.6) directly, the approaches in Banerjee et al. (2008) and Friedman et al. (2007) consider the dual

$$\max_{W \succ 0}\{\log \det(W) - p : \text{s.t. } \|W - S\|_\infty \leq \rho\}. \tag{17.17}$$

The subproblems solved at each iteration of the BCD methods in Banerjee et al. (2008) and Friedman et al. (2007) are constructed as follows. Given a positive definite matrix $W \succ 0$, W and S are partitioned conformally as

$$W = \begin{pmatrix} \xi & y^\top \\ y & B \end{pmatrix} \text{ and } S = \begin{pmatrix} \xi_S & y_S^\top \\ y_S & B_S \end{pmatrix},$$

where $\xi, \xi_S \in \mathbb{R}$, $y, y_S \in \mathbb{R}^{n-1}$, and $B, B_S \in S^{n-1}$. It follows from (17.9) that $\log \det W = \log(\xi - y^\top B^{-1} y) + \log \det B$, and B is fixed, so the BCD subproblem for (17.6) becomes the quadratic program

$$\min_{[\xi; y]} y^\top B^{-1} y - \xi, \quad \text{s.t.} \quad \|[\xi; y] - [\xi_S; y_S]\|_\infty \leq \rho, \; \xi \geq 0. \qquad (17.18)$$

Note that (17.18) is separable in y and ξ. The solution ξ is equal to $\xi_S + \rho$. In fact, the first-order optimality conditions of (17.6) and $X \succ 0$ imply that $W_{ii} = S_{ii} + \rho$ for $i = 1, \dots, n$. Hence, problem (17.18) reduces to

$$\min_y y^\top B^{-1} y, \quad \text{s.t.} \quad \|y - y_S\|_\infty \leq \rho. \qquad (17.19)$$

The BCD method in Banerjee et al. (2008) solves a sequence of constrained problems (17.19). After each step the duality gap for (17.6) at the current iterate W^k can be obtained as $\langle (W^k)^{-1}, S \rangle - p + \rho \|(W^k)^{-1}\|_1$. The BCD method proposed in Banerjee et al. (2008) for solving (17.17) is outlined in algorithm 17.1.

Algorithm 17.1 Block coordinate descent method for (17.17)

1:   Set $W^1 = S + \rho I$, $k := 1$, and $\varepsilon \geq 0$.
2:   **while** $\langle (W^k)^{-1}, S \rangle - p + \rho \|(W^k)^{-1}\|_1 \geq \varepsilon$ **do**
3:       **for** $i = 1, \cdots, p$ **do**
4:           Set $B := W^k_{i^c, i^c}$, $y_S = S_{i^c, i}$, and $B_S = S_{i^c, i^c}$.
5:           Solve (17.19) to get y.
6:           Update $W^k_{i^c, i} := y$ and $W^k_{i, i^c} := y^\top$.
7:       **end for**
8:       Set $W^{k+1} := W^k$ and $k := k + 1$.
9:   **end while**

Each instance of (17.19) is solved by applying an interior-point quadratic programming solver. Also, the inverse of W^k is computed at each iteration. This makes each iteration of the BCD costly. Moreover, the sparsity of the primal solution $X^k = (W^k)^{-1}$ is not exploited, and is obtained only in the limit. Since the BCD approach typically does not produce accurate solutions, this sparsity is hard to recover accurately.

The glasso method proposed in Friedman et al. (2007) is based on algorithm 17.1 except for the method of solving (17.19). Specifically, it can be easily verified that the dual of (17.19) is

$$\min_x \; x^\top B x - y_S^\top x + \rho\|x\|_1, \tag{17.20}$$

which is also equivalent to

$$\min_x \; \left\| B^{\frac{1}{2}} x - \tfrac{1}{2} B^{-\frac{1}{2}} y_S \right\|_2^2 + \rho\|x\|_1. \tag{17.21}$$

If x solves (17.21), then $y = Bx$ solves (17.19).

Problem (17.21) is equivalent to the Lasso problem (Tibshirani, 1996). The Lasso problem is then solved using a coordinate descent algorithm, which does not require computation of either $B^{\frac{1}{2}}$ or $B^{-\frac{1}{2}}$. Indeed, if we recall the subproblem (17.16) which is generated and solved at each coordinate descent step for (17.14) we see that to compute the scalars r_i and η_i we require only elements of B and y, since $B_i^{\frac{1}{2}} B_i^{\frac{1}{2}} = B_i$ and $B_i^{\frac{1}{2}} B_i^{-\frac{1}{2}} y_i = y_i$.

For each BCD iteration, the resulting solution of the Lasso subproblem (17.21) has the same nonzero pattern as the corresponding row of $X^k = (W^k)^{-1}$, which can be viewed as the current estimate of the inverse of the covariance. Hence, the sparsity pattern is explicitly available and is exploited by the glasso algorithm. This makes glasso significantly more efficient than COVSEL (developed in Banerjee et al. (2008)). In terms of CPU time glasso is currently the most efficient block coordinate descent approach to the SICS problem. On the other hand, it is unclear if the convergence results of BCD apply to glasso due to the way the subproblems are addressed in the implementation.

It is easy to extend the BCD approach to the more general formulation (17.7). One needs to consider the appropriate partitioning of M,

$$M = \begin{pmatrix} \xi_M & y_M^\top \\ y_M & B_M \end{pmatrix}.$$

The bound constraints in (17.19) become $-y_M \le (y - y_S) \le y_M$ and $\xi = \xi_s + \xi_M$. Both glasso and COVSEL easily extend to this formulation.

The BCD approach cycles through each row/column of W one by one. Hence the nonzeros in the inverse covariance matrix generated by *glasso* are generated by rows. This may introduce small nonzero elements, which are undesirable in the solution. It is possible to modify this method so that only one element in each row is updated at a time, thus bringing the whole approach closer to the simple coordinate descent used for Lasso. However, it is unclear how to efficiently select the next working variable to be updated. A simple cycling rule would be equivalent to the approach in Friedman et al. (2007) which we just described. The method in the next subsection is a primal greedy coordinate descent, which addresses the issue of the working variable selection.

17.2.2   Primal Greedy Coordinate Descent

We now describe an algorithm which addresses the primal problem directly and also uses coordinate descent,[1] which naturally preserves the sparsity of the solution. The method is referred to as SINCO (Sparse INverse COvariance) and is introduced in Scheinberg and Rish (2009). Unlike the dual BCD approach of COVSEL, glasso, and the general RBR, SINCO optimizes only one diagonal or two (symmetric) off-diagonal entries of the matrix X at each step. The advantages of this approach are that only one nonzero entry (discounting the matrix symmetry) can be introduced at each step, and that the solution to each subproblem is available in closed form as a root of a quadratic equation. Computation at each step requires a constant number of arithmetic operations, independent of p. Hence, in $O(p^2)$ operations a potential step can be computed for *all pairs* of symmetric elements (i.e., for all pairs (i, j)). Then the step which provides the *best* objective function value improvement can be chosen, which is the essence of the greedy nature of this approach. Once the step is taken, the update of the gradient information requires $O(p^2)$ operations. Hence, overall, each iteration takes $O(p^2)$ operations. Note that each step is also suitable for massive parallelization.

In comparison, glasso and COVSEL (and RBR) require solution of a quadratic programming problem whose theoretical and empirical complexity varies depending on the method used, but always exceeds $O(p^2)$. These algorithms apply optimization to each row/column consecutively; hence the greedy nature is lacking. On the other hand, a whole row and a whole column are optimized at each step, thus reducing the overall number of steps. As is shown in Scheinberg and Rish (2009), SINCO, in a serial mode, is comparable to glasso, which is orders of magnitude faster than COVSEL, according to the results in Friedman et al. (2007). Also, SINCO may lead to a lower false-positive error than glasso since it introduces nonzero elements greedily. On the other hand, it may suffer from introducing too few nonzeros and thus misses some of the true positives, especially on dense networks.

Perhaps the most interesting consequence of SINCO's greedy nature is that it reproduces the regularization path behavior while using only one value of the regularization parameter ρ. We will discuss this property further

---

1. We should call it "coordinate ascent" since we are solving a maximization problem; however, we use the word "descent" to adhere to standard terminology.

after we describe the algorithm.

17.2.2.1 Algorithm Description

The main idea of the method is that at each iteration, the matrix X is updated by changing one element on the diagonal or two symmetric off-diagonal elements. This implies the change in X that can be written as $X + \theta(e_i e_j^\top + e_j e_i^\top)$, where i and j are the indices corresponding to the elements that are being changed. The key observation is that given the matrix $W = X^{-1}$, the exact line search that optimizes the objective function of problem (17.7) along the direction $e_i e_j^\top + e_j e_i^\top$ reduces to a solution of a quadratic equation, as we will show below. Hence, each such line search takes a constant number of operations. Moreover, given the starting objective value, the new function value on each step can be computed in a constant number of steps. This means that we can perform such line search for all (i, j) pairs in $O(p^2)$ time, which is linear in the number of unknown variables X_{ij}. We then can choose the step that gives the best improvement in the value of the objective function. After the step is chosen, the dual matrix $W = X^{-1}$ and, hence, the objective function gradient are updated in $O(p^2)$ operations.

We now describe the method. For a fixed pair (i, j) consider now the update of X of the form $X(\theta) = X + \theta(e_i e_j^\top + e_j e_i^\top)$, such that $X \geq 0$. Let us consider the objective function as the function of θ:

$$f(\theta) = \log\det(X + \theta e_i e_j^\top + \theta e_j e_i^\top) - \tag{17.22}$$

$$\langle S, X + \theta e_i e_j^\top + \theta e_j e_i^\top \rangle - \sum_{i,j=1}^{p} M_{ij}|X + \theta e_i e_j^\top + \theta e_j e_i^\top|. \tag{17.23}$$

We use the property of the determinant (17.10) and the Sherman-Morrison-Woodbury formula (17.11) to obtain

$$\det(X + \theta e_i e_j^\top + \theta e_j e_i^\top) = \det(X + \theta e_j e_i^\top)(1 + \theta e_j^\top (X + \theta e_j e_i^\top)^{-1} e_i)$$
$$= \det(X)(1 + \theta e_i^\top X^{-1} e_j)(1 + \theta e_j^\top X^{-1} e_i - \theta^2 e_j^\top X^{-1} e_j (1 + \theta e_i^\top X^{-1} e_j)^{-1} e_i^\top X^{-1} e$$
$$= \det(X)(1 + 2\theta e_i^\top X^{-1} e_j + (\theta e_j^\top X^{-1} e_i)^2 - \theta^2 e_i^\top X^{-1} e_i e_j^\top X^{-1} e_j).$$

Given the dual solution $W = X^{-1}$, we can write the above as

$$\det(X + \theta e_i e_j^\top + \theta e_j e_i^\top) = \det(X)(1 + 2\theta W_{ij} + \theta^2(W_{ij}^2 - W_{ii}W_{jj})).$$

We define the function $g(\theta)$ as

$$g(\theta) := 1 + 2\theta W_{ij} + \theta^2(W_{ij}^2 - W_{ii}W_{jj}).$$

Recalling that W and S are symmetric, but M is not necessarily so, maximizing the objective function $f(\theta)$ over θ is equivalent to

$$\max_{\theta} \ \log(g(\theta)) - 2S_{ij}\theta - (M_{ij} + M_{ji})|X_{ij} + \theta|. \qquad (17.24)$$

This problem can be rewritten as

$$\max_{\theta} \ \min_{u \in \mathbb{R}:\, |u| \leq M_{ij} + M_{ji}} \ \log(g(\theta)) - 2S_{ij}\theta - u(X_{ij} + \theta). \qquad (17.25)$$

Swapping the min and the max, we have as the inner problem

$$\max_{\theta} f_u(\theta) = \log g(\theta) - 2S_{ij}\theta - u(X_{ij} + \theta), \qquad (17.26)$$

which can be solved by setting the derivative of the objective with respect to θ to zero.

$$f_u'(\theta) = \frac{g'(\theta)}{g(\theta)} - 2S_{ij} - u = \frac{W_{ij} + \theta(W_{ij}^2 - W_{ii}W_{jj})}{\theta^2(W_{ij}^2 - W_{ii}W_{jj}) + 1 + 2\theta W_{ij}} - 2S_{ij} - u.$$

To find the maximum of $f(\theta)$, we need to find θ for which $f'(\theta) = 0$. Letting a denote $W_{ii}W_{jj} - W_{ij}^2$, this condition can be written as

$$W_{ij} - 2S_{ij} - u - (a + 2W_{ij}(2S_{ij} + u))\theta + a(2S_{ij} + u)\theta^2 = 0.$$

If we know the value of u, we can obtain the optimal θ from the above quadratic equation. From the optimality conditions for the primal-dual problem (17.25) we know that for optimal θ

$$
\begin{aligned}
X_{ij} + \theta &\geq 0 \quad \text{if } u = M_{ij} + M_{ji} \\
X_{ij} + \theta &\leq 0 \quad \text{if } u = -M_{ij} - M_{ji} \\
X_{ij} + \theta &= 0 \quad \text{if } -M_{ij} - M_{ji} < u < M_{ij} + M_{ji}.
\end{aligned}
$$

Hence, considering the three different scenarios, one can find the optimal solution to (17.25) via solving quadratic equations in a constant number of steps, which does not depend on p. In Scheinberg and Rish (2009) the details of solving the quadratic equation are given along with the proof that the solution corresponding to the maximum of $f(\theta)$ always exists.

Once the optimal θ is computed, the objective function improvement is easily obtained from $f(\theta) - f(0)$, where $f(0)$ is the objective function value from the previous iteration. Since for each (i, j) pair the optimal θ and the objective function value improvement can be computed in a constant number of operations, then in $O(p^2)$ operations one can compute the (i, j) pair, which gives the largest objective function improvement. Once such a pair is determined, the appropriate update to X can be performed and the iteration is completed.

The inverse $\bar{W}$ of $\bar{X} = X + \theta e_i e_j + \theta e_j e_i$ is obtained, according to the Sherman-Morrison-Woodbury formula, in $O(p^2)$ operations as follows:

$$\begin{cases} \bar{W} &= W - \theta(\kappa_1 W_i W_j^\top + \kappa_2 W_i W_i^\top + \kappa_3 W_j W_j^\top + \kappa_1 W_j W_i^\top) \\ \kappa_1 &= -(1 + \theta W_{ij})/\kappa \\ \kappa_2 &= \theta W_{jj}/\kappa \\ \kappa_3 &= \theta W_{ii}/\kappa \\ \kappa &= \theta^2(W_{ii}W_{jj} - W_{ij}^2) - 1 - 2\theta W_{ij} \end{cases} \qquad (17.27)$$

We outline the steps of the SINCO method in algorithm 17.2.

Algorithm 17.2 Coordinate descent method for (17.7)

1:   Set $X^1 = I$, set $k := 1$, $\theta^k = 0$, and $\varepsilon \geq 0$. Compute $f(X^1)$.
2:   **while** $f(\theta^k) - f(X^k) \geq \varepsilon$ **do**
3:       **for** $i = 1, \cdots, p$, $j = i, \cdots, p$ **do**
4:           Compute $\theta_{ij} = \arg\max_\theta f(\theta)$ where $f(\theta)$ is defined by (17.22)
5:           If $f(\theta^k) > f(\theta_{ij})$, $\theta^k = \theta_{ij}$, $(i, j)^k = (i, j)$
6:
7:       **end for**
8:       Update W^k according to (17.27) for $(i, j) = (i, j)^k$.
9:   **end while**

The overall per-iteration complexity of SINCO is $O(p^2)$. Moreover, this algorithm lends itself readily to massive parallelization. Indeed, at each iteration of the algorithm the step computation for each (i, j) pair can be parallelized, and the procedure that updates W involves simply adding to each element of W a function that involves only two rows of W. Hence, the updates can be also done in parallel, and in very large-scale cases the matrix W can also be stored in a distributed manner. The same is true of the storage of matrices S and M (assuming that M needs to be stored, that is, not all elements of M are the same), while the best way to store the X matrix may be in sparse form.

The convergence of the method follows from the convergence of a block coordinate descent method on a strictly convex objective function, as is shown for the RBR method in Wen et al. (2009). The only constraints are box constraints (nonnegativity), and they do not hinder the convergence. In the case of SINCO we extensively use the fact that each coordinate descent step is cheap and, unlike the glasso algorithm, we select the next step based on the best function value improvement. On the other hand, we maintain both the primal matrix X and the inverse matrix W, while glasso method does not. However, none of these differences prevent the convergence result for RBR in Wen et al. (2009) to apply to both methods.

17.2.3   Regularization Path

One of the main challenges in sparse inverse covariance selection is the proper choice of the weight matrix M in (17.6). Typically M is chosen to be a multiple of the matrix of all ones as in the formulation (17.6). The multiplying coefficient ρ is called the regularization parameter. Clearly, for large values of ρ as $\rho \to \infty$, the solution to (17.6) is likely to be very sparse and eventually diagonal, which means that no structure recovery is achieved. On the other hand, if ρ is small as $\rho \to 0$, the solution X is likely to be dense and eventually approach S^{-1} and, again, no structure recovery occurs. Hence, exploration of a regularization path is an integral part of the sparse inverse covariance selection.

Typically, problem (17.6) is solved for several values of ρ in a predefined range and the best value, according to some criteria, is selected. The reason only a scalar parameter ρ is usually considered is that it is expensive to explore solutions along a multi-dimensional grid.

The work in Krishnamurthy and d'Aspremont (2009), also presented in this volume, addresses an algorithm for the SICS problem that computes the entire regularization path by a path-following method.

In this section, we concentrate on computing the regularization path (or some parts of it) by solving the SICS problem for a finite range of values of ρ rather than the complete path. All methods described in this chapter, including the alternating linearization method described below, are very well suited for the efficient computation of the regularization path, since it directly exploits warm starts. When ρ is relatively large, a very sparse solution can be obtained quickly. This solution can be used as a warm start to the problem with a smaller value of ρ and, if the new value of ρ is not too small compared with the previous value, then the new solution is typically obtained in just a few iterations, because the new solution has only a few extra nonzero elements.

The regularization path is evaluated via the ROC curves showing the trade-off between the number of true positive (TP) elements recovered and the number of false positive (FP) elements. Producing better curves (where the number of TPs rises fast relative to FPs) is usually an objective of any method that does not focus on specific ρ selection. An interesting property of SINCO is that it introduces nonzero entries into the matrix X as it progresses. Hence, if one uses looser tolerance and stops the algorithm early, then a sparser solution is obtained for any specific value of ρ. What we observe, as seen in figure 17.1, is that if we apply SINCO to problem (17.6) with ever tighter tolerance, the ROC curves obtained from the tolerance solution path match the ROC curves obtained from the regularization path. Here we show several examples of the matching ROC curves for various random networks (see Scheinberg and Rish (2009), for details of the experiments). The ROC curves of the regularization path computed by glasso are very similar to SINCO's ROC curves (Scheinberg and Rish, 2009). Note that changing tolerance does not have the same affect on glasso as it does on SINCO. The numbers of TP and FP do not change noticeably with increasing tolerance. This is due to the fact that the algorithm in glasso updates a whole row and a column of C at each iteration while it cycles through the rows and columns, rather than selecting the updates in a greedy manner.

Our observations imply that methods like SINCO can be used to greedily select the elements of the graphical model until the desired trade-off between FPs and TPs is achieved. In the limit SINCO solves the same problem as glasso and hence the limit numbers of the true and false positives are dictated by the choice of ρ. But since the real goal is to recover the true nonzero structure of the covariance matrix, it is not necessary to solve problem (17.6) accurately. For the purpose of recovering a good TP/FP ratio, one can apply the SINCO method without adjustments to ρ.

For details on the numerical experiments presented here, see Scheinberg and Rish (2009).

17.3 Alternating Linearization Method

Efficient alternatives to BCD methods are the gradient based methods which we discussed briefly in section 17.1.2. These methods have higher per-iteration complexity (typically $O(p^3)$), but they usually converge in fewer iterations than a BCD method. Moreover, several of these methods, including the one presented here, have variants with provable complexity bounds.

We discuss here the alternating linearization method (ALM) introduced in Goldfarb et al. (2009) for solving (17.6) which we write here as

$$\min_{X \in S^p_{++}} \quad F(X) \equiv f(X) + g(X), \tag{17.28}$$

where $f(X) = -\log \det(X) + \langle S, X \rangle$ and $g(X) = \rho \|X\|_1$. An effective way to approach an objective function of this form is to "split" f and g by introducing a new variable, that is, to rewrite (17.28) as

$$\min_{X, Y \in S^p_{++}} \{f(X) + g(Y) : X - Y = 0\}, \tag{17.29}$$

and to apply an alternating-direction augmented Lagrangian method to it.

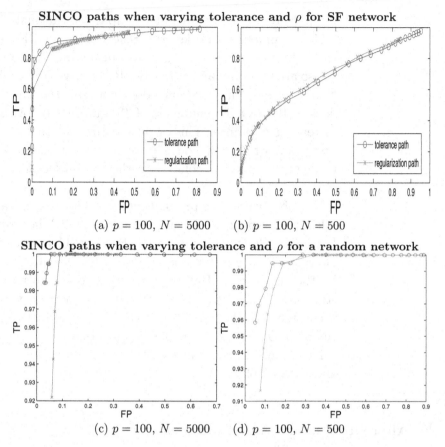

Figure 17.1: Random and scale-free networks: SINCO paths when varying tolerance and ρ.

Given a penalty parameter $1/\mu$, and an estimate of the Lagrange multiplier Λ, at the kth iteration the augmented Lagrangian method minimizes the augmented Lagrangian function

$$\mathcal{L}(X, Y; \Lambda) := f(X) + g(Y) - \langle \Lambda, X - Y \rangle + \frac{1}{2\mu} \|X - Y\|_2^2,$$

with respect to X and Y, that is, it solves the subproblem

$$(X^k, Y^k) := \arg \min_{X, Y \in S_{++}^p} \mathcal{L}(X, Y; \Lambda^k) \tag{17.30}$$

and updates the Lagrange multiplier Λ via

$$\Lambda^{k+1} := \Lambda^k - (X^k - Y^k)/\mu. \tag{17.31}$$

Minimizing $\mathcal{L}(X, Y; \Lambda)$ with respect to X and Y jointly is difficult, while doing so with respect to X and Y alternatingly can be done efficiently, as we will show below. Moreover, minimization over Y does not have to include the constraint $Y \in S^p_{++}$.

The following alternating-direction version of the augmented Lagrangian method (ADAL) is often advocated (see, e.g., Fortin and Glowinski (1983); Glowinski and Le Tallec (1989)):

$$
\begin{cases}
X^{k+1} & := \arg\min_{X \in S^p_{++}} \mathcal{L}(X, Y^k; \Lambda^k) \\
Y^{k+1} & := \arg\min_Y \mathcal{L}(X^{k+1}, Y; \Lambda^k) \\
\Lambda^{k+1} & := \Lambda^k - (X^{k+1} - Y^{k+1})/\mu.
\end{cases}
\tag{17.32}
$$

In Goldfarb et al. (2009) the following symmetric version of the ADAL method is considered:

$$
\begin{cases}
X^{k+1} & := \arg\min_X \mathcal{L}(X, Y^k; \Lambda^k_Y) \\
\Lambda^{k+1}_X & := \Lambda^k_Y - (X^{k+1} - Y^k)/\mu \\
Y^{k+1} & := \arg\min_Y \mathcal{L}(X^{k+1}, Y; \Lambda^{k+1}_X) \\
\Lambda^{k+1}_Y & := \Lambda^{k+1}_X - (X^{k+1} - Y^{k+1})/\mu.
\end{cases}
\tag{17.33}
$$

Let us assume for the moment that $f(X)$ is in the class $C^{1,1}$ with Lipschitz constant $L(f)$,[2] while $g(X)$ is simply convex. In this case, from the first-order optimality conditions for the two subproblems in (17.33), we have

$$
\Lambda^{k+1}_X = \nabla f(X^{k+1}) \quad \text{and} -\Lambda^{k+1}_Y \in \partial g(Y^{k+1}),
\tag{17.34}
$$

where $\partial g(Y^{k+1})$ is the subdifferential of $g(Y)$ at $Y = Y^{k+1}$. Substituting these relations into (17.33), we obtain the following algorithm for solving (17.28), which we refer to as an alternating linearization minimization algorithm.

Algorithm 17.3 Alternating linearization method (ALM)

1: **Input:** $X^0 = Y^0$
2: **for** $k = 0, 1, \cdots$ **do**
3: 1. Solve $X^{k+1} := \mathrm{argmin}_{X \in S^p_{++}} Q_g(X, Y^k) \equiv f(X) + g(Y^k) - \langle \Lambda^k, X - Y^k \rangle + \frac{1}{2\mu}\|X - Y^k\|^2_2$;
4: 2. Solve $Y^{k+1} := \mathrm{argmin}_Y Q_f(X^{k+1}, Y) \equiv f(X^{k+1}) + \langle \nabla f(X^{k+1}), Y - X^{k+1} \rangle + \frac{1}{2\mu}\|Y - X^{k+1}\|^2_2 + g(Y)$;
5: 3. $\Lambda^{k+1} = \nabla f(X^{k+1}) - (X^{k+1} - Y^{k+1})/\mu$.
6: **end for**

Algorithm 17.3 can be viewed in the following way: at each iteration we construct a quadratic approximation of the functions $g(X)$ at the current

---

2. This does not hold for our $f(X)$ if we consider $X \in S^p_{++}$, but it holds in a smaller feasible set.

iterates Y^k and minimize the sum of this approximation and $f(X)$. The approximation is based on linearizing $g(X)$ (hence the name "alternating linearization method") and adding a "prox"-term $\frac{1}{2\mu}\|X - Y^k\|_2^2$. Then the linearization step is applied to $f(X)$ at the next iterate X^{k+1}. The purpose of these linearizations is to replace one of the functions with a simple linear function with a prox-term to make optimization easy. It is important, however, that the resulting functions $Q_g(X, Y^k)$ and $Q_f(X^{k+1}, Y)$ provide a good approximation of $F(X)$. For theoretical purposes it is sufficient that at the obtained minimum $Q_g(X, Y^k) \leq F(X)$ $(Q_f(X^{k+1}, Y) \leq F(Y))$, which means that the reduction in the value of $F(X)$ achieved by minimizing $Q_g(X, Y^k)$ in step 1 and $Q_f(X^{k+1}, Y)$ in step 2 is not smaller than the reduction achieved in the value of $Q_g(X, Y^k)$ $(Q_f(X^{K+1}, Y))$ itself. For the case of $Q_f(X^{k+1}, Y)$, when μ is small enough $(\mu \leq 1/L(f))$, the quadratic function $f(X^{k+1}) + \langle \nabla f(X^{k+1}), Y - X^{k+1} \rangle + \frac{1}{2\mu}\|Y - X^{k+1}\|_2^2$ is an upper approximation to $f(X)$, which means that condition $Q_f(X^{k+1}, Y) \leq F(Y)$ is guaranteed to hold. For the case of $Q_g(X, Y^k)$, however, condition $Q_g(X, Y^k) \leq F(X)$ may fail for any $\mu > 0$, since $g(X)$ is not smooth. In this case, if the condition fails, one can simply skip the step in X by assigning $X^{k+1} = Y^k$. This leads to the following ALM with skipping steps (algorithm 17.4).

Algorithm 17.4 Alternating linearization method with skipping step

1: **Input:** $X^0 = Y^0$
2: **for** $k = 0, 1, \cdots$ **do**
3:    1. Solve $X^{k+1} := \arg\min_X Q_g(X, Y^k) \equiv f(X) + g(Y^k) - \langle \Lambda^k, X - Y^k \rangle + \frac{1}{2\mu}\|X - Y^k\|_2^2$;
4:    2. If $F(X^{k+1}) > Q_g(X^{k+1}, Y^k)$, then $X^{k+1} := Y^k$.
5:    3. Solve $Y^{k+1} := \arg\min_Y Q_f(X^{k+1}, Y) \equiv f(X^{k+1}) + \langle \nabla f(X^{k+1}), Y - X^{k+1} \rangle + \frac{1}{2\mu}\|Y - X^{k+1}\|_2^2 + g(Y)$;
6:    4. $\Lambda^{k+1} = \nabla f(X^{k+1}) - (X^{k+1} - Y^{k+1})/\mu$.
7: **end for**

Algorithm 17.4 is identical to 17.3, and hence to the symmetric ADAL algorithm (17.33) as long as $F(X^{k+1}) \leq Q_g(X^{k+1}, Y^k)$ at each iteration. If this condition fails, then the algorithm simply sets $X^{k+1} \leftarrow Y^k$. Algorithm 17.4 has the following convergence property and iteration complexity bound (Goldfarb et al., 2009).

Theorem 17.1. *Assume ∇f is Lipschitz continuous with Lipschitz constant*

$L(f)$. For $\mu \leq 1/L(f)$, algorithm 17.4 satisfies

$$F(y^k) - F(x^*) \leq \frac{\|x^0 - x^*\|^2}{2\mu(k + k_n)}, \forall k, \tag{17.35}$$

where x^* is an optimal solution of (17.28) and k_n is the number of iterations until the kth for which $F(x^{k+1}) \leq Q_g(x^{k+1}, y^k)$. Thus, algorithm 17.4 produces a sequence which converges to the optimal solution in function value, and the number of iterations needed is $O(1/\varepsilon)$ for an ε-optimal solution.

The iteration complexity bound in theorem 17.1 can be improved. Nesterov (1983, 2004) proved that one can obtain an optimal iteration complexity bound of $O(1/\sqrt{\varepsilon})$, using only the first-order information. His acceleration technique is based on using a linear combination of previous iterates to obtain a point where the approximation is built. This technique has been exploited and extended by Tseng (2008), Beck and Teboulle (2009), and many others. Specifically for SICS a method with complexity bound $O(1/\sqrt{\varepsilon})$ was introduced by Lu (2009). A similar technique can be adopted to derive a fast version of algorithm 17.4 that has an improved complexity bound of $O(1/\sqrt{\varepsilon})$, while keeping the computational effort in each iteration almost unchanged. However, we do not present this method here, since when it is applied to the SICS problem, it does not appear to work as well as algorithm 17.4.

Note that in our case $f(X) = -\log \det(X) + \langle S, X \rangle$ does not have a Lipschitz continuous gradient in general. Moreover, $f(X)$ is defined only for positive definite matrices while $g(X)$ is defined everywhere. These properties of the objective function make the SICS problem especially challenging for optimization methods. Nevertheless, we can still apply (17.33) to solve the problem directly. Moreover, we can apply algorithm 17.4 and obtain the complexity bound in theorem 17.1 as follows. As proved in Lu (2009), the optimal solution X^* of (17.28) satisfies $X \succeq \alpha I$, where $\alpha = \frac{1}{\|S\| + p\rho}$ (see proposition 3.1 in Lu (2009)). Therefore, the SICS problem (17.28) can be formulated as

$$\min_{X,Y} \{f(X) + g(Y) : X - Y = 0, X \in \mathcal{C}, Y \in \mathcal{C}\}, \tag{17.36}$$

where $\mathcal{C} := \{X \in S^n : X \succeq \frac{\alpha}{2}I\}$. We know that the constraint $X \succeq \frac{\alpha}{2}I$ is not tight at the solution. Hence, if we start the algorithm with $X \succeq \alpha I$ and restrict the stepsize μ to be sufficiently small, then the iterates of the method will remain in the domain where the gradient of $f(X)$ is Lipschitz continuous, and we can apply algorithm 17.3 and theorem 17.1.

Note, however, that the bound on the Lipschitz constant of the gradient of $f(X)$ is $1/\alpha^2$, and hence can be very large. It is not practical to restrict μ in the algorithm to be smaller than α^2, since μ determines the stepsize at each iteration. Below is a practical version of our algorithm applied to the SICS problem.

Algorithm 17.5 Alternating linearization method for SICS

1: **Input:** $X^0 = Y^0$, μ_0.
2: **for** $k = 0, 1, \cdots$ **do**
3:     0. Pick $\mu_{k+1} \leq \mu_k$.
4:     1. Solve $X^{k+1} := \arg\min_{X \in \mathcal{C}} Q_g(X, Y^k) \equiv f(X) + g(Y^k) - \langle \Lambda^k, X - Y^k \rangle + \frac{1}{2\mu_{k+1}}\|X - Y^k\|_F^2$;
5:     2. If $F(X^{k+1}) > Q_g(X^{k+1}, Y^k)$, then $X^{k+1} := Y^k$.
6:     3. Solve $Y^{k+1} := \arg\min_Y f(X^{k+1}) + \langle \nabla f(X^{k+1}), Y - X^{k+1} \rangle + \frac{1}{2\mu_{k+1}}\|Y - X^{k+1}\|_F^2 + g(Y)$;
7:     4. $\Lambda^{k+1} = \nabla f(X^{k+1}) - (X^{k+1} - Y^{k+1})/\mu_{k+1}$.
8: **end for**

17.3.1   Solving Subproblems of ALM for SICS

We now show how to solve the two optimization problems in algorithm 17.5. The first-order optimality conditions for step 1 in algorithm 17.5, ignoring the constraint $X \in \mathcal{C}$, are

$$\nabla f(X) - \Lambda^k + (X - Y^k)/\mu_{k+1} = 0. \tag{17.37}$$

Consider $V \operatorname{Diag}(d) V^\top$, the spectral decomposition of $Y^k + \mu_{k+1}(\Lambda^k - S)$, and let

$$\gamma_i = \left(d_i + \sqrt{d_i^2 + 4\mu_{k+1}} \right) /2, i = 1, \ldots, p. \tag{17.38}$$

Since $\nabla f(X) = -X^{-1} + S$, it is easy to verify that $X^{k+1} := V \operatorname{Diag}(\gamma) V^\top$ satisfies (17.37). When the constraint $X \in \mathcal{C}$ is imposed, the optimal solution changes to $X^{k+1} := V \operatorname{Diag}(\gamma) V^\top$ with

$$\gamma_i = \max\left\{ \alpha/2, \left(d_i + \sqrt{d_i^2 + 4\mu_{k+1}} \right) /2 \right\}, i = 1, \ldots, p.$$

We observe that solving (17.37) requires approximately the same effort $(O(p^3))$ as is required to compute $\nabla f(X^{k+1})$ itself. Moreover, from the solution to (17.37), $\nabla f(X^{k+1})$ is obtained with only a negligible amount of additional effort, since $(X^{k+1})^{-1} := V \operatorname{Diag}(\gamma)^{-1} V^\top$.

The first-order optimality conditions for step 2 in algorithm 17.5 are

$$0 \in \nabla f(X^{k+1}) + (Y - X^{k+1})/\mu_{k+1} + \partial g(Y). \tag{17.39}$$

Since $g(Y) = \rho\|Y\|_1$, the solution to (17.39) is given by

$$Y^{k+1} = \mathrm{shrink}(X^{k+1} - \mu_{k+1}(S - (X^{k+1})^{-1}), \mu_{k+1}\rho),$$

where the "shrinkage operator" $\mathrm{shrink}(Z, \rho)$ is defined as $\mathrm{shrink}(Z, \rho)_{ij} = \mathrm{shrink}(Z_{ij}, \rho)$. It is trivial to see that to apply algorithm 17.5 to the extended formulation (17.7), one only needs to modify the shrinkage step $\mathrm{shrink}(Z, M)$, which is then defined as $[\mathrm{shrink}(Z, M)]_{ij} = \mathrm{shrink}(Z_{ij}, M_{ij})$.

Note that the $O(p^3)$ complexity of step 1 which requires a spectral decomposition, dominates the $O(p^2)$ complexity of step 2 which requires a simple shrinkage. There is no closed-form solution for the subproblem corresponding to Y when the constraint $Y \in \mathcal{C}$ is imposed. One can attempt to impose this by a line search on the value of μ_k. Imposing such a constraint in practice limits the stepsize too much, and the performance of the algorithm deteriorates substantially. Thus, resulting iterates Y^k may not be positive definite, while the iterates X^k remain so. Eventually, due to the convergence of Y^k and X^k, the Y^k iterates become positive definite and the constraint $Y \in \mathcal{C}$ is satisfied. Relaxing the positive definiteness constraint during the course of the algorithm appears to be desirable for the overall performance.

17.4 Remarks on Numerical Performance

A number of numerical comparisons of optimization methods for SICS have been presented in the literature. See, for instance, Duchi et al. (2008); Goldfarb et al. (2009); Scheinberg et al. (2010); Friedman et al. (2007); Lu (2009); Scheinberg and Rish (2009) for the comparison of methods discussed in this chapter. The comparison we discuss here is based solely on the time and iteration efficiency of the algorithms to achieve comparable solutions in terms of their objective function values. The sparsity patterns recovered by these methods (aside from COVSEL) appears to be comparable.

In summary, the first-order methods in Duchi et al. (2008); Goldfarb et al. (2009); Scheinberg et al. (2010); Lu (2009) outperform glasso in Friedman et al. (2007), which substantially outperforms [3] COVSEL in Banerjee et al. (2008) and somewhat outperforms SINCO in Scheinberg and Rish (2009). Most of the tests were performed on instances up to the size $p = 2000$. On very sparse structured large matrices SINCO outperforms glasso. In

---

3. The Fortran implementation of glasso has some faults and occasionally breaks down, especially in large-scale cases. We believe it is an issue with the implementation rather than the algorithm so we do not elaborate on this further.

all experiments in Scheinberg et al. (2010) ALM outperforms the projected gradient method in Duchi et al. (2008), and the smooth accelerated gradient method in Lu (2009), in terms of CPU time and accuracy of the solution. The first-order methods do not exploit the solution sparsity, in that regardless of the sparsity, the per-iteration complexity remains $O(p^3)$. The per-iteration complexity of glasso and SINCO is empirically smaller, but the number of iterations required to achieve comparable accuracy is larger than that of the first-order methods.

17.5   References

O. Banerjee, L. El Ghaoui, and A. d'Aspremont. Model selection through sparse maximum likelihood estimation for multivariate gaussian for binary data. *Journal of Machine Learning Research*, 9:485–516, 2008.

A. Beck and M. Teboulle. A fast iterative shrinkage-thresholding algorithm for linear inverse problems. *SIAM Journal of Imaging Sciences*, 2(1):183–202, 2009.

A. d'Aspremont, O. Banerjee, and L. El Ghaoui. First-order methods for sparse covariance selection. *SIAM Journal on Matrix Analysis and its Applications*, 30 (1):56–66, 2008.

J. Duchi, S. Gould, and D. Koller. Projected subgradient methods for learning sparse Gaussians. *Proceedings of the 24th Conference on Uncertainty in Artificial Intelligence*, 2008.

M. Fortin and R. Glowinski. *Augmented Lagrangian methods: applications to the numerical solution of boundary-value problems*. North-Holland, 1983.

J. Friedman, T. Hastie, and R. Tibshirani. Sparse inverse covariance estimation with the graphical lasso. *Biostatistics*, 9(3):432–441, 2007.

R. Glowinski and P. Le Tallec. *Augmented Lagrangian and Operator-Splitting Methods in Nonlinear Mechanics*. SIAM, Philadelphia, 1989.

D. Goldfarb, S. Ma, and K. Scheinberg. Fast alternating linearization methods for minimizing the sum of two convex functions. Technical report, Department of IEOR, Columbia University, 2009.

D. Heckerman. A tutorial on learning Bayesian networks. Technical report, Microsoft Research, 1995.

V. Krishnamurthy and A. d'Aspremont. A pathwise algorithm for covariance selection. *preprint*, 2009. arXiv:0908.0143.

S. Lauritzen. *Graphical Models*. Oxford University Press, 1996.

L. Li and K.-C. Toh. An inexact interior point method for l_1-regularized sparse covariance selection. *Mathematical Programming Computation*, 2(3–4):291–315, 2010.

Z. Lu. Smooth optimization approach for sparse covariance selection. *SIAM Journal on Optimization*, 19(4):1807–1827, 2009.

Z. Lu. Adaptive first-order methods for general sparse inverse covariance selection. *SIAM Journal on Matrix Analysis and Applications*, 31(4):2000–2016, 2010.

N. Meinshausen and P. Buhlmann. High dimensional graphs and variable selection with the Lasso. *Annals of Statistics*, 34(3):1436–1462, 2006.

A. Nemirovski. Prox-method with rate of convergence $O(1/t)$ for variational inequalities with Lipschitz continuous monotone operators and smooth convex-concave saddle point problems. *SIAM Journal on Optimization*, 15(1):229–251, 2005.

Y. E. Nesterov. A method for unconstrained convex minimization problem with the rate of convergence $\mathcal{O}(1/k^2)$. *Doklady Akademia Nauk SSSR*, 269:543–547, 1983.

Y. E. Nesterov. Introductory lectures on convex optimization: A basic course. 87, 2004.

Y. E. Nesterov. Smooth minimization of non-smooth functions. *Mathematical Programming, series A*, 103:127–152, 2005.

K. Scheinberg and I. Rish. SINCO - a greedy coordinate ascent method for sparse inverse covariance selection problem. 2009. Preprint available at http://www.optimization-online.org/DB_HTML/2009/07/2359.html.

K. Scheinberg, S. Ma, and D. Goldfarb. Sparse inverse covariance selection via alternating linearization methods. In *Advances in Neural Information Processing Systems 23*, 2010.

L. Sun, R. Patel, J. Liu, K. Chen, T. Wu, J. Li, E. Reiman, and J. Ye. Mining brain region connectivity for alzheimer's disease study via sparse inverse covariance estimation. *Proceedings of the 15th ACM SIGKDD International Conference on Knowledge Discovery and Data Mining*, 2009.

R. Tibshirani. Regression shrinkage and selection via the lasso. *J. Royal. Statist. Soc B.*, 58(1):267–288, 1996.

P. Tseng. On accelerated proximal gradient methods for convex-concave optimization. *submitted to SIAM Journal on Optimization*, 2008.

M. Wainwright, P. Ravikumar, and J. Lafferty. High-dimensional graphical model selection using ℓ_1-regularized logistic regression. In *Advances in Neural Information Processing Systems 20*, pages 1465–1472. 2007.

C. Wang, D. Sun, and K.-C. Toh. Solving log-determinant optimization problems by a Newton-CG primal proximal point algorithm. *Preprint*, 2009.

Z. Wen, D. Goldfarb, S. Ma, and K. Scheinberg. Row by row methods for semidefinite programming. Technical report, Department of IEOR, Columbia University, 2009.

H. Wolkowicz, R. Saigal, and L. Vandenberghe, editors. *Handbook of Semidefinite Programming*. Kluwer Academic, 2000.

M. Yuan and Y. Lin. Model selection and estimation in the gaussian graphical model. *Biometrika*, 94(1):19–35, 2007.

X. Yuan. Alternating direction methods for sparse covariance selection. 2009. Preprint available at http://www.optimization-online.org/DB_FILE/2009/09/2390.pdf.

18   A Pathwise Algorithm for Covariance Selection

Vijay Krishnamurthy   kvijay@princeton.edu
ORFE, Princeton University
Princeton, NJ 08544, USA

Selin Damla Ahipaşaoğlu   sahipasa@princeton.edu
ORFE, Princeton University
Princeton, NJ 08544, USA

Alexandre d'Aspremont   aspremon@princeton.edu
ORFE, Princeton University
Princeton, NJ 08544, USA

Covariance selection seeks to estimate a covariance matrix by maximum likelihood while restricting the number of nonzero inverse covariance matrix coefficients. A single penalty parameter usually controls the tradeoff between log-likelihood and sparsity in the inverse matrix. We describe an efficient algorithm for computing a full regularization path of solutions to this problem.

18.1   Introduction

We consider the problem of estimating a covariance matrix from sample multivariate data by maximizing its likelihood while penalizing the inverse covariance so that its graph is *sparse*. This problem is known as covariance selection and can be traced back at least to Dempster (1972). The coefficients of the inverse covariance matrix define the representation of a particular Gaussian distribution as a member of the exponential family; hence sparse maximum likelihood estimates of the inverse covariance yield sparse representations of the model in this class. Furthermore, in a Gaussian model, zeros in the inverse covariance matrix correspond to *conditionally* independent variables, so this penalized maximum likelihood procedure simultaneously stabilizes estimation and isolates *structure* in the underlying graphical model Lauritzen (1996, see).

Given a sample covariance matrix $\Sigma \in \mathbf{S}_n$, the covariance selection problem is written as

$$\text{maximize} \quad \log \det X - \mathbf{Tr}(\Sigma X) - \rho \, \mathbf{Card}(X)$$

in the matrix variable $X \in \mathbf{S}_n$, where $\rho > 0$ is a penalty parameter controlling sparsity and $\mathbf{Card}(X)$ is the number of nonzero elements in X. This is a combinatorially hard (nonconvex) problem and, as in Dahl et al. (2008), Banerjee et al. (2006), and Dahl et al. (2005), we form the convex relaxation

$$\text{maximize} \quad \log \det X - \mathbf{Tr}(\Sigma X) - \rho \|X\|_1, \tag{18.1}$$

which is a convex problem in the matrix variable $X \in \mathbf{S}_n$, where $\|X\|_1$ is the sum of absolute values of the coefficients of X here. After scaling, the $\|X\|_1$ penalty can be understood as a convex lower bound on $\mathbf{Card}(X)$. Another completely different approach, derived in Meinshausen and Bühlmann (2006), reconciles the local dependence structure inferred from n distinct ℓ_1-penalized regressions of a single variable against all the others. Both this approach and the convex relaxation (18.1) have been shown to be consistent in Meinshausen and Bühlmann (2006) and Banerjee et al. (2008), respectively.

In practice, however, both methods are computationally challenging when n gets large. Various algorithms have been employed to solve (18.1) with Dahl et al. (2005), using a custom interior-point method, and Banerjee et al. (2008), using a block coordinate descent method where each iteration required solving a Lasso-like problem, among others. This last method is efficiently implemented in the Glasso package by Friedman et al. (2008), using coordinate descent algorithms from Friedman et al. (2007) to solve the inner regression problems.

One key issue in all these methods is that there is no a priori obvious choice for the penalty parameter. In practice, at least a partial regularization path of solutions has to be computed, and this procedure is then repeated many times to get confidence bounds on the graph structure by cross-validation. Pathwise Lasso algorithms such as LARS (Efron et al., 2004) can be used to get a full regularization path of solution using the method in Meinshausen and Bühlmann (2006), but this still requires solving and reconciling n regularization paths on regression problems of dimension n.

Our contribution here is to formulate a pathwise algorithm for solving problem (18.1) using numerical continuation methods (see Bach et al. (2005) for an application in kernel learning). Each iteration requires solving a large structured linear system (predictor step), then improving precision using a block coordinate descent method (corrector step). Overall, the cost of moving from one solution of problem (18.1) to another is typically much lower than that of solving two separate instances of (18.1). We also derive a coordinate descent algorithm for solving the corrector step, where each iteration is closed form and requires only solving a cubic equation. We illustrate the performance of our methods on several artificial and realistic data sets.

The paper is organized as follows. Section 18.2 reviews some basic convex optimization results on the covariance selection problem in (18.1). Our main pathwise algorithm is described in Section 18.3. Finally, we present some numerical results in Section 18.4.

In what follows, we write $\mathbf{S}_n$ for the set of symmetric matrices of dimension n. For a matrix $X \in \mathbb{R}^{m \times n}$, we write $\|X\|_\mathrm{F}$ its Frobenius norm; $\|X\|_1 = \sum_{ij} |X_{ij}|$, the ℓ_1 norm of its vector of coefficients; and $\mathbf{Card}(X)$, the number of nonzero coefficients in X.

18.2 Covariance Selection

Starting from the convex relaxation defined above

$$\text{maximize} \quad \log \det X - \mathbf{Tr}(\Sigma X) - \rho\|X\|_1 \tag{18.2}$$

in the variable $X \in \mathbf{S}_n$, where $\|X\|_1$ can be understood as a convex lower bound on the $\mathbf{Card}(X)$ function whenever $|X_{ij}| \leq 1$ (we can always scale ρ otherwise). Let us write $X^*(\rho)$ for the optimal solution of problem (18.2). In what follows, we will seek to compute (or approximate) the entire regularization path of solutions $X^*(\rho)$ for $\rho \in \mathbb{R}_+$. To remove the nonsmooth penalty we can set $X = L - M$ and rewrite Problem (18.2) as

$$\begin{aligned} \text{maximize} \quad & \log \det(L - M) - \mathbf{Tr}(\Sigma(L - M)) - \rho\mathbf{1}^T(L + M)\mathbf{1} \\ \text{subject to} \quad & L_{ij}, M_{ij} \geq 0, \quad i, j = 1, \ldots, n, \end{aligned} \tag{18.3}$$

in the matrix variables $L, M \in \mathbf{S}_n$. We can form the following dual to problem (18.2) as

$$\begin{aligned} \text{minimize} \quad & -\log \det(U) - n \\ \text{subject to} \quad & U_{ij} \leq \Sigma_{ij} + \rho, \quad i, j = 1, \ldots, n, \\ & U_{ij} \geq \Sigma_{ij} - \rho,, \quad i, j = 1, \ldots, n, \end{aligned} \tag{18.4}$$

in the variable $U \in \mathbf{S}_n$. As in Bach et al. (2005), for example, in the spirit of barrier methods for interior-point algorithms, we then form the following (unconstrained) regularized problem

$$\min_{U \in \mathbf{S}_n} - \log \det(U) - t \left(\sum_{i,j=1}^{n} \log(\rho + \Sigma_{ij} - U_{ij}) + \sum_{i,j=1}^{n} \log(\rho - \Sigma_{ij} + U_{ij}) \right)$$

$$(18.5)$$

in the variable $U \in \mathbf{S}_n$, and $t > 0$ specifies a desired tradeoff level between centrality (smoothness) and optimality. From every solution $U^*(t)$ corresponding to each $t > 0$, the barrier formulation also produces an explicit *dual* solution $(L^*(t), M^*(t))$ to Problem (18.4). Indeed, we can define matrices $L, M \in \mathbf{S}_n$ as follows

$$L_{ij}(U, \rho) = \frac{t}{\rho + \Sigma_{ij} - U_{ij}} \quad \text{and} \quad M_{ij}(U, \rho) = \frac{t}{\rho - \Sigma_{ij} + U_{ij}}.$$

First-order optimality conditions for Problem (18.5) then imply

$$(L - M) = U^{-1}.$$

As t tends to 0, (18.5) traces a central path toward the optimal solution to Problem (18.4). If we write $f(U)$ for the objective function of (18.4) and call p^* its optimal value, we get (as in Boyd and Vandenberghe (2004, §11.2.2)),

$$f(U^*(t)) - p^* \leq 2n^2 t.$$

Hence t can be understood as a surrogate duality gap when solving the dual Problem (18.4).

18.3   Algorithm

In this section we derive a predictor-corrector algorithm to approximate the entire path of solutions $X^*(\rho)$ when ρ varies between 0 and $\max_i \Sigma_{ii}$ (beyond which the solution matrix is diagonal). Defining

$$H(U, \rho) = L(U, \rho) - M(U, \rho) - U^{-1},$$

we trace the curve $H(U, \rho) = 0$, the first-order optimality condition for Problem (18.5). Our pathwise covariance selection algorithm is defined in Algorithm 18.1.

Typically, in Algorithm 18.1, h is a small constant, $\rho_0 = \max_i \Sigma_{ii}$, and U_0 is computed by solving a single (very sparse) instance of problem (18.5) for

---

Algorithm 18.1 Pathwise Covariance Selection

Input: $\Sigma \in \mathbf{S}_m$
1: Start with (U_0, ρ_0) s.t $H(U_0, \rho_0) = 0$.
2: **for** $i = 1$ to k **do**
3:     *Predictor Step.* Let $\rho_{i+1} = \rho_i + h$. Compute a tangent direction by solving the linear system

$$\frac{\partial H}{\partial \rho}(U_i, \rho_i) + J(U_i, \rho_i)\frac{\partial U}{\partial \rho} = 0$$

    in $\partial U/\partial \rho \in \mathbf{S}_n$, where $J(U_i, \rho_i) = \partial H(U, \rho)/\partial U \in \mathbf{S}_{n^2}$ is the Jacobian matrix of the function $H(U, \rho)$.
4:     Update $U_{i+1} = U_i + h\partial U/\partial \rho$.
5:     *Corrector Step.* Solve problem (18.5) starting at $U = U_{i+1}$.
6: **end for**
Output: Sequence of matrices U_i, $i = 1, \ldots, k$.

---

example.

18.3.1   Predictor: Conjugate Gradient Method

In Algorithm 18.1, the tangent direction in the predictor step is computed by solving a linear system $Ax = b$ where $A = (U^{-1} \otimes U^{-1} + D)$ and D is a diagonal matrix. This system of equations has dimension n^2, and we solve it using the conjugate gradient (CG) method.

18.3.1.1   *CG iterations*

The most expensive operation in the CG iterations is the computation of a matrix vector product Ap_k, with $p_k \in \mathbb{R}^{n^2}$. Here, however, we can exploit problem structure to compute this step efficiently. Observe that $(U^{-1} \otimes U^{-1})p_k = \text{vec}(U^{-1}P_kU^{-1})$ when $p_k = \text{vec}(P_k)$, so the computation of the matrix vector product Ap_k needs only $O(n^3)$ flops instead of $O(n^4)$. The CG method then needs at most $O(n^2)$ iterations to converge, leading to a total complexity of $O(n^5)$ for the predictor step. In practice, we will observe that CG needs considerably fewer iterations.

18.3.1.2   *Stopping criterion*

To speed up the computation of the predictor step, we can stop the conjugate gradient solver when the norm of the residual falls below the numerical tolerance t. In our experiments here, we stopped the solver after the residual decreases by two orders of magnitude.

18.3.1.3 *Scaling and warm start*

Another option, much simpler than the predictor step detailed above, is warm starting. This means simply scaling the current solution to make it feasible for the problem after ρ is updated. In practice, this method turns out to be as efficient as the predictor step, as it allows us to follow the path starting from the sparse end (where more interesting solutions are located). Here, we start the algorithm from the sparsest possible solution, a diagonal matrix U such that

$$U_{ii} = \Sigma_{ii} + (1 - \varepsilon)\rho_{\max}I, \quad i = 1, \ldots, n,$$

where $\rho_{\max} = \max_i \Sigma_{ii}$. Suppose, now, that iteration k of the algorithm produced a matrix solution U_k corresponding to a penalty ρ_k. Then, the algorithm with (lower) penalty ρ_{k+1} is started at the matrix

$$U = (1 - \rho_{k+1}/\rho_k)\Sigma + (\rho_{k+1}/\rho_k)U_k,$$

which is a feasible starting point for the corrector problem that follows. This is the method that was implemented in the final version of our code and that is used in the numerical experiments detailed in Section 18.4.

18.3.2 Corrector: Block Coordinate Descent

For small problems, we can use Newton's method to solve (18.5). However, from a computational perspective, this approach is not practical for large values of n. We can simplify iterations by using a block coordinate descent algorithm that updates one row/column of the matrix in each iteration (Banerjee et al., 2008). Let us partition the matrices U and Σ as

$$U = \begin{pmatrix} V & u \\ u^T & w \end{pmatrix} \quad \text{and} \quad S = \begin{pmatrix} A & b \\ b^T & c \end{pmatrix}.$$

We keep V fixed in each iteration and solve for u and w. Without loss of generality, we can always assume that we are updating the last row/column.

18.3.2.1 *Algorithm*

Problem (18.5) can be written in block format as

$$\text{minimize} \quad -\log(w - u^T V^{-1} u) - t(\log(\rho + c - w) + \log(\rho - c + w))$$

$$-2t\left(\sum_i \log(\rho + b_i - u_i) + \sum_i \log(\rho - b_i + u_i)\right),$$

$$(18.6)$$

in the variables $u \in \mathbb{R}^{(n-1)}$ and $w \in \mathbb{R}$. Here $V \in \mathbf{S}^{(n-1)}$ is kept fixed in each iteration. We use the Sherman-Woodbury-Morrison (SWM) formula

---

Algorithm 18.2 Block coordinate descent corrector steps

Input: U_0, $\Sigma \in \mathbf{S}_n$
1: **for** $i = 1$ to k **do**
2:     Pick the row and column to update.
3:     Solve the inner problem (18.6) using coordinate descent (each coordinate descent step requires solving a cubic equation).
4:     Update U^{-1}.
5: **end for**
Output: A matrix U_k solving (18.5).

(see e.g., Boyd and Vandenberghe, 2004, Section C.4.3) to efficiently update U^{-1} at each iteration, so it suffices to compute the full inverse only once, at the beginning of the path. The choice and order of row/column updates significantly affect performance. Although predicting the effect of a whole ith row/column update is numerically expensive, we use the fact that the impact of updating diagonal coefficients usually dominates all others and can be computed explicitly at a very low computational cost. It corresponds to the maximum improvement in the dual objective function that can be achieved by updating the current solution U to $U + w e_i e_i^T$, where e_i is the ith unit vector. The objective function value is a decreasing function of w and w must be lower than $\rho + \Sigma_{ii} - U_{ii}$ to preserve dual feasibility, so updating the ith diagonal coefficient will decrease the objective by $\delta_i = (\rho + \Sigma_{ii} - U_{ii})U_{ii}^{-1}$ after minimizing over w. In practice, updating the top 10 percent row/columns with the largest δ is often enough to reach our precision target, and very significantly speeds up computations. We also solve the inner problem (18.6) by a coordinate descent method (as in (Friedman et al., 2007)), taking advantage of the fact that a point minimizing (18.6) over a single coordinate can be computed in closed form by solving a cubic equation. Suppose (u, w) is the current point and that we wish to optimize coordinate u_j of the vector u. We define

$$\begin{aligned}
\alpha &= -V_{jj}^{-1} \\
\beta &= -2u_j (\textstyle\sum_{k \neq j} V_{kj}^{-1} u_k) \\
\gamma &= w - u^T V^{-1} u - \alpha u_j - \beta u_j^2.
\end{aligned} \tag{18.7}$$

The optimality conditions imply that the the optimal u_j^* must satisfy the cubic equation

$$p_1 x^3 + p_2 x^2 + p_3 x + p_4 = 0, \tag{18.8}$$

where

$$p_1 = 2(1+2t)\alpha, \ p_2 = (1+4t)\beta - 4(1+2t)\alpha b_j$$
$$p_3 = 4t\gamma - 2(1+2t)\beta b_j + 2\alpha(b_j^2 - 2\rho^2), \ p_4 = \beta(b_j^2 - \rho^2) - 4t\gamma b_j.$$

Similarly, the diagonal update w satisfies the following quadratic equation:

$$(1+2t)w^2 - 2(t(u^T V^{-1} u) + c(1+t))w + c^2 - \rho^2 + 2tc(u^T V^{-1} u) = 0$$

Here too, the order in which we optimize the coordinates has a significant impact.

18.3.2.2   *Dual Block Problem*

We can derive a dual to Problem (18.6) by rewriting it as a constrained optimization problem to get

$$
\begin{aligned}
\text{minimize} \quad & -\log x_1 - t(\log x_2 + \log x_3) - 2t\left(\sum_i (\log y_i + \log z_i)\right) \\
\text{subject to} \quad & x_1 \le w - u^T V^{-1} u \\
& x_2 = \rho + c - w, \ x_3 = \rho - c + w \\
& y_i = \rho + b_i - u_i, \ z_i = \rho - b_i + u_i,
\end{aligned}
\tag{18.9}
$$

in the variables $u \in \mathbb{R}^{(n-1)}, w \in \mathbb{R}, x \in \mathbb{R}^3, y \in \mathbb{R}^{(n-1)}, z \in \mathbb{R}^{(n-1)}$. The dual to Problem (18.9) is written

$$
\begin{aligned}
\text{maximize} \quad & 1 + 2t(2n-1) + \log \alpha_1 - \alpha_2(\rho + c) - \alpha_3(\rho - c) \\
& - \sum_i \left(\beta_i(\rho + b_i) + \eta_i(\rho - b_i)\right) \\
& + t\log(\alpha_2/t) + t\log(\alpha_3/t) + 2t\left(\sum_i \left(\log(\beta_i/2t) + \log(\eta_i/2t)\right)\right) \\
\text{subject to} \quad & \alpha_1 = \alpha_2 - \alpha_3 \\
& \alpha_1 \ge 0,
\end{aligned}
$$

$$\tag{18.10}$$

in the variables $\alpha \in \mathbb{R}^3, \beta \in \mathbb{R}^{(n-1)}$ and $\eta \in \mathbb{R}^{(n-1)}$. Surrogate dual points then produce an explicit stopping criterion.

18.3.3   Complexity

Solving for the predictor step using conjugate gradient as in Section 18.3.1 requires $O(n^2)$ matrix products (at a cost of $O(n^3)$ each) in the worst case, but the number of iterations necessary to get a good estimate of the predictor is typically much lower (see experiments in Section 18.4). Scaling and warm start, on the other hand, have complexity $O(n^2)$. The inner and outer loops of the corrector step are solved using coordinate descent, with each coordinate iteration requiring the (explicit) solution of a cubic equation.

Results on the convergence of the coordinate descent in the smooth case can be traced back at least to Luo and Tseng (1992) or Tseng (2001), who focus on local linear convergence in the strictly convex case. More precise convergence bounds have been derived in Nesterov (2010), who shows linear convergence (with complexity growing as $\log(1/\varepsilon)$) of a randomized variant of coordinate descent for strongly convex functions, and a complexity bound growing proportionally to $1/\varepsilon$ when the gradient is Lipschitz continuous coordinatewise. Unfortunately, because it uses a randomized step selection strategy, the algorithm in its standard form is inefficient in our case here, as it requires too many SWM matrix updates to switch between columns. Optimizing the algorithm in Nesterov (2010) to adapt it to our problem (e.g., by adjusting the variable selection probabilities to account for the relative cost of switching columns) is a potentially promising research direction.

The complexity of our algorithm can be summarized as follows.

- Because our main objective function is strictly convex, our algorithm converges locally linearly, but we have no explicit bound on the total number of iterations required.
- Starting the algorithm requires forming the inverse matrix V^{-1} at a cost of $O(n^3)$.
- Each iteration requires solving a cubic equation for each coordinatewise minimization problem to form the coefficients in (18.7), at a cost of $O(n^2)$. Updating the problem to switch from one iteration to the next, using SWM updates, then costs $O(n^2)$. This means that scanning the full matrix with coordinate descent requires $O(n^4)$ flops.

While the lack of a precise complexity bound is a clear shortcoming of our choice of algorithm for solving the corrector step, as discussed by Nesterov (2011), algorithm choices are usually guided by the type of operations (projections, barrier computations, inner optimization problems) that can be solved very efficiently or in closed form. In our case here, it turns out that coordinate descent iterations can be performed very fast, in closed form (by solving cubic equations), which seems to provide a clear (empirical) complexity advantage to this technique.

18.4   Numerical Results

We compare the numerical performance of several methods for computing a full regularization path of solutions to Problem (18.2) on several realistic data sets: the senator votes covariance matrix from Banerjee et al. (2006), the *Science* topic model in Blei and Lafferty (2007) with 50 topics, the covariance matrix of 20 foreign exchange rates, the UCI SPECTF heart dataset (diagnosing of cardiac images), the UCI LIBRAS hand movement dataset, and the UCI HillValley dataset. We compute a path of solutions using the methods detailed here (Covpath) and repeat this experiment using the Glasso path code (Friedman et al., 2008), which restarts the covariance selection problem at $\rho + \varepsilon$ at the current solution of (18.2) obtained at ρ. We also tested the smooth first-order code with warm start ASPG (described in Lu (2010)) as well as the greedy algorithm SINCO by Scheinberg and Rish (2009). Note that the latter only identifies good sparsity patterns but does not (directly) produce feasible solutions to problem (18.4). Our prototype code here is written in MATLAB (except for a few steps in C), ASPG and SINCO are also written in MATLAB, and Glasso is compiled from FORTRAN and interfaced with R. We use the scaling/warm start approach detailed in Section 18.3 and scan the full set of variables at each iteration of the block-coordinate descent algorithm (optimizing over the 10 percent most promising variables sometimes significantly speeds up computations but is more unstable), so the results reported here describe the behavior of the most robust implementation of our algorithm. We report CPU time (in seconds) versus problem dimension in Table 18.1. Unfortunately, Glasso does not use the duality gap as a stopping criterion, but rather lack of progress (average absolute parameter change less than 10^{-4}). Glasso fails to converge on the HillValley example.

Dataset	Dimension	Covpath	Glasso	ASPG	SINCO
Interest Rates	20	0.036	0.200	0.30	**0.007**
FX Data	20	**0.016**	1.467	4.88	0.109
Heart	44	**0.244**	2.400	11.25	5.895
ScienceTopics	50	**0.026**	2.626	11.58	5.233
Libras	91	**0.060**	3.329	35.80	40.690
HillValley	100	**0.068**	-	47.22	68.815
Senator	102	**4.003**	5.208	10.44	5.092

Table 18.1: CPU time (in seconds) versus problem type for computing a regularization path for 50 values of the penalty ρ, using the path-following method detailed here (Covpath), the Glasso code with warm-start (Glasso), the pathwise code (ASPG) in Lu (2010) and the SINCO greedy code by Scheinberg and Rish (2009).

As in Banerjee et al. (2008), to test the behavior of the algorithm on examples with known graphs, we also sample sparse random matrices with Gaussian coefficients, add multiples of the identity to make them positive semidefinite, then use the inverse matrix as our sample matrix Σ. We use these examples to study the performance of the various algorithms listed above on increasingly large problems. Computing times are listed in Table 18.2 for a path of length 10, and in Table 18.3 for a path of length 50. The penalty coefficients ρ are chosen to produce a target sparsity around 10 percent.

Dimension	Covpath	Glasso	ASPG	SINCO
20	**0.0042**	2.32	0.53	0.22
50	**0.0037**	0.59	4.11	3.80
100	**0.0154**	1.11	13.36	13.58
200	**0.0882**	4.73	73.24	61.02
300	**0.2035**	13.52	271.05	133.99

Table 18.2: CPU time (in seconds) versus problem dimension for computing a regularization path for 10 values of the penalty ρ, using the path-following method detailed here (Covpath), the Glasso code with warm-start (Glasso), the pathwise code (ASPG) in Lu (2010) and the SINCO greedy code by Scheinberg and Rish (2009) on randomly generated problems.

Dimension	Covpath	Glasso	ASPG	SINCO
20	**0.0101**	0.64	2.66	1.1827
50	**0.0491**	1.91	23.2	22.0436
100	**0.0888**	10.60	140.75	122.4048
200	**0.3195**	61.46	681.72	451.6725
300	**0.8322**	519.05	5203.46	1121.0408

Table 18.3: CPU time (in seconds) versus problem dimension for computing a regularization path for 50 values of the penalty ρ, using the path-following method detailed here (Covpath), the Glasso code with warm-start (Glasso), the pathwise code (ASPG) in Lu (2010) and the SINCO greedy code by Scheinberg and Rish (2009) on randomly generated problems.

In Figure 18.1, we plot the number of nonzero coefficients (cardinality) in the inverse covariance versus the penalty parameter ρ, along a path of solutions to problem (18.2). We observe that the solution cardinality appears to be linear in the log of the regularization parameter. We then plot the number of conjugate gradient iterations required to compute the predictor in Section 18.3.1 versus number of nonzero coefficients in the inverse covariance matrix. We notice that the number of CG iterations decreases significantly for sparse matrices, which makes computing predictor directions faster at the sparse (i.e., interesting) end of the regularization path. Nevertheless, the complexity of corrector steps dominates the total complexity of the algorithm, and there was little difference in computing time between using the scaling method detailed in Section 18.3 and using the predictor step. Hence the final version of our code and the CPU time results listed here make use of scaling/warm start exclusively, which is more robust.

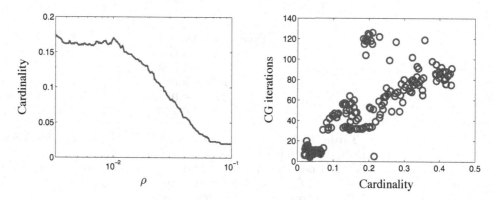

Figure 18.1: *Left:* We plot the fraction of nonzero coefficients in the inverse covariance versus penalty parameter ρ, along a path of solutions to Problem (18.2). *Right:* Number of conjugate gradient iterations required to compute the predictor step versus number of nonzero coefficients in the inverse covariance matrix.

Finally, to illustrate the method on intuitive data sets, we solve for a full regularization path of solutions to Problem (18.2) on financial data consisting of the covariance matrix of U.S. forward rates for maturities ranging from 6 months to 10 years from 1998 until 2005. Forward rates move as a curve, so we expect their inverse covariance matrix to be close to band diagonal. Figure 18.2 shows the dependence network obtained from the solution of Problem (18.2) on this matrix along a path, for $\rho = .02$, $\rho = .008$, and $\rho = .006$. The graph layout was formed using the yFiles–Organic option in Cytoscape.

The string like dynamics of the rates clearly appear in the last plot. We also applied our algorithm to the covariance matrix extracted from the correlated topic model calibrated in Blei and Lafferty (2007) on 10 years of articles from the journal *Science*, targeting a graph density low enough to reveal some structure. The corresponding network is detailed in Figure 18.3. Graph edge color is related to the sign of the conditional correlation (green for positive, red for negative), while edge thickness is proportional to the correlation magnitude. The five most important words are listed for each topic.

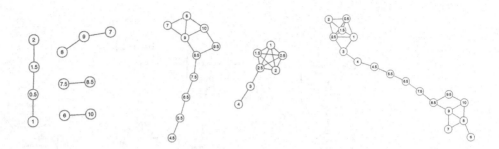

Figure 18.2: Three sample dependence graphs corresponding to the solution of problem (18.2) on a U.S. forward rates covariance matrix for $\rho = .02$ (left), $\rho = .008$ (center), and $\rho = .006$ (right).

18.5 Online Covariance Selection

In this section we will briefly discuss the *online* version of the covariance selection problem. This version arises if we obtain a better estimate of the covariance matrix after the problem is already solved for a set of parameter values. We will assume that the new (positive definite) covariance matrix $\hat{\Sigma}$ is the sum of the old covariance matrix Σ and an arbitrary symmetric matrix C. With such a change, the "new" dual problem can be written as

$$
\begin{array}{ll}
\text{minimize} & -\log\det(U) - n \\
\text{subject to} & U_{ij} \le \rho + \Sigma_{ij} + \mu C_{ij}, \quad i,j = 1,\ldots,n, \\
& U_{ij} \ge \Sigma_{ij} + \mu C_{ij} - \rho,, \quad i,j = 1,\ldots,n,
\end{array}
\tag{18.11}
$$

in the variable $U \in \mathbf{S}_n$, where ρ is a parameter value for which the corresponding optimal solution is already calculated with the old covariance matrix Σ. The problem is parameterized with μ, so that $\mu = 0$ gives the original problem whereas $\mu = 1$ corresponds to the new problem.

For many applications, one would expect C to be small and the optimal solution U^* of the original problem to be close to the optimal solution of the new problem, say $\hat{U}^*$. Hence, regardless of the algorithm, U^* should be used as an initial solution instead of solving the problem from scratch.

In the spirit of the barrier methods and the predictor-corrector method that we have devised in this chapter, we can develop a predictor-corrector algorithm to solve the online version of the problem fast, as follows. We form a parameterized version of the regularized problem

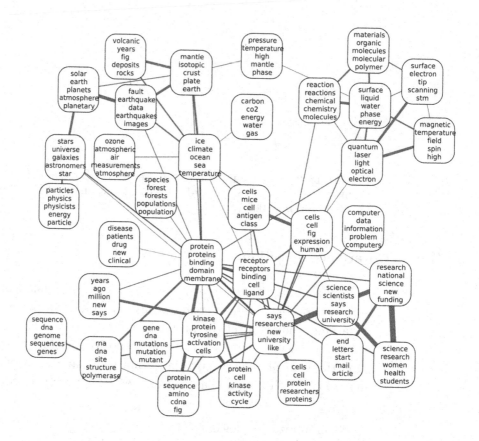

Figure 18.3: Topic network for the Science Correlated Topic Model in Blei and Lafferty (2007). Network layout using cytoscape. Graph edge grayscale is related to the magnitude of the conditional correlation while edge thickness is proportional to the correlation magnitude.

$$\min_{U \in \mathbf{S}_n} \quad -\log \det(U) - t \sum_{i,j=1}^{n} \log(\rho + \Sigma_{ij} + \mu C_{ij} - U_{ij})$$
$$-t \sum_{i,j=1}^{n} \log(\rho - \Sigma_{ij} - \mu C_{ij} + U_{ij}) \tag{18.12}$$

in the variable $U \in \mathbf{S}_n$, and $t > 0$ is the tradeoff level as before. Let us define matrices $\hat{L}, \hat{M} \in \mathbf{S}_n$ as follows:

$$\hat{L}_{ij}(U, \mu) = \frac{t}{\rho + \Sigma_{ij} + \mu C_{ij} - U_{ij}} \quad \text{and} \quad \hat{M}_{ij}(U, \mu) = \frac{t}{\rho - \Sigma_{ij} - \mu C_{ij} + U_{ij}}.$$

As before, optimal $\hat{L}$ and $\hat{M}$ should satisfy $(\hat{L} - \hat{M}) = U^{-1}$, and Problem (18.12) traces a central path toward the optimal solution to Problem (18.11) as t goes to 0.

Defining

$$\hat{H}(U,\mu) = \hat{L}(U,\mu) - \hat{M}(U,\mu) - U^{-1},$$

we trace the curve $\hat{H}(U,\mu) = 0$, the first-order optimality condition for problem (18.12), from the solution for the original problem to one for the new problem as μ goes from 0 to 1. The resulting predictor-corrector algorithm is Algorithm 18.3, which solves the online version efficiently.

Algorithm 18.3 Online Pathwise Covariance Selection

Input: $\Sigma, U^* \in \mathbf{S}_m$, $\rho \in \mathbb{R}$, and $c \in \mathbb{R}^{n \times r}$.

1:   Start with (U_0, μ_0) s.t $\hat{H}(U_0, \mu_0) = 0$, specifically, set $\mu_0 = 0$ and $U_0 = U^*$.

2:   **for** $i = 1$ to k **do**

3:       *Predictor Step.* Let $\mu_{i+1} = \mu_i + 1/k$. Compute a tangent direction by solving the linear system

$$\frac{\partial \hat{H}}{\partial \mu}(U_i, \mu_i) + J(U_i, \mu_i)\frac{\partial U}{\partial \mu} = 0$$

in $\partial U / \partial \mu \in \mathbf{S}_n$, where $J(U_i, \mu_i) = \partial \hat{H}(U, \mu)/\partial U \in \mathbf{S}_{n^2}$ is the Jacobian matrix of the function $\hat{H}(U, \mu)$.

4:       Update $U_{i+1} = U_i + (\partial U/\partial \mu)/k$.

5:       *Corrector Step.* Solve Problem (18.12) for μ_{i+1} starting at $U = U_{i+1}$.

6:   **end for**

Output: Matrix U_k that solves Problem (18.11).

As for the offline version, the most demanding computation in this algorithm is the calculation of the tangent direction, which can be carried out by the CG method discussed above. When carefully implemented and tuned, it produces a solution for the new problem very fast. Although one can try different values of k, setting $k = 1$ and applying one step of the algorithm is usually enough in practice. This algorithm, and the online approach discussed in this section in general, would be especially useful and sometimes necessary for very large datasets, as solving the problem from scratch is an expensive task for such problems and should be avoided whenever possible.

Acknowledgements

The authors are grateful to two anonymous referees whose comments significantly improved the chapter. The authors would also like to acknowledge support from NSF grants SES-0835550 (CDI), CMMI-0844795 (CAREER), CMMI-0968842, a Peek junior faculty fellowship, a Howard B. Wentz Jr. award, and a gift from Google.

18.6 References

F. Bach, R. Thibaux, and M. Jordan. Computing regularization paths for learning multiple kernels. In *Advances in Neural Information Processing Systems*, volume 17, pages 73–80. MIT Press, 2005.

O. Banerjee, L. El Ghaoui, A. d'Aspremont, and G. Natsoulis. Convex optimization techniques for fitting sparse Gaussian graphical models. In *Proceedings of the 23rd International Conference on Machine Learning*, 2006.

O. Banerjee, L. El Ghaoui, and A. d'Aspremont. Model selection through sparse maximum likelihood estimation for multivariate Gaussian or binary data. *Journal of Machine Learning Research*, 9:485–516, 2008.

D. Blei and J. Lafferty. A correlated topic model of science. *Annals of Applied Statistics*, 1(1):17–35, 2007.

S. Boyd and L. Vandenberghe. *Convex Optimization*. Cambridge University Press, 2004.

J. Dahl, V. Roychowdhury, and L. Vandenberghe. Maximum likelihood estimation of gaussian graphical models: numerical implementation and topology selection. *UCLA preprint*, 2005.

J. Dahl, L. Vandenberghe, and V. Roychowdhury. Covariance selection for non-chordal graphs via chordal embedding. *Optimization Methods and Software*, 23 (4):501–520, 2008.

A. Dempster. Covariance selection. *Biometrics*, 28(1):157–175, 1972.

B. Efron, T. Hastie, I. Johnstone, and R. Tibshirani. Least angle regression. *Annals of Statistics*, 32(2):407–499, 2004.

J. Friedman, T. Hastie, H. Höfling, and R. Tibshirani. Pathwise coordinate optimization. *Annals of Applied Statistics*, 1(2):302–332, 2007.

J. Friedman, T. Hastie, and R. Tibshirani. Sparse inverse covariance estimation with the graphical lasso. *Biostatistics*, 9(3):432–441, 2008.

S. Lauritzen. *Graphical Models*. Clarendon Press, 1996.

Z. Lu. Adaptive first-order methods for general sparse inverse covariance selection. *SIAM Journal on Matrix Analysis and Applications*, 31(4):2000–2016, 2010.

Z. Q. Luo and P. Tseng. On the convergence of the coordinate descent method for convex differentiable minimization. *Journal of Optimization Theory and Applications*, 72(1):7–35, 1992.

N. Meinshausen and P. Bühlmann. High dimensional graphs and variable selection with the Lasso. *Annals of Statistics*, 34(3):1436–1462, 2006.

Y. Nesterov. Efficiency of coordinate descent methods on huge-scale optimization problems. *CORE Discussion Papers*, 2010/2, 2010.

Y. Nesterov. Barrier subgradient method. *Mathematical Programming, Series B*, 127:31–56, 2011.

K. Scheinberg and I. Rish. SINCO—a greedy coordinate ascent method for sparse inverse covariance selection problem. 2009.

P. Tseng. Convergence of a block coordinate descent method for nondifferentiable minimization. *Journal of Optimization Theory and Applications*, 109(3):475–494, 2001.

Printed in the United States
by Baker & Taylor Publisher Services